Buffon

CORNELL HISTORY OF SCIENCE SERIES

L. Pearce Williams, Editor

Georges-Louis Leclerc, comte de Buffon (1707–1788). From *Oeuvres complètes,*
I, frontis.

BUFFON

A LIFE IN NATURAL HISTORY

by JACQUES ROGER

Translated by Sarah Lucille Bonnefoi
Edited by L. Pearce Williams

CORNELL UNIVERSITY PRESS

Ithaca and London

The publisher gratefully acknowledges the assistance of the French Ministry of Culture in defrayng part of the cost of translation.

French edition, *Buffon, un philosophe au Jardin du Roi,* © Librairie Arthème Fayard, 1989.

Translation copyright © 1997 by Cornell University

First published 1997 by Cornell University Press.

Printed in the United States of America.

Cornell University Press strives to utilize environmentally responsible suppliers and materials to the fullest extent possible in the publishing of its books. Such materials include vegetable-based, low-VOC inks and acid-free papers that are also either recycled, totally chlorine-free, or partly composed of nonwood fibers.

Library of Congress Cataloging-in-Publication Data

Roger, Jacques, 1920–1990.
  [Buffon. English]
  Buffon : a life in natural history / by Jacques Roger ; translated by Sarah Lucille
Bonnefoi ; edited by L. Pearce Williams.
      p.   cm.—(Cornell history of science)
  Includes bibliographical references (p.   ) and index.
  ISBN 0-8014-2918-8 (cloth : alk. paper)
    1. Buffon, Georges Louis Leclerc, comte de, 1707–1788.   2. Naturalists—France—
Biography.   3. Natural history.   I. Williams, L. Pearce (Leslie Pearce), 1927-   .
II. Title.   III. Series: Cornell history of science series.
  QH31.B88R5713   1997
  570'.92—dc21
  [B]   96-37948

Cloth printing 10 9 8 7 6 5 4 3 2 1

to my wife

Foreword

George-Louis Leclerc, comte de Buffon (1707–1788) was one of the giants of the French Enlightenment—that *siècle de lumières* that we now consider, rightly, to have given birth to the modern world. Buffon was unusual, even among his contemporaries, in the breadth and depth of his intellectual interests and abilities. The article in the *Dictionary of Scientific Biography* classifies him simply as an expert in natural history. This was, of course, the field in which he made his most significant contributions. In his youth, however, he showed proficiency in mathematics and published on probability theory. Later in his life, he set up and ran an industrial forge that relied on his knowledge of metallurgy. His major works were in natural history, but that is a very large field. He was concerned with questions of taxonomy and his discussions of the problems of classification contributed significantly to the beginnings of scientific classification of animals and plants. It is no exaggeration to suggest that Buffon's ideas stimulated much of the scientific work in natural history that culminated in the Darwinian synthesis. Even when he was wrong, as he often was, he was precise enough in his errors to permit others to see beyond his vision. That, as Sir Karl Popper pointed out long ago, is a major achievement.

Buffon, however, was not just a *savant;* he was also an administrator and an arbiter of literary taste as well as a subtle and experienced politician in the Old Regime. His territory was the Royal Botanical Garden, later to become the Museum of Natural History during the French Revolution. As its director, he had to wheedle and coax and plead with royal administrators to squeeze money out of the increasingly petrified royal finances. He was eminently successful and managed to turn the Royal Garden into one of the major scientific institutions of Europe.

As a member of the French Academy, that body that jealously guards the French language and influenced then as now the canons of literary taste, Buffon played an important role in the development of modern French literature. It was he who maintained, in a famous discourse, that style was the man. It was not Buffon's fault that later authors often sacrificed substance for style. Similarly, as a member of the Academy of Sciences, Buffon was able to influence the direction that natural history was to take after his death. It was his assistant, Lamarck, who coined the French word *biologie* at the beginning of the nineteenth century.

To undertake a biography of such a man demands courage. Jacques Roger was the perfect biographer for Buffon. His massive and authoritative *Les sciences de la vie dans la pensée française du XVIIIe siècle* (1963) revealed his mastery of all the fields that Buffon mastered. The present volume is the fruit of Professor Roger's deep immersion in the works of Buffon and his contemporaries. He paints a picture of the man, his world, and his work that permits us to see Buffon in all his complexity, genius, and humanity.

I should like to end this foreword on a personal note. For many years Jacques Roger was Andrew Dickson White Visiting Professor at Cornell University. I was proud to count him a friend as well as a colleague. I have been honored to serve as the editor of this, his final book.

[Editor's note: I have worked very closely with Ms. Bonnefoi, the translator. From the beginning, we agreed that clarity was our primary concern. Literary style, like a delicate wine, does not travel well. Deviations from the literal text were justified, we thought, in the name of clarity. As editor, I take full responsibility for such deviations.]

L. PEARCE WILLIAMS

Ithaca, New York

Contents

Illustrations

Illustrations credited to *Oeuvres complètes* are from *Oeuvres complètes de Buffon,* ed. M. A. Richard, Paris, Baudouin Frères, 1826.
Photographs courtesy L. Pearce Williams.
Those credited to *Histoire naturelle* are from Buffon, *Histoire naturelle, générale et particulière,* Amsterdam, J. H. Schneider, 1766–1785.
Reproduced from *368 Animal Illustrations from Buffon's "Natural History,"* Mineola, N.Y., Dover, 1993. Courtesy of Dover Publications, Inc.

Preface

In elementary school before World War II, students learned what a scientist should be: a modest man in a white lab coat, experimenting among his flasks and test tubes, without any other ambition than to discover the secrets of nature for the good of humanity, and ignoring the temptations of the world: lust, money, and power. Suspicious of fashions in style, he writes soberly and only for his knowledgeable colleagues. A secular monk, he is a priest of the religion of science: in short, he is the legendary Pasteur, who has given his name to so many streets in France.

Buffon was the opposite of this traditional image. He loved money and became rich. He loved power, and he frequented those in power. The best portrait we have of him shows him, according to the great skeptical philosopher David Hume, costumed as a marshal of France. He loved women, and not just for their beautiful souls. His laboratory experiments were few and his assumptions often questionable. He let his imagination go well beyond the facts. He wrote too well and seduced the general public. To say that such a man was a great scientist is almost immoral.

Indeed, some have tried to show that he was not a great scientist. The naturalists of his time were admirable observers, capable of describing with great precision the smallest details of an insect's structure. When Buffon told them that in studying details so closely they were forgetting nature, they revolted—especially when they realized that the nature about which Buffon was talking must replace God the creator who, for them, explained the order of the world. Buffon's great success with the public was an added insult. It was easy for devout Christians such as Réaumur, devout Voltaireans such as d'Alembert, and those devoted to minute detail like Guettard to maintain that Buffon was not a scientist. He was the "great wordsmith"; so said d'Alembert, whose style was not brilliant. "Still more buffonades," complained Guettard, who disliked generalized ideas.

The best scientists of Revolutionary France, before they were ennobled by Napoleon or Louis XVIII, considered it impossible to be both a count and a true scientist. They glorified Daubenton, who had no generalized ideas but held no noble titles either and who happily presented himself as a victim of Buffon's tyranny. After 1830, things changed. The Romantic imagination liked broad perspectives on nature and history, and the change in the intellectual climate, like developments in natural history, allowed a clearer view of what had been foreseen in Buffon's theories. The count's face was fading into the past, but editions of his work, curiously enough, were multiplying, signed by the greatest scientific names of the time, from Cuvier to Flourens.

The scientists of the Third Republic were divided. Count Buffon in his legendary lace cuffs did not attract them much. In the age of Darwin and the battle against the Church, it was helpful to find in prerevolutionary France a sort of forerunner of evolution and a probable materialist. Nevertheless, the lace cuffs were embarrassing. In 1878, however, Félix Hémon published his *Éloge de Buffon* [Eulogy of Buffon], one of the best studies of the century. In spite of this, in 1941 Émile Guyénot was still irritated by the "Lord of Montbard," and his "flowery and often pompous style." The intelligent and critical sympathy of Jean Rostand remained the exception.

Then came the epistemologists, and catastrophe. In *La Formation de l'esprit scientifique* [The Formation of the Scientific Spirit] (1938), Gaston Bachelard quoted Buffon several times as an example of a "pre-scientific" mind. In the name of a "psychoanalysis of objective knowledge," he denounced "the influence of pre-scientific concepts of unconscious origins" and gave as an example the famous saying "The lion is the king of beasts." Bachelard continues, "The lion is the king of beasts because it is fitting to someone in favor of order that all beings, even beasts, have a king." The problem is that the saying he criticized is found in the article "Du Lion" in the *Histoire naturelle* [Natural History] only in the following sentence: "One must admit that the strength of this king of beasts does not suffice in the presence of a Hottentot or a Negro who often dares to attack it face to face with fairly light weapons." A paltry royalty, that. In reality, Buffon is repeating a saying traditional since La Fontaine, which affirms man's superiority over animals. Here, historical explanation is perhaps of greater help than psychoanalysis.

What is worse, Bachelard's remark is not at all original; it is part of a venerable tradition that dates back to the Revolution. In his course offered in the early years of the École Normale, Daubenton quoted this saying and attributed it to Buffon without actually naming him. And he concluded, to the applause of his students, "Citizens, there is no king in nature!" After which, the saying passed from book to book without ever referring, of course, to the original text. Buffon's great mistake was to have published thirty-six quarto volumes.

Today we no longer view scientists the way we did fifty years ago. We find it normal that they are active in public life and in constant contact with political powers, and we think it necessary that they address the general public and not keep their discoveries within the limited circle of their peers. We also know, since Darwin and Poincaré, Einstein and de Broglie, that scientific theories play a role in scientific progress that is just as essential as discoveries and the verification of experiments. Finally, and above all, the history of science has become a discipline in its own right and the historian of science has learned that the role of any historian is to understand and not to judge—as Lucien Febvre says, the only capital sin in history is anachronism.

In addition, since 1952 we have been party to a revival of Buffon studies, marked not only by a return to texts that were little used earlier but also by a willingness to take seriously the work of a man who was one of the greatest minds of his time. I have participated in these studies, but clearly this book would not have been possible without the work of all the French and foreign scholars who are listed, incompletely no doubt, in the bibliography.

This book, however, is a biography. I believe that the biography of a scientist must be as much a history of his thoughts as of his life and actions, which drove me to analyze the texts as precisely as possible. I also want to situate Buffon once again in a historical, political, and intellectual context, without which it is impossible to understand either the ups and downs of his career or the complexity of his works. Buffon is a man of his century, and although his century may fail to explain him fully, he nonetheless cannot be understood outside his era. In all this I have, like Buffon, respected my reader. I have rejected easy simplifications, believing that one can be clear without being simplistic.

The biography of a scientist, however, poses another and more difficult problem. Marcel Proust asked literary critics not to confuse the man with the novelist, the works with the author. Things are not that simple. The psychology of a scientist does not explain a scientific theory, yet an intellectual temperament expresses itself in all acts of being: in ways of loving, of leading one's life, of writing and thinking. All Buffon's actions indicate the same mixture of sensuality and intellectualism and the same passion to possess. Whether it be money or nature, he wished both to enjoy things and to arrange them, in his accounts as in natural phenomena. This passion was cool and patient, more a passion of intelligence than of the heart, and a passion to rule in the name of Reason. But Buffon was also driven to assure his independence within society as much as within science, to affirm his liberty and his critical spirit. It will perhaps be surprising to see how much Buffon resembles Jean-Jacques Rousseau, even though their fates were so different.

His realism permitted him to take advantage of the political situation of France under Louis XV to feed his ambition, to be his own master and to depend on no one. Protected by the minister Maurepas and Mme. de Pompadour, he understood that the royal administration was more protective of intelligent technicians than defenders of religion. Provided that one respected the proprieties of society, one could achieve honor and fortune, that is to say independence and freedom of thought. He had no illusions about the social order of his time—indeed he criticized it in his own way—yet he knew how to use it. He passed for a conservative in politics, even though he was seen as a revolutionary in science. A "philosophe" more extreme than many, also an intendant in the Royal Botanical Garden, seigneur of Montbard, finally comte de Buffon through the prince's grace, above all untouchable by virtue of his scientific and literary eminence—it is understandable that he attracted suspicion, even hatred, from all sides.

For me, Buffon is first of all the author of the *Natural History*. I have tried to illustrate how, in this immense work, he transformed the way of understanding nature. He did not give new answers to traditional questions but asked new questions; he opened new fields of research; he suggested a new way of conceiving natural history by his audacity and often by his mistakes. I have not ignored his inadequacies, knowing they added to his genius and to a system of thinking in which one must find the internal logic; in addition, had he been a "model scientist" he would have been imprisoned in what Thomas Kuhn calls "normal science," and he never would have had so many new ideas.

Allow me to cite two objective witnesses to support my views, two naturalists of our century who both love the history of their discipline. Nils Nordenskiöld of Finland wrote in 1928 of Buffon, "In the purely theoretical sphere, he was the foremost biologist of the eighteenth century, the one who possessed the greatest wealth of ideas, of real benefit to subsequent ages and exerting an influence stretching far into the future." This judgment was completely upheld in 1982 by Ernst Mayr, one of the greatest living naturalists. Considering the importance of the eighteenth century in the history of natural history, this is like saying that Buffon was the most important naturalist between Aristotle and Darwin. And so it is well worth the effort to dedicate a book to him.

I decided to show Buffon the writer only briefly and in the final pages. Buffon has too often been seen only as a writer, and I wanted to react, perhaps excessively, against that tendency. I have, however, quoted him often, and readers will discover for themselves that Buffon was a great writer. I will surely be reproached for not being critical enough and for systematically looking for favorable explanations to excuse the weaknesses of my author. Perhaps. Yet here once again, I wanted to challenge a tradition of ignorance and disdain especially among my scientific colleagues which is not yet dead. In any case, a historian is never neutral. Let me admit therefore that I would not have written this biography without a secret sympathy for my author, and not just for his ideas and how he expressed them: in the last two centuries many attitudes and theoretical questions have been transformed in scientific, philosophic, or religious thought. Buffon belongs to history. But I have admired his intellectual daring, his willingness to understand and to extend the empire of human reason as far as possible, and, behind the facade of conformity, his freedom of spirit and his rejection of ready-made ideas.

This book, as I have said, would not have been possible without the studies dedicated to Buffon over the last thirty years. Buffon worked in so many areas that I was obliged to ask for help. In particular I want to thank Pierre Costabel, historian of mathematics, and Denis Woronoff, historian of metallurgy, who helped me with their expertise but bear no responsibility for the way in which I have used the information that they kindly furnished. I also owe much to the young researchers who have been working with me for many years and who have become colleagues and friends: in particular, Mme. Rey, Mme. Bitbol-Hesperiès, and M. Blanckaert. Finally, I must thank the members of the Alexandre-Koyré Center and the Centre International de Synthèse, in particular Mlle. Illic, Mme. Bilodeau, Mme. Groult, and Mme. Biard, who willingly helped me in my research and from whose competence and care this work benefited.

JACQUES ROGER

Sury-en-Vaux
October 5, 1989

PART ONE

THE AMBITIONS OF AN INTELLIGENT BOURGEOIS

CHAPTER I

The Rise of the Family and Personal Ambitions

The Family Environment and Childhood Years

If a man's destiny were written in his origins or his heredity, Buffon would have died president of the Burgundy parlement.* He would have been, at worst, a local celebrity unknown to the outside world and, at best, a second-class writer known only to specialists. That he had a passion for the sciences and became the greatest French naturalist is a sort of joke of nature, the result of a personal calling, and ultimately inexplicable. The historian must admit, with regret, that a human destiny cannot be reduced to a game of cause and effect.

Georges-Louis Leclerc, who later became the comte de Buffon, was born on September 7, 1707, in Montbard, Burgundy, the eldest son of Benjamin-François Leclerc and Anne-Christine Marlin.[1] "Montbard, a small town," says the *Encyclopédie* of Diderot and d'Alembert. If we can believe the old engravings, it was more like a large burg at the foot of a hill crowned by a feudal castle. The *Encyclopédie* adds, "There is a royal manor house, a constabulary, a salt warehouse." Montbard was, therefore, a small administrative center where junior government workers, the "king's officers," administered justice, collected taxes, maintained order, and more or less tyrannized their neighbors.

For several generations the entire Leclerc tribe had lived there with their allies, the Nadaults and the Daubentons. They intermarried, aided one another, and disputed their inheritances. They were thrifty and patient. And above all, they were not in a hurry to die: three of Buffon's direct

---

*Ed. note: The French parlement should not be confused with the English Parliament. The first was a court of royally appointed magistrates with limited administrative responsibilities; the second was the elected legislative body of Great Britain with limited (but important) judicial powers.

ancestors lived to nearly ninety or more. That was sufficient time to become wealthy, even if it were done slowly. And that was what they did.

In the beginning, the Leclerc family most likely consisted of well-off laborers. One day, a son left his plow and after an apprenticeship with a barber-surgeon settled in Montbard. This was Buffon's great-grandfather. He bled people, bandaged wounds, and cared for the peasants and the humble folk too poor to pay for the luxury of seeing a doctor. Thanks to his savings, his son succeeded in becoming a doctor. He went to college, learned Latin, and came back to Montbard to establish his practice. The son of the doctor was able to buy the office of "counselor to the King, judge of the court of the city of Montbard." He even served as the mayor and governor of Montbard from 1695 to 1697, until the official holder of the office, his cousin Jean Nadault, was old enough to take over. The son of the judge, Benjamin-François Leclerc, was the father of our Buffon. He was a lawyer in the Burgundy parlement, and he bought the office of agent for the salt warehouse, which meant that he collected the salt tax, the most unpopular tax of all.

The slow and steady rise of the Leclerc family to positions that ensured ease and notability was typical of France of the Ancien Régime. Buffon owed his entire philosophy of life, as well as his physical vigor, to his family. "Genius," he would say, "is only a greater gift of patience." This active patience, which was directed toward the future but ready to seize every occasion that arose in the present, was used by his ancestors, and he used it in turn for his career as well as his works. It was in unexpected ways, but thanks to these same persistent virtues, that he topped off the Leclerc family's wealth.

Later in his life, Buffon would stay loyal to the allies of the family. One of the Daubentons, a notary of Montbard, had presided over his father's marriage contract. When the notary's son Pierre Daubenton, who was the intendant's subdelegate of Dijon, needed his help, Buffon intervened. In all likelihood, Buffon also made Pierre a collaborator on the *Encyclopédie*. Again it was he who chose Pierre's younger brother, Louis Daubenton, a rather obscure doctor in Montbard, as his assistant, opening up a fine scientific career for him. Finally, in 1793, Buffon's son married Betzy Daubenton after his divorce. The Nadaults did not desert him either, even beyond the grave; Nadault de Buffon, a magistrate of the Second Empire, would edit his correspondence.

Montbard's horizons were quite limited. In 1706, however, Buffon's father had married a woman a little older than he, Anne-Christine Marlin, and she brought "expectations" with the marriage. She had an uncle, Georges Blaisot, who had made his fortune as a tax collector for Sicily for the duke of Savoy. Georges Blaisot had no children. It was certainly not a coincidence that the young couple asked him to stand godfather to their first son, whom they named Georges.

Uncle Blaisot kept his promises. He provided his niece with a large dowry and died well-off in 1714. After the death of his widow in 1717, his considerable wealth passed to the young Georges-Louis. As a wise and experienced father, Benjamin-François used this money to acquire the holding of Buffon, a little village several kilometers north of Montbard, along with the lord's rights to the manor of Montbard—being a tax collector certainly did not repel him. Then in 1720 he bought the office of counselor to the Burgundy parlement. The family left Montbard and came to settle in Dijon, in a house worthy of their new fortune. Benjamin-François had managed his affairs quite efficiently.

The Leclerc family already had ties with the Burgundy parlement. Becoming a member, however, meant joining a powerful body and an influential provincial aristocracy. As magistrates, the primary responsibility of members of parlement was to administer justice, which, in a country that was as bureaucratic as France of the Ancien Régime, gave them considerable authority. In addition, the parlements exercised political power that Louis XIV had broken but that they had recovered after his death: royal orders had the force of law in a jurisdiction only if they were duly registered by the magistrates, who sometimes did so reluctantly. The members of parlement, as owners of an office for which they had paid dearly, defended the interests of their class, were proud of their prerogative, and were often unbearably arrogant. They reigned in Dijon. Although they fought one another, despite complicated family ties, they ruled the surrounding countryside where they owned vast holdings. They dominated the social and intellectual life of the city, for these ennobled bourgeois eagerly cultivated literature and, as the eighteenth century progressed, philosophy and science. To live in Dijon was to have access to the life of the intellect.

We know little of the Leclercs' family life. The father was never a great magistrate. The young Georges-Louis, who soon had two brothers and two sisters, seems never to have been very close to his father, and violent conflicts of interest later would come between them. What relationship did he have with his mother? She had been his first tutor during the years at Montbard, and she is said to have had a quick intelligence and great moral strength. Was she very religious? It is interesting to note that Buffon's two brothers entered the Cistercian order and the only sister to survive became mother superior in the Ursuline convent of Montbard. At that time, however, sending children into religious orders was often a way of establishing them while keeping the majority of the inheritance intact for the eldest child. Buffon would later say that he owed his intellectual and moral qualities to his mother. He even made the generalization that children resemble their mothers more than their fathers. A psychoanalyst could certainly draw many conclusions from that remark.

Buffon, consciously or not, seems to have wanted to repress all memories of his childhood. "A man must look at his first fifteen years as being nothing," he wrote in 1749. "Everything that happened to him, everything that occurred in this long interval of time, is erased from his memory or at least has so little relationship to the goals and the things that have busied him since that he is no longer at all interested in it; it is not the same succession of ideas or, as it were, the same life."[2] We will never know what was behind this desire to erase his childhood. Yet it was perhaps during these first years that his tenacious desire for independence, which so strongly marked his life and thought, took root.

As soon as he arrived in Dijon, he had to spend most of his time at the Collège de Godrans, run by the Jesuits. This college, founded in 1587, had a good reputation. Its classes included all the sons of good family from the city and, since tuition was free, other boys from Dijon. It was here that Buffon became friends with the future abbé Le Blanc, the son of the prison warden. These day students were very busy and did not have the time to become bored.[3] In addition to classes, the teachers organized theatrical productions of pious tragedies, often in Latin, or sessions of verbal sparring in Latin or French, and many other public productions for the parents who came to admire their children's success.

The college held an important place in the city's social life. It was even seen as being too important in the eyes of certain bourgeois, scandalized by the behavior of the older students who did not shy away from brawls or frequenting cabarets and other places of ill repute. Nevertheless, the students took their studies seriously. Study of the classics predominated, with detailed analysis of the great Latin writers, which was appropriate for future lawyers, who would be using the art of oratory all their lives. Bossuet had been a student in this college, and it is possible that the young Buffon owed a slightly archaic conception of style, which was more suitable to the reign of Louis XIV than Louis XV, to his teachers and his close study of Latin orators. As in other schools of the Society of Jesus, however, scientific education was honorably represented by regular classes in mathematics and physics.

Buffon does not seem to have been a brilliant student. His peers remembered a boy who was somewhat slow, quiet, and with more ability in athletics than real intellectual prowess. Nonetheless, he was fascinated by mathematics, reading Euclid and the *Analyse des infiniment petits* [Analysis of the Infinitely Small] by the marquis de l'Hospital outside of class. This served as a classical introduction to the study of calculus. As a recent discovery of Leibniz and Newton, calculus cannot yet have been in the mathematics curriculum of the schools, but Buffon was particularly interested in it.

If Buffon owed his taste for science and his qualities as a writer to his Jesuit teachers, there is one area where his teachers failed, and that is his religious education. The Jesuits wanted to form a Christian elite for Catholic countries. In order to measure their success, one would have to follow their thousands of students during their entire lives. The failure was complete with Buffon, and even more so with another of their famous students, François-Marie Arouet, or Voltaire. Nothing seems more naturally foreign to Christianity than Buffon's thoughts, and I will speak of this matter again. He did not fight religion, he ignored it. It was only in his social behavior that he attempted to keep up appearances.

Finishing school at the age of sixteen in 1723, Buffon enrolled in the law school of Dijon, which had just been established. There he encountered one of his classmates from school, Charles de Brosses, and he met another son of a member of parlement, Richard de Ruffey, who had graduated from the prestigious school of the Parisian Jesuits, Louis-le-Grand. The three would remain friends all their lives. Richard de Ruffey would become president* of the Burgundy parlement and one of the great participants in the intellectual life of Dijon. De Brosses would also become president of the Burgundy parlement and would make a name for himself in literature with his famous *Lettres d'Italie* [Letters from Italy], along with a *Histoire des navigations aux terres australes* [History of Sea Voyages to Southern Lands] and, more inopportunely, by bloody battles with Voltaire that I will discuss later.

It is easy to understand why Ruffey and de Brosses went to law school, but why Buffon? It is possible that his father forced him to, thinking that his son would succeed him in his office. Fortunately, classes must have left time for other activities. The three young men, with the inseparable abbé Le Blanc, were admitted into the circle of Bouhier, the parlementary president. It was a distinguished honor, but easily explained: Ruffey and de Brosses came from important parlementary families, and the young LeClerc was already known to Bouhier. Bouhier had held the title of lord of Buffon for a while and was one of the heirs of the president Jacob, who had sold the lordship of Buffon to Benjamin-François. In short, they were all among friends.

Bouhier, a lawyer and respected humanist—well enough known to have been elected to the Académie Française without living in Paris—was a great figure in the intellectual life of Dijon. His library was famous; upon his death, it held thirty-five thousand books and two thousand manuscripts. Every week, for three hours, he was visited by a limited number of carefully chosen friends and scholars. This old-fashioned humanist had become interested in mathematics during his youth; he knew the English way of thinking, admired Locke, and was capable of discussing the philosophy of Leibniz. It was perhaps here that Buffon discovered those "philosophers" who would become the masters of thought of the eighteenth century. It is said that Bouhier encouraged Buffon in his taste for the sciences and

*Ed. note: A president was a chief justice.

philosophy. One thing is certain: in 1726, when Buffon left law school with his diploma in hand and barely nineteen years old, he decided to abandon law and the magistracy to dedicate himself to science.

The Years of Apprenticeship

It is easy to imagine that the young man's decision was not well received by his family. That a young magistrate discussed Cartesian vortices in a gathering of intellectuals was acceptable. That he entertained himself by collecting plants, stuffed crocodiles, or other exotic curiosities was also acceptable. But that the son of a magistrate envisioned science as his profession was inadmissible. In the early eighteenth century, science had not yet acquired the intellectual stature and the bourgeois respectability that it would have in the nineteenth, and for a simple reason: except for medicine and the teaching profession, science in 1725 led nowhere. At that time, for the son of a magistrate to become a doctor or a teacher was to fall beneath his station. Such professions were good for sons of the petit bourgeois. Even a teacher at the Royal College or in the Royal Botanical Garden did not have high social rank; this was too much like a common school pedant. The pay was also pitiful. The scientific community, with its internal hierarchy, was still a closed world inside French society. The only other possibility was the Royal Academy of Sciences, which was respected although no one really seemed to know what it was. To hope to enter it was a completely unrealistic ambition for a twenty-year-old from the provinces with no connections. One can imagine that Benjamin-François, after all that he had done to rise to his position, was beside himself when he learned of his son's decision, and that his colleagues shook their heads in disapproval. For several years, Buffon must have felt quite sharply this blind hostility from Dijon society. That he did not go back on his decision proved his strength of character and his intellectual independence.

Top priority, however, was to decide where he should continue his studies. In mathematics, Buffon had already attained the current level of research. He later said that by the age of twenty he had himself discovered Newton's "law of binomials." Even at the Royal College, he would not have found what he needed; the occupant of the chair of mathematics, François Chevallier, devoted his lectures to the art of fortification. The only way to make any progress was to contact active mathematicians directly. Buffon did this in 1727 by starting a correspondence with Gabriel Cramer, who, though barely older than Buffon, was already a professor at the Geneva Academy. We do not know how Buffon knew Cramer; perhaps it was through a friend in Dijon, Loppin de Gemeaux, himself enamored of mathematics. His early letters are lost, but we know that thanks to Cramer, he began to be known in mathematical circles.

It is clear that Buffon was suffocating in Dijon, where he passed for an eccentric or even for the delinquent son of a good family. He was undoubtedly earning some income, for in 1728 he decided to leave for Angers. Why Angers? Perhaps he wanted to rejoin Father de Landreville, a mathematics professor of the Oratory.* In any case, he found a more suitable environment in Angers. "You only need to leave home in order to value something and be appreciated and loved at the level that you deserve," he wrote to Richard de Ruffey the following year. He added, "I will do everything possible to stay away from Dijon as long as I can."[4] In Angers he studied mathematics, read Newton perhaps for the first time, and discovered the *Élements de la géometrie de l'infini* [Elements of Geometry of the Infinite] by Fontenelle, which will be discussed later. In addition, he botanized and took several courses at the medical school. Here he also quarreled with someone and fought a duel with him. A love rival? With an Englishman or an officer of the Royal Croatian Regiment? Did he kill or only wound his adversary? We know nothing for sure. In any case, our apprentice scholar was forced to flee and return unceremoniously to Dijon with a well-established reputation as a hothead. Only his friends and the aged Bouhier continued to trust him.

Were they wrong? In Dijon, our young man met a young English lord, the duke of Kingston, whom he had perhaps already met in Angers,[5] and became friendly with him. Four years younger than Buffon, Evelyn, the marquis of Dorchester, Viscount Newark, Baron Pierrepont, had lost his father at the age of two and his mother at the age of nine. The death of his grandfather had made him the second duke of Kingston at the age of nineteen. As his family did not know what to do with him, they sent him to the Continent. His aunt, Lady Mary Wortley Montagu, did not have a very high opinion of him: "The Duke of Kingston has hitherto had so ill an education 'tis hard to make any judgment of him. . . . As young noblemen go 'tis possible he may cut a good figure amongst them."** As an extra precaution, the family had chosen as the duke's mentor Nathaniel Hickman, a man who had vague claims to a medical education and was barely his elder. Hickman was an avid pipe smoker, which horrified Buffon; he had a slight bent as a naturalist and above all was a connoisseur of paintings, engravings, and objets d'art, in which he happily trafficked to earn pocket money. For all that, Hickman was a likable man; when he returned to England with his pupil, Buffon wrote of him, "He is missed by honest people; as for the prostitutes and the painters, they are in mourning."[6] Which makes one wonder about the company Hickman kept.

When Kingston and Hickman left Dijon on November 3, 1730, Buffon left

---

*Ed. note: A religious order devoted to teaching.

**Letters from the Right Honourable Lady Mary Wortley Montague, 1709 to 1762, Everyman ed. (London, 1906), p. 237.

with them. It is difficult to consider it an educational trip! Our travelers went to Nantes, disembarked leisurely near Bordeaux, and then made their way to Toulouse and Montpellier, reaching Lyons by May 1731. There, Buffon had to leave his friends and return to Dijon for "family matters." Perhaps he had learned that his grandmother was seriously ill; she died on August 1.

In October, Buffon was in Geneva, where he was finally able to speak face to face with Cramer, with whom he had faithfully corresponded. Then he left with Kingston and Hickman. The trio went through Turin, Milan, Genoa, Pisa, Florence, and finally arrived in Rome, in time for Mardi Gras. After that, the three friends separated; Kingston and Hickman enrolled in the University of Padua in March, while Buffon returned to Dijon. For him, serious work was about to begin.

We have two series of letters written by Buffon during his travels, and they show very different aspects of his character. He wrote to Richard de Ruffey of the cities that he passed through, their cleanliness and charm, the beauty of the women, the quality of the wine and cuisine, the entertainment, and above all the theater and opera. He wrote about Montauban: "The city is small but charming because of its situation, its architecture, and the pure air that one breathes." "Toulouse is a large and beautiful city . . . at least six times bigger than Dijon. Women are quite beautiful, and except for the old ones, I don't remember having seen any that were ugly." There must have been temptations because Buffon, ill at the beginning of the trip, was now in full health, and he could no longer do anything but complain "of having too much for a young man, and especially for a young traveler who does not have the same opportunities as you, sirs, permanent citizens of a good city, to get rid of surplus energy." He went on to ask for romantic gossip about Dijon. In Rome, it was the theater that enchanted him: "Four magnificent serious opera houses and as many for comic operas, without counting several small theaters, provide the ordinary pleasures for the Roman people; I have to admit that for me they are extraordinary because of the music's excellence and the dances' absurdity, the magnificence of the decorations and the metamorphosis of the eunuchs who play all the female roles. . . . This difference is so unremarkable for the Roman people that it is common to ogle and to speak of the beauty of these geldings in the same way that we would discuss that of a pretty actress; they have kept their ancestors' taste so much that they have perverted everything else."

Buffon was not unhappy doing what we now call sociology. In Bordeaux, "half the people are uncouth, and the other are dandies, but they are dandies five hundred leagues from Paris, which means they are completely wasted. You would laugh to see them, with their red heels and without swords, walking on tiptoe in the streets, where two or three inches of mud always covers the cobblestones. . . . They never are anything less than counts or marquis, even if they possess only a field or a farm and they are only knights of industry. . . . Gambling is the only occupation here and the only pleasure of these people; they gamble grandly and, during the carnival [of Mardi Gras], wearing masks. The most common game is *les trois dés;* and even more remarkable is that each mask[ed person] brings his own dice and his own cup. You have to be quite stupid to fall into such a trap."

Unlike the dandies of Bordeaux were the bourgeois of Nantes: "The inhabitants are all merchants, coarse people so scorned in our region, but whose way of life seems to me the most reasonable. They do not take on airs of preferring an *ordinaire* [a public coach] to a barouche headed by braided livery or a six-horse chariot, and they prefer to be rich and bourgeois than poor and noble. . . . As for me, I cannot say that they are wrong." The praise of Nantes' merchants turned to severe criticism of Dijonais vanity: in Nantes "everything shows the wealth that commerce produces, whereas in Angers as in Dijon, everything is scrimped and saved. Everyone does more there than he can [afford]; pride and poverty walk hand in hand, legitimate daughters of a ridiculous contempt for trade that is present there. I did not have a bad opinion of my homeland before I was outside of it; but now that I can judge it through comparison . . . I cannot help but see the faults in the picture. . . . I therefore end by being against it, and I do so without reserve."[7] It is easy to understand why Buffon never considered settling in Dijon.

In these letters, one can see, looking back, the psychological roots of some of Buffon's scientific ideas. The "surplus" that must "disappear" foreshadows his theory of organic molecules and his criticism of celibacy as the source of mental troubles. The game of *les trois dés* reminds us that Buffon worked on the theory of games of chance. Above all, however, we see the expression of a particular temperament with a taste for tangible reality and real wealth, along with a rejection of noble pretensions that were not upheld by great wealth. Before becoming a count, Buffon wanted to be rich, and rich he would be. That he saw nothing in Italy other than the bad roads, the risks of the mail, and the operas in Rome with their castrati is hardly surprising: he never had much taste for the fine arts. Curiously, nature does not seem to have impressed him either, and nothing in his letters brings us to think, as Condorcet would later affirm, that it was in Italy that Buffon discovered a calling as a geologist. Even Montesquieu and De Brosses, the two most famous French travelers in Italy of that era, showed more curiosity than he for natural phenomena. No text authorizes the belief that Hickman, because of his membership in the Royal Society of London, could have communicated a taste for natural history to Buffon.

Perhaps the four letters addressed to Cramer during these months of travel allow us to understand why he was so uninterested in nature at that time: as far as the sciences were concerned, Buffon was still fascinated only by mathematics. His letters to the young Calvinist scholar speak neither of women nor of gastronomy, but only of problems to be solved. Nothing was mentioned about his trip, other than some books he bought and some meetings with Italian mathematicians. As for the rest, there were only curves and equations. Buffon was inconsistently successful in finding the solutions to the problems that Cramer proposed to him, and we do not know if he knew how to solve those that he proposed to Cramer in turn. At least it is apparent that he either was familiar with the mathematical research of his time or was gradually discovering it. Clairaut, a mathematician of great standing, had respect for him. Clairaut had been elected to the Academy of Sciences at the age of sixteen, against the Academy's rules, and at eighteen was officially accepted into this illustrious company. Everything thus encouraged Buffon to persevere in this path.

Of all these letters there is one that merits a closer look, a letter that even Buffon judged to be important enough to publish in 1777 in his *Essai d'arithmétique morale* [Essay on Moral Arithmetic]. It discusses a problem concerning games of chance, which had first been brought up by a mathematician from Basel, Nicolas Bernouilli, and which would later become famous under the name of the Saint Petersburg paradox. The game was that of heads or tails (at the time it was called "cross or tails"), where one of the players, Pierre, tosses a coin into the air "as many times as is necessary for it to land heads." "If that were to happen on the first toss," the other, Paul, would give him a crown. He would give him two if it happened on the second toss, four if on the third, eight if on the fourth, sixteen if on the fifth, etc. "The question is, therefore, how much Pierre must give to Paul to make up for what Paul loses, or, in other words, what is the amount equal to Pierre's expectations, as he can only win."[8]

The answer given by a calculation of probabilities is quite simple, and Buffon explained it very clearly: Pierre had one chance out of two to win a crown, one chance out of four to win two, one chance out of eight to win four, and so on. For each toss, his expectation was therefore a half-crown "because the number of crowns increases at the same rate as the number of probabilities decreases; therefore the sum of all expectations is an infinite sum of money, and consequently Pierre must give Paul the equivalent of half an infinity of crowns."

Let us leave Buffon's comments of 1777 and come back to the 1731 letter to Cramer. The mathematical calculation is indisputable: "However, far from giving an infinite equivalent, no one with any sense would want to give twenty crowns, or even ten. The reason for this disparity between the mathematical calculation and common sense seems to me to consist in the difference between money and the pleasure that money can buy. A mathematician, in his calculations, sees the importance of money only in its quantity, that is to say, in its numerical value; but a moral man values it otherwise: in the benefits or pleasure that he can obtain with it." A man is no happier with a billion than he is with a hundred. Buffon introduced two ideas here that according to a modern historian of mathematics, René Taton, are "two fundamental notions of the modern theory of probability: the principle of mathematical expectation and the concept of utility."[9] That was impressive for a beginning amateur.

The first conclusion to be drawn from this example is that mathematics, for all its logical rigor, is incapable of representing reality—in this example, a psychological reality. Buffon came back to this idea and developed it in the beginning of the *Natural History*. The followup to this letter went further. A man, said Buffon, would be wrong to gamble with his basic necessities. He must gamble only with that which is superfluous to him. But what is defined as necessary? "The sum that is needed to continue living as one has always lived. . . . I suppose that the basic necessities cannot obtain new pleasures for us, or to be more precise, I would count as nothing the pleasures or benefits that we have always had, and then I would define as superfluous that which could bring us other pleasures or new benefits; I would add that the loss of basic necessities would be deeply felt; as such it cannot be compensated by any expectations, and to the contrary, the feeling of the loss of what is superfluous is limited, and consequently it can be compensated; I believe that a man feels that truth himself when he gambles, because a loss, especially a large one, will always cause more pain than a gain of an equal amount will give pleasure."

This statement expressed an entire philosophy of existence, made up of bourgeois caution and a taste for reasonable pleasures, as well as a philosophy of ambition. If the basic necessities permit a man "to continue to live as he has always lived," then all lasting increase in wealth, which at the outset might be considered superfluous, incorporates itself so to speak into these basic necessities. A man cannot limit himself; he must always progress. A reasonable man should not even risk that which is superfluous, "unless . . . he sees the expected sum to be necessary for success in an undertaking that gives him proportionately more pleasure than that same sum with which he gambles, and it is because of this way of looking at future happiness that one cannot make any rules; there are people for whom the expectation in and of itself is a greater pleasure than the enjoyment obtainable by spending what they gamble." Pascal had already said, "A man spends his life happily by gambling a little every day. Give him in the morning the money that he could win every day, on the condition that he not gamble, and you will make him unhappy."[10] For Pascal, this was a mark of weakness. For Buffon, happiness boiled down to the pleasure of "succeeding in an intention." One only needs to take reasonable risks.

If mathematical logic is but little help in these matters, a concrete estimate of basic necessities could be useful. "Therefore, in order to reason more accurately about all these things, it is necessary to establish several principles. . . . The basic necessities for a king will be, for example, an income of ten million (because a king who had less would be a poor king); the basic necessities for a man of good station will be an income of ten thousand livres (because a man of good station who had less would be a poor lord); the basic necessities for a peasant will be five hundred livres, because unless he is in total misery, he cannot spend less to live and feed his family." Buffon noted without comment the inequality of social conditions, which is not surprising. At least he knew that below a certain "vital minimum" the peasant was destitute. As for himself, he did not want to be "a poor lord." And he proved it immediately.

His father, widowed for a year, had taken it into his head to remarry "at the age of fifty . . . a girl of twenty-two, who has almost nothing to offer other than her youth. You see, Sir, the wrong that this matter would do to me; you can also judge it from the force with which I oppose it." That Buffon spoke of the problem to Richard de Ruffey[11] was due not just to their old friendship. Ruffey was at this point the president of the Burgundy parlement, and Buffon undoubtedly had already thought of a possible lawsuit. This precocious remarriage of a widower with children was too common at this time for Buffon to see any offense to the memory of his mother. It was his personal interests that were in play. The marriage took place anyway; on December 31, 1732, Benjamin-François wed the young Antoinette Nadault, who was of course a relative. Buffon had ostensibly left for Paris and did not attend the wedding. He then threatened his father with a lawsuit in order to obtain the fortune left him by his mother; the situation was even more serious because his father had not been a good administrator. In 1729 Benjamin-François had been forced to sell the holding of Buffon at the very same moment that his son began to be called Leclerc de Buffon. Buffon mobilized all of his friends from parlement: Ruffey, Bouhier, maybe even De Brosses. We do not know if the lawsuit actually took place, but Buffon was compensated; he could buy back the holding of Buffon and have at his disposal the inheritance from his uncle Blaisot. Not for one moment did he doubt the legitimacy of his action, which justified, he said to Ruffey, "the discontent of a well-born son, caused by a father who was either heartless or carried away by passion." The relationship between father and son remained strained for a long time.

Since July 1732, however, Buffon had settled in Paris in order to continue his scientific career. By the spring of 1733, at the age of twenty-six, he had already become what he would remain for his entire life: a scientist in Paris and a landowner in Burgundy. He knew how to handle two careers at once, and he played one against the other with an astonishing gift for seizing the moment, using his friends, and putting his scientific talents at the service of his wealth and his wealth at the service of his science.

CHAPTER II

A Gentleman Academician

The Conquest of the Academic Establishment

In July 1732 Buffon settled in the Faubourg Saint-Germain in Paris, in the home of Gilles-François Boulduc, apothecary to the king. This residence was certainly not chosen because of the community's charm or Buffon's intention to enter a profession that required apprenticeship with a master. How did Buffon know Boulduc? We do not know, but we know that Boulduc was not just any master-apothecary. He was first apothecary to the king, and since 1716 he had also been a member of the Royal Academy of Sciences. In 1729 he had just succeeded his father, Simon Boulduc, also an academician, as professor of chemistry at the Royal Botanical Garden. In 1695 the apothecaries at the Garden had obtained the creation of a "secondary chair" for chemistry that was reserved for them and gave a sorely needed scientific luster to their guild.

It is easy to understand why Buffon wanted to introduce himself into the academic world: he wanted to prepare his candidature for the Academy of Sciences, which for a man of twenty-seven was hardly extraordinary. Founded by Colbert in 1666 and reorganized in 1699, the Royal Academy of Sciences was not the conservatory of national scientific glory that it has become today.[1] It was possible to enter it quite young—at the age of twenty—as an "assistant," the lowest rank, and at twenty-five at the higher ranks of "associate" and "pensioner." The Academy was divided into six classes: geometry, astronomy, mechanics, anatomy, chemistry, and botany. Each class had a limited number of spaces. When a vacancy occurred the Academy would elect to a vacant position of assistant within a certain specialty. After that the assistant would rise more or less quickly up the hierarchy to become a pensioner and would finally retire as a "veteran."

Only the pensioners received a *pension,* or salary, and the veterans were allocated a retirement pension. The assistants and associates enjoyed not only a very honorable title but also all the privileges of the Academy, which were quite significant: the freedom to pursue works of their choice, grants for their research and publications, exemption from presenting their works before the censor providing that they had the Academy's approval, and access to official positions that were quite lucrative. The elections, which were for all practical purposes in the hands of the pensioners, followed a complicated ritual: the concerned "class" proposed an order of preference among the candidates; the entire Academy chose and ranked two names that were then presented to the king, or in reality to the appropriate government minister, who then made a definitive choice. No particular title was required in order to be a candidate for the position of assistant; it was enough to have demonstrated a promising talent. In fact, the Academy, which functioned as a national center for scientific research, was largely made up of young men devoted to research.

Buffon's friends were convinced that he had promising talent. Now the academicians needed to be convinced. So why choose a chemist to introduce him? Without inventing some devious strategy, one can assume that Buffon was not very sure of his ability to make a career of mathematics; perhaps he understood that its abstract and purely logical nature did not fit with either his intellectual temperament or his deep desire to understand reality. A science of matter, like chemistry, might satisfy him more. Maybe he was just simply unable to choose, not being, as Condorcet would later say, "trained for any science in particular" because "all that stimulated his thought or expanded his understanding charmed him." This tendency was quite noble, but it was not enough to get him elected to the Academy. The groundwork had to be laid, and the candidate had to prove himself.

For the groundwork, Buffon already had some allies in place: Boulduc, Clairaut, and probably also Maupertuis, a good friend of Clairaut's. In 1732 Maupertuis, almost thirty-four, was a brilliant mathematician and astronomer. He had just published a revolutionary work, the *Discours sur les différentes figures des astres* [Discourse on the Different Shapes of the Heavenly Bodies], in which he openly sided with Newton over Descartes, and whose appearance brought him into contact with Voltaire and Mme. de Châtelet, themselves whole-hearted Newtonians. It so happened that Mme. de Châtelet's husband, with whom she had remained on good terms though they no longer lived together, was governor of Semur-en-Auxois, less than five leagues from Montbard. Since Buffon was both a Newtonian and from Montbard, he had ample opportunity to become acquainted with Maupertuis. Moreover, he now had protectors in high places. Was he sure enough of his success to announce his coming election to his friends? In any case, Bouhier felt confident enough to tell one of his correspondents on January

23, 1733, that Buffon "was at the point of being accepted at the Paris Academy of Sciences."

Buffon still had to write something, and he wisely chose to stay within mathematics, where he felt more at ease. He thought up an original subject, one that combined in a novel way the two disciplines he knew best, probability and differential calculus. At the time, anyone could ask the Academy's opinion of a new scientific work. Buffon made use of this opportunity and most probably in early 1733 presented to the scholarly gathering his "Mémoire sur le jeu de franc-carreau" [Memoir on the Game of *Franc-carreau*]. It was a work that was already fairly polished, and it is possible that he had already written it and even had it approved by mathematician friends at the time Bouhier wrote of Buffon's success.

The original text of the paper has disappeared, and we know of it only through contemporary accounts, in particular that of Fontenelle in the *Histoire de l'Académie des Sciences* [History of the Academy of Sciences] for 1733, and in the text that Buffon published in 1777 in the *Essay on Moral Arithmetic*. Apart from a few corrections, this text probably corresponds to the original.[2]

The game of franc-carreau is very simple: "In a room with a parquet floor or tiled with equal-sized tiles, of whatever shape, a coin is thrown into the air: one of the players bets that the coin will fall franc-carreau, that is, on one single tile; the other bets that the coin will fall on two tiles, that is, onto one of the cracks that separates them; a third bets that the coin will cover two cracks; a fourth bets that the coin will cover three, four, or six cracks. The question is: what is the outcome for each of these players?"[3] Buffon, however, immediately moved on to a slightly different problem: what must the proportions between the side of the tile and the diameter of the coin be such that the game will be fair, that is, so that all the players will have an equal chance of winning?

The significance of this problem, as Buffon indicated, was the application of geometry to the calculus of probability, a new idea since until then probabilities had been examined only in "relation to discrete quantities," that is to say, numbers. In the beginning, it was a matter of fairly simple geometry, and Buffon was only "amusing himself by doing the calculations" for tiles of simple shapes: squares, equilateral triangles, or hexagons. Notice in passing that Buffon differentiated between geometrical figures, where the lines have no thickness, and the real conditions of the game in a tiled room where "the cracks between the tiles have some width, giving the advantage to the player who bets for the crack." Here again, as for the Saint Petersburg paradox, the mathematical conclusions must be corrected in order to correspond to reality.

After this simple geometry, Buffon went on to more complicated problems, first by supposing that the coin used was not round but "a coin of another shape, like a square Spanish pistole." The 1733 text perhaps studied this situation at greater length, but the 1777 text passed directly to what has become the "needle problem", a classic problem to which Buffon's name is now linked and for which he is considered the founder of geometric probability.

"Imagine that in a room, in which the floor is divided simply by parallel cracks, someone throws a stick into the air, and one of the players bets that the stick will cross none of these parallels, and that another in opposition bets that the stick will cross some of these parallels: what is the outcome for these two players? It is possible to play this game on a checkerboard with a sewing needle or a pin without a head." In a great number of cases, the needle will cross or not cross one of the parallel cracks according to the angle that it forms with these cracks, and this angle will vary in a continuous manner. Buffon must therefore have had recourse to calculus and brought into play the formula for the area of a cycloid, established by Pascal in the seventeenth century.

This last point brings up the interesting question of the relation between Buffon's mathematical work and Pascal's. Newton's binomial, which was Buffon's first discovery, is directly linked to the arithmetic triangle, to which Pascal had dedicated several treatises written about 1654; one of the uses of the triangle, according to Pascal himself, was to "find the powers of binomials."[4] Pascal therefore could have served as a common source for both Newton and Buffon.[5] It is quite likely that Buffon also knew of Pascal's work on games of chance and probability, a connection that I have established in passing. No doubt the properties of this curve were common knowledge at the time, but Buffon's mention of it brings us back once again to Pascal. Moreover, when Buffon, having explained the needle problem, raised other "curious and even useful" questions that could be treated in the same way, the first that he quoted was the following: "How much would a man risk to cross a river on a more or less narrow plank?" This brings to mind Pascal's famous text: "The greatest philosopher in the world is on a plank wider than necessary above a precipice; even though his reason assures him of his safety, would not his imagination prevail?"[6] We find here, however, a difference in perspective between the two authors that we noticed earlier in relation to the passion for gambling. For Pascal, human weaknesses were beyond remedy. They were the obvious mark of man's contradictory nature and therefore of the original sin that separates him from God. Buffon had no nostalgia for a lost paradise. He always looked at things in terms of relationships, and he sought only to suggest through calculation reasonable conduct and to correct our excessive fears and hopes. There is no common ground between these two visions of the world.[7]

All things considered, Buffon's paper was not, mathematically speaking, perfect. There were several errors of calculation; the reasoning was flawed and often skipped over obstacles without always really seeing them. Even in comparison with the norms of mathematics of the eighteenth century, which were less rigorous than ours today, Buffon was not always attentive and precise enough. Neither the originality nor the strength of his ideas, however, can be denied. In previously published works on probability, nothing could have suggested to Buffon the type of problem he attacked. After him, aside from a brief passage by Laplace, who corrected an error in Buffon's calculation without even deigning to name him, it was not until the second half of the nineteenth century that these questions were taken up again. Today, they are found in unexpected applications and have even given birth to a new discipline, stereology,* which claims Buffon as its distant ancestor.[8] Events went his way, and this apparently arbitrary interest in applied mathematics led to methods that permitted the application of mathematics to geology, metallurgy, and cytology. What hurt Buffon's mathematical career was surely not a lack of competence or imagination but more likely a certain impatience that did not adapt itself well to the meticulousness of the discipline.

The Academy received Buffon's paper and entrusted it to two reviewers, who happened to be Clairaut and Maupertuis. This was lucky, if indeed it was luck. Their report, presented on April 25, 1733, was very complimentary and ended by noting, "beside much knowledge of geometry, there is much ingenuity on the part of the author." Even more worth remarking is that the paper was read by Clairaut to the Academy at the following meeting. Such a distinction was rare. In his *History* of the Academy for 1733, Fontenelle published a summary of the paper and spoke of the "first-class geometry to which this subject guided M. Le Clerc, who was happy to be led." In short, this first essay was a success, which augured well for the future.

While his friends were working for him in the Academy, Buffon had already returned to Montbard, where he had decided to keep his primary residence. Why not Dijon? Perhaps because his father was still living there, perhaps also because the diversions of the parlementarians did not seem interesting to him, judging from this excerpt of a January 29, 1733, letter he wrote to Ruffey in Paris: "Here is the news from the country: several days ago some intelligent young people who were playing blind man's bluff at the ball, whipped the young M. de la Mare; his mother, who was present, came forward and tried to protest; they answered by making fun of her, saying that she was silly and that it all was nothing more than a joke. At the Sunday concert, the Councilor Malteste met Mme. Jolivet on the stairs and placed something on her, to which you ask, what? . . . his hand down her bosom to her navel. She turned around and, rightly incensed, slapped him

---

*Stereology is the science of constructing a three-dimensional structure from a thin cross-section taken from a rock, metal, or cell.

violently. He responded with atrocious insults; we still do not know how this will turn out. . . . Another adventure: a young treasurer, whom you know well, was the recipient of a slap at the ball, which he apparently received calmly; there were fortunately only two ladies and five p[rostitutes] there. The first two had to go outside, as everyone was using the others behind their backs." The distractions of these noble magistrates were not very refined. Buffon was not actually shocked and rather enjoyed stories that were a bit risqué, but the idea of staying in a city where he would not be considered a parlementarian's equal could not have pleased him. He liked his Dijonnais friends, and he would, if need be, help them, but from a distance.

In Montbard, on the contrary, he was the master, practically the lord. He very quickly demolished the modest house where he was born, bought some neighboring buildings, and had built in their place a vast mansion that he would call with pride "the castle." He also quickly took possession of the knoll that dominated it, razed some of its medieval fortifications, and built in their place a great park with terraces, where he installed a menagerie, a laboratory, and his workplace. Some historians are scandalized to see one person take possession of a public domain in such a manner. But it is very unlikely that Buffon acted without the tacit permission of the authorities, particularly the prince de Condé, who was the governor of Burgundy, and his protégé Claude Monseigneur, "Architect and Inspector of State Buildings." The province must have had quite a number of unusable feudal buildings not to be upset at getting rid of one of them. It was not the practice at the time to raise a scandal over the demolition of "gothic" buildings.

Above all, in Montbard, Buffon was close to his land, which he would use to advance his scientific and political career. From 1733 on, he responded eagerly to the demands of the prince de Condé, who hoped to put together a mineralogical collection in his castle in Chantilly. Buffon had the Montbard area searched in order to send interesting specimens to the prince. In doing so, he was working for both himself and his family. First of all, he gained the good graces of the Burgundy prince-governor, who would soon be very useful to him. At the same time, he also earned those of his representative, M. de Montigny, the general treasurer of the States* of the province. His cousin, Pierre Daubenton, Montbard's subdelegate, was under the command of Montigny: since Daubenton was not very efficient in the handling of public finances, he needed the indulgence of his superior. Buffon would plead his cause and all would work out. The marvels and constraints of provincial diplomacy!

Much time passed before Buffon became interested in mineralogy again. Another opportunity to attract the favors of an important individual

---

*Ed. note: The provincial governing body.

steered him toward the research that perhaps determined his future career. This time it was Jean-Frédéric Phélypeaux, comte de Maurepas.
Maurepas, the descendant of a dynasty that had provided the secretaries of
state since 1610, had, as it were, inherited in 1715 at the age of fourteen
the office of secretary to the King's House, to which in 1723 had been
added the Department of the navy. As the secretary to the King's House, he
was the supervisory minister to the Academy of Sciences. Intelligent and
active, he understood immediately that the navy needed geometers and
astronomers and that the academicians needed the navy for their voyages
of exploration and astronomical observation. It was he who in 1736 sent
out two great scientific expeditions, one to Lapland, with Maupertuis and
Clairaut, and the other to Peru, with two other academicians, Bouguer and
La Condamine. The aim of these expeditions was to measure a degree of
the meridian, one from near the North pole and the other from near the
equator in order to determine definitively whether the earth is a sphere
flattened at the poles, as Newton and Maupertuis upheld, or flattened at
the equator, as Descartes and his disciples claimed. The trip to Lapland,
from whence Maupertuis would return with two charming Lapp women,
was relatively easy, despite the mosquitoes. The trip to Peru, where the
academicians had to cross the Cordillera of the Andes to make their la-
borious observations, was a long, painful, and perilous enterprise. As it
turned out, it seemed that Newton was correct. Not the least paradox of
this whole affair was that the Royal Academy of Sciences proved Newton's
superiority over Descartes.

The French navy, however, like the other European navies of the time,
needed not only astronomers but also wood to build its ships, and from
Colbert's time this problem had obsessed the ministers of the Navy.
Maurepas therefore turned to the academicians and in 1731 asked them to
study methods that would increase the strength and longevity of wood.
With great regret, the Academy had to answer that it did not have the
means to do the necessary experiments. Buffon, with the forests that he
owned near Montbard, had these means, and in May 1733, before even
entering the Academy, he started his experiments. We shall return to this
research, which he continued for many years, indeed for almost his entire
life, since in 1774 Buffon examined trees that he had planted forty years
earlier. His experiments attracted the attention of Maurepas, who would be
one of Buffon's most faithful protectors until Maurepas' disgrace in 1749.
They also contributed to guide Buffon toward botany, or more precisely
toward plant physiology, and perhaps even determined his future.

In the autumn of 1733 Buffon returned to Paris. The Academy of Sciences had not forgotten him: he was invited there on November 25 to read
"a written work on geometry"—in fact a study of a problem in mechanics.
On December 12 an opening for an associate astronomer was announced.
The Academy proposed two names: Maraldi, who was already an assistant,

and Buffon, who had done nothing in astronomy. Maurepas made it known that the king had chosen Maraldi and had ordered that Grandjean de Fouchy, an assistant in mechanics, should take Maraldi's former position in the astronomy class and that therefore an assistant position was available in mechanics. The election took place the following week. The members of the mechanics class proposed three names: "M. Le Clerc, Meynier, and the abbé Sauveur." The Academy proposed the first two to the king. We know that the king, or more truthfully Maurepas, had chosen Buffon on January 9, 1734. The maneuver had been elegant: to prefer Buffon over Maraldi would have been an injustice. Grandjean was more an astronomer than a student of mechanics, and neither Maynier nor Saveur, obscure mathematicians, would ever enter the Academy, even though the latter was an academician's son.

Buffon accepted his triumph modestly. "They have honored me a thousand times more than I deserve," he wrote to Bouhier. "They hastened the vacancy of the place that I fill at the Academy; they preferred me over distinguished candidates." This statement went perhaps a bit far, but after a success it is inappropriate to speak badly about one's unfortunate rivals. Suffice it to say that the election had been well prepared and that Bouhier was well informed when he wrote to the abbé Le Blanc, "A long time ago, I knew that they were preparing a place in the Academy of Sciences for M. Le Clerc. I am pleased that he has obtained it; he is most worthy of it."[9] For his part, Le Blanc had approved of the election by saying, "I knew very well that he had a good mind." This leads one to believe that not everyone in Dijon was as convinced of that fact.

After January 16, 1734, Buffon attended the meetings at the Academy, as stipulated by the regulations. The honor of belonging to this august company, however, was not enough for him to change his way of life. As early as the spring of 1734 he left for Montbard and did not return before the fall. This absence was not normal: the academic vacation lasted only from September 8 to November 11. Buffon had a good excuse: he worked in Montbard. It was there, however, that he also felt at home and lived the life that pleased him, with his beautiful brand new house and the woods where he directed the work on his logs. Clearly the election to the Academy was for him a means and not an end. He showed and would always show as much independence with regard to the Academy as to his place of origin. He used institutions but did not put himself into their service. For this some of his colleagues would not forgive him.

The Slow March toward Natural History

Buffon pursued with more energy than ever the completion of his "castle" and the improvements of his park. Le Blanc, who was free at that moment and claimed to know architecture, advised him and directed the workers. Meanwhile Buffon paid his respects to the local squires of the area, who must have taken an unfavorable view of the luxurious expenditures of this upstart bourgeois. "He just left to see several neighboring gentlemen," wrote Le Blanc, "and he will not take great pleasure in it because among the nobles, at least those around here, as you know,

> One does not find any of them tractable.
> Whether they are old or not, rich or lacking,
> How to live among them is beyond understanding.
> To make a long story short, they are unbearable.

I speak of them knowingly, I have seen several examples. Recently he has made me superintendent of his buildings, and I have just put down the trowel to write you, for as you may imagine, I have here in my castle thirty men to supervise, as many masons as laborers, and in order to have it done well, I have to set an example."[10]

The big task was the demolition of the old feudal castle, of which Buffon would keep only the outer walls, the dungeon and the "Saint Louis Tower," where Saint Bernard's mother was born. Terraces were built with stones from other buildings, which leveled off the summit of the mound. Through a series of stairs and monumental gates, Buffon could walk directly from his house into the park, which he had planted in the French style but with many exotic trees. Tons of plant soil had to be brought in, and the work would last for many years.

In 1734 Buffon undertook the creation of a tree nursery for Montbard. Scientifically, this nursery fit in very well with his research on trees, but Buffon found a way to make an advantageous financial operation out of it. In 1735 the triennial meeting of the Burgundy States was held in Dijon, presided over by the prince de Condé. Buffon came to pay his respects to the prince, and he was very well received. The following year, with Condé's influence, he sold his tree nursery to the province at a higher price, naturally, than what it had cost him, and he was named director with an annual salary of 1,200 livres. The trees had made a decidedly good return. Buffon, however, was not ungrateful. That same year, 1736, the prince had a son. Buffon celebrated the happy event by organizing a large celebration at Montbard, with the firing of cannons and musketry, concerts in the park, a fountain of wine in front of his house, which "ran abundantly and without interruption until midnight," and finally a large supper, with lights and fireworks, "which had been prepared several weeks beforehand, in anticipation of the happy news" and which "lasted more than an hour, to the noise of cannons and muskets, and the sound of all the instruments and the greatest number of echoes." To end it all, there was a ball and a light meal. In short, a princely entertainment.

Still, publicity was needed to make this local event known to Paris. Buffon saw to it and had Daubenton write a lovely letter to an imaginary poet, which was duly published by the *Mercure*.[11] In passing, this letter praised Buffon for the transformation at Montbard: "The chaos of the old castle has been sorted out: the god of gardens has regarded the placement with a favorable eye, and things are disposed in such a way as to attract the Muses and the Graces even by your song." At the end of the century, even Court-épée, the stern author of a *Description du duché de Bourgogne* [Description of the Duchy of Burgundy], congratulated Buffon for knowing how "to spread taste and charm in the ruinous masses of this vast site, as irregular as it is."[12] From 1734 on, he was no longer M. Leclerc, nor even Leclerc de Buffon, but simply M. de Buffon.

"If you were to cover my gardens with six-franc coins," Buffon said one day, "that would still not be the price that they cost me." Buffon used local labor; family lore has left us with an image of a paternal Buffon, preferring as workers the poor and the unemployed—up to two hundred of them, it was said. "It is a useful way to give charity without encouraging laziness," he affirmed, and he made sure that the workers were not overcharged for the baskets in which they transported the earth.[13] We will see that all of Montbard's inhabitants did not consider Buffon a paternal lord, but on this point, as with many others, the testimonies are contradictory, and his character is not easy to discern.

At Montbard Buffon was not only busy with his castle and his park. He worked with prodigious energy and in several directions at the same time; and what was most surprising was that all the research areas that he explored would one day, sometimes much later, appear in the *Natural History*. Daubenton's letter to the *Mercure* quoted above curiously points out that "M. de Buffon is at the Academy of Sciences and works on chemistry." In fact, in 1737 Buffon asked a Parisian correspondent to send him several chemistry books. On the other hand, he was already involved through Bouhier's intermediary in a controversy with Louis Bourguet, a professor at Neuchâtel, on the theory of generation. What most occupied him were the trees on which he continued his experiments and more generally, plant physiology, which at the time was called plant physics. He planted seeds in different soils, such as clay, sand, and gravel, in order to know which ones were best for different species. He stripped the bark from standing trees before cutting them to see if that would increase the solidity of the wood. As less bark was gathered this way than by traditional methods and the bark was used to tan leather, Buffon experimented to see if leather could be tanned with oak wood instead of bark. Above all, he started to experiment systematically on the resistance of pieces of wood to breaking. All of this would be communicated to the Academy, to which we will return later.

Buffon was not satisfied with conducting experiments in order to enrich agricultural technology rather than science. He did not want just to observe, he wanted to explain, and for this he needed a scientific knowledge of plant physiology. He looked toward England for this knowledge. Perhaps he had been attracted to England ever since his encounter with the duke of Kingston. He was not the only one attracted to England, since at this time a veritable anglomania reigned in France, of which Montesquieu and Voltaire were but two examples among many. Through Kingston, whom he had encountered again in Paris, Buffon had made many English friends. He encouraged the abbé Le Blanc to leave for England, which resulted in his publication in 1743 of *Lettres d'un Français sur les Anglais* [A Frenchman's Letters on the English], which would have a great success. The marquis d'Argenson, an unfortunate politician but a good representative of the "enlightened" French aristocracy, wrote of the English, "Everything about the English shows freedom of thought, with a deepness of thought that is stimulated by freedom."[14]

Buffon surely thought the same, and his anglomania was so notorious that legend tells of him spending a year in England. If he ever intended to make such a trip, he seems not to have realized it. But he did learn English, perhaps spoke it with Kingston and other Englishmen in passing, used it more or less successfully in fragments of letters, and knew it well enough to read and use works by English scholars. In forestry, he cited the fundamental work by Evelyn, of which the fifth edition had just been published in 1729. In 1735 he published a French translation of an important book by Stephen Hales, the *Statique des végétaux et l'analyse de l'air* [Vegetable Staticks; Or, An Account of some Statical Experiments on the Sap in Vegetables . . . An Attempt to Analyse the Air*], the original English version having been published in 1727.

At this time, when more and more scholars were abandoning Latin in favor of their native tongues, translation of scientific works became a necessity. There were as yet no professional scientific translators. To publish a translation was very worthwhile and was not considered beneath a scholar. It was even a way for a novice to become known. Buffon had not chosen his author by chance.

A member of the Church of England, like many other English scholars of his time, Stephen Hales (1677–1761), nonetheless dedicated his existence to scientific research. When he was still young, he had studied blood circulation and had measured the pressure of blood in the arteries. After that, in accordance to a widely accepted analogy of the time, he studied the circulation of sap in plants. As a mechanist and devoted Newtonian, he sought to explain the phenomena of plant physiology through the forces exerted on liquids in movement in living bodies. This work led him to the problem of

---

*Ed. note: This is a slight abbreviation of the actual title.

the role of air in the life of plants. This role, which he saw primarily from a physical and mechanical point of view, led him to think about the chemical composition of air. Following suggestions made by Newton himself, certain chemists explained the reactions between bodies by the fundamental physical forces of attraction and elasticity. Hales made a distinction between a fixed and attracting air incorporated within plant matter and a free and elastic air that was given off by it. He thus cleared the way for research in the area of plant respiration, which developed during the course of the century, and also the field of "air analysis," in which the new chemistry of Lavoisier later took root.

Buffon's translation bears witness to an excellent command of the English language and an exceptional attention to stylistic problems.[15] In his preface, which was his first published text, Buffon praised his author's taste for experiments. "It is through these keen, reasoned, and sustained experiments that Nature is forced to show her secrets: all other methods have never succeeded. . . . It is a question of knowing what happens, and recognizing what is presented to our eyes; knowledge of the effects will lead us imperceptibly to that of the causes, and we will no longer fall into the absurdities which seem to characterize all systems. . . . This is the method that my author has followed; it is that of the great Newton; . . . it is that which the Academy of Sciences has made a law to adopt; . . . in short, it is the path that has led great men for all of time, and which still leads them today."[16]

Hales's book was indeed a series of carefully described experiments with only a few comments. Hales himself described his method in these terms: "It is consonant to the right method of philosophising, first, to analise the subject, whose nature and properties we intend to make any researches into, by a regular and numerous series of Experiments: And then by laying the event of those Experiments before us in one view, thereby to see what light their united and concurring evidence will give us."[17] Hales was not satisfied with just doing experiments, however; he also proposed theories that his experiments did not always support.[18] He wanted to introduce Newtonian physics into physiology and therefore insisted on the roles of heat, attraction, and fermentation as fundamental forces of life. Buffon for the most part remained faithful to this model. More important, Hales preoccupied himself less than his contemporaries, and even completely ignored, botanical classification and the organic structure of plants. What interested him were the life processes of plants and the forces that governed them.

It seems Buffon immediately adopted this viewpoint, which gave to his work its originality. The scientific language of the era, faithful to that of Aristotle, distinguished between the "history" of a natural object and its "physics." History was content to describe, in the most precise manner possible. Physics sought to explain the function of an object, its laws and causes. For many naturalists of the time, this work was premature, even suspect. Fontenelle, the "Perpetual Secretary" of the Academy of Sciences, thought that natural history was "perhaps the only physics within our grasp," and in 1740 the *Journal des Savants* commented, "One must wait until there is a complete natural history before creating systems of physics."[19] Hales showed Buffon that a "plant physics" could be started immediately. Buffon therefore knew what he was doing when he chose to translate "an English work of physics . . . whose discoveries have struck me so forcibly and so greatly overshadow that which is seen in other works of its kind."[20]

Clearly Buffon had written this emphatic praise of experimentation in his preface in order to disarm critics. Others were saying much the same thing, but the experiments of which naturalists spoke were often no more than observations. It was within physics that "experiments," in the modern sense of the word, were carried out. Was Buffon aware of the difference and was he playing with the word, or was he unconsciously being ambiguous? It is impossible to know. We see here for the first time a difficulty that we will encounter again, for it is connected with the thought of a man who often had deep intuitions but who was not always capable of finding the words to express them.

CHAPTER III

From the Academy to
the Garden's Door

A Well-Organized Life

Starting in 1735, Buffon adopted the unchanging rhythm of life that he would keep up until his death. In the spring he left for Montbard; in the fall he returned to Paris. No event, no new function, no official responsibility, no criticism, be it open or veiled, would make him abandon his customary long stays in Montbard. Those who worked with him or were under his orders had to adapt to his lifestyle. And everywhere, the same rule was in force: do not waste time.

In Montbard the day started early. Buffon, however, enjoyed his sleep, and early mornings were painful to him. "I loved sleep in my youth," he said of himself, "it relieved me of a lot of time." He tells how, because he was "unhappy with himself," he had asked Joseph, an elderly servant, to wake him before six o'clock, promising him a crown each time he succeeded. One morning, having run out of arguments, Joseph pulled off the bedclothes and poured a bowl of cold water on his master. He received his crown, and Buffon ends the story by saying, "I owe ten to twelve volumes of my works to poor Joseph."[1] Buffon was, and would remain until his death, a formidable machine for work: fourteen hours a day for forty years.

As soon as he got up, he dressed, ate frugally, and then started to work. He read, wrote, and prepared the papers that he would present to the Academy of Sciences, soon followed by his translation of Newton. He was not a man, however, to stay shut up in his office all day. He took great pleasure in surveying his lands and forests. "I have often spent entire days in the woods," he wrote, "where it is necessary to be called from a distance and to listen attentively in order to hear the sound of the horn and the voices of dogs or men."[2] Nature was not then as tame as she is today, and

[28]

the forests were still mysterious and slightly frightening. For a naturalist, walking alone in them was a memorable experience.

Above all, it was in nature that he could be active. During a passing illness, Buffon wrote, "I'm furious to be kept back in my room and to not be able to fell trees and do my experiments."[3] He was in his element in the fresh air, among his workers, having the irrigation ditches dug and the trees planted in the nursery of Montbard, overseeing the development of the park, stripping trees of their bark or cutting them down to test their solidity. A rural man and a man of action, he was not an abstracted intellectual. The knowledge that he wanted to gather had to be useful, first of all to him, and then to others. In Montbard, he thought and acted; and he was happy. He was the master because he was at home. He decided, he paid, he was free. His will was written in his buildings, his terraces, the thousands of trees that he had had planted in regular rows which grew slowly, writing into the countryside the order desired by human reason, the bearers of future profit. Future, because he already knew that "the great worker of Nature is time," and it was with nature that he worked; it was she that he wanted to put into his service. He could not, like an engineer, who works on inert material, immediately realize a project drawn up on paper. He could not live in the immediate moment. He incessantly prepared for the future. He lived and thought in time and with time. For him, too, time was also a great worker.

Meanwhile his park was being built, and here he soon set up his workplace, a small building placed on a rampart at the top of the hill, a single naked room furnished with two tables of dark wood, one for him and one for his secretary. The engravings of the *Histoire naturelle des oixeaux* [Natural History of Birds] that can be seen there today were not hung until much later. For quite a long time, there was only one engraving on the wall, a portrait of Newton. No books, few papers. The distant tops of the hills were visible through the two windows. It was here that he would write the majority of his works, in the solitude of his thoughts but also, in exchange, with the freedom of his solitude. Here again he was his own master.

For all that, he was not cut off from the world. He started receiving his Parisian friends, in particular Helvétius, the young *fermier general** already fascinated by philosophy and whose book *De l'esprit* [On the Mind] created a scandal in 1758. His visits were not for social reasons. "We are here three friends who lack only beards to be philosophers," wrote the inevitable Le Blanc. "I put M. de Buffon at the head, as both our host and our mentor. The second is M. Helvétius. . . . He is indeed a philosopher right down to his fingertips. He is also very talented with poetry. . . . Yours truly is the third. . . . We work in this manner, each one on his own: one solves prob-

*Ed. note: A tax collector who paid a fixed sum to the king prior to collecting taxes, and whose income was the difference between the taxes collected and the fixed sum.

lems, the other composes verses, the third writes about morals and the customs of nations. We live almost like three hermits, we hardly see each other except at meals, and the only difference is that we stay at the table a little longer and we speak there more of Newton or Descartes, Virgil or Racine, than of the Desert Fathers."[4] Buffon and Helvétius had many things in common, most important a great admiration for Locke's philosophy. In any case, Helvétius was very happy at Montbard, where he spent more than two months in the fall of 1739.[5] Buffon, however, was not only occupied with Newton. He was looking for a good wine for the duke of Kingston, and he did not forget his friends from Dijon, especially since the nursery of Montbard required him to correspond with provincial authorities.

He had to return to Paris, however, at the latest by Saint Martin's Day, or November 11. That day marked the solemn public reassembly of the Academy of Sciences, and it would have been improper not to attend. Still, Buffon was not always there. He did not like Paris. In February 1738 he wrote to Le Blanc, "I am delighted when I think that you rise every day before dawn; I would like to imitate you; but this miserable life in Paris is contrary to these pleasures. I supped very late yesterday, and I was kept up until two hours after midnight. That is not the way to rise before eight in the morning, and one's head is not very clear after six hours of rest! I sigh for the tranquility of the country. Paris is hell, and I have never seen it so full and so stuffed. I, unfortunately, do not have a taste for sticky involvements; at any moment, there are many endlessly going on. I would rather spend my time making water flow and planting hops than to waste it here with useless errands, and more uselessly paying court."[6] Let us be clear about this: this was no longer the reign of Louis XIV, and no one scratched any more at the door of the king's chamber. It was his ministers, or at least the senior civil servants and the people in high places, that Buffon needed to see. Why? Simply because one never knew, and it was important not to be forgotten. Buffon probably did not have any specific ideas, but knew that he had to be known.

Friends were not always reliable either. Take, for example, the duke of Kingston. Once again resident in the capital since 1732, he had become a very Parisian figure. "I dined with Milord Quinston [Kingston]," wrote the abbé Prévost. "He is an extremely likable man whose intelligence is astonishing, for no matter what subject is put before him, he reasons with it as if he were the most accomplished master of the art." The Burgundy wine that Buffon had obtained for the duke had perhaps gone to the abbé's head. Naturally, Kingston, who had conformed to the customs of his compatriots, had become the official lover of a young woman at the Opéra, Mlle. Carton. Then he met Mme. de La Touche, and things became more serious.

Mme. de La Touche was the youngest of three natural daughters of the wealthy financier Samuel Bernard and his recognized mistress, Mme. Fontaine. At the age of seventeen she had married a M. Vallet de La Touche, whose least fault—or greatest virtue—was that he was always traveling. Mme. de la Touche therefore lived with her mother in a beautiful house that her father had built in Passy for his mistress and their children. There were always many in Mme. Fontaine's house: young aristocrats, actors, literary people, artists. Some came to pay their respects to the financier, others to his daughters, one of whom would become Mme. d'Arty, the mistress of the prince de Conti, and another, Mme. Dupin de Francueil, the wife of the financier Dupin de Francueil and Jean-Jacques Rousseau's protector. All three daughters were very beautiful. Kingston fell in love with the youngest, and his passion was returned: he was surely not the first.

Returning from Italy in 1736 after a two-year absence, M. de La Touche discovered the situation, was not happy, and made it known. Kingston persuaded his mistress to flee with him to England. He left for London and she left Paris nine days later, toward Holland. No one was taken in, and all Paris had a good laugh over it. Samuel Bernard, beside himself, swore that he would "have the nobleman's head cut off, at least in effigy." This was understandable: had not some practical joker pasted a poster on his door that said, "Lost: a beautiful bitch, brunette, with a dog from England named Milord. One hundred louis reward"?[7] He and his mistress were sullied by the scandal. An anonymous song said:

> What! That La Touche gone off
>   With Kingston
> Would an entire distraught family
>   Yell thief?
> From that where can
>   So much surprise come?
> It is a fountain that will be
>   Lost in the Thames.

M. de La Touche lodged a formal complaint. M. d'Arty, the brother-in-law of the guilty party, and Buffon, the duke's friend, organized a counterattack by spreading "a huge exposé" of the husband's conduct, that "ugly little man."[8] Buffon intervened all the more willingly, since Le Blanc had joined the fugitives in England. The lawsuit dragged on and M. de La Touche never won, but he had his revenge in another fashion. The arrival of his wife did not please everyone in England. In a satire imitating Horace, and later in his *Dunciad,* Alexander Pope, the famous and formidable poet who was also on bad terms with the Montagu family, made transparent and insulting allusions to this affair. More prosaically, Lord Bathurst wrote to Swift, "I do not need foreign furniture. My neighbor, the duke of Kingston, has imported a piece, but I don't believe that it is worth the transportation costs." The two lovers would separate in 1750. Mme. de La Touche returned to Paris in 1757 and died in solitude in 1765 without ever seeing her husband again.[9]

When Buffon intervened and tried to silence M. de La Touche, he did so mostly because of his friendship with Kingston and Le Blanc, for he was always very mindful of the duties of friendship. At the same time, however, he made some important contacts. One of the rare salons that he must have frequented was that of Mme. Dupin de Francueil, the sister of Mme. de La Touche. Jean-Jacques Rousseau has left a dazzling description of this salon: great ladies and lords were seen there along with many literary figures, among them Fontenelle, the abbé de Saint-Pierre, Voltaire, and many others including Buffon, who was still only a novice.[10]

He was no doubt more at ease with the young Newtonians from the Academy of Sciences, who often got together: all were involved in the same battle against the Cartesian old guard, and life had not yet separated them. The youngest and the most talented was Clairaut, but the senior member was Maupertuis, whose trip to Lapland had made him famous. Voltaire was not there: he was in Lorraine, at Cirey castle, with Mme. du Châtelet. There these two studied Newton and carried out experiments in physics. At the end of 1738 Buffon planned to visit them.[11] Before going, he saw Mme. Denis, Voltaire's niece, who was angry at being buried in the country and preferred to be in Paris. Buffon, who in 1736 had so admired the *Épître à Uranie* [Epistle to Urania], was in 1738 much less enthusiastic about *Elements of Newton's Philosophy.* "I had thought that Voltaire would succeed very poorly in commenting on Newton," he wrote.[12] Did Voltaire's analysis of Newtonian science seem insufficient to him? Or was it Voltaire's declared deism and his conception of a science necessarily limited by the free intervention of God? We do not know, but it would certainly appear that these two men did not share the same philosophy.

Voltaire ignored him, or pretended to. As someone who could never stay long in one place and who always felt ill and passed from enthusiasm to depression, he seems to have envied Buffon's strength and balance. "That one travels to glory by different paths," he wrote to Helvétius, "but he also arrives at happiness, and he carries himself marvelously. The body of an athlete and the soul of a sage, that is what is needed in order to be happy."[13] The preceding year, when Helvétius was in Montbard—and Voltaire knew that Buffon would read his letter—Voltaire had written to Helvétius, with a perfectly false modesty, "I am the lost child of a party of which M. de Buffon is the head."[14] And further, "He pleases me so much that I would like to please him."[15]

Voltaire liked to please and Buffon did not mind being admired, especially by such a great man. But Voltaire's comment had perhaps a more precise motive. He was at the time truly interested in the sciences. At Cirey, Mme. de Châtelet and he had hoped to compete for a prize offered by the Academy of Sciences on "The nature and the spread of fire." The two lovers-become-rivals had, independently, experimented, consulted friends, and sent in a paper without communicating the results to the other, but neither one received the prize. In 1738 Voltaire had written an "Ode à Messieurs de l'Académie des Sciences qui ont été sous l'équateur et au cercle polaire mesurer des degrés de latitude" [Ode to the Men of the Academy of Sciences who have been to the equator and the polar circle to measure the degrees of latitude]. The poem did not add much to his literary glory; he sang the praises of both Maupertuis, who was still his friend, and the Academy. In 1741 Voltaire sent the Academy his paper "On motive forces." There is little doubt that Voltaire would have liked to enter the Academy of Sciences.[16] In the fall of 1739, when Voltaire sent Helvétius these letters so full of respect and admiration for Buffon, Helvétius had just been named an associate in the Academy, which gave him the right to vote for the election of a new member. This perhaps explains Voltaire's correspondence.

Buffon accepted the praise but was not taken in by it. In 1750, while Voltaire was in Berlin, they continued to write each other.[17] Five years later, Buffon prudently wrote to Ruffey, "I am delighted that you are in touch with Voltaire; he is indeed a great man, and also a very nice man."[18] The open breach in their relationship would come later.

Did Buffon frequent those salons that set the tone of intellectual life in Paris, and especially the salon of Mme. du Deffand? Apparently not, and it does not seem as if he would have been comfortable there. The conversation of the time, rapid, brilliant, superficial, in which a Fontenelle or a Montesquieu would excel, did not suit him any more than it would Jean-Jacques Rousseau—that would be something that the two men had in common. Even the theater, the great passion of the time, interested him little. In 1732 he had gone to see Voltaire's *Zaïre* and concluded, "I should have left instead of being suffocated there."[19] In literature, he judged that Marivaux was good only for "small intellects and affected people."[20] His tastes were vastly different from those of his century, and this made him aloof.

There remained the Academy of Sciences. Buffon occupied a subordinate position, which forced him to be respectful. He worked a great deal, presented papers often, and during the winter made up for his absence during the spring and summer. He built up solid friendships, which would serve him when the time came. An assistant in Mechanics, he had not yet had to confront the naturalists. In short, he bothered no one. Yet it was already evident that he was not suited for teamwork.

So much for his work, friendships, and contacts. What do we know of

Buffon's emotional life? Nothing, because there is nothing to know. Buffon, said a contemporary, "almost always reduces love to simple physics." No question of wasting time. No question, therefore, of playing the romantic lover. In speaking of his youth, Buffon would later say, "I had a little mistress whom I adored, and, well, I forced myself to wait until six o'clock before going to see her, often even at the risk of no longer being able to find her." Another contemporary would translate this poetically by saying that Buffon "was looking for his daily sexual pleasures in a class of woman . . . who would not take more time than the two minutes they say that angels take to cover themselves with their wings so as not to be jealous of our pleasures."[21] From this personal conception of love he would construct a philosophy. The combination of an active intelligence and a strong sensuality would also define his intellectual temperament.

Goodbye to Mathematics

During the month of March 1736, Buffon presented a second paper on the game of franc-carreau to the Academy, a paper that has not been preserved but that showed his desire to continue his research in mathematics. He had just taken up his correspondence with Cramer again by writing to him, "I am going to deliver myself entirely over to my taste for mathematics."[22]

It was not the Academy that benefited from these projects. In 1740 Buffon published a translation of a work by Newton, *La Méthode des fluxions et des suites infinies* [The Method of Fluxions and Infinite Series]. The original text, written in Latin, was quite old, and Buffon pointed out why it had never been published; he dated it around 1664 to 1671, which has been debated.[23] Only in 1736 had John Colson, a mathematics professor, published an English translation, accompanied by commentaries that Buffon judged not worthy of translation. Newton's text was an introduction to calculus for beginners, and Buffon was the first to say that it contained nothing new and would not teach mathematicians much. So, why the translation? Buffon answered this in his preface. He wanted to take a stand on three contemporary debates: the nature of infinity, the invention of the calculus itself, and the attack that Berkeley had just launched against the calculus. Buffon expressed his opinion on these three points clearly, even violently.

Let us start with the second question: who had invented calculus, Newton or Leibniz? The quarrel between the two men, or more precisely between Newton's disciples, secretly driven by their master, and Leibniz, was already legendary. The stakes, which greatly surpassed those of a simple argument of priority, were at the same time metaphysical and political. In

London, the Newtonians occupied key places in the scientific and intellectual community, with the result that Newtonian philosophy had become the scientific justification for a certain political and religious position and for a restoration of the Church of England.[24] Leibniz's philosophy opposed that of Newton in many areas, and this factor threatened Newton's influence. In addition, in 1714, with the death of Queen Anne, the elector of Hanover had risen to the English throne under the name of George I. Would he bring Leibniz, a Hanoverian, to London? The Newtonians feared so and started a campaign. From 1715 to 1716, a famous correspondence between Leibniz and the Newtonian Samuel Clarke brought out the differences between the two philosophies. The Newtonians had already accused Leibniz in 1711 of purely and simply plagiarizing Newton with regard to the calculus, and in 1712 the Royal Society of London published a collection of documents titled *Commercium epistolicum* which clearly upheld their position. Leibniz's death in 1716 did not put an end to the debate, since the Continental mathematicians generally sided with Leibniz, whose method seemed better to them, as in fact it was. We know today that the two discoveries were independent and were the almost inevitable outcome of the development of mathematics in the seventeenth century.

In his preface, Buffon rewrote the history of the calculus—drawing inspiration largely from a book that Fontenelle had published in 1727, *Eléments de la géométrie de l'infini* [Elements of the Geometry of the Infinite]— in which he sided strongly with Newton. He was rightly criticized for his lack of objectivity, and he thus became closely tied, as we shall see, with English scholars, whose point of view he blindly adopted. In France, furthermore, he became involved with Clairaut, Maupertuis, and Voltaire in a battle in defense of Newton. His translation and preface must be viewed from this perspective; historical objectivity was not his main concern. What did he really know about Leibniz? Maupertuis and Mme. du Châtelet would soon be interested in Leibniz, and Buffon himself would come across him again. Leibniz was only an obscure metaphysician and a jealous mathematician.

The debate on infinity tells us something about Buffon's intellectual temperament. It is an old debate, which focused on the existence of a real or, as philosophers said, an "actual" infinity. Everyone admitted that a series of numbers could be extended to infinity or that a quantity could be divided infinitely. But that was only a possibility. The question was to know if an infinite series of numbers really existed. Medieval theologians, considering God's infinity, wondered if an infinite universe was not the necessary consequence of a creator's infinite power, and at the end of the sixteenth century Giordano Bruno had come to this conclusion. At the end of the seventeenth century a lengthy evolution of ideas had led to the Newtonian conception of an infinite time and space and, therefore, an infinite universe.[25] For all that, the metaphysical question of actual infinity was not resolved. Descartes, a mathematician and philosopher, remained cautious: "Whenever I have dealt with infinity, it has been to submit to it," he had written to Mersenne, "and not to determine what it is and what it is not."[26] The idea of an infinite universe could not be rejected either, as it would "stipulate the limits of God's works." "To us who affirm infinity," he added, "it is not our responsibility to resolve the contradictions that normally are put forward on this subject, but we are freed of all difficulties by the ingenious and very true confession that we recognize our intellect as not being infinite, and therefore it is not able to grasp that which touches infinity."[27]

Calculus gave a new topicality to this philosophical debate, since it raised the question of whether the infinitely small quantities manipulated by the new calculus really existed. Leibniz did not believe so, assuming "that infinity and the infinitely small can be taken as fictions, similar to imaginary roots, these fictions being useful and based in reality."[28] In 1727 Fontenelle defended their real existence, and Buffon seemed at first to have accepted his argument.[29] He now attacked Fontenelle without naming him.[30]

As always, Fontenelle's thoughts were very subtle. He recognized the purely human character of mathematics: "The only things that exist in Geometry are what we put there, so to speak." Buffon later adopted this idea as his own. Precisely because of this point, however, Fontenelle carefully distinguished between metaphysical infinity and geometrical infinity: "It is not right to draw from Metaphysical Infinity objections to the Geometrical one." Geometry treats a reality that is solely its own: "All that is postulated as real is necessarily derived from the reality that is postulated in its object. The infinity that it demonstrates is therefore as real as the finite, and the idea that geometry has of it is nothing more than a supposition, which is only a matter of convenience that must disappear as soon as one makes use of it." In consequence, "an infinite number has as real an existence as finite numbers."[31]

Buffon rejected Fontenelle's conclusion, mainly because he did not differentiate between geometrical and metaphysical infinities. "The idea of infinity," he said, "is only an idea of absence, and has no concrete representation." Even "space, time, and duration are not real Infinities." Likewise, "there is no number that is at present Infinite or infinitely small, or smaller or bigger than an Infinity, etc." Because "Numbers are no more than representations, and never exist independently of the things they represent," they do not have a "real existence," and things themselves cannot be infinite. Buffon then concluded: "The majority of our errors in Metaphysics come from the reality we give to ideas of privation. . . . We must therefore consider Infinity, be it infinitely small or large, only as a privation, a retreat from the idea of the finite, which we can use as a supposition that in a few cases can help to simplify ideas and permit us to generalize the results in the practice of the Sciences; thus all the art is reduced to using that supposition by trying to apply it to the subjects considered. Whatever merit it has is thus in the application or, in short, in the use that one makes of it."[32]

There is what Buffon said about "the nature of this infinity, which in enlightening men seems to have dazzled them." What he did not see was that at the same time as he was criticizing Leibniz by following Fontenelle, his ideas were actually much closer to Leibniz than to Newton. His reasons, however, differed from those of Fontenelle because his commentary had its roots in Locke, the master of all the new philosophers. That is why Voltaire, and later d'Alembert, expressed themselves in an analogous fashion. These few paragraphs, however, tell us much about Buffon's philosophy. They tell us much, first of all, about the relationship between reason and the reality of things. Experience offers us only finite objects. It is through mentally suppressing this finitude that we conceive of the idea of infinity, but that mental operation does not teach us anything about the nature of things. Descartes had affirmed that the very existence of the notion of infinity in our thoughts proved the existence of an infinite being outside of ourselves, that is, God. Buffon, following Locke, rejected Cartesian idealism. This implied, contrary to what Descartes believed, that human reason was not necessarily capable of understanding Nature. The science that men have constructed is a purely human science. Buffon later returned to this point at length.

The rejection of actual infinity also implied the rejection of infinity itself, that is, a rejection of the absolute. The Buffonian conception of science was rooted in a philosophy of relativism. If indeed the absolute exists, man can know nothing of it with certitude. Curiously, Buffon supported Pascal's ideas in this instance against Descartes, but he did away with all their religious associations. The direct consequence of this philosophy was that mathematics does not teach us anything about reality. More precisely—and here Buffon distanced himself radically from Fontenelle—mathematics does not have its own reality. Fontenelle gave an intellectual reality to numbers and geometrical figures, independent of all physical and metaphysical reality. For Buffon, there was only physical reality. Thus, mathematics was only a tool, practical, even indispensable, but nothing more. We shall see these ideas developed at greater length again in 1749. We know that Buffon had them already in 1739.

The last argument in which Buffon intervened was the one that the idealistic philosopher Berkeley had provoked by attacking the metaphysical foundations of calculus.[33] This hardly deserves mentioning here, because it is clear that Buffon addressed it only to defend his friend the English doctor and mathematician James Jurin. Regardless of what he said, Buffon certainly had not read Berkeley's book attentively. Otherwise he would have seen that Berkeley's criticisms of the status of the infinitely small corresponded exactly to his own, although they were based on an extremely different metaphysics. As with Leibniz, the fundamental philosophical differences prevented Buffon from recognizing what they had in common. His attack on Berkeley was more satire than philosophical discussion. By intervening so lightly into a serious debate, Buffon exposed himself to criticism. The interesting thing about this episode is that it shows his friendship for Jurin and suggests that it was Jurin who had advised him in the Newton-Leibniz controversy.

After his translation of Newton, Buffon did not publish anything on mathematics until his *Essai d'arithmetique morale* [Essay on Moral Arithmetic] in 1777, in which he once again took up and developed several of the texts and ideas from his youth. In discussing this work, we must therefore also examine the contents of a paper, "Sur les mesures" [On Measurement], that Buffon presented to the Academy of Sciences in December 1738 and February 1739, of which the original text is lost and of which we are aware only because of the *Essay*.[34] At least Buffon left mathematics honorably. His translation of Newton was generally well accepted, his manner of deciding between Newton and Leibniz did not attract criticism, except on Cramer's part, and his ideas on infinity even received a certain degree of approval in the *Mémoires de Trévoux*, the Jesuit journal.

There remains the often-asked question: Was Buffon a mathematician? The answer of historians of mathematics is generally negative, but it is perhaps prejudiced. Technically he was not above reproach, but no mathematician is. The greatest criticism that can be given is that he had not considered mathematics to be an end in itself, nor had he accepted its logic as definitive proof, and he never lost sight of the realities behind the symbols. In a word, it was his realism that prevented Buffon from really being a mathematician.

Forestry and Work on the Strength of Wood

Even before his election to the Academy of Sciences, Buffon, in order to attract the good graces of Maurepas, undertook experiments to find a way of increasing the strength of wood for construction. From 1737 to 1744 he presented to the Academy, either alone or in collaboration with Duhamel du Montceau, an entire series of papers on this problem. It would seem that nothing could be further from mathematical research. Buffon worked in many fields, and he applied himself with equal enthusiasm to them all.

The advantage of his experiments with wood was that they were done in the forests surrounding Montbard, where Buffon liked to work, and they both increased his wealth as a landowner and advanced his career as an

academician. By working to improve his woods, Buffon followed the example of the large Burgundian landowners and also the great English lords
who were fascinated by new technologies for agriculture and breeding. At
this time, it was impossible to apply to actual practice the theoretical knowledge of a science still in its infancy. Buffon did what was done in many
fields during the eighteenth century: apply a rational and experimental
intelligence to purely practical research. It was a scientific spirit rather than
scientific knowledge that he used. In working with his loggers, he used
traditional methods and tested them through systematic experiments.

It will be recalled that Maurepas, minister of the Navy, had asked the
Academy of Sciences to study ways of improving the strength of wood for
construction, and that Buffon, although not yet belonging to the Academy,
had started his research in 1733. Another academician a little older than
Buffon, Henri-Louis Duhamel du Monceau (1700–1782), had started the
same research the year before. As a botanist, he had become known
through his work on the cultivation of saffron. He had a chateau near
Pithiviers but did not own any forests, and had to carry out his experiments
in the woods of Orléans, which were part of the royal domain. Here he had
less freedom than Buffon, who worked on his own property. The experiments that he had to do were not allowed by Colbert's edicts, and even
Buffon had to defend himself in court against the attacks of the supervisory staff of the administration of the Waters and Forests of Avallon.* The
judiciary power refused to bow down before the authority of the minister,
and the King's Council had to intervene. Thirty years later, Buffon was still
involved in lawsuits against the administration of the Waters and Forests.[35]

Maurepas had asked the two men to collaborate, which they did, and
together they presented their first papers before the Academy. Their relationship soured afterward. Duhamel had the feeling, perhaps justified, that
Buffon took credit for discoveries made by both. At the end of a presentation by Buffon, Duhamel said to him, "My dear colleague, you have a good
memory." To which Buffon replied, "I know how to take advantage of a
good thing wherever I find it."[36] The rivalry between the two savants wors-
ened, we shall see, and Duhamel ended up by joining the anti-Buffon camp
led by Réaumur in the Academy.

The ill feeling between the two men allows us to understand why Buffon
attracted so much hostility among his colleagues, a hostility from which his
reputation has suffered until today. Conscientious, precise, modest, without literary or scientific genius, Duhamel was the model of the average
savant, who advanced knowledge methodically and usefully without being
tempted by other intellectual adventures. For someone of that nature,
Buffon's penchant for hasty assertions, grand theories, and new ideas was
an offense to the proper scientific method. Add to it his political, literary,

*Ed. note: This institution had local jurisdiction over hunting, fishing, woods, and rivers.

and social successes, and the offense became personally and morally intolerable. Still it is the bold theorists who open new paths in science that scrupulous practitioners are incapable of imagining.

With or without Duhamel, Buffon presented to the Academy a series of papers on forestry, studying the best methods of seeding or planting trees, of choosing the most appropriate soils and exposures according to the species, of cultivating young trees, of protecting them from frost, and so forth. He knew most of what had been written on these subjects. He drew little from plant physiology theorists, who could not help him much, and was much more inspired by the treatises on forestry, above all those of English authors, whose concerns corresponded more closely to his own.[37] It was, however, primarily experimentation that guided him. He reported his tests, failures as well as successes, did not pretend to have made "admirable discoveries," and advised that "on the contrary, these are commonplaces, but their usefulness could make them important."[38] He also underlined the fact that these experiments were expensive, and that it was necessary not only to produce the best wood but also lower the cost of useless products. Pure science was not at issue here.

It is clear, furthermore, that Buffon was passionate about his trees. He had thought to write a major treatise on forestry, and when he published the outline for his *Natural History*, he announced a *Traité d'agriculture* [Treatise on Agriculture]. These projects were later abandoned, but they bore witness to his interest in his research. Buffon's originality here is considerable. He was the first to consider a forest not as a collection of trees but rather as an entity in itself, a whole in which the individuals maintained specific relationships and acted upon one another; in short, something which was a precursor of what today is called an ecosystem. He noted the relationships between the different types of trees according to the way they were grouped, and how the copse played a role in the growth of trees for timber. He even noticed the role of birds in the scattering of seeds and that of field mice making their winter provisions. In short, forestry became a new discipline. Buffon, it is true, later abandoned this research, and it was Duhamel who continued it and became the recognized authority in this area, but he must share the credit for beginning it with Buffon. Buffon never forgot his forests, not only because he continued to work in them for his own personal profit but also, and above all, because the *Natural History* would reflect his work there, both in Buffon's scientific theories and in his scientific and poetic images of nature—including the place that man, originally lost in the primeval forest, had won for himself in it and must still win.

The experiments on the strength of wood were of an altogether different character, even though they were part of the same project.[39] Here, Buffon was in an area investigated by the greatest scholars of the seventeenth century: the question of the strength of materials. Galileo, Robert Hooke,

Leibniz, and Mariotte, as well as architects and engineers and many others, had elaborated theories of physics and mathematics to solve a problem that fascinated theorists of the elasticity of bodies. In this area where theory and practice intersect, two intellectual tendencies came together: that of experiment and that of mathematical theory.

As expected, Buffon took the side of the experimenters. In 1729 a famous military engineer, Bernard Forest de Belidor, had published an important work, *La Science des ingénieurs* [The Science of Engineers], in which he attacked the practitioners who believed only in experimentation and upheld the absolute necessity of mathematic theorizing. Buffon, on the contrary, started by showing that the texture of wood itself was too variable to make mathematics useful. The results would differ depending on which part of the tree a sample was taken from. In addition, no mathematical extrapolation from the results obtained from a sample of small dimensions could be applied to serious predictions on beams of great length. Buffon, who boasted of having done more than a thousand experiments, compared the resistance of bars of small dimensions taken from the center of the trunk, near the bark, and at half the distance between the two. In addition, he measured the force necessary to curve and then break beams of different sections and lengths. He measured the density of wood by weighing it, and eliminated the question of the effects of drying by always working with green wood.

From all this work, he produced comparative tables, which he then had to interpret. It was here that he ran into difficulties, and here also that Buffon's mental character showed itself. The same experiment, repeated several times, predictably gave different results. Statistical methods for the calculation of errors did not exist at that time, so Buffon had to chance choosing the measurement that seemed best to him or take an arithmetic average, as was the current practice. He established proportions between weight and resistance, between a beam's strength, on one hand, and the section and the length on the other. He believed that these would allow him to correct Galileo's formulas, and above all to predict with reasonable certainty the resistance of a given beam. Practical use prevailed over mathematical formalization, but Buffon was already convinced that mathematics could not be applied to such a complex physical reality. It would be up to others to establish more satisfying mathematical formulas from Buffon's tables. In this sense, the work on the strength of wood marked Buffon's farewell to mathematics.

England and Free Thought

Buffon already had privileged ties with England: he was a friend of the duke of Kingston, he received Englishmen passing through Montbard, and he had translated Hales and Newton. He refrained from writing in English to the abbé Le Blanc, who was in London, because, he said, "I fear all that which makes me waste time, and I hardly like what mortifies self-esteem,"[40] but he wrote a few letters in Newton's language, preferably to French friends . . . In 1739 he became a member of the Royal Society of London, of which Newton had been president and whose scientific prestige was equaled only by that of the Academy of Sciences of Paris. Well before that, however, he was in contact with English scholars and had had books sent from England. As one of his friends said, he was "all English." He had hoped to go to England, like Maupertuis, Cramer, Le Blanc, and many others, and he had no doubt abandoned the idea with regret.[41]

England was very fashionable in France among the philosophes, natural philosophers, and aristocrats, but for contradictory reasons. A freedom of thought and expression unknown to the subjects of Louis XV reigned in England, particularly in political and religious matters. The Church of England had lost many of its powers. Aristocratic life was brilliant. Great aristocrats who were very rich associated with scholars and intellectuals, divided their time between London and their vast holdings in the country, and took an interest in philosophy, science, agriculture, and animal husbandry. It was they who created the "English thoroughbred"[42] and new races of livestock. They were exactly what Buffon would have liked to be. The abbé Le Blanc knew this fact well: in his *Letters* on England published in 1745, a third of the nineteen letters addressed to Buffon were devoted to agriculture, gardens, and the architecture of country houses.[43] In England, trade was not a victim of archaic prejudices, and aristocrats could earn money. England meant wealth and freedom. Foreign visitors, with Voltaire and Montesquieu leading the pack, little cared that behind this brilliant facade the political regime was corrupt and that social inequality was just as pronounced as in France.

Buffon established more and more contacts with English scholars. I have already mentioned Jurin, and there were many others, of whom the most distinguished was probably Martin Folkes, the president of the Royal Society of London, with whom Buffon corresponded for a long time. It was an Englishman, the microscopist John Turberville Needham, who collaborated most closely with Buffon in his research on reproduction. Buffon never had the same difficulties with English scholars that he had with his French colleagues. It is true that the English never had to live with him.

Buffon was not only interested in English science. Included in the books that he had had sent over for himself and his friends were many philosophical works, often rather unorthodox. Among these were Matthew Tindal's deist tract, *Christianity as Old as Creation,* which had scandalized England, the works of the pantheist John Toland, those of the deist Shaftesbury, which would enchant Diderot, and even Woolston's *Discourse,* which ridiculed the miracles of the gospel and also had caused a scandal. All this literature, which had inspired Voltaire, did not seem to have shocked Buffon. President Bouhier's circle was not very orthodox, and De Brosses, a friend forever, could only with difficulty pass as being devout.[44] Well before 1750, it would seem, irreligion or at least skepticism had made much progress among the aristocracy and the high bourgeoisie.

As to the religious opinions that Buffon himself held at this time, we have a unique testimony, a letter to his friend Étienne-François Dutour de Salvert, the former inspector of the farms in Avallon, who was at this time the tallage collector at Riom. Dutour was interested in the sciences and in philosophy, and Buffon had several of the works mentioned above sent to him. Was it through caution that Buffon wrote to him in English to inform him of the books' arrival?[45] Be that as it may, Dutour sent to Buffon his *Système de l'âme* [System of the Soul], and Buffon answered him, "Surely you did not write this work in two weeks, but to make Systems with that material is to build on sand. . . . Many people believe like you that the soul is material, that thought is a result of a particular organization, much like sound. . . . One seeks in vain how thought is created, because in the same way that the tip of a finger cannot touch itself and the eye cannot see itself, thought cannot understand itself; this is in fact an important proof of its materialism because it follows, in that respect, the nature of bodies."[46] Here Buffon clearly expressed the materialism of the eighteenth century, specifically the refusal to consider the human soul as a spiritual substance. We should not forget this text when we read the *Histoire naturelle de l'Homme* [Natural History of Man].

By the beginning of 1739, the essential traits of Buffon's intellectual personality, his lifestyle, and his taste both for scientific research and practical activity were definitively fixed, perhaps more so than he himself imagined. Still only an assistant in Mechanics at the Academy, he concerned himself with various important scientific topics. He lacked only an opportunity to assert himself in an important investigation and the necessity to enlarge his knowledge enough to be able to make a synthesis. This was about to come to him all at once.

CHAPTER IV

Entry into the
Royal Botanical Garden

In the spring of 1739 Buffon's academic career suddenly took a new turn. On March 18 he left the Mechanics section of the Academy and was transferred, probably at his own request, to the Botany section. His intellectual evolution had thus been recognized by the administration, and this change of position, which might seem unimportant, may have determined his future. The fortunate accidents that followed would have come to nothing were it not for this initial decision.

Eight days after this transfer a rather obscure botanical pensioner, Michel-Louis Reneaume, died. The Academy proposed three candidates as his successor, including Buffon and Bernard de Jussieu. Buffon was still only an assistant, and Jussieu was already an associate. At the end of May, the king named Jussieu to take Reneaume's place and Buffon to take Jussieu's. It was a first step up.

Buffon's work to improve forests and the strength of wood had caught the eye of the minister and even the king. Agriculture, especially botany, interested Louis XV, and he asked Bernard de Jussieu to create a botanical garden at the Trianon: it was in this garden that Jussieu established a new plant classification that he was too timid to publish; it would later be successfully brought to light by his nephew, Antoine-Laurent. It has been said that Louis XV invited Buffon to come to Fontainebleau, asked him how the woods that surrounded the castle and the town could be improved, and offered him the superintendency of all the woods in the royal domain. After much thought, Buffon refused.[1] Showing no anger at the refusal, the king granted him a pension of two thousand livres, which Buffon had not solicited, to compensate him for the expenses of his research and "to put him in a position to continue his experiments on wood."[2]

[44]

An Unexpected Nomination

On July 16, 1739, Charles de Cisternay du Fay, the intendant of the Royal Botanical Garden, died of smallpox. He was forty-one years old, and no one had foreseen his death. When Buffon heard of it, he was at Montbard. If several people immediately told him the news, it was certainly not only because he was in the Academy; he was surely seen as a possible successor. He had dreamed of succeeding Du Fay and had mentioned it at least to his old friend De Brosses. From Montbard, Buffon wrote immediately to the abbé Le Blanc in Paris, another old friend who was in the process of paving his own way to the Court, and to his colleague Jean Hellot, also a sure friend. This last letter is worth quoting:

"I was going, my dear friend, to answer your first letter, when I received the second. I already knew of the death of poor Dufay, which truly grieved me. We are losing a lot at the Academy: in addition to the honor that he gave the body through his worth, he was so well known in the world and the Court that he obtained many things and spared others many troubles; he did astonishing things for the Royal Botanical Garden, and I admit to you that he has put it on such good footing that it would be a great pleasure to succeed him in that place; but I imagine that his position will be greatly coveted. Even though I would have more reasons to claim it than another, I would not dare ask for it; I know M. de Maurepas fairly well, and in turn he knows me well enough that he would give it to me without solicitation from me. I will pray my friends to speak for me, to say loudly that I am suitable for that position; that is all that I can reasonably do at present. As for what you tell me, that M. de Maurepas is determined to keep the Royal Botanical Garden in the Academy, I do not have any difficulty in believing it; but even if he did not look at Maupertuis unfavorably, I do not believe that he would give him that position. There are other people at the Academy, however. Tell me if you hear about someone being named; in short, tell me what you find out.

"You could let fall a few words of M. le comte de Caylus's wishes to M. de Maurepas. There are some things in my favor, but there are also many against, especially my age; however, upon reflection, one might realize that the intendancy of the Royal Botanical Garden needs a young, active man who can brave the sun, who knows plants and the way to multiply them, who is somewhat knowledgeable in all areas that are asked of him, and above all who understands the buildings. . . . [I]t appears to me thus that I am what they are looking for; but so far I do not have any great expectations, and consequently I will not be greatly grieved to see this position filled by another."[3]

It takes great self-mastery, dignity, and cleverness to write a letter like this under these circumstances, when the possibility of such an extraordinary career change arises. Buffon was quite right to downplay these unreasonable expectations. Among the "other people at the Academy" was Duhamel du Monceau. He had the same skills, if not more, and the same social and intellectual profile. Duhamel was seven years older than Buffon and had entered the Academy five years before him. Above all, it was believed that Du Fay had promised him the succession or, as it was called, his "survival." Although this detail has been widely accepted, because of Du Fay's relative youth I cannot believe it without reservations. At this precise moment, Duhamel was studying in London, but the two Jussieu brothers, Bernard and Antoine, had reminded Du Fay of his promise.

What Buffon did not know was that Hellot had not waited for his letter before taking action. We have an excellent account of Du Fay's last days from Fontenelle: "He fell ill last July, and as soon as we realized it was smallpox, he did not wait to hear death spoken of in well-prepared euphemisms; he resigned himself and courageously asked for the last rites, which he received completely conscious."[4] Hellot was a great friend of Du Fay, who named him as executor of his will. Hellot was therefore very close to Du Fay during his last moments, and if we can believe Nadault de Buffon, had him sign a prepared letter that recommended Buffon's nomination.[5] According to Fontenelle, Du Fay "wrote his will, which was almost a part of a letter written to M. de Maurepas, to indicate who he thought was best suited to succeed him in the Intendancy of the Royal Garden."[6] Let us therefore not be too quick to accept the malevolent rumor of a signature snatched from a semiconscious dying man. People died more lucidly years ago than today.

Was Du Fay's letter decisive? Or had Maurepas already judged that Buffon, more enterprising and active than Duhamel, was more fitted to the office? Whichever, "choosing M. de Buffon was such a good proposal that the King did not want to make any other," Fontenelle said. Buffon's candidature was presented to the king on July 25, and the next day Buffon was named, "at a salary of 3,000 livres per year."[7]

How much time did he fret in Montbard while waiting for the news? We don't know. Maybe at least until August 10: the abbé Le Blanc prided himself on being the first to tell him of it.[8] His friends rejoiced, but the nomination was much talked about. "All the medical world and all the Academy fought for that position," wrote Le Blanc. "It is worth 1,000 crowns in salary, one of the most beautiful residences in Paris, and the right to make nominations for all the positions which depend upon it. Many traveled by coach to Compiègne to solicit it; M. de Maurepas, on his own initiative, conferred it on M. de Buffon, who was fifty leagues from Paris and who did not dream of it any more than his father had. This choice was a surprise, but everyone nonetheless approved of it, and it was agreed to praise both the Minister who made such a choice and the Academician who earned it."[9]

Le Blanc had a slightly idyllic view of things: many academicians, those who were friends of Duhamel or were simply bound by the privileges of seniority, were scandalized. Buffon's success was attributed solely to the favor of the minister and must have created lasting enmities for him. Maurepas appeased Duhamel by naming him inspector general of the Navy, a position that he filled with honor but without ever forgiving Buffon for having stolen the office that he thought of as rightfully his, and which was in fact one of the highest scientific positions in France under Louis XV.

The Royal Botanical Garden

The Royal Garden of Medicinal Herbs, which in 1793 became the National Museum of Natural History, is the oldest scientific institution created by the monarchy, not counting the Collège Royal (which became the Collège de France).[10] The two establishments had in common the fact that their courses were public, neither awarded diplomas, and they needed to change only their names in order to survive the Revolution, which had suppressed all universities and academies.

The two institutions answered the same need; they compensated for the inadequacy of the University of Paris. François I had created the Royal Lecturers to teach Greek, Arabic, and Hebrew, which the university refused to teach, and soon the teaching of science came to be added to philology. Louis XIII transformed the Royal Lecturers into the Royal College and, inspired by Richelieu, also created the Royal Botanical Garden. Thus began a tradition of French administration, which still endures: every time the government deems that the university has refused to adapt itself to the requirements of new scientific theories, it creates a new institution. Thus, starting in the seventeenth century, the *grandes écoles* of civil and military engineers came into being: the École polytechnique during the Revolution, the École pratique des hautes études under the Second Empire, and the École nationale d'administration after World War II. Under each regime, the administration judged itself more competent and innovative than the professors. It was not always wrong.

Under the reign of Louis XIII, medical studies were the problem. The Faculty of Medicine of Paris was an independent, "republican," and egalitarian body. The only authority it recognized was that of its board and its dean. All those whom it had promoted to the level of regent-doctor had the right and even the duty to teach. It was a thankless and poorly paid duty, which took time that would have been more fruitfully spent in medical practice, and which was carried out with great reluctance. The college was traditionalist in its habits and dress, an adamant defender of its liberties and privileges, and officially very Catholic—even if certain of its members, like Guy Patin, were strongly heretical. It was equally traditionalist in its teaching. It swore only by the Galenic and traditional pharmacopoeia. One of its great masters, Jean Riolan le Jeune, otherwise a distinguished anatomist, had triumphantly refuted Harvey's heresies on blood circulation, which the Faculty refused to accept until 1672.

In opposition to the Parisian Faculty was the Medical Faculty of Montpellier. Its professors were appointed by the king, and, at least theoretically, after a competitive examination. Henri IV had taken it under his protection. Starting with Louis XIII, all the "King's physicians"—there were many, under the orders of the "head physician"—were from Montpellier. They had the right to a private practice in Paris, which enraged the Parisian physicians. In addition, several professors were Protestants, and they had introduced into France a new pharmacopoeia based on mineral chemical products developed in Germany by Paracelsus, a revolutionary doctor of the sixteenth century. The most famous remedy of this "chemical medicine" was antimony, a metallic and violently vomitory product, used for making another famous remedy, "emetic wine."

While Paris and Montpellier exchanged arguments and more often insults, the royal administration, in this case Richelieu, sided with the chemists. Richelieu supported Théophraste Renaudot, the creator of the *Gazette* (an early news sheet) and also a physician from Montpellier. In Paris, Renaudot established a free clinic and medical consultation, where he practiced chemical medicine under the protection of the minister; the Faculty sought in vain to have it suppressed. It was in this atmosphere that Richelieu created the Royal Botanical Garden in 1635, in accordance with the project that Guy de La Brosse had presented to him. In reality, the creation had already taken close to ten years: the decision was made in 1626, the land was bought in 1633 in the suburb of Saint-Victor, and the definitive edict for the foundation was published in 1635. Then the buildings were constructed, and the Garden finally opened its doors in 1640. Meanwhile the Faculty of Medicine had used all its powers to make the project fail.

Guy de la Brosse was a strange character. It is not even sure that this "ordinary physician" of Louis XIII was a medical doctor. He passed for a "libertine," that is, a free thinker, and he was accused of taking advantage of the remoteness of the site, far away from inhabited neighborhoods, to arrange orgies in the Royal Botanical Garden. He had, however, built a chapel in the Garden and held mass there. Guy Patin, the quick-tempered dean of the Faculty of Medicine who despised him, said that "one day he was showing his house to some women, and when he came to the chapel, he told them, 'Here is the salting tub where we will put the swine when he dies,' indicating himself."[11] Patin could have perhaps pardoned him for being an unbeliever; he could not excuse him for being a "chemist."

In fact, officially the project was to create a *jardin des simples*, or a garden of medicinal plants, based on the model in Montpellier built by Henri IV, under the pretext that the Faculty did not teach "how to perform pharmaceutical operations." The royal edict that created the Garden, however, was somewhat ambiguous. There were to be three "demonstrators": three physicians "to demonstrate to the students the interior of plants and all the medicines, and to work on the composition of all kinds of drugs through simple [herbal] and chemical means."[12] "Chemical": the word was let loose. In fact, Guy de la Brosse had a laboratory of chemistry built and equipped in order to work on the new medicines,[13] and the teaching of chemistry was immediately made a part of the establishment's program. Plants were nevertheless not neglected: two thousand three hundred were already being cultivated in 1640. Among other things, the edict specified: "We wish that in the cabinet of the aforementioned house there shall be kept a sample of all drugs, be they simple or composed, together with all rare things found in nature." Thus, the "Cabinet of the King" was created.

While the lawsuits brought by the Faculty of Medicine were multiplying, which it invariably won in the Parlement of Paris, its old accomplice, and then just as invariably lost in the King's Council, the new institution started to function under the direction of Guy de la Brosse and under the surveillance of a superintendent who was necessarily the head physician to the king. Chemistry and botany were taught for their medical uses only. La Brosse died in 1641, and the Garden survived as well as could be expected. In 1648 the Faculty attempted to ban the chemistry courses of the new intendant, William Davisson, a physician of Scottish origin and a wholehearted Paracelsian. In 1673 Louis XIV had to intervene personally in order to crush the opposition of the Faculty and of Parlement, which tried to prevent Dionis from "demonstrating" blood circulation to the Garden.

Things changed at the end of the century with Guy-Crescent Fagon, who in 1693, after a brilliant and skillful career, became the head physician to the king and the Garden's intendant.[14] A Parisian doctor, Fagon knew how to create a truce in the war between the Parisians and the Montpellierans, as well as in the war between the Faculty and the Garden. He had been born in the Garden in 1638, for his mother was the niece of Guy de la Brosse. From 1665 on he had taught chemistry there. Combining the functions of intendant and superintendent, he developed the quality of teaching and chose valuable "demonstrators": Joseph Pitton de Tournefort and Antoine de Jussieu in botany, Claude-Joseph Geoffroy and Boulduc in chemistry, Joseph-Guichard Duverney and François de La Peyronie in anatomy and surgery. He also had an amphitheater of six hundred seats built. The Garden became a truly modern scientific institution.

After the death of Louis XIV in 1715, Fagon, as was customary, lost his position as head physician but kept the superintendency of the Garden. Thereafter, the tie between the two positions was permanently broken. Fagon had been a member of the Academy of Sciences since 1699. Upon his death in 1718, the position of superintendent was abolished and the intendant became the sole director of the institution. It was obvious that he must belong to the Academy. Fagon's successor, Pierre Chirac, Intendant of the Royal Garden (the medicinal plants had disappeared), was once again a physician, but he was above all an academician. Charles-François de Cisternay du Fay, who succeeded Chirac in 1732, was not even a physician by training. Du Fay came from a military family and became a lieutenant of the Picardy regiment at fourteen. He was fascinated by electricity and physics and had educated himself in chemistry. He had been a member of the Academy since 1723. He was thirty-three when he took over the direction of the Garden. The Ancien Régime was not a gerontocracy.

Chirac, and in particular Du Fay, opened up the Garden to contemporary developments in botany, which had ceased to be a science secondary to medicine and had become a discipline in its own right. Chirac placed the Garden of apothecaries of Nantes under the authority of the Royal Botanical Garden, thanks to which many exotic plants were introduced in France; he also started to keep close contacts with travelers, who sent specimens to the Garden. But he neglected the buildings. In contrast, Du Fay, after having visited the botanical gardens of England and Holland, built hothouses for the cultivation of tropical species and took care of any necessary repairs. The description that Fontenelle gave of his activity foreshadowed that of Buffon:

"All possible activity would not have sufficed him to execute in so little time all his plans in the Garden with just the funds allocated for this establishment. He often needed to obtain extraordinary grants from the Court. Fortunately, he was very well known to the Ministers and he had access to them and a kind of freedom and familiarity which came more easily to a Military man or a man of the World than to just a simple Academician. He also knew how to behave around Ministers, how to prepare his requests well in advance, how to ask for them only when the time was right, and when they had been almost granted, how to wipe away the inevitable first refusal with good grace, how to come back to the problem during the periods that were clear of clouds; finally, he had the gift of pleasing them, which is a great part of the art of persuasion. The Ministers also knew that they had nothing to fear from his art, which only was used for ends useful to the Public and glorious for themselves."[15]

Would a president of a university or a director of a laboratory today find things have changed that much since Louis XV?

In 1729 the old Cabinet of Drugs became the Cabinet of Natural History. The doctors and surgeons who left for the colonies received very precise instructions for sending interesting specimens to the Garden. The number of visitors increased rapidly. Foreigners in particular recognized, as Fontenelle said, that "the Royal Garden [is] the most beautiful in Europe." In 1738 the young Swedish botanist Linnaeus, already famous, came to see the Garden. It is said that he melted incognito into a group of students who were following a "demonstration" by Bernard de Jussieu. When the professor asked the identity of an unknown plant, he heard a foreign voice that answered in Latin "Facies americana." "Tu es diabolus aut Linnaeus,"* answered Jussieu in the same language. The two botanists became friends, botanizing together in the Parisian region, and Jussieu had Linnaeus named as a corresponding member of the Academy of Sciences. Much later, Linnaeus would establish a list of "Officers of the Army of Flora." He put himself at its head, with the rank of head general, and Bernard de Jussieu as his second officer, with the rank of major general.[16] From 1739 on, Bernard de Jussieu preferred Linnaeus's classification to that of Tournefort.[17] Linnaeus had the example of the Parisian garden in mind when he reorganized his own Garden of Plants at Uppsala in 1745.

French or foreign, students flocked to classes in botany, anatomy, and chemistry. It was the only place in France where the professors taught the results of their own research and the latest progress in their field. Buffon himself had certainly followed the chemistry courses. He knew well the establishment for which he was responsible and that he would direct for almost fifty years. He could get right to work.

The Last Papers Presented to the Academy

The direction of the Royal Botanical Garden turned Buffon only slowly away from his academic activities. For several years, he continued to make reports on papers presented to the Academy or to present papers himself on problems of physics, demography, or the theory of sensations.[18] These subjects seemed to interest him less as he gave more attention to preparing his *Natural History*. They appear again later, however, since they touched on fundamental problems that were, for Buffon, related to the very interpretation of nature and knowledge. Later on he even reprinted several of them in the *Supplement* to the *Natural History*, often after reworking and developing them.

Among the reports of the works presented to the Academy were those that he dedicated in 1745 and 1746 to a work by Deparcieux, "Sur les

*Ed. note: "You are either the devil or Linnaeus."

probabilités de la vie" [On Probabilities in Life] a problem that he would take up again in the *Essay on Moral Arithmetic* and in other essays of volume IV of the *Supplement* in 1777. Another report was devoted to the method that Jacob Rodrigues Pereire invented for educating deaf-mutes; he returned to it later in the *Natural History of Man*. As for the original papers he presented, one dealt with this same problem of defects in sense organs ("Dissertation sur les causes du strabisme" [Paper on the Causes of Squinting], 1743), another studied the problem of physiological optics and raised analogous concerns ("Dissertation sur les couleurs accidentelles" [Dissertation on Accidental Colors], 1743),[19] and three dealt with the problem of the generation of animals, treated at length in volume II of the *Natural History*. Finally, three papers or groups of papers deserve a closer examination.

The first, written in 1740, was devoted to skyrockets.[20] The technical problem, it seems, was the shape that must be given to a rocket's powder charge so that combustion occurs completely and rapidly. There was, however, a theoretical problem as well: how much acceleration can one expect to give a rocket? Newton had explained that a projectile given sufficient speed would escape the earth's attraction. Buffon did some calculations and experiments and concluded that a rocket could be given any acceleration desired. He surely was not thinking of the artificial satellites of today, which are based on the same principles and on an analogous technology. He did, however, draw on this work for his theory of the formation of the planets, which he presented in 1749, and about which he was perhaps already thinking.

Another research topic led to several papers, of which the last was fairly late (1752): the invention of a "burning mirror." In Greek history, Archimedes set fire to Roman vessels that were attacking Syracuse by using concave mirrors that concentrated the sun's rays. According to Descartes, these mirrors "had to be extremely large, or more likely mythical."[21] Not allowing that opinion to influence him, Buffon built several square concave mirrors made up of smaller, slightly curved mirrors. The largest mirror, which measured 6 feet on one side (about 1.8 meters) was made of 360 small mirrors. With it, Buffon was indeed able to set fire to buildings made of wood at a distance of 10 to 200 feet (from 3 to about 65 meters). At a distance of 10 feet, he could melt iron. These experiments were based more on physical than geometrical optics. Buffon studied the loss of light due to reflection and experimented with photometry. As with the strength of wood, he showed that geometry did not exclude the need for experimentation. Theoretically, a mirror of 800 feet on the side was needed to burn an object at a distance of 200 feet: in practice, a 6-foot mirror was enough. His method of construction for large concave mirrors was not geometrically perfect; it did, however, have a promising future, and it is still used today.[22]

Aside from these problems of optics, Buffon was very interested in the relationship between light and heat, which he considered to be the same material of differing particle size. By submitting metals to a particular form of heat or light, he hoped to discover their composition by this "kind of calcination."[23] For him, this was just as much a problem of chemistry as physics, the boundary between the two sciences being fairly uncertain, and Mme. de Châtelet agreed with him on that point.[24] Later on Buffon returned to these problems, which played a large part in his general conception of nature. He hoped to measure heat in a manner other than by the temperatures given by thermometers, which he considered arbitrary. He wanted to find a true measure of heat and be able to talk about a heat that was double or triple another. Here again, he was inspired by Newton,[25] and it was with a particular satisfaction that he communicated the results of his experiments to Martin Folkes.[26]

In a century already fascinated by "experimental physics," such spectacular experiments interested a large public, and Buffon did not miss any occasion to use the publicity for himself. He gave several public demonstrations of the power of his burning mirrors. One was organized at the Chateau de la Muette, in the presence of the king himself, and Buffon offered Louis XV the mirror that he had just used. The "new Archimedes" was extolled in the papers. The *Mercure* published a poem from the president de Ruffey:

> Buffon! There is nothing that does not cede
> To your ingenious efforts.
> What! The miracles of Archimedes
> Are merely the games of a studious leisure for you.

At the end of about twenty lines, the reader could understand, thanks to explanatory notes, that Buffon was an honored member of the Royal Society of London and had just received the congratulations of Frederick II, who was perhaps curious to know if the Prussian army could use the new weapon.[27] Buffon was becoming famous.

The Controversy over the Law of Attraction

In 1748 and 1749, on the eve of publishing the first volumes of the *Natural History*, Buffon launched into a confusing academic controversy, which became an important episode in the history of celestial mechanics in the eighteenth century.

Around 1740, parallel progress in the methods of calculation and in observations in astronomy allowed the refinement of the Newtonian model of the movement of the planets. We know that Newton's law is simple: celestial bodies attract one another in direct relation to their masses and inversely as the square of their distance. The calculation of the relative motion of a planet in relation to the sun would also be simple, if the reciprocal attraction of the two bodies were the only consideration: all relative motions would be elliptical, conforming to Kepler's model. The planets, however, attract one another at the same time, and the complexities of these mutual attractions complicated the model's simplicity. It was therefore necessary to solve the "three body problem," by means of successive approximations, to calculate the relative motions of three bodies that all attract one another. A particularly clear case is that of the moon: its reciprocal attraction with the earth is very strong, since it is very close to it. But it also is affected by the attractive force of the sun, which is much farther away but whose mass is enormous. That attraction has a "perturbing" effect. This effect is seen by a relatively slow but perceptible movement of the line of the apsides of the ellipse, calculated as a first approximation (see diagram).

The relative movement of the moon's apogee was precisely the topic for

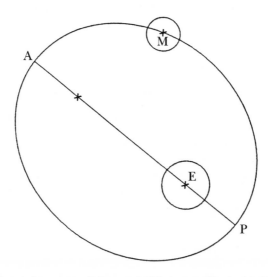

The moon (M) revolves around the earth (E) in an elliptical orbit, of which the earth is one of the two foci. The apsides are the extremes of the long axis of this ellipse: P is the perigee (the point where the moon is the closest to the earth), A is the apogee (the point where it is the farthest). The long axis AP turns slowly around the center of the earth (E).

a competition organized by the Academy of Sciences for the year 1747. Observations attributed a period of about nine years to the revolution of the apogee. According to calculations, it should be eighteen years: a huge difference. Euler in Berlin and Clairaut and d'Alembert in Paris, the greatest mathematical astronomers of the time, attempted to solve the problem in an atmosphere of feverish rivalry. As soon as the two Parisians thought they had found an important result, they hurried to deposit it at the Academy in a sealed envelope, which the perpetual secretary initialed and dated. Both shivered at the thought that Euler might find the solution and reveal it in the paper that he was expected to submit for the contest.

The first to bring the discrepancy between observation and calculation to public notice was Clairaut, on November 15, 1747, in a paper whose title showed its importance: "Du système du monde dans les principes de la gravitation universelle" [On the System of the World according to the Principles of Universal Gravitation].[28] The paper expressed a deep sense of helplessness on the part of this early Newtonian. If the law of gravitation did not account for the movements of the moon, "it must from now on be condemned without appeal." Clairaut had redone all the calculations: he had found eighteen years instead of nine. "A result so contrary to the principles of M. Newton made me at first entirely abandon attraction." Clairaut already knew that Euler had done the same calculations and wondered if he did not need to go back to Cartesian vortices.[29] A heartbreaking revision, but understandable from rigorous mathematicians.

In order to understand better Clairaut's position and Buffon's reaction, it is necessary to know that Clairaut, like many other scientists of the time such as Stephen Hales, wanted to explain many of the "phenomena that happen under our eyes, such as the roundness of drops of liquid, the ascent and the descent of fluids in capillary tubes, the curving of light rays, etc." by means of universal attraction. For all these phenomena, an inversely proportional attraction had to be assumed, not to the square but rather to the cube of the distances, or even to "higher powers." This required the assumption that there are in nature as many laws of gravitation as there are phenomena to be explained. This is what Buffon rejected.

To come back to the moon, Newton's law agreed with too many phenomena for it to be rejected lightly. Finding it "as difficult to reject as to accept," Clairaut proposed modifying it by adding to the classic equation a supplementary part that would leave the effects of attraction at large distances practically unchanged but would slightly increase it for bodies that are very close, like the moon and the earth. In place of the traditional formula m_1m_2/d^2 (where m_1 and m_2 are the masses and d the distance), one would have the formula $m_1m_2/d^2 + m_1m_2/d^4$. In that way the unreconcilable could be reconciled, bringing together agreement of Newton's law with the motion of the planets and its disagreement with that of the moon.

The first to contradict Clairaut was d'Alembert, who in turn wrote a paper with a title just as grandiose: "General method to determine the orbits and motions of all the planets, taking into consideration their mutual action."[30] D'Alembert refused to change the law of gravitation and proposed adding to the attractive force of the earth on the moon a supplementary force, whose nature he did not indicate, that would not be calculated in inverse relation to the square of the distances. In his correspondence with Euler, he insisted on the irregularity of the moon's shape and on the lack of homogeneity of its matter, and envisioned the possibility of an additional magnetic attraction already proposed by Newton. Like Buffon, he did not see any reason "to change for one single phenomenon a law that agrees with all the others and that is very simple. I have difficulty assuming that the attraction of an atom depends on anything other than its mass and its distance, and in this case the attraction can only be the product of the mass by a simple power of the distance, otherwise it would be necessary to enter a parameter—and I admit I don't know where to find one—into the expression of the attractive force."[31] Euler, on the other hand, found his ideas to be closer to those of Clairaut.

On January 20, 1748, Buffon entered into the debate with "Réflexions sur la loi d'attraction" [Thoughts on the Law of Attraction], which he read before the Academy.[32] The version printed in the Academy's *Mémoires* is only one part of the original paper, and it is only through Clairaut's response to the paper that we know of the technical part.[33] Like d'Alembert, Buffon refused to abandon Newton's law so quickly and suggested the possibility of a supplementary magnetic attraction. His main objections, however, as he recognized himself, were "metaphysical" in nature. First of all, the gravitational force "must be measured, like all qualities which come from a central point, using the inverse relation to the square of the distance, the way quantities of light, smell, etc. are measured." Then, if the law were modified in order to adapt it to all specific phenomena, as Clairaut proposed, this "would open it up to all possible and imaginable laws: a law in Physics is a law only because its measure is simple and the scale that represents it is not only always the same but also unique." If not, "there is no longer any law in Physics." If the expression consists of two terms, it is because it designates two forces. Clairaut's conclusions "are based only on calculation." But "one can represent anything with a calculation, and accomplish nothing; if one is allowed to put one or several terms at the end of an expression of a law of physics, as with that of attraction, it is . . . arbitrary instead of representing reality." We find here once again the theme of the divorce between mathematics and reality.

Clairaut soon answered both Buffon's technical and "metaphysical" criticisms.[34] In passing, however, he affirmed the supreme validity of mathematics: "Authors who construct Systems on vague reasons can lose everything through new observations, but those who begin with mathematical principles never work in vain, even when the actual observations destroy their suppositions; they always have the means to use their first research for discovering new truths." After that he refuted the analogy Buffon established between the diffusions of attraction and light: light is composed of corpuscles, and we don't know what attraction is. How can we devise laws for it? "If we make attraction depend on some metaphysical virtue that God gave to matter for reasons that are impenetrable, by what means shall we explain that attraction if it is not by facts?" In particular, we do not know if gravity is the effect of one or two causes: "Is it right for us to attempt to decide if the Creator gave an attractive virtue to matter with two different decrees or if He endowed it with two forces at the same time by a single act of will?" Clairaut, in fact, stayed loyal to Newton's position on the unknown nature of attraction, a "gift of God" to matter. In contrast, attraction for Buffon was an "essential property" of matter, and it had to be simple.

Finally, Clairaut explained that a simple curve could be expressed by a complex function:[35] "Without a doubt Nature is simple for one who sees it with a single glance and through the true relation of things, but [she] can seem composite to us who only know her through unconnected facts; and . . . one does not make a supposition contrary to the essence of things when laws are accepted that follow functions rather than powers: if we cannot express them simply, it is the fault of Algebra, which, as a language, has its imperfections." It is paradoxical that Clairaut was the one who emphasized that mathematical expressions do not necessarily take into account the simplicity of reality, while Buffon looked for an exact correspondence between the two.

On May 17, 1749, Clairaut presented a "Notice" to the Academy which created a sensation. After having considered the problem "once more from a point of view that has not been considered by anyone," he had "almost exactly reconciled observations made on the motion of the apogee of the Moon with the theory of attraction without supposing any other attractive force than that which follows the inverse proportion to the square of the distances."[36] He did not say what this "new point of view" was, which he left in a sealed envelope at the Academy and sent in the same form to the Royal Society of London. The debate should have been closed. The "Notice" was, however, followed by a brief note, which refuted an "Addition" that Buffon had made to his own paper. Buffon answered the note, and Clairaut answered the answer. The controversy, now purely theoretical, treated only the relationship between mathematics and reality.

As a matter of fact, Clairaut was furious because Buffon stated that it was because of his objections that Clairaut had given up his complex formula.[37] Clairaut also suspected the intervention of Le Monnier and the Royal

Society of London in support of Buffon. He wrote to Euler: "Several non-geometer Newtonians believe all is lost if forces other than those of M. Newton are introduced, and they have overwhelmed me with questionable objections, which I should have scorned but which I had the weakness to answer. Since my retraction, I have had many petty quarrels because I had to counter the impudence of people who thought to triumph over something that should have honored me." He added: "I would have convinced my antagonists more if I had drawn the English over to my side."[38] The Royal Society, indeed, had not acknowledged receiving Clairaut's paper.

This controversy irreparably compromised the friendly relationship between Buffon and Clairaut. Euler and d'Alembert, however, had been taken unaware by Clairaut's sudden about-face, which was so spectacular that it threw d'Alembert's work into shadow. Consequently, his relationship with Clairaut became more and more polemical. As for Euler, he was not convinced by Clairaut's conversion and waited until 1751 to return to the classic formulation of Newtonian attraction. This controversy marked Buffon's definitive break with mathematics and mathematicians.

After 1752, Buffon presented no further papers to the Academy of Sciences. In 1744 he had become its permanent treasurer, an office that was worth an additional three thousand livres. Later, because of his long stays in Montbard, he was given an assistant named Thillet. He therefore participated in the administrative life of the institution, but the intellectual climate of the Academy suited him less and less, and he felt increasingly ill at ease there. We shall come back to this progressive distancing, in which the public success of the *Natural History* and the jealousies that it aroused certainly played a part. Buffon was not a model academician, and the others let him know it.

The Early Years of the Intendant of the Royal Botanical Garden

When Buffon entered the Garden, it was still only one part of the current holdings of the Museum. To the south it was bounded, as it is today, by rue du Jardin-du-Roi, which became the rue Geoffroy-Saint-Hilaire. From there, it stretched toward the north in the direction of the Seine, from which it was separated by the vast lands of Saint-Victor abbey. All of it covered a little more than eight hectares [a little less than twenty acres]. To the west, more land belonging to the abbey separated it from the rue de Seine-Saint-Victor, today the rue Cuvier. To the east, it reached part way to where the rue de Buffon lies today. There was no hope of expansion on the side of the abbey. The possessions of the Church were inalienable and the monks did not have the right to sell, even had they wanted to. Elsewhere several purchases were possible, but that meant money, and Buffon did not yet have the same contacts at the Court that Du Fay had had. He would look into this only later, when he had the means to use a method that was sharply criticized.

As for the buildings, they were made up mostly of the "chateau," constructed in the seventeenth century along the rue du Jardin-du-Roi, which Du Fay had repaired and enlarged. It contained the intendant's apartment and the Cabinet of Natural History. Next to the chateau and its chapel, where La Brosse had been buried, a building constructed or renovated by Fagon contained an amphitheater destined mainly for the teaching of chemistry. Another building served for the teaching of anatomy. Space was lacking everywhere. Money also.

It was, however, primarily the men who worked there who made up the Garden. There were five, who shared the teaching of botany, chemistry, and anatomy. Their title was "Demonstrator," and also sometimes "Professor," and they did not necessarily teach what their title indicated. All were members of the Academy of Sciences; all were older than Buffon, sometimes much older, with more seniority in the Academy; they were more competent in their respective disciplines than he, and they knew it. Only the chemist Boulduc was an old friend. The two Jussieu brothers had campaigned for Duhamel. Buffon was very careful, therefore, not to interfere with the teachers' activities in the Garden. He agreed when the botanists asked him to replace, both in the plantations and on specimen labels, Tournefort's old classification with Bernard de Jussieu's "natural" classification and Linnaeus's binary nomenclature, despite the little sympathy that he felt for the Swedish botanist.

It was up to Buffon, however, to choose the new professors. A position frequently would be vacated because of the death of its holder: people did not retire in the eighteenth century. An aged professor could be helped by a "deputy." Buffon often respected this choice when selecting a successor: in such a way, Antoine-Laurent de Jussieu succeeded his uncle Bernard. From the beginning, Buffon's choices were excellent. In 1742 he named the anatomist Jacques-Bénigne Winslow as the successor of Hunauld. Winslow was a Jansenist, which did not appeal to Buffon, but he was the best anatomist of his time. In 1743 Buffon had to replace the two chemists. The rather colorless Bourdelin succeeded Louis Lémery, but it was Rouelle, original and brilliant, who replaced Boulduc. He entered the Academy the following year. Diderot and Lavoisier would later take his classes, which attracted many people despite the risks to the class of explosions caused by his legendary absentmindedness. In 1745 Buffon had to name a chief gardener. He chose an unknown, Jean-André Thouin, who would prove very satisfactory. All of his fortunate choices implied a deep knowledge of scientific circles, and probably some discreet consultations as well.

The task that Buffon attacked most vigorously was the reorganization

*Table of the personnel at the Garden
at the time of Buffon's arrival**

The offices are classified in the order of their creation. The first title is the official title of the office.

- "Demonstrator of the interior of plants with the title of professor of botany": Antoine de Jussieu (1686–1758). Named in 1710, Academician since 1715.
- "Demonstrator and pharmaceutical operator": Louis Lémery (1677–1743). Named in 1730, Academician since 1712.
- "Demonstrator and operator of pharmaceutical operations": François-Joseph Hunauld (1701–1742). Named in 1729, "Demonstrator-operator of the interior of plants with the title of professor of anatomy and surgery." Academician since 1725.
- "Sub-demonstrator of the exterior of plants": Bernard de Jussieu (1699–1777). Named in 1722, Academician since 1725.
- "Miniature painter": Magdeleine-Françoise Basseporte (1701–1780). Named in 1735. She was the only woman of the group.
- "Gardener": Bertamboise. Named in 1721, died in 1745.
- "Demonstrator of chemistry with the title of professor of chemistry": Gilles-François Boulduc (1675–1742). Named in 1729, Academician since 1716.
- "Guardian of the Cabinet of Drugs": Bernard de Jussieu. Named in 1732.

*(According to Laissus 1986, pp. 324–41.)

and enrichment of the Cabinet of Natural History. Upon his arrival, the Cabinet was administered without enthusiasm by Bernard de Jussieu, who still had the old title of "Guardian of the Cabinet of Drugs." In 1745 Buffon had replaced him, probably with his permission, by Louis-Jean-Marie Daubenton, who at the age of twenty-nine was an unknown practitioner of medicine in Montbard but, as we recall, part of the Leclerc clan. Daubenton received the title of "Guardian and Demonstrator of the Cabinet of Natural History," and it was up to him to order the chaos of the collections. His first contribution to the *Natural History* would be the "Description of the Cabinet of the King," which opens volume III.

Buffon expanded the space reserved for the Cabinet right away.[39] He had some work done on the "chateau" and sacrificed part of the intendant's apartment. He immediately continued enriching the collections that Du Fay had started. On the day following his nomination, he was forced to depart from his routine, leaving Montbard for Paris in the middle of August to receive the crates of plants sent to his predecessor.[40] In 1742 he tried to sustain the zeal of Arthur, the King's physician in Cayenne, who sent specimens of plants but who wanted to see an increase in his salary.[41]

He developed his network of correspondents and had Maurepas create a certificate titled "Correspondent of the Royal Botanical Garden," a purely honorary title that he distributed only sparingly. In the *Natural History*, he made it a point of honor to cite his informants, which redoubled their efforts. In the meantime, he used his contacts at the Court in order to put his hand on everything that interested him. In 1744 he had the collection of anatomical specimens of Bonnier de La Mosson given to him. In 1748 he received from the king a magnificent Italian marble table, which he had noticed in the Salle des Gardes of the Louvre. Inlaid with rare stones and pieces of coral forming a decoration of birds, flowers, and insects, the table stands today in the gallery of mineralogy in the Museum. The Cabinet grew so quickly that it soon needed more room.

Buffon, who was very circumspect when dealing with the professors, felt more at ease when it came to the Cabinet or the buildings. He could decide and act, and that is what he liked. He was, however, not in control of his own expenses; he was obliged to go to Versailles to "pay his respects." He probably had direct access to the king, and Mme. de Pompadour, the king's favorite, finally accepted him within her circle. He would not, therefore, suffer too much from Maurepas' disgrace, who was dismissed in 1749 because of a campaign of slightly pointed epigrams directed at the favorite. He did not depend any less on the minister despite his closeness with the king. He was part of the system, knew it well, and used it. He became in his own way one of the "great assistants" who made up the strength of the royal administration; the fact that he directed a scientific institution protected him from the hazards of politics, since the administration of Louis XV had learned to respect scholars. We shall see later that certain precautions were needed. Within this system of "enlightened monarchy," he worked both for the Garden and for his personal fortune. He was, clearly, a man of the Ancien Régime.

For all that he did not give up his freedom. He left for Montbard as often as he could, for that was where he was really his own master and the master of his work. He left to his Parisian collaborators the task of watching over any work in progress, with the condition that they write him to ask his advice and instructions. Now that he had a project worthy of him, he was even more zealous in keeping his hours of work. Maurepas, in naming him as director of the Garden, had asked him to publish a "Description of the Cabinet of the King." He assigned this task to Daubenton. As for himself, he would write a *Histoire naturelle, générale et particulière* [Natural History, General and Particular]. He would need ten years of work before the first three volumes appeared in 1749.

PART TWO

A NEW
NATURAL
HISTORY

CHAPTER V

The Preparation of the *Natural History*

Why the *Natural History?*

The seventeenth century had been the century of geometers. Both mathematics and rational mechanics, itself a purely mathematical science, had made great progress in the seventeenth century. At the end of the century, the invention of calculus and Newtonian celestial mechanics crowned more than one hundred years of uninterrupted efforts. The mathematicians of the eighteenth century used and perfected the instruments they had inherited, giving Newton's cosmology its definitive form: they did not invent anything completely new.

Mathematics is the masterpiece of the human intelligence. Its rules are those of reason itself, and nothing, in theory, can resist them. Seeing the descent of bodies or the movements of the planets submit to the law of an equation must have given men a feeling of triumph, a sort of intoxication. To quote Descartes, he felt "like the master and owner of nature." Men had already exercised mastery over machines, which they had invented and built with more and more fascination since the fifteenth century. Leonardo da Vinci has left so many examples, both real and imaginary. Men understood these machines because they had made them and because they knew the laws of their workings. Since Archimedes it had been known that these laws obeyed geometry.

In order to understand nature, it was enough to consider it to be a huge machine, and that was exactly what the "mechanical philosophy" created by scholars of the seventeenth century was.[1] They decided that everything was made of pulleys and levers, hydraulic circuits and gears, which transmitted the movements by which other bodies were put into motion and acted according to the laws that they themselves understood. They imagined that

[65]

in the secret depths of matter, where their eyes could not reach, infinitely hard corpuscles pressed against and struck and bounced off one another, always obeying the universal laws of motion. They had not seen them and would never see them, but, thanks to them, they could build intelligible models that, at least theoretically, could account for the phenomena they observed. Let us not make fun of them: the cleverest theories of physics in the nineteenth century would be founded on the hypothetical existence of atoms or molecules whose existence could not be proved but which were required by mathematical calculations. The real atom is a discovery of our century.

In the second half of the seventeenth century, therefore, nature was understood to be a clock. Or as Fontenelle said, "Nature is only on a large scale what a watch is on a small one." And since we know how to build watches, we should be able to understand nature. Perhaps not right away, because the number of gears was gigantic, but with time. It was enough to take apart the mechanism patiently and little by little, as with an immense puzzle, all the pieces would find their place. The "mechanical philosophy" thus became a "research program," as the philosophers of science say today, and it is to this program that the new scientific institutions created at the end of the century, the Royal Society of London and the Royal Academy of Sciences, committed themselves.

The mechanical philosophy had very precise intellectual requirements. A clock does not build itself, and it works only when it is completely built. A clock thus supposed a clockmaker, as Voltaire never tired of saying, and Nature's clock required an intelligent Creator, a "Supreme Geometer," who built it in all its least details before starting it up. This led to at least three consequences. Between the geometry of the Creator and ours, there could be a difference of degree but not of kind. Our reason functions like the divine reason, and that is why we can understand nature. In addition, the world had to be perfect and complete from the moment of its creation. Otherwise, it would not have been able to work. In the Bible, the book of Genesis spreads out the creation of the world over six days. This is five days too many. Fortunately, Saint Augustine wrote, "He who lives in eternity created everything at the same time."* And, of course, nothing had changed since the beginning. A clock does not evolve. Finally, nature's innumerable gears only transmit the motions that they received in the

*This statement, very often quoted by scholars at the end of the seventeenth century, is a quotation from one of the last books of the Old Testament, the Ecclesiasticus or the Wisdome of Jesus the sonne of Sirach, XVII, 18.+ The important words, "at the same time" (in Latin: *simul*), give a false meaning to the Latin *Vulgate* of the original Hebrew. But this false meaning fits the theology of Saint Augustine and, even more so, the mechanical philosophy of the late seventeenth century. +Ed. note: In the King James Bible (1611 or first edition), this book appears in the *Apocrypha* with the title The Wisdome of Jesus the sonne of Sirach or Ecclesiasticus. I can find no sentence in the passage cited that refers to the subject quoted.]

beginning, just as the gears of a clock transmit only the motion that the weights or the springs that make them work transmits to them. Matter is passive and does not create either force or motion. In order, therefore, to understand nature, it was enough to know the laws of the transmission of motion and the exact structure of things, or what was called their "figure." All was done "through figure and motion." The science of the seventeenth century was a science of structure.

The internal logic of the mechanical philosophy was perfectly coherent and satisfying to the intellect. It was born in the second third of the seventeenth century, when enthusiasm for knowledge was accompanied by the glorification of the will. "I am the master of myself as well as the universe," one of Corneille's heroes said, and Descartes saw fit to adopt this motto. Through the power of his reason, man, with the help of God, put himself in the center of the world. The new science would not only let him understand nature, it would also permit him to act on it and use nature for his profit. The beginnings were still timid, but they were promising. The philosophers Bacon and Descartes decided that science was useful, and governments began to believe so as well. Everyone did not share this optimism, but even the libertine skeptics or the austere Jansenists let themselves be led by the success of the new science.

In all this, natural history had been forgotten. It had flourished in the sixteenth century, when the wealth of nature, the diversity of living forms, and the discovery of the New World with its exotic products enchanted the mind. The greatest variety of beings, whether they were real or imaginary, normal or monstrous, so numerous, so different that it was impossible even to classify them except alphabetically, jostled one another in the books of Gesner, Belon, Rondelet, and Aldrovandi. Sometimes the naturalist had seen them himself and had them drawn from nature. Sometimes he had only heard about them from a traveler or had read the description of one of his colleagues. Whether he was sure of their existence or doubted it a little, he did not hesitate to make reference to them, fearing being incomplete more than being inaccurate. With an implacable erudition, the naturalist told all that he knew, all that he had seen, all that he had read. An article would not have been complete without mention of the use of an animal in heraldry, its medicinal virtues or cooking recipes. It is easy to criticize these scholars' bulimic scholarship, as well as their credulity. What they were seeking was an inventory of an inexhaustible Nature, full of mysterious forces, unpredictable in her powers and products, and before which human reason had to recognize its own impotence. Everything could be true because everything was possible.

By ruining this vision of nature, the mechanical philosophy of the seventeenth century had ruined natural history. The infinite diversity of living forms did not lend itself to the rigor of geometers, who preferred to ignore it. Save some lost travelers in far-off countries or a few doctors seeking new remedies, the immense world of minerals, plants, and animals no longer interested anyone. Only man and the products of his intellect deserved attention.

At the end of the century everything started to change. This movement was too general not to have had deep causes. Under the reign of Louis XIV or in England during the Restoration, the political storms of the Fronde or the Civil War belonged to the past. Great ambitions were dead. Politics had become an administrative affair. Everything became middle class, despite appearances, or rationalized itself, which amounts to the same thing. In France, this country subjected to a cult of a unique hero, the courtiers were bored and great ladies could not scheme any longer, except about minuscule problems of precedence. The world of Saint-Simon was the successor to that of the Cardinal de Retz. Pessimism reigned, heroism was passé, and the new moralists, Boileau and La Bruyère, were morose bourgeois who meticulously cataloged the oddities of the human heart almost like entomologists.

From this was born a new passion, curiosity. Now people looked for everything that was different, everything that disoriented them, everything that came from somewhere else. It did not matter from where or how. The arrival of Turkish or Siamese ambassadors became an event. "Chinoiseries" invaded chateaus and mansions. *A Thousand and One Nights* became the fashion in reading, then *L'espion turc* [The Turkish Spy] and the *Lettres persanes* [Persian Letters]. Louis XIV had children and dogs painted on the walls between the windows of Versailles. Real life was elsewhere, not inside boring palaces. Why should not savants,* who studied nature, have new diversions to offer?

For this very reason, the savants started to encounter difficulties. They had exhausted the joys of pure theory. Cartesian science had opened up dizzying, but more and more uncertain, perspectives. It was possible to discuss vortices and small corpuscles until one was blue in the face. But what was really true? Natural philosophers realized that it was necessary to experiment, to examine things as they truly were, to describe them carefully first, and then to reason afterward. To use the language of the time, it was necessary to find the "history" of nature before building a "physics," to describe phenomena before seeking the causes. Certainly, mathematicians continued to practice mathematics. But there were now other things to do.

Experiments can be done in public. Austere geometers of the middle of the century spoke almost exclusively among themselves. Experiments attracted crowds. This fashion of science was an important phenomenon. In

*Ed. note: The word "scientist" was invented in 1840 and has come only recently into French. "Savant," or natural philosopher, will be used here for those who made a profession of the study of nature in the eighteenth century.

1658 the Cartesian Rohault gave public lectures, for which he charged admission, which were very successful because he illustrated them with experiments. Even anatomy would soon be all the rage. Duverney would come to be nicknamed "the anatomist of courtiers"—not because he dissected them but because they came to watch him dissect. Dionis, at the Royal Botanical Garden, would have up to five hundred in the audience. The shrewd Fontenelle hurried to write his *Entretiens sur la pluralité des mondes* [Conversations on the Plurality of Worlds], in which Copernicus's system was put within the reach of the marquises. In short, there was great demand for what a newspaper of the time called "that kind of curiosity."[2] Molière and Boileau, as good bourgeois, could poke fun at women savants. The women did not let themselves be concerned by these facile mocking remarks, because the example came from on high: it was before the queen, Madame la Dauphine, and "some other ladies of first quality" that Dionis in 1681 showed the anatomy of a dead woman with her fetus. When in 1704 a graduate of the Medical College of Paris defended a thesis on the "spermatic animalcules," which the Dutch microscopist Leeuwenhoek had discovered in 1677 and which we call today spermatozoa, the public rushed to the examination. The thesis was of course in Latin, but, as Fontenelle tells us, it "so piqued the curiosity of the Ladies, and Ladies of the highest station, that it was necessary to translate it into French to initiate them into the mysteries of which they had not even the theory."[3]

The microscope, indeed, opened up an unknown and completely unexpected world. That the smallest drop of water contained thousands of living things was beyond the imagination. The word that is found most often under the pen of new observers of the microscopic world was "marvel." But the marvelous had been banned from scientific vocabulary. A universe subject to geometry was too reasonable to have surprises. Geometric rationalism was taken by surprise, caught red-handed by ignorance and presumption in its attempts to put limits on nature's richness. As Antonie van Leeuwenhoek, the most talented microscopist of the time, said, "we can only be seized by amazement, for we are incapable of conceiving the extreme smallness of the parts that animals are composed of, so that the only thing left to us is to exclaim: O unfathomable depth of supreme wisdom!" The first microscopic discoveries provoked many criticisms and incredulities, but it was necessary to bow before the facts.

Microscopes were not easy to handle at that time. Investigators had to make them themselves, which was tiresome work. Yet, a magnifying glass alone was enough to study insects, which soon became a new passion. What is more astonishing than an insect? The observers progressively discovered the cycle of their metamorphoses, the extraordinary precision of their behavior, the care that they took to preserve a progeny they would never see, the marvelous and infinite variety of their structure. Finally, it was

Human curiosities, including dwarfism and partial albinism. From *Oeuvres complètes*, XII, p. 12.

man's place in nature that was put into question. Since Aristotle at least, it had been accepted that the anatomy of the human body was the most perfect that could be conceived. In 1712 Fontenelle had to admit: "It seems in general that the most admirable of all animals, as far as mechanism is concerned, are those that we resemble the least." The detailed study of insects, an entirely new science, made great progress in the first half of the eighteenth century. Réaumur produced a perfect model of it in the six volumes of his *Mémoires pour servir à l'histoire des insectes* [Memoirs concerning the History of Insects], which he published from 1734 to 1742.

Another branch of natural history, however, started to develop at the end of the seventeenth century, but in another direction: botany. Unlike insects, plants were not a discovery. It had been known for a long time that they were very numerous and varied, but their actual number and diversity were better realized after travelers had discovered exotic flora. The difficulty, then, was to put the plant world into order, or to create classifications. The Greeks and the botanists of the sixteenth century had begun to think about this problem. At the end of the seventeenth century the work was taken up again, in particular by Mayol in France and John Ray in England. A botanical classification must satisfy contradictory requirements. On the one hand, it must allow one to find in nature a plant that one is looking for and, inversely, to identify a plant that one finds. This requirement was all the more necessary given that plants were still used as the basis for medicinal remedies: it was absolutely necessary not to confuse hemlock with parsley. On the other hand, however, plants seemed to classify themselves into natural families, in which the members resembled one another by their general appearance. For a long time, botanists had identified such families, and a classification ought not disassemble them. Botanists during the entire eighteenth century would seek compromises between a practical but arbitrary classification and a natural but impractical one. The classification of Tournefort (1656–1708) was the most popular in the beginning of the century. After 1735 Linnaeus's classification conquered the world of the botanists. But that classification, which was founded on the structure of the flower, was artificial without being clearly presented as such. This produced an entire series of debates, to which we shall return.

Observers and Classifiers

To describe and classify was the dual occupation of naturalists in the first half of the eighteenth century. In reality, the two groups of scholars more or less ignored each other. The "observers," who were primarily fascinated by insects, were barely interested in classification, and Réaumur did not hesitate to place in the "class of insects" anything that was not a quadruped, a bird, or a fish. Slugs, starfish, snakes, or lizards were all insects. A crocodile was a "furious insect," but an insect all the same. Inversely, the "classifiers," who worked mostly on plants, did not like detailed descriptions. What they analyzed with care were the distinctive characteristics that they chose for their classification, for example, the structure and placement of the reproductive organs of a flower. The rest were secondary, as they were of little use for classifying. The ambitions of the two groups also differed. The observers dedicated months of work to the study of a species of insect. They knew very well that the number of known species was already enormous and that there were surely many others that were not yet known. What they hoped was to get "their" species into the domain of known things and to make its "history" a definitive acquisition of human science. Others were doing just as much for other species, and progressively an exact knowledge of nature would be built. As for the classifiers, they wanted to place all plants, those that were known and those that were not yet known, into categories that they defined. They had a rationalist ambition that was foreign to the observers, who were full of humility before the diversity of beings.

Observers and classifiers did have several points in common. First of all, they were mainly interested in the structure of things. For the classifiers, this structure was what furnished the distinctive characteristics that permitted classification. For the observers, this structure was interesting because of its diversity and its perfect "suitability" to the lifestyle and the environment of each species. The structure of the mosquito's proboscis, the wasp's stinger, the beetle's wing sheath was a marvel of "mechanics." The observer was also enchanted to see each insect use to perfection the tool that nature had given it to do what it needed to do to survive and reproduce. Neither the observers nor the classifiers were, however, very interested in "physics," that is to say, plant or animal physiology. The functioning of the living creature was, so to speak, a necessary consequence of its structure. Stephen Hales, who studied plant physiology, was not a classic naturalist. Réaumur experimented on digestion, but it was marginal to his work in entomology. We shall see how the natural history of the time practically ignored the prickly problem of reproduction. By definition, according to the terms used at the time, "physics" was excluded from natural history.

From whence came these structures that were being described or classified? The answer was simple: they came from God. If natural historians broke, at least partially, with the rationalism of the mechanical philosophy, they retained the conviction that the world had been created by an intelligent Being who had foreseen everything so that the world functioned, that is, so that each species survived and reproduced, always resembling itself. Fixist creationism reigned practically without obstacle. Observers and classifiers shared this conviction, even if they interpreted it differently. For the classifiers, it was a question of finding the order that God had put into nature. It was because this order exists and has been imposed on nature by the Supreme Wisdom that it must be possible to create a natural classification. In classifying the plant kingdom, Linnaeus had the feeling of "walking in the steps of God." The observers, always attentive to the work of Providence, discovered it in the delicate and complicated structure of each insect and in its perfect adaptation to its needs. God was for them the Supreme Artisan who made each insect perfectly. Entomology, like cosmology and anatomy, was a proof of God's existence.

In this vein, the naturalists joined forces with the theologians and the apologists of Christianity. The seventeenth century, like the sixteenth century, had been one of religious controversy. Protestant and Catholic opposed each other fiercely, and both of them fought against the incredulity of the libertines. But violent controversies also raged within Catholicism, Jansenism, and, later, Quietism. In the Protestant countries, particularly England during the Civil War, the seventeenth century witnessed the birth of the most diverse and radical sects. Puritans, Quakers, Socinians, and many others attacked the official Church. In all these controversies, whether they were about proving the existence of God or defending a certain dogma, a wealth of abstract reasoning and intellectual subtleties went to waste, since no one convinced anyone of anything. At the end of the century, there was a general feeling of fatigue after so many useless efforts. The same movement that drove the scientists to distance themselves from abstract theories in order to cultivate observation and experimentation drove the theologians to seek evidence for the existence of God in the marvels of created nature. Jacques Abbadie, a Protestant, and François de la Mothe Fénelon, the archbishop of Cambrai, both offered the same type of apologetics, and many others imitated them.

The movement started in England. Nehemiah Grew, John Ray, George Cheyne, William Derham—all naturalists, some famous, and all men of the cloth—published scholarly works in which Newtonian cosmology, anatomy, and especially natural history served to multiply the proofs of the existence of an intelligent Creator and to point out the folly of those who believed that the order of the world was the result of chance.[4] From England the fashion passed to the Continent, first in the Protestant and then the Catholic countries. In France the great representative of this apologetic through natural history was the abbé Noël Pluche, the author of *Le Spectacle de la Nature* [The Spectacle of Nature], one of the greatest literary successes of the century, whose nine volumes, appearing between 1732 and 1750, were repeatedly republished. Réaumur, who had furnished Pluche with some of his evidence, was a little wary of the naive enthusiasm of those who believed to have understood everywhere the Creator's intentions. But he was quite convinced that he found God in the insects that he was studying: "I have often had real pleasure in seeing in detail some of the marvels that only He who knows how to make marvels has conceived to vary the species of insects so prodigiously and to perpetuate them. It was gratifying to me to present these pleasures on a level for those who could appreciate them. Excellent moral lessons can be drawn from these sweet and calm pleasures that elevate the soul toward the Being of beings, with whose existence we are too rarely concerned."[5] It was in nature and no longer in cloisters that it was possible to take refuge in order to leave the world behind and find God. Pluche's work takes the form of a conversation between worldly people during their vacation in the country, in a comfortable chateau, far from the noise of the city and their frivolous amusements. While pretending to teach how to discover nature, he preached a philosophy of life.

Pluche did not write only for adolescents at a time when colleges ignored natural history, and Réaumur did not write only for his entomologist colleagues. Both of them wanted to touch the sophisticated public; they wanted to entertain them, but entertain them in a useful way. People who were bored could only be distracted by piquing their curiosity without demanding too much intellectual effort from them. Since they liked to be astonished, they would be presented with marvels, but "true marvels." Pluche left to others the difficult and thankless task of "penetrating to the very foundation of Nature"; he was content to present the "spectacle," which Fontenelle had compared earlier to a production at the Opéra. Pluche took up this metaphor: "We believe that it is more suitable for us to describe the exterior decoration of this world and the effect of the machines that produce the show. We are allowed to see this much. We even see that it has been rendered so brilliant only in order to pique our curiosity. But, content with a representation that fills our senses and our intelligence, we need not ask that the engine room be opened to us."[6]

Was the "engine room" open to man? There was more and more doubt at the time, and the abbé Pluche himself made fun of the astronomers who claimed "measuring rod in hand, to measure the distances of the planets and to calculate the effects of motive forces." It was better to be satisfied with admiring nature and, rather than asking a lot of useless questions, to put to active use—through industry and commerce—those resources that nature offered us, since Providence has given them to us. Réaumur, who knew the difficulty of understanding what happened in the "engine room," did not try to have worldly people penetrate it either. He saved them the dryness of real science, since he had little confidence in their capacity for attentiveness. Instead he sought to present them with this "true marvel," which might distract them and bring them to God.*

The incredible success of this "experimental theology"—Pluche's first

---

*Bringing himself to speak of classifications, Réaumur pointed out that what he would say would not interest "those who are only touched by the agreeable offerings of this science [entomology]." He added, "Those sciences that from the outside are the most pleasant are dry and arid, once one examines them closely: he who wishes only to find that which is agreeable must be satisfied to touch them lightly" (Réaumur 1732, I, 41).

volume had three editions in six months—showed that the strategy was a good one. Not everyone was brought back to God, of course, but sophisticated people were not insensitive to the nostalgia for the simple life in an idyllic countryside and its "sweet and tranquil pleasures," well before Jean-Jacques Rousseau, a great admirer of the abbé Pluche, wrote *La Nouvelle Héloïse* [The New Héloïse]. They were incapable of leaving the city, so they made nature come to them. The fad of "cabinets of curiosities," which would be all the rage in the second half of the century, was already present, and some people started to hoard fossils and other "singularities of nature."[7] When Buffon started to think seriously of writing a natural history, he knew that there already was a public ready to accept it.

Another Ambition

It is certain that Buffon did not dream of walking in the steps of Réaumur or the abbé Pluche. Had he never looked at an insect? It's possible. Even the title of his work, *Natural History, General and Particular,* makes one wonder. We know what a "particular" natural history was. It was exactly what Réaumur was doing. It was what the abbé Lelarge de Lignac, a student of Réaumur's and soon Buffon's enemy, was doing when he published in 1749 a *Mémoire pour servir à commencer l'histoire des araignées aquatiques* [Memoir to Introduce the History of Water Spiders]: it would be too ambitious to pretend to have written a complete history of these interesting creatures. But what is a "general" natural history? As soon as the title of the work appeared, Réaumur showed his skepticism and his concern: Buffon was not a naturalist; his philosophical opinions, as far as we can tell, were not reassuring; above all, it was far too early to undertake a "general" history when so few "particular" histories were worthy of confidence.

Réaumur was right to be worried. In order to write a "general" history of nature it was necessary to enter the "engine room," to go beyond the description of details in order to see the "big" picture, in short, to try and reduce nature's operations into systems. This means to restore nature herself, no longer as a "spectacle" put on by Providence in order to distract people and to lead them to God but as the play of forces capable of producing the most complex phenomena. A "general" history necessarily implied another philosophy of nature, a philosophy in which God was no longer omnipresent. It is indeed doubtful that Réaumur truly foresaw the extent of the disaster. Buffon already knew what he wanted to do.

He had no intention either of writing only for his colleagues. In fact, except for some chosen friends, it was not at all for his colleagues, or what he called the "common savant," that he decided to write. Rather his intended audience was the large cultivated public, whose new philosophical tendencies he already knew or guessed, who would be more capable of accepting his ideas. Buffon did not hope to distract his readers so much as to make them think. He certainly knew their tastes, their flightiness, their frivolity, and he knew how to flatter them. But he also believed them capable of reasoning. Basically, he despised them less than Réaumur or the abbé Pluche. And it is from this public that he expected the success that the naturalists of the Academy of Sciences would certainly not give him.

This situation is sometimes found in the history of science. When the scientific community is practically unanimous in the idea it has of a science, its principles, and its methods, and when this community shares what philosophers of science, since Thomas Kuhn, have called a "paradigm," the researcher who rejects this paradigm can hardly expect to be understood by his colleagues. His only hope is to be heard by a wider audience that does not share the habits of thinking and the prejudices of professionals and is free of the sociological and intellectual inertia of the scientific community. Their ideas evolve faster and they are less sensitive to technical difficulties. This permitted them to believe that an innovative investigator could detect new truths better than his colleagues. The process is risky because it is necessary, in the final analysis, that the ideas of an innovative scholar be accepted by the scientific community. If not, he will be thrown back into the category of the fringe that interests no one.

Surrounded by Aristotelian academics, Galileo wrote his works in Italian. He meticulously showed the absurdity of the old physics, earning at the same time the hatred of other scholars and the admiration of the intelligent lovers of knowledge. Darwin, who was frightened by the hostility that he knew he would meet from the professional naturalists, composed *On the Origin of Species* like a lawyer's plea, carefully avoiding all technical vocabulary. The majority of the specialists criticized him sharply, but a large public, ready to accept his theory of evolution and the philosophy that it represented, devoured his book. Buffon found himself in the same situation. He felt that there was a public ready to receive his philosophy of nature and the type of natural history that he wanted to do. It was to this public that he would address himself, and his success, which would be enormous, could not be pardoned by his colleagues. Think today of the indignation of honest and painstaking academics when one of them, cleverer or a better writer, obtains honors on television or in weekly magazines.

The fact remains that in a few decades a new scientific community was formed, which adopted Galileo's ideas. Twenty years after the publication of *On the Origin of Species* all biologists, or almost all, were evolutionists. Did Buffon succeed in this last phase of the process? Did he succeed in giving natural history a new direction? This is a difficult question, one that I shall try to answer at the end of my study.

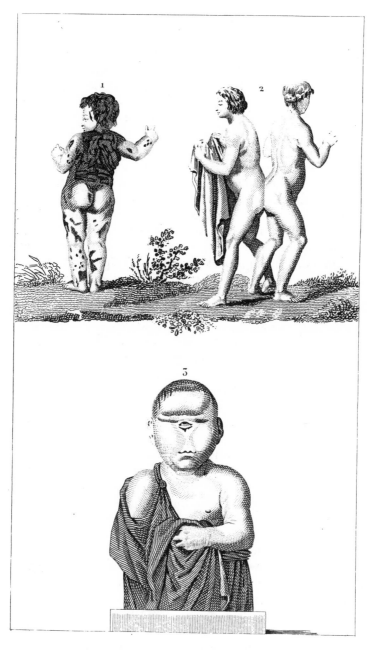

Human curiosities, including Siamese twins. From *Oeuvres complètes*, XII, p. 488.

The Preliminary Research

The first three volumes of the *Natural History*, which appeared in 1749, covered immense and very different subjects: the history and the theory of the earth, the formation of the planets, problems that we would now call "general biology," in particular the theory of reproduction, and finally the "natural history" of man, which included studies of physiology, psychology, and ethnology. We do not know how much Buffon knew about these subjects in 1739. Probably not much. We also do not know when he decided to write the *Natural History*, though it was probably very soon after his nomination to the Royal Botanical Garden, if we judge from what we know of the composition of the first three volumes.

It is certain that he had a lot to learn. We know little, unfortunately, of his apprenticeship, except that he was essentially bookish. In many areas the mass of works quoted is impressive. We know nothing precise about his method of working. It is probable that he was helped by "excerpts" done for him of works that interested him. All this documentation was later destroyed, for Buffon always feared being "buried under papers."

The *Histoire and théorie de la Terre* [History and Theory of the Earth] has a date: "Montbard, October 3, 1744." This was therefore the first of the finished texts, in a field that Buffon had never studied before. Nonetheless, he put the "final touches" on it in 1746, quoting in it a dissertation by Pierre Barrère published that year. We have no trace of any preparatory work, other than a report before the Academy in 1739 on a map of lands in the southern hemisphere presented by Pierre Buache.[8] Then came the text "De la formation des planètes" [On the Formation of the Planets], dated "Buffon, September 20, 1745." This text was therefore already written when Buffon became involved in the controversy with Clairaut, and indeed some problems discussed by Clairaut are found here, in particular that of the "figure of the earth," that is to say, the measurement of the flattening of the globe at the poles. One also finds here problems linked to the study of skyrockets, which dates from 1740. We do not have any traces of the preliminary work on the *Natural History of Man* either, except for the two papers from 1743 on squinting and the color of shadows. The only topic on which we know that Buffon worked for a long time is that of reproduction.[9] As early as 1733, he discussed it with the Swiss naturalist Louis Bourguet through the intermediary of Bouhier, reproaching Bourguet, incidentally, for holding ideas that he later adopted. In 1741 he was the first to announce to Martin Folkes the discovery of Trembley's "polyp," which impressed him a great deal.[10] In 1744 he spoke with Maupertuis, and we shall see the importance of these discussions.

The first five chapters of *Histoire des animaux* [History of Animals] are dated "At the Royal Botanical Garden, February 6, 1746." The date is, however, perhaps false, or marked the beginning of his research. A little later, Buffon met an English microscopist recently arrived from Paris, the abbé Needham, and undertook with him and several other collaborators—Daubenton, Dalibard, and Guéneau de Montbeillard—a series of observations on "seminal fluid," which would last from March to May 1748. On May 18, Buffon deposited the result of this research at the Academy in a sealed envelope,[11] before the research was completed and before presenting it orally to the Academy on December 14. At the end of his lecture he announced, "I will give an account of everything in a work that I am proposing to publish, but I thought I owed you this extract from it; I have been delighted to pay homage to the Academy with such an interesting discovery." Everyone already knew that the *Natural History* was about to appear.

The Publication of the First Three Volumes

In October, the *Journal des Savants* had announced Buffon's project to the public: a complete natural history in fifteen volumes, from minerals to man. It deliberately specified that the work had "been done according to the views and orders of M. le comte de Maurepas." That should quiet the critics.

The first volume was already in press. It contained a description of the Royal Botanical Garden and its Cabinet, a treatise on the method for studying natural history, and a theory of the Earth. Eight volumes would follow, dedicated to the natural history of animals and of man, "considered as an animal with his habits depending on race and climate." Three volumes would treat plant physiology, agriculture, and botany, and the last three volumes would treat minerals. "We promise to give the public several volumes each year."[12] In fact, the text of the first volume, half printed at the end of May, was finished at the beginning of September, and then the second volume was in press.[13] The plates, however, needed to be engraved, which held things up. In August 1749 the printing was done, but the work could not be sold until the minister had drawn up the list of people to whom it needed to be sent, as books published at the Royal Press had to be sent to the king, the royal family, and certain other great personages. It is possible that the three volumes were offered for sale in September. One thousand copies were printed.

These three quarto volumes, magnificently printed, were very impressive. On the title page, two angels blowing on the trumpets of Fame held up a globe decorated with three fleurs-de-lys and topped with the royal crown. Above was the title of the work: *Natural History, General and Particular, with the Description of the Cabinet of the King. Volume One.* The address, "A Paris, De l'Imprimerie Royale" [Paris, From the Royal Press], endowed the book with the character of an official publication. One thing is surprising: there is no author's name. Following the title page is the dedication, "To the King." It is signed "Buffon, Intendant of your Botanical Garden" and "Daubenton, Guardian and Demonstrator of your Cabinet of Natural History." Following this is a "Table of what is contained in this Volume," ending with the words "By M. de Buffon." In the following volumes, the contributions of each of the two authors would be identified in this same way.

Looking at these pages no one could know that Buffon was a member of the Academy of Sciences. As modesty was not his forte, one can ask why he renounced adorning himself with a prestigious title. Since there are no documents about this point, we can at least put forward a hypothesis. The volumes published by the Royal Press were not exempt from the censor. Publications of academicians, however, were not subjected to the administrative censor, provided that they had received official approval from the Academy, and it was also on this condition that an academician could present his academic standing on the title page of a book. Nothing leads us to believe that Buffon had submitted his work for approval from his colleagues, to whom he did not even announce the title before December 14, 1748, while the *Journal des Savants* had already made it public. Finally, the book bore no approval of a royal censor, and therefore had not been examined by the normal censor.

Everything leads us to believe that Buffon had played a shrewd game and succeeded in publishing his work without its being examined by anyone. It is difficult to believe that d'Argenson, the minister responsible for the Academy of Sciences and literary work as well as for the War Ministry, was taken in. Yet without a doubt more was involved than just a trick. The only position that Buffon wanted to claim was what would follow his signature at the end of his dedication to the king: "Intendant of your Garden." He could not have established his distance from the Academy any better, or better marked his official position and the privileges that accompanied it.[14]

CHAPTER VI

A New Discourse
on Method

The "First Discourse,"* which opens the first volume of the *Natural History,* does not have a provocative title: "How to study and treat natural history." The professional naturalists, Réaumur for example, could very well question what authority Buffon claimed to have to give lessons in this area. But introducing a work with a summary of the goals and methods was not unusual, and Réaumur himself had dedicated a summary of this kind in his "First Memoir," which opened his *Memoirs concerning the History of Insects.*

It is interesting to note that there are several common points between Buffon's "First Discourse" and Réaumur's "First Memoir": both begin by showing the discouraging immensity of the task that awaits the naturalist, and both speak of the history of the discipline. Right from the start, there are differences, however, and we quickly see that Buffon would not be satisfied to repeat the commonplace. His real subject was the knowledge that man could have of nature. This was strictly a philosophical question that went beyond the limits of natural history, unless we admit, as Buffon appeared to be ready to believe, that all knowledge of nature comes from natural history.

This is an old question, dating from the Greek philosophers, with Plato and Aristotle in the lead, and discussion of it continues today. The scientific revolution of the seventeenth century gave it a new topicality. Starting with Bacon and Descartes, philosophers and scholars discussed the power of

*The expression "First Discourse" is found in the Table of Contents of Volume I of the *Natural History.* In the same way the *History and Theory of the Earth* is titled "Second Discourse." The use of this word to designate all texts produced by Buffon became standard. A possible explanation of the archaic use of this word, which does not imply any "oratorical" sense, is found at the end of this chapter.

human reason and the method appropriate for the sciences. By defining the power of human reason, man and his place with respect to nature and to God were also defined. This leads immediately to metaphysics. Buffon knew it, but he also knew that his metaphysics was not very orthodox and that he therefore should not expose it directly. Moreover, he pretended not to consider himself a philosopher. It was only through immediate problems that were met with in the practice of natural history that he would discuss these subjects. This explains the curious structure of his text, which starts with pedagogical remarks and ends with the question, What is truth?

At the time that Buffon entered the debate, there were two main philosophical traditions concerning the powers of human reason. One came from Descartes and was based on mathematical practice. It affirmed that reason could attain truth, if it was based on "very simple and very obvious principles," such as axioms of geometry, and if it followed a logical and rigorous method such as that of mathematics. If the "clear and distinct" ideas that our reason found obvious were false, that would mean that God himself was tricking us, which was impossible by definition. Therefore, even in physical matters, we must follow the deductive method of mathematics to arrive at the truth unless in checking our deductions through experience we find that certain notions, like the notion of infinity, are above our powers. Fundamentally, however, Cartesian rationalism affirmed that human reason could understand nature.

The other tradition, represented in France by Gassendi and Pascal among others, did not give as much power to human reason. It said that experience was the only source of our knowledge, and that it only allowed us to attain knowledge of the surface of things—their appearance—or, to use the Greek word, the "phenomena." God does not guarantee the truth of our findings. We could build a science, but that science would never be more than "our" science. In putting experience first, this tradition took up the ideas that Francis Bacon had presented in the beginning of the century and that, after the whirlwind of Cartesian science, would come back into favor with scholars in London as well as in Paris.

This same tradition took on a new philosophical worth with John Locke's work. In 1690 Locke published his *Philosophical Essay concerning Human Understanding*. In this book, which would influence all thought in the eighteenth century, Locke explained that our ideas derive only from a combination of sensations and that the human mind has no other power than that of bringing about that combination. This "sensualist" philosophy, as it would be called, excluded the possibility of man's ever truly understanding nature. But it also risked putting the existence of man's spiritual soul in doubt and allowed the belief that thought was no more than a property of organized matter. When Voltaire presented Locke's ideas in his *Philosophical Letters* of 1734, he raised this matter openly and caused a scandal.

To doubt the ability of human reason to understand nature is not blame-worthy in itself. Réaumur and the abbé Pluche, as we have seen, did not think us capable of entering into the "engine room," but they did not doubt the existence of the soul. Buffon was Locke's disciple, and he only partially hid this fact. As for the human soul, his ideas were hardly ortho-dox. He would therefore have to be careful.

What to do, then, faced with "all the objects presented to us by the Universe," with "this prodigious multitude of Quadrupeds, Birds, Fish, Insects, Plants, Minerals, etc."? First of all, they must be brought together and exhibited. Natural history in the eighteenth century was not carried out in the wild. It took place in the "Cabinet of the King" and in similar institutions. This was not out of laziness but because that was how the system worked. It was necessary to gaze at length at "the diverse produc-tions of the different climates" placed side by side. "In seeing them often, and, so to speak, without a drawing, [these objects] little by little form lasting impressions, which soon are linked in our intelligence through fixed and invariable relationships, and from there we reach more general views." Let us therefore be open to our sensations: they combine by them-selves in our intelligence. It was enough to let the mind "work by itself and . . . alone form the first links that represent the order of its ideas."[1] It was good pedagogy, but it was also Locke's own philosophy.

Buffon, however, was not a man to expect everything to come from a passive mind, nor did he believe that in order to see, it was only necessary to look. It was necessary to be talented, to have "that first spark of genius, that seed of judgment" that leads to the study of nature. Not everyone has it. "The love of the study of Nature assumes two qualities in the mind that seem to be opposites: the broad view of a fervent genius that embraces everything in a single glance, and the attention to small details of a la-borious instinct that looks at only one detail at a time."[2] This was because "the true and only science is the knowledge of facts."

It was not enough to accumulate facts, however, to "make exact descrip-tions." In other words, it was not enough to do what Réaumur did. "It is necessary to try to rise to something that is greater and more worthy of our time; it is necessary to combine observations, generalize the facts, link them together through the power of analogies, and try to arrive at that high degree of knowledge where we can see how particular effects depend on more general effects, where we can compare Nature with herself in her great operations, and where we can finally find ways to perfect the different parts of Physics." Description required only "a good memory, assiduity, and caution." To think, more was needed: "It is necessary to have general views, a firm glance, and reason formed more through reflection than study; finally, it is necessary to have that quality of mind that permits us to grasp distant relationships, fit them together, and form a body of rational ideas, after having justly appreciated their plausibilities and weighed their probabilities."³

From the beginning, we are at the heart of a fundamental contradiction between a philosophy and a temperament. The human mind can only combine sensations: Buffon admitted this, because he was Locke's disciple. Yet man can also build a science: Buffon was sure of this because he felt that he himself was capable of it. Up until the end of his life and his work, Buffon would have to live with that contradiction and, as Pascal said, "hold onto both ends of the chain."

Is There Order in Nature?

To build a science is to discover an order in nature. Let us be careful here. "Since we know only one way to arrive at an end, we convince ourselves that Nature makes and operates everything through the same means." Thus, we are "naturally inclined to imagine a kind of order and uniformity in everything." Do we have this right? "Doesn't this amount to importing the abstractions produced by our limited intelligence into the reality of the Creator's works?"⁴ Formulated in such a way, the question was perfectly orthodox. Réaumur or the abbé Pluche would be the first to recognize that the wealth of Creation exceeded our understanding, and the deist Voltaire would have agreed. Yet even if we cannot discover the order of the world, we know that it exists, since this world is the work of God.

Buffon seemed to be less sure. In a language admirably calculated at least outwardly to respect the divine character of Creation, he asked the dangerous question: Is there order in nature? We are overcome not only by the prodigious number of her creations, but also by the infinite variety of her "resources" and her "very disorderliness." "Too small for this immensity, overwhelmed by the number of marvels, the human intelligence succumbs: it seems that everything that can be is; when the hand of the Creator opened, it seems, it was not to bring into being a certain limited number of species; it has thrown down, it seems, a number of related and unrelated beings all at once, an infinity of harmonic and contrary combinations, and a perpetuity of destructions and renewals."⁵

If "everything that can be is," if there is as much contradiction as harmony and as many destructions as renewals, if beings have been "thrown" almost negligently by the Creator's hand, it is necessary for us to renounce once and for all the image of a harmonious and peaceful nature that an attentive Providence had lovingly built for the happiness and the edification of man. If it is necessary to find an order, and it is indeed necessary to find one, it is to be found neither in the structure of beings nor in their supposed adaptations to their conditions of life. Réaumur's admiration was badly founded.

The pretensions of the classifiers did not have a better foundation. With a scandalous violence, Buffon ridiculed the Linnean classification. In botany, Linnaeus, "in accordance with his system . . . put together into the same classes blackberry bushes and nettles, tulips and barberry, elms and carrots, roses and strawberries, oaks and bloodwort."[6] In zoology, it was even worse. In the *Anthropomorpha* order, Linnaeus grouped "man, monkeys, sloths, and scaly lizards. One must really be obsessed with classifying to put such different beings together." The hedgehog, the mole, and the bat were put into the order of "ferocious beasts," *Ferae.* The hippopotamus and the shrew were in the order of "*Jumenta,* or beasts of burden." "A grouping . . . that is so unjustified and bizarre as if the author had intended to make it such."[7] With a cheerful ferocity, Buffon played on the weaknesses of the Linnean nomenclature and refused to enter for an instant into the logic of a system that he challenged.

He did so precisely because he criticized it for being an arbitrary logic which some naturalists wished to impose on nature. To establish their divisions, the botanists took "the liberty . . . of arbitrarily choosing a single part in the plants to be the specific characteristic." They claimed "to judge the whole from one part," which was "a metaphysical error." If you looked at the entire plant, the clear-cut divisions faded away. "Nature moves through unknown gradations and consequently she cannot be a party to these divisions, because she passes from one species to another species, and often from one genus to another genus, by imperceptible nuances."[8] All methods of classification were destined to fail: they were a requirement of the human mind, not of nature's reality.

In fact, "that way of knowing is not a science, it is only a convention, an arbitrary language." "It must be used as agreed-upon signs are used by us to understand one another." Every method is "only a dictionary where one finds words put into an order . . . as arbitrary as alphabetical order." These signs are not useless: it is necessary to have a common language in order to "understand one another."[9] They do not teach anything about the nature of things, and a method of classification is not a method of reasoning. It is not that true scientific method for which "the greatest Philosophers felt the necessity" without being capable of defining it:

"In this same century when the Sciences seem to be cultivated with care, I believe that it is easy to perceive that Philosophy is neglected, maybe more so than in any other century; the so-called scientific Arts have taken its place; almost everyone is taken up with methods of the Calculus and Geometry, those of Botany and Natural History, the formulas, in short, and the Dictionaries; everyone believes that he knows more because the number of symbolic expressions and scholarly phrases has been increased, and no one cares that all these arts are no more than scaffolding for the sciences, and not science itself."[10]

This was a surprising condemnation of an entire century, but here we find a constant theme in Buffon's thoughts, dating from his discussion of the idea of infinity and continuing up to his quarrel with Clairaut. The calculus, the mathematical expression of the law of gravitation, was only a product of these "so-called scientific Arts [that is, techniques]." As for the methods of natural history, they were only systems of arbitrary signs; they teach us nothing about nature, the only object of a true science.

At the end of the text, when he took up the fundamental question, What is truth? Buffon carefully separated mathematical truths from physical truths. Mathematical truths were purely a product of our intelligence: "We have made suppositions, we have combined them in many ways; this body of combinations is mathematical science; there is, therefore, nothing more in this science than what we have put there." Mathematics is no more than a gigantic tautology: "It is for this reason that it has the advantage of being always exact and provable, yet abstract, intellectual and arbitrary."[11] And therefore useless.

For many of Buffon's contemporaries, mathematics was entirely written in God's wisdom, and man, in progressively enunciating new theorems, did no more than laboriously discover the fragments of an eternal Truth. And since God was the Eternal Geometer, it was not surprising that the laws of nature could be expressed in mathematical formulas. To say that mathematics was a pure creation of human intelligence was almost blasphemous, and also posed another problem that we have not resolved today: specifically, what is the relationship between mathematics and nature? We will see that Buffon could not avoid addressing it.

By renouncing the methods of classification and mathematics at the same time and for the same reasons, Buffon, leaving aside the question of philosophy, was going against the strongest prejudices of his time. If it is true, as it was said,[12] that the science of the eighteenth century relied on the conviction that words could represent things, Buffon was not from that century. And his refusal was not accidental. It was founded on a philosophy of nature and man.

We therefore have to come back to reality. That is what the ancient great naturalists did, Aristotle in particular, of whom Réaumur spoke with disdain and Buffon praised at length. Aristotle knew how to avoid the two defects of modern science: he did not encumber himself with either arbitrary classification or useless descriptions. He set the example of what a "general" natural history could be by using the method of comparison and generalization. He demonstrated the relationship among animals by consideration of the most important vital functions, thus ascertaining the essential facts without neglecting the details. "The outline of the work, its organization, the choice of examples, the soundness of the comparisons, a certain turn of the ideas, which I would willingly call philosophical character"; all this was exemplary in Buffon's eyes. He was certain that Aristotle the naturalist was the model he wanted to follow and not Aristotle the philosopher deformed by scholasticism.

Above all, Aristotle took man as the starting point for all his comparisons, "more because he is the best known animal than because he is the most perfect." This is an important detail, which radically distinguished Buffon from the abbé Pluche and the providentialists. Buffon, too, placed man in the center of nature, and more completely than Aristotle had, primarily because man is a knowing subject. It was he who made science, and that science was his: it could no longer claim to resemble an imperfect image that God had of His creation. The real is therefore what man discovers around him, as if spontaneously: "Let us imagine a man who has actually forgotten everything or who wakes up innocent of everything around him; place that man in a countryside where the animals, the birds, the plants, the stones present themselves successively to his eyes. In the first moments this man will not distinguish anything and will confuse everything: but let his ideas be confirmed little by little by the reiterated sensations from these same objects: soon he will form a general idea of animated matter, he will distinguish it easily from non-animated matter . . . and naturally he will arrive at the first great division, *Animal, Vegetable, and Mineral.*"

We recognize our "innocent man": he is the novice student at the start of the text. But he is also a first appearance of an image that Buffon would come back to, along with many others: an image of a man who awakes in nature and whose sensations progressively organize themselves. This would be Condillac's and Charles Bonnet's famous statue, an image symbolic of sensualist philosophy.

As easily as he distinguished the three great "kingdoms" of nature, our "innocent man" would differentiate quadrupeds, birds, and fish within the animal kingdom, and trees and plants within the vegetable kingdom. "Here is what simple observation must necessarily give him. . . . Here, too, is what we must see as real, and what we must respect as a division given by Nature herself."[13]

The logical conclusion of this text was that there was no "chain of beings." The idea of a "chain" or "ladder of beings" was very old,[14] but it had found new life thanks to Leibniz, and it would soon be eloquently defended by the Genevan naturalist Charles Bonnet. In short, it held that all created things, from stones to angels, constituted an uninterrupted series, passing imperceptibly from one degree to another. There were therefore "organized stones" (for example, asbestos) and "lithophytes" (or stone-plants) that "form a transition" between mineral and vegetable. There were even "zoophytes" (or animal-plants, like the sponge) which serve as a bridge between plants and animals. If the division among the three king-doms was "made by Nature herself," as Buffon said, there was no longer a continuous chain between beings.

The difficulty was that Buffon had just affirmed that if man put himself "at the head of all created beings, he would see with astonishment that one could descend by almost imperceptible degrees from the most perfect creature to the most shapeless matter, from the most organized animal to the crudest mineral; he would recognize that these imperceptible nuances were the greatest work of Nature."[15] An excellent definition of the chain of being. Which text is to be believed? In fact, both of them, and Buffon explained why. We do not know of any lithophytes or zoophytes. Therefore the three kingdoms of nature were well separated. And in any case, "our general ideas were only made up of specific ideas," our categories were only valid for a "rough idea of things." In the classes of natural beings, "we clearly perceive only the middle . . . the two extremities recede and increas-ingly escape our considerations."[16] Note that the two arguments are contra-dictory and that the difficulty is not really removed.

Let us come back to our "innocent man," and let us lend him all the knowledge that we have. How would he classify animals? He would do so according to "the relationships that they would have with him; those that are the most necessary to him, and most useful, will occupy the first rank; for example, he will show a preference to the horse, dog, ox, etc. in the animal order." Then he would pass on to wild animals that live in the same country and finally to "animals of foreign climates." In this way he would study the "productions of Nature . . . in proportion to the use that he can obtain from them . . . and he would arrange them in his head relative to that order of his knowledge, because it is indeed the order according to which he acquired them, and in which it would be important for him to keep them."[17] Of course, that order led to "putting very different things together." That, however, was of little importance: "Is it not better to ar-range objects in the order and position where they are normally found, not only in a treatise of Natural History, but even in a painting or anywhere else, rather than to force them together because of a supposition? Is it not better to have the horse, which is *soliped* [single-hoofed], followed by the dog, which is *fissiped* [divided feet, i.e., toes] and who indeed traditionally follows the horse, rather than by a zebra, which is rarely known to us and perhaps has no other relationship to the horse than being single-hoofed?" Or if the Linnean classification were adopted: "Would you find that a lion resembles a bat more than a horse resembles a dog?" Since all classification is arbitrary, it is just as well "only to follow the order of the relationship that these things appear to have with ourselves in our distributions."[18]

These texts scandalized the naturalists of the nineteenth century. Notice first of all that they should not have scandalized Réaumur, or at least not for the same reasons. The abbé Pluche began his discovery of nature "with the flowers and the greenery of our gardens." From there he passed to "our vegetable and fruit gardens," and then on to "our workable earth," "our vineyards," and finally, to "our forests."[19] Pluche made an inventory of the gifts of Providence. Buffon followed the order of a "natural discovery" of nature. What they had in common was in not believing man to be capable of grasping the order of the world. The taxonomists of the nineteenth century believed that this order was knowable, but the theory of evolution upset everything, and even today "evolutionary" and "cladistic" taxonomists argue over whether crocodiles should be put with birds or with reptiles.[20] Nature does not let herself be so easily grasped.

What resulted was that the order Buffon claimed to follow, "the natural order in which all men are used to seeing and considering things," reduced nature to the French countryside of the eighteenth century. It was in the farmyard that the dog "traditionally" followed the horse. Buffon wrote for French readers, and in any case this is unimportant, since we are dealing with the arbitrary. Also, the word "natural" is ambiguous: would it be "natural"—the comparison is significant—to put a zebra next to a horse "in a painting"? Finally, we should not rule out a certain taste for provocation in Buffon. To scandalize the Linneans must not have displeased him.

Biology before the Term Existed

If all this was without real importance, it was because the goal of the *Natural History* lay elsewhere. Above all, it was necessary to give "the exact description and the true history of each thing," "not the history of the individual but that of the entire species"; to start by describing an animal from the outside and inside, without entering into useless details that belonged more to the area of comparative anatomy, but not to be limited by shapes. If classifications were futile, it was above all because they were founded on details of morphology. The natural history of an animal species must include much more: "It must include their generation, the duration of pregnancy, birth, the number of young, the care of the fathers and mothers, their type of rearing, their instinct, the place of habitation, their diet, the way that they obtain it, their habits, their tricks, their way of hunting, and then the services that they can render us, and all the uses or commodities that we can obtain from them."[21]

Here we are very far from the cabinet of the Royal Botanical Garden or Linneaus's *System of Nature*. The knowledge of the animal of which Buffon spoke was that of the man of the countryside, of the huntsman who would locate the deer for the hunt the next day, of the breeder who must lead cows to the bull or plan the birth of lambs. These were the real "relationships" between nature and man. At the same time, however, natural history breaks apart the "tables" that claim to "represent" the order of things by putting names in boxes. Behind each name there is now a living animal, that hunts or flees the hunter, that occupies a territory, reproduces, raises its young. Natural history now embraced an entire realm of new knowledge, which the classifiers had excluded.

It was not therefore through classing morphologies but in systematizing our knowledge of living beings as they live, through comparing their physiologies, their "habits," according to the climates in which they live, that we could reach "that high degree of knowledge where we can see how particular effects depend on more general effects, where we can compare Nature with herself in her great operations."[22] That was the true goal of the *Natural History*. If there was an order to the world, it was not an order of structures, those structures that the taxonomists classified. It was the order of the "operations" of nature, an order of the processes that give rise to life and its perpetual renewal, an order of the forces that animate the living world and of the laws that govern them. If we understood this order, we would be capable of creating a "Physics" of the living and explaining effects through causes and laws. Buffon's *Natural History* sought to be a biology before the word existed.

Are we able to go so far? If mathematical truths were no more than a production of our mind, what were physical truths? They could not be arbitrary, "they are based only on facts." How to pass from fact to law? "A series of like facts or, if you wish, a frequent repetition and an uninterrupted succession of the same events, make up the essence of physical truth: what one calls physical truth is thus no more than a probability, but a probability so great that it equals a certainty."[23] A little before the Scottish philosopher David Hume, of whom Buffon did not yet know, Buffon asserted that we could not be *absolutely* certain of the regularity of natural phenomena. Hume would speak of "belief." Buffon spoke of "probability," but it was a subjective probability, which he would later try to quantify. It seems to us exceedingly probable that the sun will come up tomorrow, because in man's memory the sun has always come up. In fact, nothing *proves* to us that it will come up.

Mathematical truths are obvious, but they tell us nothing about the reality of things. Physical truths are only probable. For good measure, Buffon quickly mentioned moral truths, "in part real and in part arbitrary," which "only have proprieties and probabilities as a goal and end." In order for all these "truths" to be really true, a guarantor would be needed, which could only be God, and God was totally absent. Man was alone with himself, facing nature with no guarantee of being able to understand it.

And yet it was necessary to construct a science, and a science that went beyond the simple description of things, a science that discovered the "great operations" of nature. If, after having established the facts we seek the causes of them, "we will find ourselves suddenly stopped, reduced to trying to deduce the effects from more general effects, and obliged to admit that the causes are and will be perpetually unknown to us, because our senses, being themselves the effects of causes that we do not know, can give us ideas *only of the effects,* and never the causes; it is therefore necessary to limit ourselves to calling cause a general effect, and to give up on knowing more." We will therefore discover an order, but that order will be "relative to our own nature rather than suitable to the existence of things."

This would nonetheless be an order: "These general effects are for us the true laws of nature; all phenomena that we recognize as conforming to these laws and depending on them will be that many facts explained, that many understood truths."[24] To understand nature was no longer to discover the suitability of structures; it was to discover the order of laws. By applying mathematics to physics, we can measure the relationship between cause and effect, and if "the result agrees with the observations, the probability that you have guessed correctly increases so strongly that it becomes a certainty." That was the strength of Newton's system. But outside of astronomy and optics, which treat simple objects, physics can rarely use mathematics in such a way. If you try to introduce it by force into subjects that are too complicated, you risk "stripping the subject of most of its qualities, creating an abstraction of it that does not resemble the real thing," and arriving "at a conclusion that is just as abstract." This is the source of "scholarly errors that one often accepts as the truth." If we were aware of this danger, "it would be possible to agree on the Metaphysics of the Sciences" and to discover "the true method of using one's mind in research": to observe, combine, generalize advisedly, to link the essential facts through analogies, "to base one's plan of explanation on the combination of all the relationships and present them in the most natural order."[25]

"We will give examples of this method in the following discourses, the THEORY OF THE EARTH, the FORMATION OF THE PLANETS and the GENERATION OF ANIMALS."[26] By concluding his "First Discourse" thus, Buffon imposed on his reader a comparison that already suggested "the true method of using one's mind." Descartes also had followed his *Discours de la méthode* [Discourse on Method] with three "essays," of which the first two were divided into "discourses." It is significant that Buffon had chosen the same word as Descartes, even though this word, which had the meaning of an organized presentation, had become completely archaic. In presenting his "essays" after his "discourse," Buffon emphasized his intention of imitating Descartes, or more likely of refuting him.

There was no great merit or originality in refuting Descartes in 1749. Everyone agreed that he was too daring a theorician, a fabricator of "physics novels." He had been useful in his time, pulling down the empire of scholasticism, but his time had passed. Thanks to Newton, it now was known that a science could only be built on experience without "feigning hypotheses."

Buffon, however, did not name Descartes anywhere and seems to have calculated his discourse in such a way as to make the most enemies possible in the learned world. He told mathematicians that mathematics had no practical uses. He reproached meticulous observers for not knowing how to go beyond their descriptions. He denounced classifiers for the arbitrariness and the vanity of their methods. That was a lot of people. More precisely, that was a lot of academicians.

The philosophers were not treated any better. For the benefit of Christian, Cartesian, and Platonic idealists, Buffon explained that mathematics was a creation of man and human reason would never understand nature. For the skeptics, be they Christians or deists, he affirmed the possibility of creating a science. In short, he contradicted everyone.

Above all, however, with arrogance and aggressiveness, he placed man at the center of nature. He made of him both an animal immersed in nature and a reasoning being placed in the center of nature, one who served as his own point of reference. Of man, Buffon could say, like Pascal, "If he raises himself, I lower him; if he lowers himself, I raise him." Yet above this Pascalian man there was a hidden God. Buffon sometimes mentioned the Creator, but this Creator was only a stylistic prop. At this time everyone, not only Réaumur the Christian but also Voltaire the deist, saw God in nature. Maupertuis himself, Buffon's great friend, found God in a principle of mechanics. Buffon seemed to have been struck blind. Man, as Buffon showed him here, this man who must construct a science, was alone in nature and did not seem to be moved by it. Nothing was more foreign to him than anxiety, at least metaphysical anxiety. Already for Buffon, metaphysics was no longer anything but a philosophy of the sciences.

And yet, in an ultimate contradiction, this man without God, this "innocent man" who awoke to the "real" world somewhere in the French countryside, revived the Cartesian ambition to discover the rule of law in nature. Even though Newton served as his model, Buffon would go much further than his master, to the point of betraying him. If, without excessive modesty, he took himself as a new Descartes, it was because he defined both the ambitions and the limits of a new science.

CHAPTER VII

History and Theory
of the Earth

The title of the "Second Discourse" needs to be explained to the modern reader. The "History of the Earth" was the description of everything that could be seen, the distribution of the continents and the seas, topography, stratifications, and so forth. The "Theory of the Earth" was an attempt to explain through physical causes or past events this organization of the earth. The double title signified that Buffon, as he had warned us, had the intention of combining "physics," or the research of causes, with "history," or description. In doing so, he would more or less cover the field that today we call "geology," even though the method used was still far from that of modern science.

The organization of the text is curious. The "Discourse" itself is short, fewer than sixty pages in the spacious typography of the original edition. But it is followed by nineteen articles of "Proofs," which take up nearly five hundred pages. Was Buffon trying to spare the worldly reader from tedious details and interminable descriptions? Maybe, but the first articles of the "Proofs" are also those that contain the most original and sometimes the least orthodox ideas. It is there that a theory of the formation of the solar system is found, which we shall examine in the next chapter, along with a severe criticism of those who confuse the Bible and natural history. These are matters that he perhaps preferred not to present right up front.

Indeed, by presenting a "Theory of the Earth," Buffon entered into an area of research where physics and theology were singularly mixed. Highly daring hypotheses rubbed shoulders with precise observations. This intellectual effervescence had our terrestrial globe, its structure and history, for its subject, an earth that man thought he had always known, yet which he had just discovered.

[93]

A New Science

The "Theory of the Earth," such as Buffon understood it, was a new science, but one that had aroused passionate debates for half a century. The ancient authors gave only slight attention to the sciences of the earth.[1] The existence of fossils of sea shells in areas very far from the sea had been known for a long time, and the logical conclusion was that the ocean had at another time covered these lands. That conclusion, however, did not suggest any general upheaval or any great geological activity. For Aristotle, the earth's mass was inert, and subterranean phenomena, be they volcanoes or earthquakes, affected only its most superficial layer. On the surface, only local phenomena existed: the sea could invade the land or, on the other hand, earth could fill in a bay. All these phenomena counteracted one another, and there was no general history of the earth. The possible causes of these changes were discussed throughout antiquity. According to some, they were caused by unusual local catastrophes; according to others, by earthquakes of a normal type.

Despite the remarkable ideas on the general equilibrium of the globe held in the fourteenth century by Nicole Oresme, who believed the masses of the earth and sea could change through erosion and sedimentation, the Middle Ages did not add much to the science of antiquity, except to attribute the presence of sea shell fossils found today on terra firma to the biblical Flood. This idea appeared again at the beginning of the eighteenth century. In fact, we must wait for the astronomical revolution of the sixteenth and seventeenth centuries for the earth in its physical totality to be considered an object of science. By making the earth a simple planet which turned like the others around the sun, the new astronomy freed the earth from the weight of the cosmos, whose center it had been for so long. From then on, it could be studied for itself, and its history became independent of that of the universe. Thus it was during the seventeenth century that a new area of science appeared, the Theory of the Earth.[2] The first authors, almost drunk with their recently acquired freedom, immediately offered encompassing systems by which they tried to explain the topography of our globe, which was, after all, the only thing that could be directly observed, by means of large-scale events that had built up or modified the totality of the earth's internal structure.

The first theory of the earth—the first attempt to understand systematically the earth's structure and its actual topography—was proposed by Descartes in the fourth part of his *Principles of Philosophy*, published in 1644. The author had just explained how the universe had been created with an indefinite number of vortices, each one possessing a star at its center. The earth was originally a heavenly body of this type. Then opaque matter was distributed in concentric layers, which obscured the initial luminous nucleus, while the earth's vortex was captured by the sun's. In the end, an exterior layer of earth was formed, which surrounded a layer of water. The layer of earth then cracked and collapsed into the lower layer of water. That was how, in complete disorder, the continents with their mountains were formed, surrounded by oceans.

The collapse of the superficial layer very strongly suggested the biblical Flood, and contemporary readers generally concluded that Descartes wanted to give a physical explanation of the events told in Genesis. It is probable that that was indeed Descartes's intention and that the very idea of a history of the earth was suggested by the Bible. The idea of an irreversible history of nature was foreign to Greek thought, whereas the Bible tells the history of humanity and has a natural catastrophe, the Flood, intervene.[3]

The Cartesian theory was taken up with several modifications by an English scholar, Thomas Burnet, who published a *Sacred Theory of the Earth* in 1681, in which the biblical references were quite explicit.[4] Before the Flood, according to Burnet, the earth was perfectly smooth and without surface features. The horrible chaos of the present-day mountains date only from that Flood. It was, in a way, a consequence of sin, and sinful man did not deserve better. For the rest, like many of his contemporaries in England, Burnet announced the end of the world and its final flare-up five hundred years from then, at the most. For Burnet, God's designs were accomplished through the play of physical causes, and the history of the earth was inseparable from the history of humanity, that is, the history of salvation.

Burnet's work enjoyed a considerable success. The great Newton himself did not feel it beneath him to respond to his questions and discuss his theory. In 1696 another English scholar, William Whiston, published a *New Theory of the Earth*. Whiston was a follower of Newton, but, like Burnet, he tied the Bible to the history of the earth. Like Burnet, he thought that in the beginning there were no mountains and, in addition, that the axis of the earthly poles was perpendicular to the plane of the ecliptic. There were therefore no seasons, and antediluvian man at our latitudes enjoyed an eternal springtime. When God decided to annihilate corrupt humanity, He sent a comet, whose strong power of attraction inclined the axis of the poles and broke apart the upper strata of the earth, while the humidity contained in the tail of the comet poured out torrential rains. This explained the biblical text, "the windows of the sky were opened."*

In the "Proofs" of his "Theory of the Earth," Buffon had no trouble

---

*Ed. note: Buffon referred to the passage, so it is worth quoting here: "In the year when Noah was six hundred years old, on the seventeenth day of the second month, on that very day, all the springs of the great abyss broke through, the windows of the sky were opened, and rain fell on the earth for forty days and forty nights" (*New English Bible*, Genesis 7.11–12).

ridiculing Burnet and Whiston. At least he recognized these authors had established the idea that the earth had a history and that its existing topography was the result of physical causes, even if, in their minds, these causes were at the service of the divine will. This type of theory, however, had no future. The development of another area of research and the evolution of new ways of thinking soon led to noticeably different theoretical models.

In 1669 a Danish scholar then living in Italy, Niels Stensen, also known as Steno, had published a small volume with the complicated title *De solido intra solidum naturalitur contento dissertationis prodromus* [*The Prodromus to a Dissertation concerning Solids Naturally Contained within Solids*, Eng. trans. 1671]. This text marks a fundamental date in the history of geology. Steno came to the study of geology through the study of fossils, which were then commonly called "figured stones." The "solids naturally contained within solids" were fossils. Contrary to certain contemporaries, who saw them as "sports of Nature," Steno had no doubts about the animal origins of fossils. He knew the traditional explanation based on the Flood's action, but he also knew and agreed with the objections of those who, dating as far back as Leonardo da Vinci, opposed this explanation. The Flood had been too brief and too violent an event to explain the presence and the placement of all these petrified shells. Steno was not only interested in fossils. He was equally interested in the regular arrangement of rock beds in superimposed strata. This regular stratification implied that sedimentary rocks were slowly deposited from calm waters, which was incompatible with the violence of the Flood. Furthermore, Steno immediately understood that the upper layers must have been deposited after the lower layers, and that it was therefore possible to establish a relative chronology of the geological layers. This idea seems obvious to us, but he was the first to present it. From there, Steno proposed a historical sketch, which sought to explain not only the deposit of parallel layers through sedimentation but also the events that must have broken and tilted these layers, which often were not horizontal.

As the inventor of stratigraphy, Steno did not immediately obtain the success that he deserved. On the other hand, the problem of fossil shells, ignored by Descartes and Burnet, became one of the fundamental problems for any theory of the earth. Despite the objections that were well known at the time, the majority of scholars attributed their distribution around the world to the Flood. In 1695 John Woodward, an Englishman who very closely studied the fossils of his country, attempted a synthesis between Burnet's grand theories and the problems raised by Steno. He also had the Flood intervene but was forced to assume that the waters had been capable of dissolving rock in such a way that the shells, obeying the laws of gravity, could sink into them before they settled again so that the lighter ones came to rest on the heavier ones.

Despite the diversity of explanations, all these theories have many points in common. First of all, they all consider water to be the essential agent of geological phenomena. The Cartesian idea of a central fire, supported by Leibniz in 1693, was practically abandoned. Some authors, like Hooke in England, insisted on the role of volcanoes and earthquakes, but they were very much in the minority. It was therefore impossible to imagine that the mountains had been raised up: it was the areas around them that must have either sunk or been eaten away by erosion. In addition, the majority of the theories, with the exception of Steno's, included cataclysms that completely altered the earth. Finally, these theories generally accepted the biblical chronology as it had been calculated in the seventeenth century by Archbishop Ussher, who fixed the date of Creation as 4004 B.C.* For some, such as Descartes, what was involved was the explanation of the earth's "genesis," or the manner in which it achieved its current state. For others, like Burnet, it was truly a "history": the transformation of the earth must continue until the end of time.[5] No one envisioned processes that lasted longer than human experience, which favored the appeal to major catastrophes and thus made the role accorded to the biblical Flood more acceptable.

Burnet's or Whiston's ideas, which portrayed man as a victim of God's anger, condemned to live in the ruins of a former world whose destruction had resulted from his sins, were too strongly marked by the theological spirit of the seventeenth century to survive for long. The vogue for millenarianism disappeared. As we have seen, the new religious thought insisted on the great concern of Providence for man and on the marvels of nature that lead us to adore the Creator. By maintaining that the Flood had ruined everything and that the mountains were the horrible marks of divine anger, the goodness of God was forgotten. Also forgotten was the fact that in the beginning He had created everything and that He had done so for man's happiness. With the new religious thought there was something impious in supposing that a catastrophe, even one willed by God, could upset the divine plan and the original harmony of Creation. As early as 1693 John Ray, a scholar and theologian, explained in his *Three Physico-Theological Discourses* that the mountains were beautiful and they were useful to man.

*Biblical chronologies are founded on the succession of the generations since Adam, as described in Genesis. There have been a great number of chronologies, some of which are based on the Latin translation of Saint Jerome (the Vulgate) and others on the Greek translation called the Septuagint. This question greatly preoccupied theologians of the seventeenth century, perhaps because they considered that the end of the world was near. Newton himself established a biblical chronology. That of the Anglican archbishop James Ussher (1580–1656) was the most generally accepted one of the seventeenth century.

Well before Europe discovered the Alps and their awe-inspiring poetry, "experimental theology" showed that their existence was a gift from Providence, and therefore they must have existed from the beginning of the world. Hence it was better to reject the Flood as a major geological agent. It became preferable not to insist on this regrettable episode of relations between man and his Creator.

For all this, the taste for grandiose theories did not disappear, but the biblical inspiration no longer was dominant. In 1723 the engineer Henri Gautier published *Nouvelles Conjectures sur le globe de la Terre* [New Conjectures about the Earth's Globe], which broke with the dominant historical spirit.[6] For him, the earth was hollow and full of air. The thickness of the superficial solid crust was only 3,390 toises, or about 6,400 meters (1 toise = about 6.5 feet), and the phenomena that occurred there came from lateral pressures. This crust was supported by the exact equilibrium between attraction and centrifugal force, and Gautier suggested a mechanism whereby the oceans rose while the continents sank: the same place would be at one time covered with water, at another time dry, according to a cycle whose duration was estimated by Gautier to be 35,000 years.

In 1748 a work with the bizarre title, *Telliamed,* appeared. It is in fact an anagram of the author's name, Benoît de Maillet, the former French consul in Egypt who had died ten years earlier.[7] Probably written in the first years of the century, the book presented a sort of cyclic cosmology: exposed to the sun's rays, the earth would dry up until the time when, propelled toward the outer limits of the solar system, it would become wet again and then start its descent toward the sun, drying out once again. This parching would manifest itself through the slow lowering of the sea level, for which Maillet brought proof taken from direct observation. The presence of marine fossils in the solid earth was easily explained in such a way.

The interest of the grandiose theories of Gautier or Maillet was that they opposed a cyclical model to the historical models inspired by the Bible; a cyclical model supposed, or at least allowed for, an eternal and therefore uncreated universe. This opposition was made even clearer by the publication in 1744 of a small work by Linnaeus, the *Dissertation sur l'accroissement de la Terre habitable* [Essay on the Increase of the Habitable Earth].[8] Given the idea that God, in the beginning, created only two animals and two plants of each species, Linnaeus suggested that the habitable earth had been no more than a little island, whose surface sufficed for lodging all living beings. As these beings bred and multiplied among themselves, they needed more space. Thanks to the lowering of the sea level, the solid earth increased and offered all the room necessary. This was a typically historical model, and it was based on precise and exact observations! The Swedish physicist Anders Celsius, mainly known as the inventor of the centigrade thermometer, had placed markers on the Finnish coast. The examination of these markers, which continued after his death, clearly showed a lowering of the sea level. The observation was correct, but it was incorrectly interpreted: we know today that it is not the sea that is sinking, but the Scandinavian peninsula that is slowly rising.

The new scientific spirit, however, rejected the grandiose theories that at the beginning of the century had been the rage. Gautier's theory passed unperceived (perhaps not by Buffon), and *Telliamed* was not taken seriously. The problems remained. The most serious savants, such as Antoine de Jussieu and Réaumur, studied fossils and had to admit the former presence of the ocean on what is now solid ground. This presence was also required by the arrangement of rock layers. Jussieu or Réaumur did not mention the Flood and assumed only local and perhaps relatively recent flooding, which they did not attempt to explain. This theory was applicable to the limestone shell marl of Touraine studied by Réaumur: it was known that the Gulf of Poitou had filled in only recently, and it is possible that sometime in the historical past the Touraine had also been under water. Jussieu, however, had discovered evidence of exotic plants near Lyons: how did they get there?

All these difficulties were presented in a book to which Buffon owed a great deal, the *Lettres philosophiques sur la formation des sels et des cristaux* [Philosophical Letters on the Formation of Salts and Crystals] by Louis Bourguet published in 1729. The book ended with a "Mémoire sur la théorie de la terre" [Memoir on the Theory of the Earth], which began with a description of terrestrial topography, continued with a study of rocks, their distribution and arrangement, and ended with an attempt at an explanation in the form of a history of the earth. Without mentioning the Flood, the author introduced a revolution and a "new formation" that caused an upheaval in the structure of the earth and foresaw a future upheaval that would be caused by fire. These catastrophes were explained by physical causes and an "admirable Mechanism [which] came originally from the first Construction of the World and from the order that divine Wisdom put there."[9] We will return to the debt that Buffon owed to Bourguet. Here we simply note how Bourguet tried to reconcile the order of creation with later catastrophes.

The abbé Pluche in 1735, confronted with the same problem, did not hesitate to draw from the Bible or to mention the Flood. According to him, a large part of the existing structure of the earth, the organization of the mountains, and the distribution of the rock layers were anterior to the catastrophe and were little transformed by it. All that should be attributed to the Flood was simply the folds and the fractures of stratifications, the disorder of mountains that before had been ordered harmoniously, and above all the disappearance of the eternal springtime, which regulated our climates before the catastrophe. For God had only to incline the axis of the earthly poles, and all the rest followed:

"If God, through the displacement of the axis, shook the air and forced the surface of the earth inward, what must have been the astonishment of the children of Noah at the sight of the change that had occurred to their place of residence! Instead of the delightful valleys and hills continually carpeted with the greenery that adorned the original earth, they found only fissured landscapes and rocks tumultuously dispersed, which the universal shock had broken and thrown into the air. Most mountains bristled with snow-covered peaks or hid their summits in thick fog."[10] This was a painful vision, which reminds us that Pluche had Jansenist penchants. It was ignored in the rest of his work.

Such was the theory of the earth when Buffon started to work. Everyone admitted certain facts: marine fossils attested to the former presence of the ocean on the present-day continents; rocks were arranged in horizontal or inclined strata, and the distinction between "original mountains" and "secondary"—or sedimentary—"mountains" had not yet been made. Water was generally considered to be the only possible geological agent. Two models of interpretation confronted each other. The more commonly accepted one implied an irreversible history and used the intervention of the biblical Flood. The other, that of Gautier or *Telliamed*, was cyclic and logically implied an eternal universe. Under these conditions, any choice Buffon made would necessarily have both philosophical and scientific implications.

A Truly Natural Natural History

If the "Theory of the Earth" tried to bring us face to face with nature, three articles of the "Proofs" were dedicated to a critique of authors who had written theories of the earth before Buffon. This critique was obviously selective. Buffon reproached Steno for not having given any general theory, and he criticized Hooke for giving earthquakes too much importance. His favorite targets, however, were those who based their theories on the Flood. Woodward and Bourguet were very good observers but bad theorists. Whiston received the worst treatment. All of them had committed the same error: making the biblical Flood a major geological event and attempting to explain it through physical causes. Buffon took malicious pleasure in showing that Whiston contradicted the Holy Scripture at every turn, and he ridiculed him with enthusiasm:

"Here, therefore, is the history of creation, the causes of the universal flood, of the length of the lives of the first men, and of the shape of the earth; all this seems not to have cost our author anything, but Noah's ark seems to have worried him greatly: indeed, how to imagine that in the middle of such an awful disorder, in the middle of the confusion of a comet's tail with the great abyss, in the middle of the ruins of the terrestrial orb, and in the terrible moments when the elements of the earth were mixed up and new elements were coming from the sky and Tartarus to increase the chaos, how can anyone imagine that the ark with its numerous cargo was drifting calmly on the crest of the waves? Here our author is flailing about, making a great effort to succeed in giving a physical reason for the preservation of the ark; but since this reason strikes me as insufficient, badly thought out, and not very orthodox, I will not repeat it."[11] The conclusion was obvious: "Whenever one is rash enough to try to explain theological truths through physical reasons, permits oneself to interpret the divine text of sacred books in purely human terms, or wants to apply one's reason to the will of the Most High and to the execution of His decrees, one necessarily falls into darkness and chaos."[12]

Buffon returned to the same theme ten times. The Flood was impossible to explain physically. It had to be seen as a miracle and nothing else; "above all bad physics [should] not be mixed with the purity of the Holy Book."[13] In the shelter of this haughty orthodoxy, Buffon excluded the Flood from the geological history of the earth. He made only passing references to the traditional objections. Yet by condemning with such force "the clouds of a physical theology whose obscurity and pettiness derogate from the clarity and dignity of religion, leaving unbelievers with only a ridiculous mix of human ideas and divine facts,"[14] he broke brutally with the dominant tendency of natural history of his time.

By excluding the biblical Flood from the physical history of the earth, Buffon wanted to do more than indicate science's independence from theology. The Flood was physically inexplicable, but it was also a catastrophe, and Buffon refused, for reasons of method, to resort to such explanations: "The causes of rare, violent, and sudden effects must not concern us; they are not found in the ordinary workings of Nature. It is with effects that happen every day—the movements that succeed one another and renew themselves without interruption, the constant and always repeated operations—that we must consider our causes and our reasons."[15] In other words, "in order to judge what has happened and even what will happen, we have only to examine that which happens."[16] Well before the geologists of the nineteenth century, Buffon rejected "catastrophism." He accepted only the "current causes," those that we are able to observe at work in nature. It was by excluding both miracles and catastrophes that natural history could really be natural.

The Discovery of the Earth

The "Second Discourse" leads us quickly to a sort of overview of our planet. The oceans, the continents, and the mountains pass before our eyes. Buffon serves as our guide: "I see . . . , I perceive . . . , I note . . ." are phrases that punctuate the work. This general overview is taken up again in great detail in the "Proofs," which often take the form of long accounts of physical geography whose relevance to a theory of the earth is not always clear.

The first impression is that of great disorder, but let us not be too hasty to accept it: "In paying a little more attention here, we will perhaps find an order that we did not suspect, and general relationships that we did not see at first glance."[17] As a historian recently noted,[18] Buffon was trying to discover consistencies in the structure and placement of the Continental masses, mountains, rivers, and stratifications. But finding this structural order was not an end in itself for him. First of all, he had to show the succession of the causes that had produced it, that is, the physical processes, which were just as regular. The processes that are always at work that Buffon tried to identify directly were the following: a general movement of ocean waters and sea currents, regular winds, and the phenomenon of erosion. It was to discover this order that it was necessary to study "Nature on a large scale." General causes produced different effects in different circumstances, and only a vision of the whole could bring them to light and transform simple descriptions into true science.

In such a way, we would discover "that mountains were not laid down haphazardly and were not produced by earthquakes or other accidental causes but are in fact an effect resulting from the general order of Nature."[19] Not only do they occupy clearly defined places on the continents, the mountain chains are placed there in a regular fashion, and the valleys display that remarkable regularity discovered by Bourguet, which Buffon praised on this occasion: the salient and re-entrant angles always correspond on the two sides of the valley. Another remarkable regularity: the structure of the rock layers. Buffon differentiated between the "vitrifiable" rocks (granite, sand, etc.) and "calcinable" rocks (limestone), but he was convinced that they all had been laid down by sedimentation, and that the placement of rocks in superimposed layers was universal. Contrary to Woodward, however, he underlined the fact that the layers were not superimposed in a decreasing order of specific weight: heavier rocks are often found on lighter rocks, which proved that they were not deposited there at the same time. Finally, another general phenomenon: the presence of fossil shells. In agreement with Woodward this time, Buffon stressed that their substance was the same as that of the rocks in which they were found, which proved that they were fossilized at that place. And they are everywhere, even on the highest mountains, which also proved that all the continents had been submerged sometime in their history.

Buffon's documentation was largely literary, but it was very extensive and taken from the best sources. The "Proofs" contain a considerable number of quotations, including reference at least to the author and the title of the work. But there are also many personal observations, most often made in Burgundy, which confirmed or disproved the accounts of the authors. There was even a precise analysis of the succession of the rock layers taken from Marly-la-Ville down to a depth of about thirty meters. Curiously, geology was one of the areas in which Buffon seemed to have made the greatest number of personal observations.

The Theory and Its Difficulties

The observations have been gathered, the indisputable facts have been established, the consistencies have been recorded.

"This said, let's reason."

With these four words, isolated on one line, Buffon introduced his theory. It started with a curious remark for someone who had asserted strongly that he would admit only current causes and their observable effects: "In the first moments after creation . . . the surface of the earth must have been . . . much less solid" than today. The same causes therefore must have been able "at that time to cause great revolutions in a small number of years."[20] What bothered Buffon was that he could not, or dared not, give the earth an age incompatible with biblical chronology. Nowhere else in the text did he mention the duration of geological eras, and we will see that his theory did not let him evaluate them, or even seriously ask the question. Current causes act slowly, and it is impossible to imagine that they could have done in two or three thousand years all the work that Buffon attributed to them. Buffon dodged this difficulty at the price of a statement that contradicted his principles but shielded him from theological censure.

The rest was fairly simple. The surface of the continents was constantly subjected to erosion. The rivers carried all the matter worn from the hills and mountains to the sea. The sea itself, subject to the movement of the tides and animated by a continuous motion from east to west, ate away the eastern shores of the continents. All this matter torn from terra firma was slowly deposited at the bottom of the oceans, mixing with the debris of marine shells. In these layers of soft matter, sea currents sculpted a relief similar to that of the continents. These currents also transported sediments and deposited them on top of older ones. The older of the layers was purely chronological and the heaviest of rocks could be deposited on the lightest ones. When this sea bottom emerged, we would have a new continent, and the ocean would cover an older one.

The disposition of strata and the universal presence of fossils were thus explained without any need to imagine a flood or even a mass of water larger than the existing oceans: the continents had been submerged one after the other. The regularity of movement of the ocean and sea currents explained the observed regularities in the structure of the mountains and, in particular, the correspondence of salient and re-entrant angles. Buffon said he had already observed this during his trip to Italy (before reading Bourguet?) and that it had suggested to him his "first ideas on the theory of the earth."[21] Buffon was hence convinced that he had reached the goal set for himself: to explain the known facts by natural and observable causes. Some secondary phenomena such as volcanoes and earthquakes remained. Buffon easily rid himself of those by attributing them to subterranean fires that occurred spontaneously in the superficial layers where combustible matter accumulated.

There was, however, a question that could not be avoided: "How can it be that this earth where we live, and where our ancestors like us have lived, which from time immemorial has been a dry continent, firm and far from the oceans, having been earlier the bottom of the ocean, was now above all water and so distinctly separated from it? . . . What accident, what cause could have produced this change in the globe? Is it even possible to conceive of one so powerful as to bring about such an effect?"

The first possible answer: "The facts being certain, the manner with which they came about can stay unknown without prejudicing the judgment that we make of them." Buffon would hardly be satisfied with that. He analyzed two hypotheses. One was a catastrophe like the sinking of Plato's Atlantis. The other was simply the slow work of erosion, which leveled the existing continents and raised the bottom of the oceans, combined with the slow work of the ocean, which progressively ate away the coasts and then invaded them while withdrawing from its old territory. The idea of a catastrophe was not satisfying to the mind, but it had the advantage of not bringing up the burning question of the duration of geological eras. Buffon was satisfied, for the time being, to insinuate that the second hypothesis was preferable to him,[22] leaving in shadow, however, the purely physical difficulties it raised, which his critics took much pleasure in pointing out.

We come thus to the conclusion, or more accurately, the conclusions, as there are two of them: one at the end of the discourse itself, the other at the end of the "Proofs." The first summarized the theory and affirmed its validity. The second was more modest, at least in appearance: "We have to admit that we can only imperfectly judge the succession of natural revolutions; that we judge even less perfectly the series of accidents, changes, and alterations; that the lack of historical vestiges deprives us of the knowledge of the facts; we lack experience and time; we do not reflect on the fact that the time we lack is not lacking to Nature; we want to relate the centuries past and the ages to come to our moment of birth. Human life, stretched even as much as it can be, is only one point in the total duration, one single event in the history of God's actions."[23] An edifying thought, but it leads the reader to believe that the history of the earth had been infinitely longer than the history of humanity, which of course ruins not only the application of biblical chronology to the history of nature but also man's preeminent place in this history.

Let us compare the above text with its double meaning to the end of the first conclusion, the two being separated by almost five hundred pages: "It is the waters of the heavens which little by little destroy the work of the ocean, continually lower the height of the mountains, fill in the valleys, the mouths of the rivers and the gulfs, and bring everything to the same level, and which will one day return the earth to the ocean, which will then accept it, baring new continents traversed by valleys and mountains, just like those we inhabit today."[24]

The history of the earth was therefore not ended. In fact, it could not end, because it had neither beginning nor end: it was cyclical. The alternation of the oceans and the earth could be repeated indefinitely, and each "revolution"—here in its proper sense—erased the preceding one. These "perpetual vicissitudes" went nowhere. The question of the earth's age did not mean anything, or at least there was no possible answer, and the creative act of God was relegated to a past so far distant that all trace of it was lost. The "Theory of the Earth" thus ended with two major statements: the revolutions of the globe were cyclical, and the history of man was only one point in the history of nature. Buffon presented these two conclusions in two texts separated from each other: he would use this technique each time that he clashed head-on with the reigning orthodoxy, be it scientific or theological.

Scientifically speaking, Buffon's theory of the earth was neither better nor worse than that of his contemporaries. It was rather more reasonable. It was his philosophical range that was striking: he freed geology from the Bible and opened an unfathomable past to the imagination. Others would follow him on this path, but more discreetly. The cyclical model he presented would be taken up at the end of the century by the Scottish geologist James Hutton and in particular by Jean Baptiste de Lamarck, who owed him a great deal. As for Buffon himself, he would later abandon this model, and the importance of this 1749 text would be forgotten.

CHAPTER VIII

The Formation
of the Planets

"In the beginning, God created the heavens and the earth. And the earth was without form, and void; and darkness was upon the face of the deep. And the Spirit of God moved upon the face of the waters."[1]

Few texts have been so often quoted and commented on in the Christian world as the first two verses of Genesis. The mystery of the origin of things has always fascinated the human mind. The moment of the Universe's birth is the meeting point between nothingness and being, between time and eternity. All the ancient religions developed complex mythologies telling how the gods had created the world, and almost all the early Greek philosophers devised cosmogonies. Only Aristotle avoided the problem of declaring that the cosmos was eternal. The current popularity of the big bang theory, an initial explosion and an expanding universe, shows that the question of origins has not lost any of its mysterious power. In the evolution of the universe as well as in the evolution of life, we seek to discover our origins and make sense of our presence in this world.

From the beginning of Christianity until the nineteenth century, and sometimes even still today, every word of the biblical text has been scrutinized and interpreted in the light of the scientific knowledge of the time. One thing was certain: God had created the world, and very quickly. It was understood as a creation ex nihilo, from nothingness.[2] Yet what had He created "in the beginning"? The earth "without form and void," the darkness, the waters, all that could be interpreted as an "original chaos," which was put in order little by little afterwards. How was it put in order? Through a series of divine interventions spread over the next five days, as the rest of the text says? Or through a natural evolution, as Saint Gregory of Nicaea was already claiming in the fourth century? The two interpretations remained in opposition for a long time in the Christian tradition, until the freedom that the interpreters of the Bible once enjoyed was progressively restricted and a literal interpretation of scriptures was imposed by the Protestant and Catholic Reformations.

As we have seen, it was the "mechanical philosophy" of the seventeenth century that insisted on the idea of an instantaneous and complete creation of the universe. The world machine could only function if it were perfect from the very beginning. Only one mechanical philosopher of the seventeenth century dared to propose a natural and progressive arrangement of the universe: Descartes. In his *Principes de philosophie* [Principles of Philosophy] of 1644, just before treating the "genesis" of the earth's structure and topography, he explained how, using matter and motion directly created by God, a universe similar to our own could have been formed naturally through the play of the laws of physics alone. It is interesting to note that between Descartes and modern astrophysicists who devised the big bang theory, no one dared to undertake such an enterprise. The "cosmogonies" proposed in the meantime by Kant, Laplace, and others attempted to explain only the origins of the solar system.[3]

Descartes was careful to specify that his explanation was no more than a "fable" that could in no way "do harm to the miracle of Creation," which is certainly true when his entire philosophy is considered.[4] He was nevertheless immediately attacked; scientifically, because his explanation was considered a "physics novel," and theologically because his attempt to explain the order of the world through natural processes was seen as a threat to the Christian faith. Although he was not an atheist himself, there was the risk that he might inspire atheism in others. Christian theology, interpreted in the light of the "mechanical philosophy" of the seventeenth century, insisted that the universe was immediately, spontaneously, and completely created by God in its current state. Malebranche, whose Christian philosophy so well represented the tendencies at the end of the seventeenth century, liked to repeat Saint Augustine's saying, cited earlier: "He who lives in eternity created everything at the same time."[5] This went well beyond the requirements of the biblical text, and it forces one to consider the story of Genesis, which spread the creative act of God over six days, a poetic fiction.

This natural philosophy, part mechanistic and part Christian, achieved its most advanced form in England, first appearing in the work of Robert Boyle (1627–1691), a chemist and physicist but also a Christian philosopher. Boyle criticized Descartes for willfully ignoring "final causes," that marvelous harmony which made the structures of natural beings so perfectly adapted to their function and their needs, and which so obviously revealed divine wisdom. By explaining the wonderful structure of things through the blind interplay of physical forces, Descartes came close to reviving Epicurus's philosophy, which explained everything by chance. It was Newton, however, who dealt the death blow to Cartesian theories when

he published his *Mathematical Principles of Natural Philosophy* (known generally as the *Principia*) in 1687. For the Cartesian cosmology of vortices, he
substituted a celestial mechanics that was infinitely more precise and mathematically more satisfying. This celestial mechanics met the new requirements of natural theology much better than Cartesian theory.

We know that the Newtonian system assumed the existence of universal
gravity, a force by which the sun, planets, and their satellites attracted one
another. This force was the same as the one that pulled bodies toward the
center of the earth, and the oceans toward the moon and sun to form the
tides. It is proportional to the masses that attract each other and inversely
proportional to the square of the distance between them. It could therefore be said that the moon was constantly falling toward the earth and the
earth toward the sun. If, however, neither one were actually falling, it was
because gravity's action was constantly counterbalanced by the speeds of
the moon and the earth in their orbits. If they stopped moving, the moon
would really fall to the earth and the earth to the sun. Conversely, if gravity
suddenly stopped affecting them, their own speed would make them fly out
of their orbits tangentially, much as a stone which is swung on the end of a
string flies away when the string is released.

We see here how Newtonian mechanics required a perfect balance between gravity and the planet's orbital speeds. If one became stronger than
the other, the system would be destroyed: the planets would fall into the
sun or fly off into space. Since gravity depends on mass and distance, the
masses, distances, and speeds must have been carefully calculated, so to
speak, in advance.

But there was more. In the second edition of the *Principia*, which appeared in 1713, Newton added a "general Scholium" in which he pointed
out all the observable regularities of the solar system: all the planets turn in
the same direction in concentric circles* and almost in the same plane. All
the planets' satellites turn in the same direction as well, and almost in the
same plane as the orbit of the planets themselves. From these observations,
he arrived at the following conclusions:

"It is not to be conceived that mere mechanical causes could give birth to
so many regular motions. . . . This most beautiful system of the sun, planets, and comets could only proceed from the counsel and dominion of an
intelligent and powerful Being. And if the fixed stars are the centres of
other like systems, these, being formed by the like wise counsel, must be all
subject to the dominion of One, . . . and lest the systems of the fixed stars

*Ed. note: Newton and Jacques Roger were perfectly aware of Kepler's Law, which stated
that the planets move in ellipses. The "eccentricity" of the ellipses is quite small, so that a
purely qualitative description of planetary orbits as concentric circles provides a model sufficient for ordinary conceptual purposes.

should, by their gravity, fall on each other, he hath placed those systems at immense distances from one another."[6]

Newton was only confirming what was already known: as early as 1692 Richard Bentley had used his celestial mechanics as proof of God's existence, and Newton himself had encouraged him.[7] Thus, Newton's astronomy joined the other marvels of nature discovered by naturalists, and all the weight of his enormous scientific prestige was put into the service of the new natural theology and its creationism. In France, Voltaire called himself the apostle of this rational religion and endeavored to popularize both Newton's gravity and creationist deism. Buffon, we have seen, was little satisfied with the result. The first "Proof" of the "Theory of the Earth" lets us understand why.

A Science of Natural Causes

Voltaire was pleased that Newtonian science imposed limits on the curiosity of the human mind. In his *Philosophical Letters* he said this about gravity: "The cause of that cause is in God's bosom." And, quoting the Book of Job, he added, "You will go up to that point and you will not go any further."[8] Buffon, it seems, did not share this skepticism. Yet, in fact, Voltaire was loyal to Newton's meaning. So, what to do?

Buffon started by expounding Newton's system and the role of gravitation. He remained discreet about the nature of this inexplicable force, which was the object of many lively controversies. He was satisfied to say that gravity's existence was perfectly proven through experience and that it was "generally associated with all matter." In doing so, he imitated the discretion that Newton himself had shown in his public texts, the only texts that Buffon could have known.* In any case, he did not seem tempted to appeal to divine intervention. Like many others after him, Buffon considered gravity a given, one of those "general effects" whose causes escape us, and which a positive science must not, at least for the time being, attempt to explain.

"One single thing stands in the way and is indeed independent of the theory, and that is the force of impulse." But what force is it? Once

---

*"But hitherto I have not been able to discover the cause of those properties of gravity from phenomena, and I frame no hypotheses . . ." (*Principia, General Scholium,* in Newton 1966, II, p. 547). The sentence "I frame no hypotheses" ("hypotheses non fingo" in the original Latin) has become famous. Newton implies only that gravity does not follow the ordinary laws of mechanics, where forces communicate through contact. He developed this point at length in two letters to Bentley (January 17 and February 25, 1693, in Newton 1961, pp. 253–4) and concluded from it that the force of attraction could not be a force inherent in matter. It was exactly this idea that several philosophers of the eighteenth century adopted, including Diderot and Buffon. Buffon certainly did not know of the letters to Bentley, which were published only at the end of the century.

launched, the planets or their satellites continue to turn under the simple effect of inertia—that great discovery of mechanics in the seventeenth century. It is enough to suppose that the planets moved in a non-resisting medium. Buffon did not mention this problem, however difficult it appeared. The question he raised was that of the *impulse*, or the very cause of this orbital motion of the heavenly bodies. In other words, who launched them?

Buffon's answer deserves quoting, because it was typical of the way he presented his ideas when they were unorthodox: "This force of impulse was certainly communicated to the heavenly bodies in general by the hand of God, when He gave impetus to the Universe; but as we must abstain from recourse to causes outside Nature in Physics as much as possible, it seems to me that in the solar system we can give an explanation for this force of impulse in a plausible enough way, and we can find a cause for it."9

Descartes, in the same way, had juxtaposed God's direct intervention and the "plausible" action of a physical cause, and his philosophy allowed him this. One hundred years later, it was necessary to look for another interpretation. The simplest was that science must omit God. Buffon paid tribute to the reigning theology and creationism purely as a formality and ostensibly confined himself to the study of nature, breaking not only with the science of his time but also with Newton himself. What made his position strong was that he did not accomplish this break under the auspices of a philosophy of nature but in the name of science alone. His having spoken of the creative act of God at the outset should have sheltered him from any suspicion of atheism. Once God had been named, however, physics went to work, with its own means. We have difficulties imagining today how daring that break was, since Buffon's position has become the normal and almost spontaneous one for scientists. Buffon's step was not as innocent as he made it seem, and it was in fact a latent atheism that allowed him thus to separate God from science.

Buffon's stand not only raised religious problems. It also posed serious problems of method. In his "Theory of the Earth," Buffon wanted to take into account only the "current causes," those that could be directly observed at work. To reconstruct the earth's past, it was therefore enough to extrapolate from the present and imagine that the same causes had acted for very long periods. Seeking the cause of the planets' movement was an entirely different problem: the only currently "observable" cause, the force of inertia, was no help. It could not explain the first impulse. It was therefore necessary to use the observable present to reconstruct a past event that has left no other trace than the present situation.

Buffon started by specifying his goal. The solar system is made up of planets and comets. The comets "travel across the solar system in all sort of directions, and . . . the inclinations of the planes of their orbits differ widely among them." Nothing therefore could be said about them. On the other hand, the planetary system presents all of the regularities that Newton had already pointed out, and these Buffon later considers in great detail. These regularities allowed one to imagine "some common ground in their movement of impulsion." Buffon thus attempted to explain the "formation of the planets," and nothing more. He did not propose a "cosmogony" in the true sense of the word, as Descartes used it, and if cosmogony can be mentioned in reference to this text, it is simply because the astronomy of the time concentrated almost exclusively on the study of the solar system.

Buffon then presented his hypothesis: the sun was a sphere of matter liquified by heat. A comet might have struck it obliquely and torn away a little of its mass, say $\frac{1}{650}$. Comets were then considered to be very dense heavenly bodies, making this possible. This mass, expelled in the form of a torrent of melted matter, then condensed into spheres, which formed the different planets and continued to turn around the sun in the direction that the impact of the comet had given to their movement. The "obliqueness of the shock" gave these spheres a very rapid movement of rotation about their axes. The centrifugal force brought about by this rotation spun off a part of the planet's matter, which condensed in the form of satellites, while the planets themselves, still liquid, took the form of spheres flattened at the poles. Then they progressively cooled and solidified. One single event thus allowed the explanation of all the observed characteristics of the solar system.

Most of the text is devoted to a detailed examination of these characteristics: for example, the planets with the fastest rotation are also those that are flattest at the poles and have the greatest number of satellites. The planets farthest from the sun are the least dense, which Buffon explained in his own way, and here he came into direct conflict with Newton, for whom a planet's density was proportional to the solar heat that it would have to withstand. Buffon rejected this explanation because "that is only a final cause," and Buffon was looking for a "physical relationship." There is a long discussion about the flattening of the earth at the poles. This was very topical for, as we know, the Academy of Sciences had sent two expeditions, one to Lapland and one to Peru, precisely to measure this flattening. The measurements that were taken did not, however, coincide with the numbers proposed by Newton. Buffon tried to reconcile the fact with the theory in a way with which we are quite familiar: the theory "supposes that the earth has a regularly curved shape," that is, it did not take mountains into account. This allowed Buffon to keep the theory, which he needed for his argument, without contradicting the measurements made by his friend Maupertuis.

Nevertheless, a serious difficulty linked to celestial mechanics remained, which Buffon could not ignore: if one supposes that the comet had torn away a part of the sun's matter, this mass once launched into space should, after a complete revolution, come back practically to its starting point, that is, the sun itself. Now, the sun is located more or less in the center of the planetary orbits. Buffon found a multitude of possible explanations in order to rid himself of this objection. In the torrent of matter thus projected, "the movement of the parts in front must have been accelerated by the parts in the rear," while in turn, "the force of attraction of the front must also have accelerated the movement of the rear." The entire mass would therefore act like a rocket animated by its own acceleration, and not like an ordinary projectile. Buffon's interest in rockets is understandable, but the analogy is a vague one, and the effect does not seem proportional to the cause. It is also possible to assume that the sun was itself displaced by the shock of the comet. This displacement, however, would have brought about an equal displacement of the mass of matter that rotated around it. In short, Buffon came out of it pretty badly: he had to suppose an inexplicable change in the direction of the primary torrent of solar matter without giving any reason for it.

In fact, Buffon asserted nothing and was satisfied to present his "system" as "plausible." What seemed implausible to him was that the regular structure of the solar system could be the product of chance. He evaluated this implausibility: there was 1 possibility out of 64 that all the planets would turn in the same direction; a "7,692,624 to 1" chance for the inclination of their orbital planes to be the same. That their movements had a common origin was therefore a "probability amounting almost to a certainty." The collision of a comet with the sun was also very likely. But the explanation as a whole did not have the same degree of plausibility, and everyone was free not to accept it: "However likely what I have said up to now appears to me, so far as the formation of the planets and their satellites is concerned, since each person has his own limit, especially for estimating probabilities, and since this limit depends on the power of the mind for combining relationships that are more or less remote, I do not hope to convince those who do not want to believe any of it."[10] For Buffon, the "likelihood" of the system amounted to a certainty. Too bad for small minds that were unable to grasp the "power of analogies."

Buffon obtained this likelihood from the agreement he discovered between the necessary consequences he attributed to the impact of a comet on the sun and the regularities currently observed in the solar system. What he actually presented was what is today called a "model": a hypothetical fact is posed, the necessary consequences are deduced according to the laws of physics, and then these consequences are compared with the observed facts. This is what the creators of the big bang theory do, using an immense amount of modern knowledge. This is also what Buffon did, with infinitely more modest knowledge, and observing with much less rigor the laws of celestial mechanics of his time. This model had nothing to do with the Cartesian "fable," which claimed to be based on "very obvious principles" and an a priori definition of matter. Buffon was satisfied to start from a more simple state of the same system that he was studying. He can legitimately be reproached for his lack of rigor and his accumulation of secondary hypotheses, which he used to disguise the inadequacies of the principal hypothesis; and he can also be reproached for not trying, with good reason, to mathematize his model. But these too obvious weaknesses must not blind us to the philosophical and scientific originality of his venture, an originality that Laplace himself—who was hardly inclined to be indulgent regarding Buffon—agreed to recognize.[11]

From Model to History

The difference of method and epistemological status separating the "Theory of the Earth" and the essay on the formation of the planets raise a certain number of difficulties. The *Theory of the Earth*, we have seen, supposed cycles of events that repeated indefinitely and had immediately observable causes. The essay on the formation of the planets hypothesized a unique event in the past that created an unchanging state: the planets will never fall back into the sun. This meant that the earth itself had undergone an irreversible evolution: it was a molten sphere when it was formed, it then cooled and solidified. What happened between those first moments of its history and the beginning of its current state? Do visible traces of its primary state remain?

Buffon could ill afford to ignore these questions since he knew of a short text by Leibniz titled *Protogea*, which appeared in 1693.[12] The text argued that the earth had begun by being a star, until the time the fire that burned in it went out because of a lack of combustible matter. Sand and other vitrifiable matter were products of this primeval fire. Buffon had two objections to this hypothesis. The first was "that of saying things that are equally possible or impossible and thus to which it would be pointless to apply the rules of probability." It was a hypothesis that was neither verifiable nor refutable, as we would say today. "The great defect of this theory," Buffon added, "is that it does not apply to the present state of the earth; it explains the past, and this past is so ancient and has left us so few vestiges that one can say anything that one wants about it."[13]

This criticism has at least two good points. On the one hand, it clearly shows that the *Theory of the Earth* of 1749 in no way purported to be a "history" of the earth. It claimed to explain what exists by what is happening before our eyes, that is to say, it strove to be a science. This tension between sciences and history is characteristic of all scientific disciplines that need the testimony of the past in order to explain the present. It is found throughout the history of geology and also throughout the entire history of the theory of evolution. Scholars have most often sought, as Buffon did here, to construct a science rather than to reconstruct a history, since history is not a science and cannot claim to have the same status or prestige.

The second interesting point in Buffon's criticism of Leibniz, and of all those who had attempted to recreate a history of the earth, was Buffon's emphasis on the requirements of hypotheses. The theories he criticized could be neither proven nor refuted. More precisely, to use Buffon's own words, the "rules of probability" could not be applied. He did not mean, we can be sure, probability in the statistical sense of the word, but rather the degree of plausibility a theory could have. Buffon was convinced that his hypothesis of the formation of the planets possessed a very high degree of plausibility, because he thought he had shown that it agreed very precisely with the observed facts. According to him, such was not the case with Leibniz's theory. This was the difference between a hypothesis and a "model" like the one that Buffon wanted to establish. This must be remembered when we see Buffon, thirty years later, practically adopt all the main points of Leibniz's theory. He adopted it only because he was convinced that he had given it a status of "plausibility" that it had not had with its original author.

A last point must detain us. The collision of a comet with the sun can be considered a "catastrophe." Buffon carefully excluded catastrophes from his theory of the earth, and he made fun of those who used them wherever they fancied. Buffon did not use them, first because catastrophes were the direct interventions of God in the history of nature, and Buffon excluded God from this history. Second, he omitted them because they were catastrophes, and he wanted to know only the current causes, which always obeyed the same laws. How could he make the collision of a comet with the sun fit into his principles?

Buffon overcame this difficulty by showing that that encounter was a fortuitous, and at the same time very probable, accident. As a fortuitous accident, it escaped the general determinism of natural phenomena. It could very well not have happened, and the fact that it did happen does not at all imply that it would ever happen again. Here chance broke the chain of determinism, just as brutally as divine anger did, according to Whiston or Woodward, when it let loose the great Flood. In such a way history was born. The normal sequence of natural causes only generated an eternal repetition of the present. Chance alone could create the unique and irreversible event, after which nothing would remain as it was before.

This conception of history, so astonishingly modern, calls to mind Voltaire's, while it runs counter to that of the nineteenth century.

At the same time the encounter of a comet and the sun was not only possible but probable. Buffon insisted on the great number of comets and the fact that several of them pass very close to the sun. This fortuitous event had nothing miraculous about it and nothing that contradicted the laws of nature. Chance had been domesticated, so to speak, by probability. Buffon, however, did not go to the point of suggesting the possibility of a new encounter and, why not, the formation of new planets. If history were founded on uncertainty, it could not allow prediction. Moreover, if new planets came to be formed in the same way, they would enter into an already existing planetary system and would have to create a new configuration within it. History never repeats itself, since a new event does not erase the past but adds to it: "A throw of the dice will never abolish chance."

With the essay on the formation of the planets, history entered into Buffon's thoughts, even while his theory of the earth sought to exclude it. A philosophical will to reach as far as possible in the search for natural causes, rather than purely scientific reasons, pushed Buffon to refuse Newtonian creationism and to introduce history as if by accident. This already proved something that would be better understood in the nineteenth century, namely that it is history, not science, that is the greatest adversary of creationism. Buffon, preoccupied by other questions, did not attempt either to extend or to theorize on the role of history in his thought. We shall see that this role continued to grow and that Buffon, setting out to write a natural history, ended up by devoting himself to the history of nature. For him as for the great evolutionists of the nineteenth century such as Lamarck or Darwin, the tension between science and history became a constant theme in his work.

From Generation
to Reproduction

In the first volume of the *Natural History* some unorthodox views and daring theories had been already set forth. It was the beginning of Volume II of the *History of Animals,* however, that really incited passionate controversies across Europe, controversies that would echo until the end of the century. If Buffon had wanted to shake up the scientific world of his time, he had succeeded perfectly. In order to understand how daring and original his ideas were, we must, as in the preceding chapters, briefly recall those that were commonly accepted by naturalists in the middle of the eighteenth century.

The Generation of Animals

The *History of Animals* is in fact the treatise titled "On the Generation of Animals" announced at the end of the "First Discourse." The change of title simply indicated that Buffon was attempting to reinterpret the broad problems of animal life in light of the specific problem of "generation." This "history of animals," which became in the table of contents a "general history of animals," presaged what later would be called "animal biology"— or just "biology," in the sense that plants played a role in Buffon's thoughts.

Since the translation of Aristotle's works into Latin, the Latin word *generatio* and its French equivalent *génération* (in English: generation) have traditionally referred to the process by which living beings engender beings similar to themselves. Since antiquity the remarkable and inexplicable phenomenon of animal and plant generation had inspired countless theories and controversies.[1] It suffices to say here that most of the older theories relied on the intervention of immaterial forces: the Aristotelian "form" or "entelechy;" the "faculty" or "formative virtue" of Galen's medical disciples, the "ferments" and "archeus" (life principle) of iatrochemists. Be they agents of divine will, as some thought, or of the mysterious power of nature, as others insisted, these ambiguous powers, neither material nor really spiritual, possessed the "ability" to form without fail a chicken from an egg or a horse from the stallion's "seminal fluid," whether or not, according to different theories, it was mixed with the "seminal fluid" of the mare. Without fail, that is, except when there were "mistakes" which produced monsters.

As we can guess, the mechanical philosophy of the seventeenth century did away with all these mysterious powers and acknowledged only "matter and motion." Easier said than done. Once again, it was Descartes who, not content to explain the formation of the cosmos and the earth by the laws of physics, also wrote a treatise of embryology based on the same principles. Corpuscles of various forms, caught up in a sort of living vortex, miraculously constructed a human body. This treatise, *De la formation du foetus* [On the Formation of the Fetus], published posthumously in 1664, drew only ironic or embarrassed comments.

William Harvey, the great anatomist who discovered the circulation of the blood, also dedicated many years to the study of animal generation. He made many close observations of the development of the chicken in the egg and of the embryo in the womb of viviparous females. From these observations he first drew the conclusion that the embryo was formed progressively, that is, through "epigenesis." He also concluded, however, that the "male semen" did not furnish any matter to the embryo, that it produced a sort of "contagion"* of the eggs of oviparous animals or in the wombs of viviparous animals, and that in the womb the embryo first took the form of an egg. Thus the phrase with which he headed his work: *Ex ovo omnia,* "Everything comes from an egg." In general, however, Harvey remained an Aristotelian and did not in any way adopt the mechanical philosophy. He did not believe in the existence of a female "seminal fluid" and considered the "female testicles," soon to be called "ovaries," as simple lymph glands. The weight of his methodical observations and his scientific prestige made him an author who was not easily put aside.

From 1667 on, a series of discoveries changed the framework of the problem. Three anatomists—a Dane, Nicolas Steno, and two Dutchmen, Van Horn and Regnier de Graaf—proved that the "female testicles" of viviparous animals were actually ovaries analogous to those of oviparous animals, and produced eggs ("ova"), which descend to the womb by the fallopian tubes. In fact, they confused ova with what we call "de Graaf follicles," but that confusion did not interfere with their success. In 1668

*Ed. note: In Latin, the word means an "influence," usually bad.

the Italian anatomist Francesco Redi proved by a very simple experiment that worms found on meat that were thought to be produced by "spontaneous generation" really came from eggs laid by flies. With a surprising rapidity, the ancient belief of spontaneous generation was abandoned by the scientific world. Finally, in 1687 the Dutch microscopist Antonie van Leeuwenhoek discovered by chance the "spermatic animalcules," which would later be called spermatozoa. The modern reader might easily believe that after these three discoveries the problem of "animal generation" was close to being cleared up, if not completely resolved.

In fact it was nothing of the kind. That the "female seminal fluid" of viviparous animals must be replaced by eggs was not a problem. That the "spermatic animalcules" constituted the active part of the male semen was much more difficult to prove, and this would take until the nineteenth century. What remained inexplicable was the way an animal was formed from an egg, from sperm, or from both, or the way a plant was formed from a seed. This was especially inexplicable by means of the rules of the mechanical philosophy.

This difficulty comes from the fact that the central problem of embryology is the creation of an organization. With the naked eye, and even with the microscopic techniques then available, the "anatomists" of the beginning of the eighteenth century were incapable of discovering a structure in either the egg or seminal fluid that had any relationship to the being that eventually emerged. Cell theory and the discovery of chromosomes date from the middle and the second half of the nineteenth century. The genetic code and its mode of action on cellular physiology were discovered in the middle of our century. Molecular embryology, despite its tremendous progress in the last few years, is still far from completely explaining the mechanisms of embryonic development.

In the beginning of the eighteenth century, therefore, the problem was to explain the appearance of order, and an extraordinarily complex one at that, from nothing. It was easy to say, like Descartes, that animals were machines and that all physiology could be referred back to mechanical or hydraulic problems. To explain their generation was more difficult. In 1685 the clever Fontenelle, in his *Lettres galantes* [Gallant Letters], put his finger on the difficulty: "You say that Beasts are Machines, just like Watches? Yet if you put a Male Dog Machine next to a Female Dog Machine, a third little Machine could result from them; two Watches, however, could rest next to each other for their entire existence without ever making a third Watch. Now, Madame de B. . . and I find through our Philosophy that all things that are two but have the power to make three are of a quality elevated well above that of a Machine."[2] What to do, if Cartesian embryology were really a "physics novel" and one wanted to stay faithful to the mechanical philosophy that accepted only "matter and motion"?

As can be seen, the "generation of animals" posed the same problem as the system of the world, that is to say that of the origin of order. We saw how Newton and his contemporaries explained the order of the world through God's creative act. The naturalists shared the mechanical and creationist philosophy of their colleagues in astronomy. There was, however, only one solar system, which must have been created only once. Plants and animals, on the contrary, are born every day in front of our eyes. To say that God directly creates them one by one every minute is to reject the "unity of the time of Creation," and "to introduce God into the machine." It was impossible to go that far, since the mechanical philosophy admitted, at least in general, that once the world was created, God let it be governed by the laws of motion.

The solution to all these scientific and philosophical problems was found around 1670 in a theory often called the "preformation of germs" (or simply, "preformation"). This theory brought together two assertions, one of which dated from the first half of the century. Botanists, despairing of explaining the formation of a plant from a seed, had decided that the plant was already present, "preformed" in the seed. It was in the form of a "germ," extremely small, but completely and perfectly organized. It therefore needed only to grow. To this idea, which dealt with entirely different philosophical concerns, the naturalists of the end of the century added a second one, that of the "preexistence of germs." Not only was the plant already completely present "in germ" in the seed, but the germ itself had been directly created by God in the beginning of time, and was simply waiting for a certain moment to grow. As the learned philosopher Malebranche explained in 1674 in his *Recherche de la Verité* [Search for the Truth], if an apple tree comes from an apple seed, it is because it was already there, since its formation cannot be explained by the laws of motion. It was therefore already there with its apples and their seeds, and the apple trees contained in these seeds, and their apples and their seeds . . . and so on, not to infinity, but containing all the generations of apple trees necessary to produce apples until the end of the world.

It is not surprising to see Malebranche become the advocate of this new theory. Germ preexistence not only got rid of an insoluble scientific problem; it also confirmed the famous verse of the book of Ecclesiasticus, which has already been cited several times: "He who lives in eternity created everything at the same time." It confirmed the Malebranchian idea that nature was no more than a passive mechanism in which God alone acted. More precisely, it resurrected the Augustinian theory of "seminal reasons," adapting it to the mechanical philosophy. Saint Augustine is cited many times in our naturalists' works.

Clearly mechanical creationism guaranteed the astonishingly rapid success of the theory of preexisting germs in the scientific community at the end of the seventeenth century. It is obvious now that it was also this creationism that ensured the success of Redi's experiments refuting spontaneous generation, a theory that was even more incomprehensible for the reigning philosophy than normal generation. Savants of the time were not only philosophers; they knew how to support a new theory with scientific observations. It was the great Dutch entomologist Swammerdam who in 1669 showed that when a caterpillar is dissected, a completely formed butterfly is found inside. He was the first to draw the theory of preexistence from this observation. In a tulip bulb, Malebranche discovered a completely formed tulip. In 1672 the great Italian anatomist Marcello Malpighi communicated to the Royal Society of London an observation he had made with the microscope: while examining a freshly laid chicken egg, he had discovered the elements necessary for the embryo already visible in the cicatricule,* and from that observation he confirmed the preformation of the germ. We know today that embryonic development starts before the egg is laid, which explains this observation. It is also necessary that an egg be fertilized. Malpighi had examined an unfertilized egg, found nothing, and said so. His caution was for naught: his observation would be quoted constantly as an experimental proof not only of preformation but even of the preexistence of germs.

In a few years, the new theory was accepted by almost all the scientific world. The only serious debates concerned the way the germs were preserved before their growth. Some believed that they were dispersed in nature and had to wait to find their way into an organism of their species: the panspermia theory. Very quickly, however, it was generally accepted that germs were "boxed" one inside the other. This was Malebranche's idea, and Swammerdam had already used it to explain how men could have sinned through Adam: they were already contained in his loins. A possible objection was the difficulty in imagining the smallness of germs for distant generations: Malebranche responded that matter was infinitely divisible, and reason's greatest power was precisely to go beyond the feeble powers of our imagination.

How could this theory be reconciled with the recent discovery of eggs and spermatozoa in viviparous animals? If the germ was a complete organism, it could not be in the egg and the spermatozoon at the same time, nor half in one and half in the other. A choice had to be made. We can imagine that Leeuwenhoek opted for the spermatozoa that he had discovered. His theory, called "animalculism," had a few followers including the illustrious Leibniz. But the majority of savants preferred the "ovist" doctrine, which placed the germ in the egg. Indeed, the spermatozoon was a sort of animal in itself. As such, it was exposed to death. Besides, hundreds of thousands

*Ed. note: The chalaza.

of sperm cells are found in semen, and as a contemporary said, "that is a lot of wasted seed." In the egg the germ seemed to be more sheltered. In addition, based on experiments done by Regnier de Graaf on rabbits, it was believed that ovulation did not occur if there was not copulation: nature did nothing in vain. The egg has also always been the symbol of the origin of things from the most ancient antiquity. Finally, had not Harvey written *Ex ovo omnia?* In short, ovism won.

If the doctrine of the preexistence of germs seems absurd to us today, let's look at the mark it has left on French vocabulary. Everyday French says that something is *en germe* in something else, indicating that something is present without being visible. Scientific language still talks of embryonic "development," whereas this "development" is actually a "disenveloping," or the progressive appearance of something hidden which becomes visible little by little. It is also the primary meaning of the word "evolution," which etymologically means "unrolling" and originally denoted the unrolling of a manuscript, which in antiquity was in the form of a long scroll. The word "evolution" with its scientific meaning appeared in the work of Charles Bonnet to designate the growth of the preexisting germ. The word took on the meaning that we know in the nineteenth century because the first "evolutionists" imagined the evolution of life modeled on embryonic development, still seen as the progressive appearance of something that was already there.

These remarks about vocabulary point to the philosophical implications of the theory of preexisting germs. Everything existed from the very beginning. History, be it of the individual, or of life, or even of humanity, was no more than the progressive manifestation of a preexisting, albeit invisible, reality. This means that history does not create anything and does not produce anything either new or unexpected. History is no more than a "developer" in the photographic sense of the word: something that makes visible a hidden but already present image. Hence, it becomes possible to formulate "laws of history" in the same way as laws of embryonic development and to foresee a final state of nature or society. From Leibniz through Auguste Compte to Marx, the philosophy of time, nature, and history underlying the theory of preexisting germs played a considerable role in Western thought.

Controversial Hypotheses

The theory created many scientific difficulties, which inspired prolonged controversies in the first half of the eighteenth century, without actually leading to the rejection of the doctrine. If the preexisting germ is in the egg, of what use is the male? In 1740 a young Genevan naturalist,

Charles Bonnet, showed that in the case of aphids there are only viviparous females, at least during the summer, and at the very moment of their birth young aphids already contain very recognizable aphid embryos. This was interpreted as proof of the preexistence of germs. But if the male was useful only "to put into motion" the development of the germ, how was it that offspring resembled their sires? This problem was especially difficult when it concerned well-known hybrids, such as mules that result from the mating of a male donkey with a mare which possess, among other features, the long ears of their sire. Does that mean that God had foreseen the breeder's intervention, or that the donkey's sperm had influenced—but how?—the development of the horse's germ?

Another difficult problem: monsters. If an abnormal child is born, for example, a fetus with two heads, should we believe that God had created it as such, or were two normal germs accidentally compressed against each other to produce that result? The question arose soon and immediately took a theological turn. Could a good and wise God create monsters? This idea was repugnant to reason. Yet what right did human reason have to pretend to judge God and limit His freedom? As early as 1684, Malebranche, who believed in accidents, and Arnauld the Jansenist, who defended divine freedom, confronted each other on this question.

The debate at the heart of the Academy of Sciences continued for years. From 1733 to 1742, it took a particularly lively turn between two anatomists, Lemery and Winslow. Winslow showed that certain monsters were too well organized for them to be just simple mechanical accidents. Lemery was "revolted" by the "system of originally monstrous eggs": "Imagine a first-rate clockmaker whose honesty equals his skill: if someone, not knowing who had made some very bad watches, suddenly took it into his head to attribute them to our clockmaker and claimed, in doing so, to be praising his liberty, which did not force him to make only excellent watches and which permitted him to make monstrously bad ones, I ask what we should think of that sort of praise."[3] The difficulty with God the clockmaker was that he became directly responsible for the quality of his watches. Rattled, but not convinced, Winslow, despite the austerity of his religious faith,* finished by questioning the very theory of preexisting germs itself.

Other phenomena were even more troubling. From the end of the seventeenth century, it was observed that certain animals had the ability to regenerate parts of their bodies that had been artificially amputated. First it was lizards' tails, then the claws or feet of crabs and crayfish. At this time this phenomenon was called "reproduction." In order to interpret it within the preexisting germ theories, the existence of germs of tails and feet, or even segments of feet, had to be assumed, as the animal only regenerated

*Ed. note: A Dane, the grandson of Lutheran pastors, and a student in theology, he came to France to study anatomy and had been converted to Catholicism by Bossuet.

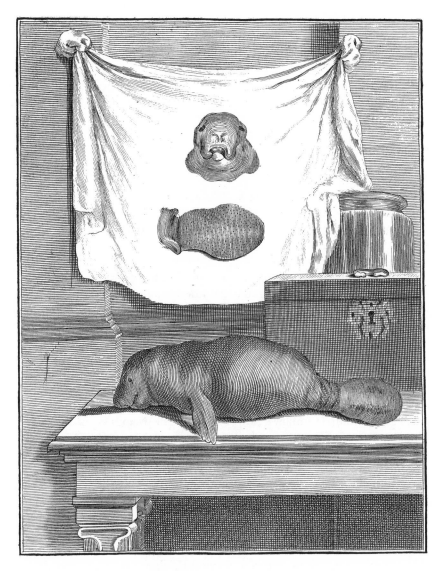

Manatee. From *Histoire naturelle*, p. 130.

that which had been taken away. The presence of many germs also had to be assumed, because if the regenerated foot was amputated, another grew in its place. As early as 1712, Réaumur expressed his perplexity, without actually questioning preexisting germs.

In 1740 Abraham Trembley, a young amateur naturalist from Geneva, who was at that time living in Holland, made an even more surprising discovery, which he hurried to communicate to Réaumur. It involved a very small animal, which he called a "polyp" because it was shaped like a tube with tentacles used to trap the particles it ate (today this animal is called a freshwater hydra). At first Trembley thought that it was a plant, but he saw that it moved and that it also had two extraordinary properties. First of all, it reproduces as do plant cuttings: something resembling a plant shoot emerges from its trunk and progressively becomes a complete polyp and finally detaches itself from the adult. More astonishing, if it is cut in two, each half regenerates into a complete animal. Moved by curiosity, Trembley cut his polyps in two, in four, in eight: each fragment regenerated a complete polyp.

Confronted with these marvels, Réaumur cautiously asked to see the animal. He saw it and did not believe his eyes. What was worse, it was soon discovered that other animal species such as starfish and earthworms possessed the same capacity. How many preexisting germs were necessary in these beings to explain these surprising "reproductions"? Or was it necessary to admit that living matter had a natural power to "reproduce"? Réaumur, incapable of entertaining that hypothesis, did not hesitate to state that in the case of these phenomena, it was "germ development [that] was occurring before our eyes," and he wondered what became of the "polyp's soul" when it was thus cut into pieces. For a man of Réaumur's quality to be so insistent, there had to be serious reasons; the main one was that it was impossible for him to imagine nature possessing the power to form an organized being.

The preexistence of germs posed a much more banal and less spectacular problem than Trembley's polyp. If the germ preexisted in the egg, by what mechanisms did its growth or "development" operate? This question was asked in 1729 by Bourguet (with whom, it will be recalled, Buffon had spoken through the intermediary of Bouhier). Bourguet had presented his ideas in his *Philosophical Letters on the Formation of Salts and Crystals,* a work in which he also set forth his theory of the earth, the importance of which we have seen for Buffon. Bourguet started with a comparison between animal growth and that of crystals. Crystals grow, however, in a mother liquor, whose molecules have the same form as that of the crystal. The growth of a living being requires a more complicated mechanism, because the molecules used for this growth must be assimilated by the organism. Bourguet therefore imagined what he called an "organic mechanism" capable of assuring this assimilation. For this purpose, he supposed that in each organ an "organic mold" existed, which gave molecules furnished by food the shape needed for their incorporation into the organ. Passing from there to the problem of generation, Bourguet suggested that molecules thus assimilated could gather in the seminal fluids of both sexes and form the elements of the embryo.

Bourguet did not abandon ovism or preexistence for all that: in order to form the embryo, a "general mold" was necessary to put the molecules into the appropriate order, and this general mold could only be a preexisting germ. In discussing these questions, Bourguet explained at length some older ideas that the germ theory had eclipsed: the idea defended long ago by Hippocratic doctors that male or female semen came from all parts of the body; the idea that nutritive molecules were assimilated by the different organs, called by the ancient doctors the "third digestion"; and finally the idea that this assimilation was accomplished by the action of a "mold." All these ideas had been upheld in opposition to preexistence at the end of the seventeenth century in France and Italy by doctors of epicurean and materialistic tendencies. Bourguet either rejected them or adapted them to his own system, but he did explain and discuss them. Buffon had no need to go elsewhere to find them.

To these ideas of material origin, Bourguet added others, which probably came from Leibniz, the most important one being the distinction between organic and brute matter. This distinction concerned not only the nature of vital mechanisms, since an "organic mechanism" differed from that of inert matter but also that of matter itself. Bourguet put forward the idea of the circulation of matter (Charles Bonnet would take up this idea), which led him to assert that "the molecules that enter organized Bodies can through union and separation form all necessary alterations. . . . They suffice for everything, while remaining what they are, by means of the Mechanism that God instituted from the beginning."[4] These molecules were therefore not the same kind as those which formed brute matter.

In 1734 Buffon wrote to Bouhier that he refused to continue his discussion with Bourguet because they could not even agree on the facts. Among the facts that Bourguet admitted and Buffon rejected was the presence of sperm "in the semen of females."[5] Bourguet's opinion was indeed unorthodox, but Buffon later claimed it as his own. Buffon owes Bourguet even more. Just as he did for his "Theory of the Earth," Buffon borrowed many of Bourguet's ideas, with which, it is true, he built a very different system.

He had not conceived this system alone. In 1741 his old friend Maupertuis was working on his *Essai de cosmologie* [Essay on Cosmology], which he would publish only in 1750, in which he fiercely attacked the naturalists' providentialism and their abuse of final causes. In 1744 he published his *Dissertation sur le nègre blanc* [Essay on the White Negro], followed one year later by the *Vénus physique* [Physical Venus]. On the pretext of explaining an albino Negro who had been introduced to Parisian salons, he in fact attacked all the problems presented by animal generation. He condemned ovism completely for being inadequately proven, as well as preexistence, which informs us only of an animal's creation whereas its formation had to be explained. He dared praise Cartesian embryology and resuscitated

Harvey's observations, which proved epigenesis over preexistence. He thus came back to mixing the two semens, male and female, each one containing the necessary parts for the formation of an embryo. How do these parts come together to form the correct order? A difficult question, but Newtonian attraction suggested the existence of affinities, as the chemists called them, or even "relationships of union," which could be an explanation.

In any case, an obscure hypothesis was better than doing away with the problem as preexistence did. The hypothesis at least had the advantage of explaining the facts of heredity, which so embarrassed the followers of preexistence, and to which Maupertuis paid special attention. Buffon, who cited Maupertuis as being "the first to start to approach the truth,"[6] certainly had long discussions with him about these problems. He would not accept all of Maupertuis' ideas, but it was perhaps from him that Buffon drew courage to present his own.

The Theory of Reproduction

The *History of Animals* of 1749 is made up of two series of texts clearly distinguished by Buffon himself. The first five chapters contain his general theory and are dated "Royal Botanical Gardens, February 6, 1746." The six following chapters present the results of experiments done later and are followed by a "Recapitulation," which is dated "Royal Botanical Gardens, May 27, 1748." It is as if Buffon were anxious to emphasize that his theory came before his experiments.

The first chapters' progression is slow and muddled. Several times Buffon found it necessary to interrupt himself and justify ideas introduced by general reflections on the way man can understand nature and the difficulties that the functioning of the human mind poses for this understanding. This slow pace has at least the merit of showing the close relationships that unite the theory of generation to Buffon's general philosophy and to his vision of man and nature.

The first idea developed at length was that of the unity of the living world. Animals and plants possess common properties, which radically separate them from inorganic matter: a "sort of living form," an "animated organization," and especially the "capacity to reproduce." This capacity was so essential that it sufficed to prove that animals and plants "are practically of the same order." There was even, "according to our way of perception," a hierarchy that descends from man to animals and animals to plants. This hierarchy descends in imperceptible degrees, and Buffon seized the occasion to show how hazy the difference was between man and animals, using his habitual method that consisted of introducing unorthodox ideas where they were least expected.

In passing, Buffon stated his definition of animal species for the first time: "One must consider as the same species, that which by means of copulation perpetuates itself and conserves the similarities of that species, and as different species, those that through the same means can produce nothing together."[7] This definition was important, modern, and new; and as we shall see, it had numerous consequences.[8]

If there were, therefore, both a living and an inanimate world, what was the difference? Were they both equally "natural," one just as much as the other? And was it not necessary to have recourse to a specific "principle" in order to explain life? In an astonishing twist in the long sentence that ends the chapter, Buffon explained that these questions were meaningless: "the Creator had not put any fixed divisions" between animals and plants, "production of an animal does not cost Nature more, and perhaps costs less, than production of a plant," and it must be believed "that in general the production of organized beings costs Him nothing, and that finally, the living and the animate, far from being a metaphysical quality of beings, are physical properties of matter."[9] Without warning, Buffon stated the fundamental principle of materialistic biology.

Thus living beings "reproduce." This very statement excluded creationism and preexistence. It was "reproduction in general," that "power to produce one's likeness," which required study, and Buffon was the first to give the word its modern meaning. It was not a matter of studying the "generation" of this or that species, but "reproduction" as a universal phenomenon of the living. In doing this, Buffon went outside of natural history and defined a field of research that had to be a science of life in general, a "biology."

"Let us gather facts to give us ideas." This statement has remained famous, and rightly so, for it was typical of Buffon's procedure. He found facts in all the observations made before him, and he called upon them in detail in a later chapter. He repeated them all, including those of Aristotle and Harvey, but he also examined them critically in the light of one another, without necessarily favoring the most recent ones. His own experiments came later, once the theory had been formed. Before that, he needed to find a "hypothesis" that took all the known facts into account. The procedure was the same as with the theory of the earth or the explanation of the planets' formation. This was not the modern experimental process, which verifies a hypothesis through experimentation, but neither was it pure speculation from a priori principles. Buffon did not ignore facts; on the contrary: he wanted to explain them all, instead of imitating certain "observers" who favored what they had seen themselves and ignored the rest. The danger, of course, lay in giving facts of unequal worth the same value. Moreover, such a method did not lead to the discovery of new facts. But that was not Buffon's goal, at least not immediately. He had made a few microscopical observations of the reproductive organs of plants, but he did not get much from them. For him the more urgent task was to reorganize natural history based on new principles, and evidently it was this reorganization that his colleagues found inexcusable.

If the facts came from elsewhere, it was Buffon who organized them. In order to study "reproduction," it was necessary to start at the simplest level, and Buffon criticized those who had studied only the reproduction of man and higher mammals, the most complicated level. The study must start with the simplest organisms, those in which a single part is able to regenerate the entire organism. This is the case with many plants, certain worms, and the polyp. It was from this model of the polyp's regeneration, later precisely called "reproduction," that Buffon attempted to form a general theory. What had been an incomprehensible exception became with him the simplest case of a universal phenomenon. The entire nineteenth century, starting with Lamarck, would study life in the same way, beginning with the least complex organisms, which after 1850 would be called "unicellular."

If a fragment of elm or a polyp could regenerate a tree or a complete animal, it was because each part of these organisms contained a "germ" of the entire organism. We would say today that each cell, possessing the totality of the genetic code of the species, is capable of producing an organism of that species, what we call a "clone." Yet Buffon was not only a victim of the ignorance of his time; he was above all a prisoner of a mechanistic imagery that he tried to blur, without being able to be rid of it. If he used the word "germ," like the partisans of preexistence, it was because, like them, he could only conceive of an organism's origin as the miniature of the organism. This required him, in the case of the elm or the polyp, to consider the individual as "an assembly of germs or small individuals of the same species." Like Bourguet, he would confirm this idea using the example of the composition of a crystal of sea salt, which "is a cube made up of an infinity of other cubes." The smallest cubes that we can imagine must themselves be composed of "primary and constituent parts" that must also be cubes.

By analogy, Buffon concluded that the elm and the polyp "are organized bodies made up of other similar organic bodies, of which the primary and constituent parts are also organic and similar." Buffon soon called these "primary and constituent" parts "organic molecules," an expression that had an enormous success and that would be widely used until the nineteenth century. Most of the naturalists who rejected Buffon's ideas attacked his "organic molecules" in particular and severely criticized his stubbornness in defending them until the end of his life. Yet, paradoxically, up to the end of the century, everyone felt obliged to discuss them.[10] These "organic molecules" represented a threat and a need at the same time.

They were a threat to all those who saw in living beings machines necessarily created by God, and who rejected the idea of active matter. They represented a need because the new discoveries had shown the limits of traditional mechanism and revealed the dynamism specific to life. Buffon avoided mechanical imagery, at least partly, by using the expression "organic molecules," because organic matter was not organized; it was life without organization. This was a terribly hazy notion whose success came precisely because it expressed a vague intuition and it could be used for vitalist thought as well as for a chemical conception of life. This notion was very close to that of "living matter," which biology would be rid of only in the twentieth century.

Buffon immediately asserted that these "living organic parts" "are common to animals and plants" and are "primary and incorruptible." As the primary matter of life, they entered into the makeup of living beings and were freed upon their death, thus circulating indefinitely. They were what assured organisms' growth: they were absorbed with food, separated from useless inorganic matter, and then assimilated by the different organs. But how? Buffon took from Bourguet the idea of the "interior mold" and added only a "penetrating force" conceived on the analogy of Newtonian attraction, which integrated organic molecules into the living substance of organs.

Let us come back to the elm and the polyp. Each of the "germs" that they contained could play the role of "interior mold," grow and develop, and become an adult organism. Reproduction was nothing more than a particular case of growth. This was another idea that influenced all of nineteenth century biology and even had its moment of glory with Ernst Haeckel around 1860. Since "to feed oneself, develop, and reproduce are . . . effects of a single and common cause," Buffon devoted a chapter to nutrition and development, which he explained by the "intussusception" (the word comes from Bourguet) of the "organic parts" that, thanks to "penetrating forces," assimilated themselves to the organs, which served as "interior molds." When the growth of a polyp was ended, the organic molecules furnished by nutrition and that could no longer be used for growth would then be used for the development of new individual germs, which emerged like buds from the adult organism.

What remained was the problem of sexual reproduction. To explain it, it was enough to suggest that the organic molecules that are no longer used for growth assemble in the reproductive organs, and "form the seminal liquor there, which in both sexes is, as can be seen, a sort of extract from all body parts."[11] In each of the two seminal fluids there were, therefore, "small organized bodies," which, however, could only develop when the two fluids were mixed. It was possible, moreover, that at first this mixture produced only "a sort of rough sketch of the animal, a small organized body in which only the essential parts are formed."[12] Here Buffon no longer spoke of "germs": he left the door open for a true "epigenesis," the mechanism of which he did not even attempt to explain. We shall come back to this point.

Finally, Buffon recalled the two great phenomena supporting his theory: that animals reproduce only after having finished growing and, in particular, the "children's resemblance to their parents," or what we today call "heredity." These facts cannot be understood "unless it is admitted that the two parents have contributed to the formation of the child's body and, consequently, that the two seminal fluids have been mixed." As Buffon tells us, it was this problem of "resemblance" that compelled him to abandon the first system he had imagined, "a mixed system where I used spermatic worms and the eggs of the females as the primary organic parts that form the first speck of life, to which I supposed, like Harvey, the other parts joined themselves in a symmetrical and proportionate order through the forces of attraction."[13]

It was doubtless this system that Buffon used in opposition to Bourguet's. If he finally had to abandon it, it was, he said, "because, having myself taken the care to observe with all the accuracy of which I am capable a large number of families, in particular the most numerous ones, I could not withstand the multiplicity of the proofs."[14] It is too bad that Buffon never published the summary of these observations, which made him, along with Maupertuis and Réaumur, one of the pioneers of what we today call "genetics," the science of heredity.

The facts of heredity created the most serious difficulty possible for the preexistence of germs. It was a well-known difficulty, as were the insoluble problems that the polyp's "reproduction" posed. What we see here is a classic phenomenon in the history of the sciences, one that Thomas Kuhn called a "change of paradigm." The new paradigm that Buffon proposed was purposely based on the most embarrassing phenomena for the paradigm of preexistence, phenomena that were known but consciously neglected or left out thanks to ad hoc hypotheses. The fact remained that the new paradigm proposed by Buffon also raised a certain number of difficulties, and the experiments that Buffon would soon do with Needham would complicate the problem. We shall see how in the following chapter.

Buffon's theory allowed the problem of "animal generation" to escape from the impasse where it had been stuck, by giving a natural and physical origin to the order of living things. This order no longer had to come directly from divine creation, nor was it imposed on matter by some mysterious power: it came directly from the organization of the parents, who "reproduced themselves" in the embryo. By thus introducing the new concept of "reproduction," Buffon changed the meaning of the phenomenon. He also made many other new ideas possible, which he either developed himself or left to others. For example, if it is the parents' bodies that are reproduced, the individual modifications undergone by these bodies before reproduction should be reproduced. This is the inheritance of acquired characteristics and was the basis of Lamarckian evolutionism. Here Buffon introduced a conception of heredity that endured until the end of the nineteenth century and the birth of genetics.

From Reproduction to
the Problem of Life

Several Problems of the Philosophy of Science

As can be imagined, there is no point in judging Buffon's system in the light of modern sciences; we must look only at the problems that it raised at the time. We shall come back to the question of "eggs" and "spermatic animalcules," which Buffon disposed of rather quickly. We shall start instead with the difficulties that Buffon was conscious of and that he tried to remove.

The first problem came from scientific logic. Buffon presented his theory only as a hypothesis, but a hypothesis that met certain criteria. There were questions, he told us, that cannot be asked of nature, those that touch upon the universal properties of matter and those that concern absolutely unique phenomena. In both cases all comparisons were impossible. Beyond these, there were "questions of fact": "why are there trees? why are there dogs? why are there fleas?" All these questions are unanswerable, for they can only be answered by final causes, which are not causes, since "a moral preference can never become a physical reason." It is therefore useless to ask "why animals and plants reproduce." It is a "question of fact," and therefore unanswerable. The question to ask was *how* they reproduce. Any hypothesis that attempts to resolve this problem must meet the same requirements:

"It is necessary to exclude . . . all [hypotheses] that imagine the thing done, for example, those that lead one to suppose that all the germs of the same species were contained in the first germ or that at each reproduction there is a new creation that it is an immediate effect of God's will, because these hypotheses are reduced to questions of fact, for which it is not possible to find answers: all hypotheses that have final causes as a goal need to be rejected . . . because these hypotheses, instead of depending on the physical causes of the effect to be explained, are based only on arbitrary relationships and moral preferences."[1]

The preexistence of germs was therefore disqualified a priori, not for scientific but rather for logical reasons. To say that "in the first creature reproduction was completely finished is not only to admit that one does not know how it happens, but also to abandon the will to think about it."[2] The theory was also disqualified because it succumbed to the illusion of a real infinity. On this point we know Buffon's position, since he referred to it in a footnote in the preface of his translation of Newton; to imagine germs infinitely boxed one inside the other "is to put the object outside the field of one's vision and then to say that it is not possible to see it."[3]

"I admit," Buffon recognized, "that here it is easier to destroy than to establish."[4] What he attempted to establish was not without problems. Is it easy to imagine, for example, that an elm or a polyp is composed of elm or polyp germs just like a cube of sea salt is composed of little cubes? Probably not, but why? Because, thanks to "a strongly established preconceived notion in men's minds, it is believed that the only means for judging a compound are through its elements and that in order to know a being's organic constitution, it is necessary to reduce it to its simple and non-organic parts."[5] What is here called a compound's elements is in reality an abstraction: "We therefore constantly take the abstract for its elements and the real for the compound. In nature, on the contrary, the abstract does not exist, nothing is simple, and everything is compound." We consider animals more compound than plants and plants more so than minerals, but in reality we know nothing. "We do not know if a globule or a cube costs Nature more or less than a germ or any organic part."[6] Instead of trusting his ideas alone, man should follow experience, and say only that a fragment of an elm generates an entire elm, as a fragment of a sea salt crystal regenerates a whole crystal when in a solution of sea water.

Another difficulty: what is an "interior mold"? The two notions are contradictory. Why? Because "our senses are only the judges of the exterior of bodies," and therefore "we understand clearly the exterior affections and the different shapes of the surfaces." If, conversely, our eyes "were made in such a way as to depict the interior of bodies to us, we would therefore have a clear idea of that interior, without it being possible for us to have, through this same sense, any idea of the surfaces." We are irremediably limited by the organization of our senses. Buffon here takes up an idea, albeit in another form, that he had already presented to the Academy of Sciences in 1738 and to which he would return in 1777: if we accept part of a straight line as a unit of measure, we cannot precisely measure a circle. Conversely, if we take a piece from a circle, we no longer can measure a straight line.[7] In nature the problem does not arise, since "these figures are only of our invention."[8]

Because of their weight, we know, however, "that there are interior qualities in bodies." We thus have the right to accept "the idea of an interior mold related to this quality."[9] This did not mean, as it was believed, that the "interior mold" was a force "analogous" to the force of attraction.[10] It was a form, almost a structure, that existed, like attraction, in the very mass of matter. We will never have a clear idea of it, but the effects of these "interior qualities" can be compared. The fact that "this expression *interior mold* appears to contain two contradictory ideas" was of little importance: "The opposition is only in the words," and "when it is necessary to represent ideas that have not yet been expressed, we are sometimes obliged to use terms that seem to be contradictory."[11] Our words correspond only to simple ideas, and our simple ideas are only "the first perceptions that come to us through the senses."[12] It is therefore not surprising that words cannot represent the interiors of objects.

It is interesting to note that Buffon, without knowing it, put forward an idea already expressed by Newton in his Letter to Bentley: "Those things w[hi]ch men understand by improper & contradictious phrases may be sometimes really in nature without any contradiction at all."[13] He certainly knew what Newton had written in the "General Scholium" that he had added to the second edition of the *Principia* in 1713: "In bodies, we see only their figures and colors, we hear only the sounds, we touch only their outward surfaces, smell only the smells, and taste their savors; but their inward substances are not to be known either by our senses or by any reflex act of our minds."[14] Newton concluded from this that it was impossible to know God's true nature. Buffon on the other hand, concluded that we can never know the true nature of nature.

Newton also underlined the fact that gravitation could not be a mechanical cause because it does not act "according to the quality of surfaces . . . (as mechanical causes do) but according to the quantity of the total matter."[15] Buffon felt the need to defend himself, and above all to defend his recourse to "penetrating forces," against "those who have taken as a foundation of their philosophy the admission of only a certain number of mechanical principles."[16] These principles were founded only on the information given by our senses, and were as limited as they were. If, therefore, we can discover "a new general effect through reflection or comparisons or measurements or experiments, we will have a new mechanical principle that we can use with as much confidence and advantage as any of the others."[17] Gravitation was now a respectable mechanical principle, and Buffon, like many others before and after him, called upon it to legitimize his idea of "penetrating forces."

So be it. But was this new mechanics enough to explain life? Does the definition of an "organic mechanism" require nothing else? And first of all, what does the adjective "organic" mean? Buffon did not define it anywhere.

The word had been the object of a controversy between Leibniz and the famous German chemist Georg Ernst Stahl. Stahl leaned toward an animist interpretation, and Leibniz insisted on mechanism. For Bourguet, a disciple of Leibniz, the "organic mechanism" assumed the existence of an "infinity of molecules . . . accommodated to particular systems, predetermined from the beginning by the supreme Wisdom, each one joined to a singular and dominant Activity or Monad."[18] Buffon was surely not Leibnizian enough to take such a definition from him. Thus he did not speak of "organic mechanism" but of "parts" or "organic molecules," which is very different. The word here had neither the vitalist sense that it would soon take on, according to which organic matter was animated by a "vital principle," nor the chemical sense that it received in the nineteenth century. For Buffon, in this first part of the text, it was no more than a word that allowed him to stay faithful to a mechanical conception of life, all the while respecting the uniqueness of life.

Did Buffon really stay faithful to the doctrine of mechanism, enriched as it was with Newtonian dynamics? It is doubtful. We should note that he challenged here, even before it was stated, the principle that would direct all modern biology: "to know the organic constitution of a being" is to "reduce it to simple and nonorganic parts." This is not the place to examine if modern biology and particularly molecular biology have attained this ideal, or to enter into a debate on "reductionism" and its limits. Let us simply say that the concept of "organic parts" rules out the pursuit of vital mechanisms even while seemingly respecting the principles of mechanism.

Buffon might well have considered this concept as legitimate. First of all, it was based on his sensualist philosophy: since our senses give us only a superficial image of things, our minds cannot hope to really understand natural mechanisms. In addition, it represented a certain methodological caution, analogous to the prudence Buffon had shown in his theory of the formation of the planets. In that theory he did not claim to go back to the absolute beginning of the solar system but only to an anterior state of the system, in which the sun was already the center of the comets' movements. Here Buffon accepted the existence of life as a hard fact. The existence of a fixed quantity of indestructible "organic molecules," which circulated indefinitely in what we today call the "biosphere," was taken here as one element of the order of the world. In the chapters that followed, Buffon tried to clarify his thoughts on the relationships between mechanism and life. It was not until much later, however, in 1777 or 1778, that he proposed one and even two explanations for the origin of organic molecules.

Let us therefore admit the existence of organic molecules and of an "interior mold" in which the molecules could integrate themselves through the action of a "penetrating force" analogous to Newtonian attraction.

Lion. From *Histoire naturelle,* p. 122.

Buffon like Bourguet called this integration "intussusception." But was it really an "assimilation," that is, a mechanism by which random organic molecules were transformed in order to be integrated into the substance of each particular organ? It does not seem so. For one thing, Buffon spoke of organic molecules as "a secondary and foreign matter that penetrates the interior, that becomes similar in shape and identical to the matter of the mold,"[19] which indeed appears to be a good definition of the process of assimilation. Elsewhere, he supposed that there were "many varieties and kinds of organic parts which differ greatly from each other" and that "each part of the organized body receives the kind that is the most fitting for it."[20] This way of conceiving assimilation was important for his conception of reproduction itself.

"Each interior mold only admits the organic molecules that are proper to it, and finally . . . when development and growth are almost completely finished, the surplus of organic molecules that before were needed here are sent from each of the individual's parts to one or several places where, finding themselves all together, they unite to form one or several small organized bodies, which must be all similar to the primary individual, since each part of this individual has contributed the organic molecules that were the most analogous to it: those that would have served for its development had it not been finished, those that through their similarity can be used for nutrition, finally those that have almost the same organic form as the parts themselves."[21]

It was not only necessary, therefore, to admit that the existence of organic molecules was a fact and also, at least for the time being, inexplicable, but in addition that these molecules were as preadapted to the diverse organisms whose development and reproduction they assured as they were to the diverse organs of these organisms. In this condition they could reproduce, in the male and female semens, "kinds of little living and organized bodies," which rather mysteriously "need one another so that the organic molecules that they contain can join together and form an animal."[22]

Given these conditions, can Buffon be considered a partisan of epigenesis? If epigenesis supposes, as Harvey thought and as modern embryology admits, the progressive and successive formation of the organs from an almost apparently inexistent structure without relationship to the animal that it reproduced, certainly not. The little organized being must come into existence immediately upon the mixture of the two semens. Buffon hesitated, however, on this important point. He had in passing evoked the possibility of this primitive embryo being still "only a sort of rough sketch of the animal, an organized body in which only the essential parts are formed."[23] He would come back and develop this idea, as we shall see at the end of this chapter. A more serious admission was that of a universal and, as it were, cosmic "preformation" of organic molecules. Here Buffon was more hesitant and even more "creationist" than the preformationists of the beginning of the seventeenth century, who at least gave adult organisms the power of really organizing the seed or the embryo.

In fact, Buffon only succeeded in attacking the preexistence of germs, a theory so metaphysical that it was easy to combat. Despite his efforts, he remained a prisoner of the mechanical images that he vainly tried to exorcise. He could not imagine how living nature could exert the power that he wanted to give to it. His timidity, however, was not only philosophical. In addition to being a prisoner of mechanical theory, Buffon was also a prisoner of the scientific literature on which he founded his theory.

Some Anatomical Problems

As has been said, Buffon devoted a long chapter to the examination of everything that had been written on the generation of animals. He started with Plato, which permitted him to criticize idealism in general and that of Malebranche in particular, and to reject once again the use of final causes. He followed this with subtle praise of Hippocrates and Aristotle, reproaching the latter, along with Descartes, for having considered the question "only from the point of view relative to their system of philosophy." Finally, he came to Harvey, and he wrote a remarkably complete, precise, and well-informed review of the observations, discoveries, and theories presented on the subject since Harvey.

As for the "system of eggs," it was easy for him to show, by relying on Malpighi and Antonio Vallisnieri, that what Regnier de Graaf and the others had taken for eggs were only "vesicles" of the ovary (called de Graaf follicles today) and that no one had really seen the eggs. This fact was correct: ova of vivaporous animals were only observed directly with a microscope by Karl Ernst Von Baer in 1827. Buffon concluded that ovarian "vesicles" contained a "female semen" and that the "system of eggs," which he necessarily associated with preexistence, "explained nothing" and was "badly founded." Note that it was Malpighi who furnished him with the essential elements of his criticism.

As for the "spermatic animals," whose existence was undeniable, Buffon took up all the objections which the "ovists" had presented against "spermist" preexistence, from which he concluded that the one was even more unlikely than the other. He therefore felt free to propose another interpretation of the nature and role played by spermatozoa, which the majority of his contemporaries considered as "parasites" rather than agents of reproduction.[24]

The essential problem, as we have seen, was that of embryonic development. This is where preexistence and epigenesis opposed each other and where so many philosophical problems originated. The representative of epigenesis was Harvey, whose observations and conclusions Buffon explained at great length. We know these conclusions appeared very seductive to Buffon; if, in the end, he did not accept them, it was because of Malpighi's authority. Buffon analyzed Malpighi's observations carefully, and he judged them superior to those of Harvey because they had been made with a microscope. What had Malpighi seen? In a chicken egg that was *fertilized* but not yet incubated, he discovered that the cicatricule already contained the rough outline of the embryo, and he concluded, "rightly" according to Buffon, "that the fetus exists in the egg even before it is incubated, and that its first rough shape had already sent down deep roots."[25] This directly contradicted Harvey.

On the other hand, in examining "with the same attention the cicatricule of unfertilized eggs that a hen produced without having had communication with the male," Malpighi saw "only a globulous body like a mola, which contained nothing organized."[26] Contrary to an entire tradition dating back to the 1670s, Buffon understood that Malpighi's observations were in fact opposed to preexistence; but the respect he had for Malpighi's work forced Buffon to admit that the embryo was almost instantaneously formed after fertilization. This was the thesis that he defended and that his theory tried to explain.

After all, Malpighi was certainly the best embryological observer of his time, and no one after him had attempted to observe systematically the development of a chicken embryo with a microscope. In 1746, when Buffon finished chapter V, which ended the first part of his *History of Animals,* he had not yet made any personal observations. He therefore relied on Malpighi's observations, which he was able to interpret correctly, thanks to his personal philosophy and his rejection of preexistence.

But he would also use the microscope, and the results of these observations would lead him to conclusions that were more complex but not necessarily more reassuring for partisans of received ideas.

Microscopic Observations

Since the theory of reproduction just examined was founded only on the discussion of other authors' works and observations, Buffon felt it necessary to submit it to the test of new observations. That is what he explained in the beginning of chapter VI, which began the second part of his text. If indeed there were "an infinity of organic parts" in living beings, and if these organic parts assembled themselves in semen for reproduction, they should be observable. More precisely, Buffon suspected that "the animalcules that are seen in male semen are perhaps these same living organic molecules, or at least they are just the first grouping or the first assembly of these molecules, but if this is true, female semen, like that of the male, must contain live organic molecules," reunited here also in the form of spermatic animals. The reproductive organs of plants must also contain some. There should even be some in the "glandulous bodies" of the ovaries, that is, in the de Graaf follicles, where many had vainly sought to discover eggs. Buffon thought about it "for more than one year," made "an outline of experiments," and determined "to undertake a series of observations and experiments that require a lot of time." Unsatisfied with his microscope's quality, and perhaps mistrusting his own inexperience with a difficult technique, he asked for the assistance of the abbé Needham, a young, already well-known microscopist.[27]

John Turberville Needham (1713–1781) was an English Catholic priest who had been seized by a passion for microscopy.[28] In 1745 in London, he had published a collection of observations of squid semen, in which he had discovered strange little mobile bodies that acted like little machines. This observation led him to wonder if the spermatic animalcules were not "phenomenally small machines" rather than "live creatures." He also believed he had seen analogous machines in the dust of plant stamens, namely pollen. Finally, he had observed wheat blight and discovered the phenomenon of "latent life": the "little eels" responsible for the disease stayed dried and inert for a long time: it sufficed to wet them with water for them to return to life.

In 1746 Needham came to Paris, recommended to Réaumur by Abraham Trembley and to Buffon by Martin Folkes, the president of the Royal Society of London, of which Needham would become a member in 1747 as the first Catholic priest admitted to this society. It seems that Réaumur, Trembley, and Buffon competed for the honor of translating Needham's first book into French, a fact that indicates the renown he enjoyed. Be that as it may, Needham was without a doubt closest to Buffon on many points; he shared the rejection of germ preexistence and was disposed to give nature more powers than other naturalists. All this did not prevent him from being a very devout Christian and being worried that Christian thought was being compromised by a natural philosophy that he considered fragile.

This is not the place to follow Needham's personal thoughts, which clearly differed from Buffon's. Let us instead follow Buffon's account of events. The experiments took place at the Royal Botanical Garden from March to May 1748, except for a few repeated in the fall. They took place in the presence of Needham, Daubenton, Guéneau de Montbeillard, and the young botanist Thomas François Dalibard. Since these observations gave rise to a controversy that lasted until the end of the century and even beyond, it is necessary to stop here for an explanation.

According to Buffon, the observations were made with a "double microscope," a term everyone, contemporaries of Buffon as well as historians, has interpreted as a "compound microscope," meaning equipped with two lenses: an objective and an eyepiece. This interpretation is confirmed by the engraving by Sève, which shows a compound microscope made by the English optician John Cuff from 1744 on. This instrument, whose magnify-

An eighteenth-century microscopic salon. From Buffon, *Histoire naturelle, générale et particulière*, Paris, Imprimerie royale, 1749–1767, II, p. 1.

ing power was around 100×, had well-known defects: large spherical and chromatic aberrations and a weak power of resolution. These defects gave rise to the criticisms of the German microscopist Ledermüller in 1756 and also especially those of Spallanzani: Needham and Buffon had not been able to see well because they had a bad instrument. If they insisted on defending their observations, it was because of pure stubbornness and the desire to uphold their theories.

A very recent study, however, no longer permits this interpretation.[29] Without going into a meticulously detailed investigation, let us say only that its conclusion was clear: Needham, in his initial observations and later in those he made with Buffon, used a *simple* microscope, also made by John Cuff, especially for Martin Folkes. Buffon spoke of a "double microscope" because the instrument had a second lens, but it served only to focus the light. With a magnifying power of more than 400× and all the optical qualities of a simple microscope, the instrument was superior to all others of its time—superior, especially, to those of Leeuwenhoek, Ledermüller, and Spallanzani. It was only in the nineteenth century, in 1826, that Adolphe Brongniart confirmed certain of Needham's observations and that Robert Brown—who discovered "Brownian motion"—armed with his own observations, did not hesitate to revive "the supposed constituent of elementary Molecules of organic bodies, first so considered by Buffon and Needham."* If our two observers defended their observations to the end,

*Ed note: R. Brown, "A Brief Account of Microscopical Observation on the Particles Contained in the Pollen of Plants," July 30, 1828, in *The Miscellaneous Botanical Works of Robert Brown . . .* , 2 vols. (London, 1866–7), 1:463–79.

even if they were not in agreement on how they should be interpreted, it was because they had good reason to believe that the observations were better than those of their colleagues. As for these colleagues, they criticized the observations because they were unable to reproduce them and because they supported theories they could not accept.

It was true that they had been led astray by Buffon himself, who had spoken of a "double microscope." At issue here was perhaps the style of the *Natural History* and the public that Buffon wanted to reach. Several times Buffon alluded to observations or experiments he had done. We have no reason not to believe him, but he never gave any details of them, and only spoke briefly of them in his text and as an additional proof of the validity of his theories. His reputation of never making any observations himself came from this practice. It so happened that in this particular case, Buffon gave a technical description of his microscope and its optical characteristics, a description that should have been enough to identify it. He did not describe it in the *Natural History*, where it would have bored everyone. It was put into the paper that he presented before the Academy of Sciences in 1748. But the *Mémoires* of 1748 appeared only in 1752, and no one, not even historians, paid attention to this text in the brouhaha provoked by the *Natural History*. No one, that is, except perhaps Réaumur, who let Lelarge de Lignac say that Buffon's observations were badly done; he was very unwilling to describe his own.

In the *Natural History*, Buffon, who claimed to be "well trained" in microscope use, limited himself to a precise description of the difficulties of this work and the precautions necessary for the correct use of the instrument. Throughout the text he indicated the practical problems he had had to solve and how, for example, in order to maintain the heat of a cow's womb brought to him from the slaughterhouse, he had it placed in a basket wrapped in warm linen on which a live rabbit was placed. He described his failures as well as his successes, and the whole gives an unquestionable impression of honesty.

The first series of observations was made on male sperm. Buffon and his colleagues examined human sperm, taken "from the seminal vesicles of a man who had died a violent death and whose cadaver was recent and still warm," probably a man condemned to death, who had just been executed. These observations were repeated on another cadaver in the same condition. Then it was the sperm of dogs, rabbits, and rams (they had to wait until October when these animals would be in heat). In all, they made several dozen observations, sometimes prolonged for several days in order to follow the changes involved. Each time they examined pure semen, then sperm diluted with water, and finally, at least for the animals, pieces of testicles put into water to "brew" in jars "scrupulously" plugged "with cork and parchment."[30] This precaution, probably proposed by Needham, suggests that "aerial germs" were already familiar.

Brown bear. From *Histoire naturelle*, p. 87

In the pure male sperm, Buffon and his colleagues observed the "filaments in motion" which had already so greatly embarrassed Leeuwenhoek, as well as "spermatic animalcules." But they interpreted these "animalcules" (Buffon avoided using the word) much differently from Leeuwenhoek. What they saw, according to Buffon, were "globules" that progressively detached themselves from the "filaments" to which they stayed attached by a "thread" for a certain time, at the end of which they oscillated like a pendulum bob on the end of its string. Finally, the thread broke and the globule freed itself, dragging the thread behind it. The globules then started to advance, always with a pendular movement of oscillation. Finally, the thread shortened and disappeared: the globules thus liberated moved more quickly and freely in every direction: "They resembled animals then more than ever," Buffon noted.[31] "I can guarantee that each of these observations was repeated a great number of times and followed with all possible precision, and I am persuaded that these threads that these bodies in motion drag behind them are neither a tail nor a limb that belongs to them and are a part of themselves." The length of these tails was very variable; they hindered the globules' movement: "We clearly see that they seem to try to rid themselves of them."[32] From then on, it was obvious to Buffon that these "globules" were not "animals." From then on, too, Buffon ran the risk of discovering similar "globules" pretty much everywhere.

When Buffon's engravings are compared with Leeuwenhoek's, it is obvious that they did not see the same thing. Leeuwenhoek saw and drew spermatozoa. Why did not Needham and Buffon see them? The error is not easy to explain. Although Buffon was admittedly a novice in microscopy, Needham was an expert, and their observations were good. The proof is furnished by the description of certain globules' movement, which "appeared to come from the same side and all together, they crossed the microscope's field in less than four seconds' time, they were lined up against each other, they moved in a line of seven or eight abreast, and followed one another without interruption, like marching troops."[33] This observation was new and, like several others accompanying it, was "one of the first exact descriptions not only of the Infusoria, but of the process whereby colonies succeeded one another and how certain species multiplied by division."[34] Even beyond the protozoans, thanks to their microscope's power, Buffon and Needham could observe bacteria and fragments of cells animated by Brownian movement. What they saw can be observed today by repeating their observations with instruments of the same power.[35] Buffon and Needham had really seen "animated globules" in their microscope. They therefore had the right to think that their observations confirmed their theories.

To see was not enough to understand. For different reasons, Needham and Buffon were more interested in the "globules" than they were in spermatozoa, which they considered as one kind of globule among others. From this point on, they would find their globules not only in all the male semen they examined but also in the female "seminal liqueur," that is, in the liquid that they gathered from "ripe" de Graaf follicles. They did not know that the ovum had already been expelled from the ovary. A crucial experiment, however, deserves mention, since it could not be explained in this way. A drop of this "female seminal fluid" was extracted from a female dog:

"I therefore examined this fluid under the microscope and from the first glance I had the satisfaction of seeing moving bodies with tails, which were almost absolutely similar to those that I had just seen in the seminal fluid of the male dog. Mr. Needham and Mr. Daubenton, who observed after me, were so surprised by this resemblance that they could not persuade themselves that these spermatic animals were not those of the male dog that we had just seen; they thought that I had forgotten to change the slide and that some of the male dog's fluid could have remained, or that the toothpick with which we had gathered several drops of the female dog's fluid could have been used before for that of the male dog. Mr. Needham therefore personally took another slide and another toothpick, and having gathered the fluid from the slit of the glandular body, he examined it first and saw the same animals again, the same bodies in motion, and he was as convinced as I was not only of the existence of these spermatic animals in the female's seminal fluid but also of their resemblance to those of the male's seminal fluid. We saw the same phenomenon again at least ten times in a row and on different drops."[36]

The only possible explanation here was that the female, "who had been in heat for five or six days," had already been covered by a male without our observers' knowledge. Needham did not have the same theoretical interest as Buffon did for this observation, and it is likely that Buffon emphasized Needham's and Daubenton's incredulity as a supplementary proof.

The import of Leeuwenhoek's experiments had to be addressed. Buffon dedicated a long chapter to these experiments, where he expressed his great respect for "this famous observer" to whom he "took great care not to compare himself." He even gave Leeuwenhoek the honor of discovering these "spermatic animalcules" despite Nicolaas Hartsoeker's claims, which the French had often taken seriously. Buffon cited Leeuwenhoek's texts, published in the *Philosophical Transactions* of the Royal Society of London, and reproduced the drawings Leeuwenhoek had published. But Buffon also skillfully showed how Leeuwenhoek had changed his mind once he had adopted his system of preexistence, and Buffon concluded by stating that as far as the "facts" were concerned, they agreed; a bold statement, at the very least.

On two essential points, therefore, the observations proved Buffon right, but the observations were faulty. The fatal confusion between spermatozoa and other microorganisms would lead him to find similar "globules" in the various infusions he and Needham prepared: infusions of male animal testicles or female ovaries as well as infusions of animal or plant tissue. From these efforts, Buffon and Needham would arrive at different conclusions.[37] For Buffon, the question was, What is the nature of these "globules" and what consequences should be deduced concerning the very nature of life?

On Mechanism and Life

Since Leeuwenhoek's "spermatic animalcules" did not have tails, they were not animals. On this last point, Buffon was right, even if his observations were faulty. We must not forget that Leeuwenhoek's spermatic animalcules were not our modern spermatozoon, a single cell produced by the male organism: it was a true animal, and the passage from animalcule to man supposed a "metamorphosis" analogous to the passage from caterpillar to butterfly. Buffon was therefore justified in asking how these animalcules reproduced themselves, especially in animals where the reproductive organs were obliterated from one year to the next.[38] It must be remembered that at that time it was thought that all animals, no matter how small, must be organized like all other known animals. In 1741 a naturalist close to Réaumur, Gilles-Auguste Bazin (1681–1754), spoke of microscopic animals "twenty-seven million times smaller than a Mite," and added, "Each one of these animals . . . has all the essential parts which make up a living animal; it has a head, a chest, intestines, nutritive organs, those for generation, veins, muscles, nerves, blood, a heart, and maybe eyes."[39] Bazin, of course, defended preexistence and the idea of germs boxed one inside the other as well as the divisibility of matter to infinity—everything that Buffon rejected. This fact must be remembered in order to understand Buffon's reaction.

After having recalled that our idea of animals was only a "general idea, formed from specific ideas that came from a few specific animals," and that many animals did not correspond to this definition, and after having also recalled that it was impossible to draw a clear line between plants and animals, Buffon wondered if there existed a delineation between plants and minerals, between the living and the nonliving. He came to "suspect" the existence of three categories of intermediary beings: "organized bodies that, not having the power to reproduce, for example, would still have a sort of life and movement," such as eggs; "other beings that, without being animals or plants, could very well enter into the composition of them both" (Buffon gave no examples here); "and finally, other beings that would be no more than the primary assemblage of organic molecules,"[40] "organized bodies [that] are more like natural machines than animals,"[41] such as all those "moving bodies" whose movement was "continuous and without any rest," which constantly change shape and which Leeuwenhoek himself had often observed. These "natural machines," or "organic parts in motion," can be seen just as easily in tooth tartar as in rainwater or plant infusions. "Organized bodies" were only the "primary assemblage" of organic molecules.

That these "moving organic parts" were not "true animals" was proved by Needham's experiments: they were found in infusions of "grilled or roasted meats," whereas true animals could not have survived the test of fire.[42]

They were therefore made up of indestructible organic molecules. This meant that for Buffon, "spontaneous generation" in the real meaning of the word, that is, the birth of something living from something nonliving, did not exist. Only "equivocal generation" existed, meaning that an organized body could emit organic molecules which then reorganize themselves into a body of another species: this was the origin that Buffon attributed to parasitic worms, which infested large organisms.[43] Nature's fecundity was immense: "There are perhaps as many beings, either living or vegetating, that are produced by the fortuitous collection of organic molecules as there are animals or plants that can reproduce by a constant succession of generation." These were a new kind of "organized beings," and they had not all been discovered.[44] These beings were born from "fermentation," which also meant that "what we call fermentation" was due only to the action of "organic parts" and had nothing to do with what happened in the case of minerals.[45] It is not the least surprising to see Buffon give in passing a quasi-Pasteurian definition of fermentation.

Finally, therefore, more than a "nuance" remained between the living and the nonliving: organic molecules were definitely living. What was established was the distinction between the living and the organized, with three distinct levels of life: the organic, the "natural machines," and the true organisms, animal or plant. Buffon thus tried to reconcile his two major assertions: the existence of a mass of primary and indestructible organic molecules, and the fact that life is not "a metaphysical level of beings." Regardless of the importance of the distinctions he thus made, it is certain that in his texts of 1749 Buffon did not succeed in finding a satisfactory reconciliation between his metaphysics and his science.

A Return to Embryology

The last chapters of the *History of Animals* are not generally read. This is a mistake because Buffon, convinced that he had established the foundations of his general theory, examined here the specifics of generation in different species, and discussed in detail two very precise problems: the quasi-instantaneous formation of the embryo at the moment of the mixture of the two seminal fluids and the development of this primitive embryo.

By studying the "variety in animal generation," Buffon sought, much as he did in his "Theory of the Earth," to establish regularities in comparisons between animal sizes, lengths of gestation, and the number of young per pregnancy, the time elapsed between onset of fertility and the age of reproduction, and between fecundity of oviparous animals and viviparous animals, etc. Probably the most interesting remarks concerned the often

neglected problem of the physiological phenomena that precede reproduction in the two sexes. Other interesting remarks dealt with the phenomenon of "latent life" discovered by Needham in the "eels" that caused ergot,
a disease in wheat and rye. It was the formation and development of the
embryo, however, that attracted his attention more.

The first point raised numerous difficulties. Why did not each seminal
fluid form an embryo before mixing with the other? How did the mixed
organic molecules arrange themselves into the necessary order? Whence
came the placenta and the different envelopes of the fetus, visibly alive and
organized, but with no equivalent in the adult organisms of the parents?
Buffon answered all these questions without much confidence. It is con-
ceivable that a "fixed base" was needed so that the mixed organic molecules could constitute a new embryo, and this fixed base could be supplied
by the molecules from the sexual parts, which obviously came from only
one of the parents. The abundance of these molecules in one or the other
parent determined the sex. The "fixed base," once formed, created a
"sphere of attraction," or several if there were several embryos, around
which other molecules arranged themselves "according to the laws of affin-
ity." Chemico-Newtonian vocabulary was not enough to make the explanation very clear, but Buffon only presented this theory as "probable." The
unused molecules formed the envelopes of the fetus, and the parent whose
sex did not prevail supplied more molecules for the rest of the body—the
reason children resemble the parent of the opposite sex.

From all this, it could at least be concluded that if the embryo's formation in the mixture was "immediate," it was not "instantaneous." Buffon
discussed this point again at length in the last chapter. Here he repeated
Malpighi's microscopic observations of the chicken egg and verified that
the whole embryo existed in the cicatricule at the time it was laid.[46] This
does not mean, however, that an embryo at that stage was a miniature
animal, first of all, because there were "fundamental and essential parts"
without which an animal could not live. Buffon seemed ready to admit, on
Aristotle's authority, that this was the case for the digestive system. But he
insisted in particular on the fact that observation proved that the first thing
visible was the spinal column, with a bulge at one end, which would become the head. Only then did the "double and symmetrical parts" appear,
which seemed to "take their origin" from that primary axis and develop
"through a sort of vegetation." Buffon's description was precise and
seemed to attest to personal observation. In addition his conclusion was
remarkable: "This first development is very different from that which follows; it is a production of parts that seem to be born and that appear for
the first time; the other, which follows it, is no more than a growth of all the
parts already born and formed on a small scale, almost as they must be
when larger."[47] It is tempting to say that Buffon came as close to true

epigenesis as both his respect for Malpighi and his theory of reproduction allowed him.

If the formation of the embryo's first outline was not instantaneous, its "development" was not a simple phenomenon either. It is possible to admit that all the parts of the embryo were present, but their "relative positions" were different from what they would become. Unfortunately, it was not known how the "double parts" were "folded on each other," and the developed form did not help. It was the same as "when you amuse yourself by folding paper" to make a boat or a crown, a notion that led Buffon to make a very original remark:

"It would be a problem that goes beyond known Geometry to determine the shapes that can result from all the developments of a certain number of given folds. All that which has an immediate relationship to position is absolutely lacking from our mathematical sciences; this Art, which Leibniz called *Analysis situs,* is not yet born, and nevertheless this Art, which would let us know the relationship of position between things, would be as useful as, and perhaps more necessary to the natural Sciences than, the Art that has as its object only the size of things; this is because we need to know the form more often than the matter."[48] The science for which Buffon wished, at the time when analysis and calculus were triumphing, is topology, which was discovered only much later and was applied, but only in our time, specifically to the study of morphogenesis.

On the embryo's formation, on monsters, on the power improperly given to maternal imagination, on birth, and on a hundred other questions that he addressed, Buffon gathered a considerable documentation, which made a veritable encyclopedia out of his last chapters. He never stopped proposing personal explanations, but for many of his readers, this mass of information, presented with neither pedantry nor jargon, must have been one of the reasons for his text's success, as was the case for the entire *Natural History.* A final "Recapitulation," dated May 27, 1748, summarized the theory, confirmed the existence of the interior mold and organic molecules, and ended on the ancient Aristotelian theme of the eternity of the species: "As long as some individuals survive, the species will always be brand new . . . all [species] will live on their own as long as they have not been annihilated by the Creator's will." This last sentence reintroduced divine intervention while rejecting it at the outset; it was by themselves that species survive.

The theory of reproduction that Buffon presented in 1749 can be summarized in a few sentences. The same thing can be done with Darwin's theory of evolution. In both cases, they are reduced to a deceptive simplicity. This is why I did not hesitate to enter into this maze of a text, at the risk of wearying the hurried reader. The theory arose from fifteen years of contemplation of a crucial problem of contemporary science, a spontaneous yet well thought out rejection of germ preexistence, an attentive reading of all available literature, along with a philosophy of nature and of knowledge. Then came the discussions with Maupertuis. Buffon's acquaintance with Needham and all the observations done with or without him followed these discussions. Finally, he raised all the related questions that also required answers. The result was terribly complex, sometimes contradictory, often hesitant. For the main parts, Buffon was sure of himself: his observations confirmed his theory. We remember only his "commanding tone"—a voice that seeks to require agreement. We forget his doubts and uncertainties. Perhaps he finished by forgetting them himself.

What is striking today, more so even than Buffon's daring or his mistakes, is his inability to realize his own ambitions. He wanted to introduce into biology a type of mechanical thought enriched by the Newtonian dynamics. This is what the "interior mold" and the "penetrating forces" represented. Yet he admitted the existence of "live organic molecules," perhaps born from an unnatural coupling of the Epicurean atom and the Leibnizian monad, but in any case, incapable of being explained by mechanical ideas. He wanted to destroy preexistence, in order to give back to Nature the powers taken from her, but he had to admit that organic molecules were preexistent and preadapted, and he could not admit a true epigenesis. Was he scrupulous in his methods or lacking in imagination? It seems likely that he was conscious of these inadequacies, even though he never spoke of them, because later he attempted to correct them.

CHAPTER XI

The Natural History of Man

Man was present from the very first pages of the *Natural History:* man seeking to understand the nature that surrounded him, man subjected to the limits imposed on him by his sensory organs, capable of building the abstract construction of mathematics thanks to his ability to reason, yet incapable of reaching, in his knowledge of nature, a certitude other than a very high degree of probability; man, finally, who willingly placed himself in the center of the world, at the top of the ladder of beings, and decided to organize his discovery of the living world in concentric circles, starting with what was closest and the most familiar to him. This decision, far from showing a monstrous pride, in reality meant that man renounced knowing nature's true order—the order that it pleased God to introduce, if God exists—and it means that man resigned himself to building only the imperfect science of which he knew he was capable.

In all this, man was regarded as a knowing subject. But as a natural being, man was an object of knowledge and, since "he must put himself in the class of animals whom he resembles by everything material he is,"[1] it was with himself that he had to start all natural history. Truthfully, there was neither great daring nor great originality in placing physical man among the animals. It was even traditional since Aristotle, who had more specifically classed man among the domestic animals.[2] In his *Systema Naturae* of 1735, Linnaeus had placed man in the class of quadrupeds and created an anthropomorphic order, where man was grouped with monkeys and sloths. This grouping provoked several protests[3] and incited an ironic comment from Buffon: "It really is necessary to be obsessed with classifying to put such different beasts as man and sloths together,"[4] all because of the arrangement of their teeth.

The fact remained that this was a delicate subject. Since antiquity, an entire philosophic tradition, be it materialist or skeptical, insisted on abolishing all differences between the nature of animals and man. Pliny, whom Buffon praised because he "communicated to his readers a certain freedom of intelligence, a boldness of thought that is the seed of Philosophy,"[5] had affirmed the existence of intermediary species between man and animals, men with dog's heads or tails. Their existence was so well accepted that Jesuit missionaries believed they had found these strange beings in California. Montaigne took wicked pleasure in showing that animal instinct was often superior to human reasoning—another point shared with this tradition, which Buffon mentions briefly.[6] Very briefly, though, because he was not at all tempted, as we shall see, to humiliate man before animals.

Opposing this philosophy, Christian thought reminded man that he had been created in God's image, a reasonable being made to adore the Creator and to rule over creation. But it could not forget that original sin had taken many of his privileges away from man and that his sinful reason might envy the innocence of brutes. The Cartesian theory of the animal-machine had also raised protests in many areas, including scholars and Christian philosophers.* The question of animals' souls was still current in 1749.[7] In short, the distance between orthodoxy and impiety was narrow, and the same statements could, according to the intentions and contexts, be pious or reprehensible.

Was it necessary to step onto this slippery slope? Would it not have been simpler either to leave man out of natural history or to limit his presence to a few anatomical and physiological descriptions? Tradition leaned toward this prudent solution. Indeed, classically anthropology (the word was already common) was made up of two distinct parts "of which one treated man's soul and the other his body."[8] This distinction was precisely the one Buffon rejected. The history of an animal species, he announced in the "First Discourse," must not only contain the description of the form of animals and their physiology but also describe "the fathers' and mothers' care and attention, their kind of education, their instinct . . . their habits, their wiles,"[9] in short what we would call today their behavior or even their psychology. There was therefore no reason, where man was concerned, to exclude from natural history the description of his "habits," that is, his intellectual, moral, and social life, and everything that modern ethnologists call his "culture." It is the same being who lives, sleeps, eats, reproduces, populates the different countries of the world, discovers the world through the intermediary of his sensory organs, reflects on what he feels,

*The existence of souls in beasts had been upheld, in opposition to Descartes, by both "skeptical philosophers," such as Gassendi and La Fontaine, and Christian scholars, such as Claude Perrault. Réaumur believed in beasts' souls because of "an interior feeling, and even a kind of spirit of justice" (Réaumur 1734, VI, lxvii). In 1749 a man named Guer published a *Histoire critique de l'âme des bêtes* [Critical History of the Soul of Beasts] in two volumes with, for the most part, a materialist tendency. The question was thus timely. Buffon's analysis only reignited the debate, of which we shall see the continuation.

speaks, makes up societies, and attempts to build a science from nature. It was this being in his entirety that Buffon wanted to treat in his *Natural History of Man*.

The project was therefore new and the title was not naive. For his contemporaries, it necessarily recalled a recent and scandalous work, the *Histoire naturelle de l'âme* [Natural History of the Soul] by the materialist physician Julien Offroy de la Mettrie, which came out in 1745. The title clearly indicated the author's intention not to let there be any doubts about the purity of his orthodoxy. From this came the double role of the first chapter, "On the Nature of Man," as well as its ambiguity: it was necessary both to dispel the suspicion of materialism and to stay loyal to the naturalist's method. The meeting point of these two preoccupations was man's superiority over animals. Buffon believed in this deeply and demonstrated it as a naturalist through arguments he judged to be irrefutable. But first he justified it by a philosophical analysis whose sincerity was more dubious; its study will require a look at other chapters of this same *Natural History of Man*.

We shall come back to this point again, since man does not disappear from the *Natural History* after the end of this treatise. We find him again and again in the *Histoire des quadrupèdes* [History of Quadrupeds], which is where Buffon discreetly presented his more daring ideas. In this sense, the *Natural History of Man* is the most visible, most careful, perhaps also the most deceiving of all the texts that Buffon devoted to this difficult subject.

"On the Nature of Man"

From the beginning, Buffon founded his philosophical analysis on Cartesian dualism, which was then the orthodox doctrine of the Church. It was therefore a problem "of clearly recognizing the nature of the two substances of which we are made."[10] To succeed, Buffon followed the Cartesian method of intellectual introspection. The only thing of which we are sure is our existence as thinking beings: "The existence of our soul is proved to us, or rather we only form a unity, this existence and us: to be and to think are the same thing for us; this truth is intimate and more than intuitive, it is independent of our senses, our imagination, our memory, and of all our other relative faculties."[11]

Therefore, we have a soul. Yet what can we say about it? "Simply to say [that it] is unextended, immaterial, immortal" only denies it that which we accord to matter, and "what kind of knowledge can we acquire on this path of negation?"[12] None, and here we find again what Buffon had said of that other negative notion, infinity. It was therefore necessary to say that "our soul has only one very simple, very general, very constant form; that form is thought. It is impossible for us to perceive our soul otherwise than through thought; this form has nothing divisible, nothing expansive, nothing impenetrable, nothing material about it. Therefore the subject of this form, our soul, is indivisible and immaterial."[13] And, consequently, it is imperishable, since that which is simple is indestructible.

Existing by itself, this soul is independent of the sensations that it receives through the intermediary of the sensory organs. This is because there is nothing in common between "the sensation that we feel" and the physical phenomena that affect our organs. What relationship is there between the sound we hear and the vibration of air that strikes our ear? Now if "the interior sensation is completely different from what causes it," it is necessary to conclude that "if things exist outside us, they are in themselves completely different from what we judge them [to be]." In the extreme, they "could very well not exist."[14] We are therefore much less sure of their existence than of that of our soul. The very existence "of our body seems doubtful, as soon as we start to think that matter could only be a mode of our soul, one of its ways of seeing."[15]

Descartes had also explained that the existence of the exterior world was less certain to us than that of our own thought. Here, however, Buffon seemed rather to follow Berkeley's argument. Berkeley had shown with much more detail than Descartes the difference in nature between our sensations and the physical phenomena which cause them, and he had concluded by denying the very existence of matter.* Berkeley was known in France, where his early works had been translated in 1734.[16] We have seen that Buffon, in the preface of his translation of Newton, had summarily criticized Berkeley's attacks on the metaphysics of calculus. It is possible that Buffon now had a greater knowledge of him after having discussed him with Needham, who in 1750 endeavored to find a definition of matter that answered Berkeley's objections.** Diderot, with whom Buffon had at this point regular contact, also knew of Berkeley's thinking, which he described in 1749 as a "system that, to the shame of human intelligence and philosophy, is most difficult to combat, even though it is the most absurd of all."[17]

---

*George Berkeley (1685–1753), an Irish Anglican priest of English origin who would become the bishop of Cloyne, Ireland, had published in 1710 a *Treatise concerning the Principles of Human Knowledge* and in 1713 *Three Dialogues between Hylas and Philonous,* in which he explained in detail his "immaterialist" philosophy.

**According to Needham, matter did exist, but it must be reduced to "simple principles," whose combination produced on our senses the impressions that drive us to attribute to it properties such as extension, solidity, impenetrability, etc., which in reality it did not possess. It is tempting to say that Needham reduced matter to energy. He thus thought he had avoided the radical conclusions to which "Malebranche, Berkeley and some others . . . carried off by a sort of metaphysical enthusiasm" had been led (Cf. Needham 1750, p. 458, and Roger 1971, p. 502 ff.).

We are therefore not sure of the existence of the exterior world. "But let us admit this existence of matter, and even though it is impossible to demonstrate it, let us give ourselves over to ordinary ideas and say that it exists."[18] By its very platitude, this conclusion was revealing. It signified that Buffon renounced battling Berkeley. Descartes, however, would have given him the means to do so, on the condition that he accept in its entirety the Cartesian analysis of the soul and its powers, because for Descartes, the soul was not just an empty substance, a power of thinking without its own contents. It directly possessed innate ideas, which owed nothing to sensations. "Thought, which knows itself, . . . finds first in itself the ideas of several things,"[19] wrote Descartes. Among these innate ideas was that of God, whose "veracity" would forevermore guarantee our knowledge of the exterior world. Buffon rejected this argument. The reason, we know, was that like Locke, he rejected innate ideas. But then what good was this whole Cartesian analysis and the entire discourse on "the two substances"?

To understand more clearly, it is necessary to leave the first chapter of the *Natural History of Man* and pass directly to the chapters in which Buffon studied the function of the sensory organs. There we leave Cartesian metaphysics and come back to the philosophical topics of the middle of the eighteenth century. In these chapters, Buffon treated the physiological problems of vision and hearing, explained nearsightedness, squinting, and deafness, analyzed the function of touch in the various parts of the body, and presented to the public a new method for the reeducation of deaf-mutes: a collection of technical information that was philosophically inoffensive.

Several incidental remarks do, however, give the reader food for thought. Buffon noted that according to Fontenelle, a deaf-mute, healed by chance, "led a purely animal life" before his recovery.[20] He noted also that "animals that have hands seem to be the most spiritual" and that, in general, animals were more or less "stupid" in inverse proportion to their ability to recognize the different shapes of objects by touch.[21] He concluded from this that if a newborn baby were denied the freedom to move its hands and touch objects, the development of its sense of touch would be retarded, "this so important sense, on which all our knowledge depends. . . . One man perhaps has more intelligence than another only for having had in his first childhood a greater and earlier use of this sense."[22] Let us then permit babies to wriggle freely and grab everything that is within their reach. And above all, let us understand that intellectual activity depends on the activity of the senses, exactly as Locke had said.

More specifically, Buffon concluded: "It is through touch alone that we can acquire complete and real knowledge, it is this sense that corrects all the other senses whose effects would be no more than illusions and would only produce errors in our mind if touch did not teach us to judge."[23] This

conclusion reunited an entire series of earlier remarks on the insufficien-
cies of vision and hearing. Vision, by itself, does not allow us to judge the
distance or the size of an object: the angle that defines a small close object
is the same as that which defines a large distant object. The same uncertainty exists for the intensity of a sound and the distance of its source. There is
more: we have two eyes and an inverted image of objects is formed on each
retina—we should therefore see everything as double and inverted. Buffon
was persuaded that this was how a newborn saw the world until touch
corrected these false impressions.[24]

With these statements, Buffon entered into one of the most famous
philosophical debates of the eighteenth century, one that touched on what
is generally called the "Molyneux problem." William Molyneux (1656–
1698) was a great Irish lord, a scholar and philosopher, the author of the
first English translation of Descartes's *Meditations* as well as a treatise on
optics. He had asked his friend Locke the following question: imagine
someone blind at birth who could tell the difference between a sphere and
a cube by touch; imagine that he recovers his sight; would he be able to
recognize the two figures by just looking at them? This question, discussed
by Locke in 1694 in a new edition of his *Essay concerning Human Understanding*, stimulated an immense literature.

The seventeenth century was fascinated not only by geometry but also by
problems of optics and vision, the most geometric of the senses. In the
Dioptrique [Dioptrics], which accompanied his *Discours de la méthode* [Dis-
course on Method] (1637), Descartes had studied the functioning of the
eye; even though, he insisted "that it is the soul that sees, and not the body,"
and that it sees through the intermediary of a "common sense" centered in
the pineal gland.[25] He explained the physiological mechanisms thanks to
which sight could give us a unique and upright image of things, as well as
an idea of their distance and size.[26] This text was fundamental and widely
known in the eighteenth century.[27] Malebranche, despite his Cartesian
convictions, had dedicated several chapters of his *Recherche de la Vérité*
[Search for the Truth] (1674) to showing that sight alone could give us an
idea neither of size nor of distance.[28] This idea was given not by sensations
but by "natural judgments" that "are made in us, without us, and even
despite us," and that in reality God "makes in us."[29] Malebranche's analyses
were also widely known in the eighteenth century.[30] Indeed, Locke's discussion of Molyneux's problem is part of the attack he led against Malebranche and his English disciples.[31]

What Locke nonetheless retained from Malebranche was the notion of a
"judgment" that unconsciously transformed the sensation: "the ideas, we
receive by sensation, are often, in grown people, altered by the judgment,
without our taking notice of it."[32] This judgment did not come from God,
as Malebranche said, but from "habit," that is, experience. Without that

experience, Molyneux's blind man would not be able to differentiate the cube from the sphere.

It was still necessary, and here lies the problem, for there to be a power capable of exercising this "judgment," capable therefore of comparing sensory impressions among themselves, of interpreting them and even modifying them. For Locke, this power existed. It was the understanding in the form of a "power to perceive."[33] The existence and the nature of this "power" would become the basis of the debate.

In 1709 Berkeley was of the same opinion as Locke, but for other reasons. In 1738 Voltaire, in his *Elements of Newton's Philosophy*, also leaned in this direction, but most important, he reported to the French public that an experiment had been carried out in 1728 by the English oculist Cheselden, who had surgically removed a cataract from a man blind from birth. The experiment confirmed Locke's analysis.[34] Condillac, however, in 1746, in his *Essai sur l'origine des connaissances humaines* [Essay on the Origin of Human Knowledge], insisted that the postoperative blind man could recognize the two shapes provided that his eye had been used a little. Condillac rejected the "unconscious judgments" that served as intermediaries between perception and the soul's conscious activity.[35]

In the spring of 1749, this theoretical debate gave way to an unexpected incident in Paris. A Prussian oculist, Hilmer, was to operate on a young blind girl, Mlle. Simoneau, and the operation was to be under Réaumur's patronage. It is easy to imagine the excitement of the Parisian intellectuals. Diderot asked Réaumur's permission to be present at the moment the "primary bandage" was taken off, when this young blind woman would see for the first time. According to Diderot himself, Réaumur refused. According, however, to his daughter, Mme. de Vandeul, Diderot was present at the removal of the bandage but quickly understood that it had already been removed a first time, in the sole presence of a friend of Réaumur's, Mme. Dupré de Saint-Maur. Diderot had protested, and Mme. de Saint-Maur had him put in prison to punish him; according to the abbé Trublet, it was Réaumur himself who had Diderot arrested.[36]

The fact was that Diderot, having "started to philosophize with some friends," of whom Buffon was probably one,[37] published his *Lettre sur les aveugles* [Letter on the Blind] in the beginning of June 1749, at the very time when Buffon was putting the finishing touches on his *Natural History of Man*.[38] Buffon took advantage of the moment to include in a note a quotation from Diderot's *Letter,* whose "very fine and very true metaphysics" he praised;[39] on July 23 this same metaphysics earned Diderot the honor of being locked up at Vincennes by *lettre de cachet.** Mme. de Saint-Maur counted for little in this affair; Diderot had already been marked by

---

*Ed. note: A royal and arbitrary warrant of arrest under the Ancien Régime.

the police as a dangerous author, and in 1749 the government had decided to crack down on wrong-thinking intellectuals. Diderot was only one of the victims of repression. We do not know what Buffon thought when he saw the author of a work he publicly praised imprisoned. In any case, he did not suppress his note.

These chapters of the *Natural History of Man,* which were published in September, arrived right in the middle of this philosophical and judicial uproar. Buffon, who sided with Locke against Condillac, attracted the lasting hostility of this fussy philosophe. But what was it really all about? Arguing with Locke, Berkeley, and Voltaire that the raw data of sight could and must be corrected by the information provided by touch boiled down to saying that the human mind was capable of correcting one sense by another, that that correction was made automatically. What we take as a raw sensation is in reality a "judgment" that includes an activity of the mind.

It was therefore about "judgment" and not sensation that Buffon wrote when he discussed our power to estimate by sight the size or distance of objects, on pages that curiously seems to owe a lot to Malebranche, perhaps through Voltaire's influence. This "judgment" gives us a "correct idea" and serves as an intermediary between the sensation and the activity of thought. But where is this judgment formed?

It was here, in two apparently hasty reflections, that Buffon was the more original and daring. In the chapter "Des sens en général" [The Senses in General], he noted that "the nerves are the direct organ of feeling," that is, with the ability to feel, and that sensations are only "different kinds of feelings." The senses thus are only "varied forms of the same substance," and consequently sensations "are not as essentially different from one another as they appear."[40]

Moreover, in the chapter "De la nature de l'homme" [On the Nature of Man], a turn of phrase seems almost to have escaped from Buffon's pen: "Our interior sense, our soul," he wrote.[41] In the context, the "interior sense" might have a Cartesian meaning. But Buffon knew Aristotle too well to ignore that, for him, the idea of an "internal sense" (which, by the way, Diderot also used) designated a common sense capable of uniting and comparing different sensations.[42] Descartes, as we have seen, admitted the existence of this common sense, which he placed in the pineal gland, in the seat of the soul—an idea that seems highly ambiguous.[43]

Is it possible that for Buffon the "interior sense," or the power to judge, resided in the nervous system itself, or at least in the brain? If such were the case, we are in the presence of a completely materialistic idea, and one that posed a formidable problem: what difference is left between man and animal?

In any case, we are very far from Cartesian philosophy. By following Locke, Buffon had, like him, displaced the problem of ideas from the domain of metaphysics to that of psychology. We now progress by successive steps from a purely physical "feeling" to a raw "sensation," from a sensation to a "judgment," and from a judgment to an idea. As with Descartes, the soul was identified with thought, but was it still an autonomous power, an independent "substance"? The question deserved to be asked, and it is a great temptation to believe that Buffon, by directly attacking the problem of the soul and by using a Cartesian method and vocabulary so conspicuously, wanted above all to disguise the possible materialistic implications of his philosophy of the mind.

We have to recognize that he succeeded perfectly. When the theologians at the Sorbonne examined the book, they did not dream of reproaching Buffon for his materialism. Quite the contrary, they judged it necessary to remind him that according to Christian faith, man has a body as well as a soul, and it was not permitted to doubt the existence of the exterior world.[44] In short, it was Berkeley's immaterialism that they denounced.

"I fear that I have already said too much about a subject that many people will perhaps see as foreign to our goal: should considerations of the soul be found in a book of natural history?"[45] The question, we have seen, was not purely rhetorical. By uniting the two traditional discourses on man, Buffon brought the soul into natural history, and it was the project itself that he defended: "Why want to take away the history of the noblest part of his being from the natural history of man? Why degrade him inappropriately and want to force us to see him only as an animal, while he has such a very different, distinguished, and superior nature to that of beasts that it would be necessary to be as unenlightened as they are in order to confuse the two?"[46] At least on this point, Buffon was perfectly sincere: he believed in the absolute superiority of man over animals. But it was as a naturalist not as a philosophe that he would prove it. The chapter "The Nature of Man," which had begun in the Cartesian mode, ended by being a chapter of natural history.

That difference implied a complete change in method. In order to compare man and animals and "to judge the nature of both of them, it would be necessary to know the inner qualities of the animals as well as we know ours," which was impossible. It was therefore necessary to make the comparison by studying man as we study animals: from the outside, by comparing only "the results of the natural operations of both of them."[47]

This point is essential, because it signified two things. The first was that all man's "operations" are "natural," including, we shall see, speech, intellectual activity, and the creative innovation that gives rise to civilization. The second was that Buffon refused to humanize animals. This ancient and venerable temptation, illustrated by touching stories of helpful dolphins or thankful lions, was one that not many evolutionist naturalists of the nineteenth century could resist. As a prime example, Darwin gave to animals in general and to his dog in particular the majority of human feelings, including shame, humiliation, and a sense of fun.[48] Buffon would never go that far. Beyond that, Buffon and Darwin clearly had different life experiences, which perhaps served as a base for their theoretical positions. For Darwin, the dog was the companion of the solitary hunter setting out with his gun on his shoulder, searching for ducks or partridge, the loyal servant who obeyed at the sound of its master's voice and fetched the bird brought down on the wing. The dog in Buffon's eyes was the pack dog, which man watched, directed, or sent forth from the height of his horse. Darwin's dog slept close to the fireplace, and the living room must have smelled like dog. Buffon's dogs slept in their kennel. The relationship could not have been the same. It is even very doubtful that Buffon "liked animals" in the ordinary sense of the word. He admired them, which is not the same thing. Sentimentality toward animals was largely unknown in the eighteenth century, especially in France.[49] It would develop in the nineteenth century, as man felt more and more assured of his domination over the animal world, and less and less satisfied with the society in which he lived. "The more I know man, the more I like my dog": the saying is caricatural but revealing. At the root of many "ecologic" ideologies there is more hatred of society than love of nature.

By deciding to look only at man and animals from the outside, by comparing only their behavior, as we would say today, Buffon contradicted all the introspective analysis that he had just done. For philosophy and morality, he substituted a science that could claim to be objective by comparing only observable behavior and by renouncing, at least in principle, all speculation about "psychological motivations" inaccessible to observation. From then on, he could well use a moral and psychological vocabulary to talk about animals: it would only serve to simplify the language. This was a frightening simplification, however, because in the long run, it was this same vocabulary that would be emptied of its substance, which has clearly happened with the sociobiologists of the twentieth century.[50] In the end, the very idea of "human nature" almost disappears.

It remained that a simple comparison of behavior was enough to show man's superiority over animals. First of all, man had domesticated animals. "Anyone would admit that the stupidest of men is capable of controlling the most intelligent of animals . . . and that it is less by force and adroitness than through superiority of nature, because he has rational purposes, a plan of action, and a series of means through which he forces the animal to obey him."[51] No animal has ever domesticated another.*

Second point: man speaks and animals do not, even though the anatomi-

---

*Ed. note: The alert reader will no doubt here ask, What about ants and aphids? The two entomologists I consulted suggested that this statement depends on one's definition of domestication.

MAMMIFÈRES. *Pl. 65.*

"L'Anonyme" (Fennec or Sahara fox), Syrian hyrax, and aye-aye.
From *Oeuvres complètes*, VI, p. 133.

cal structure of the vocal organs would allow it, as is shown by the monkey or the parrot. "It is therefore not the mechanical powers or the material organs, but it is the intellectual power, it is thought that they lack."[52]

Finally, because they do not think, animals "do not invent or perfect anything." Beavers build their dams, bees construct their hives, as beavers and bees have always done, which proves that "their operations are only mechanical results and purely material." For if it were necessary to give them a soul, "we would be required to make only one for each species, in which each individual would participate equally; this soul would therefore be necessarily divisible, and in consequence it would be material and very different from ours."[53] The merit of this subtle reasoning served at least to embarrass the defenders of animals' souls, in particular Réaumur, the admirer of bees.

"This is more than what is needed to show us the excellence of our nature, and the immense difference that the goodness of God put between man and beast." For once, it was not true that nature worked only "through nuances and imperceptible degrees": "There is an infinite distance between the faculties of man and those of the smallest animal." Buffon congratulated himself, as far as man was concerned, "for having proven the spirituality of his soul."[54]

This claim was perhaps excessive, but Buffon could at least flatter himself for having muddled the issue. Was the spiritualism he displayed sincere, or was it only the mask of a materialism that, although it certainly recognized the superiority of man and his reason over animal automatism, defined human reason merely as the product of a more complex organization and of the social practices that this organization permitted? Historians have discussed this, and although I lean toward the second hypothesis, I recognize that my opinion is based more on Buffon's total work than solely on the analysis of the chapters just discussed. As intendant of the Garden of Louis XV, required to sign his works, Buffon could not allow himself the anonymous boldness of a d'Holbach. But we do not have the right to conclude that he was a hypocrite just because he needed to be careful. At this point of our analysis, we have to suspend judgment. It is even possible, as we shall see, that the problem is not essential.

For readers of the time, however, the spiritualist conclusion of the chapter "On the Nature of Man" was perhaps of less importance than the famous text that ends the chapter "On the Senses in General." There, in a short "philosophical narrative," Buffon imagined the sensations and the successive thoughts of a "man as one can believe the first man was at the moment of creation, that is, a man whose body and organs were perfectly formed, but who wakes up innocent of himself and everything around him." In a vaguely paradisiacal decor, near a palm tree, the senses of sight, hearing, smell, touch, movement, and taste progressively aroused more and more complex ideas in him, but sleep obliterated them. Upon waking, this first Adam discovered Eve. "I put my hand on this new being, what rush of emotion! . . . I feel her move under my hand . . . I feel the birth of a sixth sense" and, despite the coming night, "I was too alive to fear ceasing to be, and it was in vain that the darkness in which I found myself reminded me of the idea of my original sleep."[55]

As a masterpiece of the "Louis XV style," this passage was very successful. Many other philosophes, including Condillac and Charles Bonnet, would use this kind of philosophical fiction in order to illustrate the awakening of consciousness and of the mind under the influence of our sensations. It is possible to wonder, however, if the reverie induced in the reader by the last lines of this text would tempt him to plumb the mysteries of the soul's spirituality and immortality.

CHAPTER XII

The Birth of Anthropology

When discussing man, Buffon first had to speak as a philosopher. He could not allow himself not to do so, as that would mean isolating himself from current events as well as refusing to participate in a debate in which he felt he had something to say. The subject lent itself to philosophy and even required it. The chapters we just studied, however, are mingled with others from which philosophy was certainly not absent but in which the scientific aspect predominated and gave a more original authority to Buffon's opinions. These analyses, whether medical or ethnographic, surely made an even greater impression on his contemporaries.

The Ages of Life

Four chapters of the *Natural History of Man* were dedicated to the ages of life. Buffon distinguished four: childhood, puberty, adulthood, and old age. This distinction was based on criteria that were more physiological than social, at a time when adolescence was just starting to be recognized as an intermediary state between childhood and adulthood.[1]

There are many banalities in these chapters, many remarks and much advice from medicine. Their only originality comes from the fact that they figured in a work of natural history. There are also a few new remarks, however, and a new way of uniting factors, traditionally separated into different areas of knowledge, in the same study of man.

On childhood, Buffon noted primarily "the disgust that can come from the details of the care that this state requires."[2] The poetry of childhood had not yet been born. In following the life of infants from birth to the appearance of language, Buffon introduced several original observations.

First of all, he emphasized the trauma of birth, in which the child passes from a liquid to an air environment; this gave him the strange idea of experimenting to see if puppies born in water could become amphibious animals.[3] Far less original was the careful description he gave of the new-born infant along with its normal dimensions (but not of its weight, which may surprise us today).

What especially interested Buffon was the progressive awakening of the sensations and the "soul." The newborn's eye is sensitive to light but sees nothing. It must train itself, as vision is educated by touch. In this stage, the child has only "bodily sensation, similar to that of animals." It is only after forty days that it laughs and cries, which is specific to man, "as laughter and tears are products of two internal sensations, both of which depend on the action of the soul" because they imply "knowledge, comparisons, and thoughts."[4] Despite these remarks, Buffon said nothing about the emotional relationships a newborn could have with those around him, primarily his mother or nurse. Like the majority of his contemporaries, he was only really interested in sensations and thought: all that lay in between eluded him.

In fact, this chapter "On Childhood" is largely a treatise of pediatrics, in which Buffon intervened in contemporary debates. If he often criticized the customs of the time, he did so by comparing them with the customs of other cultures; this appeal to ethnology is quite surprising. The Lapps, Indians, and other peoples from the North plunge the newborn into cold water, and even "the mothers bathe with their children . . . an instant after birth; we regard this custom as very dangerous, yet the women very rarely perish after birth, whereas despite all our efforts we see a great number among us perish."[5] Buffon offered the same sort of criticism for the custom of tightly swaddling children. Other peoples were satisfied by covering them or even, in warm countries, leaving them naked. Were not "the Siamese, the Japanese, the Indians, the Negroes, the Savages of Canada, those from Virginia, from Brazil," etc. more reasonable than we? For they were then able to keep the child clean, which the custom of swaddling hardly permitted.

There is some emotion in this description of early childhood: "As they are required to stay in the same position in the cradle, and they are always restrained by the binding of the swaddling clothes, this position becomes fatiguing and painful after a certain time; they are wet and often chilled by their excrements, whose acridity irritates the skin, which is fine and delicate, and consequently very sensitive. In this state, children are powerless; in their weakness they can only ask for relief by crying; we must take the greatest care to relieve them." "Only . . . maternal tenderness . . . is capable of this continuous vigilance, of these small acts of attention that are so necessary: can we expect it from mercenary and common wet-nurses?"

Buffon developed these themes with conviction,[6] and Rousseau would only have to take them up in *Émile*.

Buffon made these same kinds of remarks for nutrition, weaning, hair growth, dentition, and the normal growth of the child before and after birth. Buffon was also the first to notice that the speed of growth tended to diminish after birth, before increasing again at the time of puberty. He would later be the first to publish accurate tables on this subject.[7]

Another new preoccupation was the calculation of the life expectancy of the newborn, in a time when infant mortality was considerable. Buffon referred to English and French tables of mortality, to which we shall return later. According to tables from London, half of all children died before the age of three. The French tables were more optimistic, so to speak: half of all children survived until the age of seven or eight.[8] Buffon took advantage of these facts to condemn the placing of children "in the hospitals of big cities," where they perished in great numbers "from a sort of scurvy or other illnesses that were common to them all, to which they would not have been exposed had they been raised separately." It would be better to send them to the country.[9]

Finally, the child starts to speak. Buffon analyzed the movements necessary for the pronunciation of different sounds and thus explained why children always start by pronouncing the same ones, a phenomenon that occurs in all languages. The chapter ends with the child's initiation into reading: Buffon did not believe that this should start too early, and he was distrustful of "prodigies of four years, eight years, sixteen years of age, who became only fools or very common men at twenty-five or thirty." Was he remembering his own childhood? In any case, his conclusion was clear: "The best education of all is that which is the most ordinary, the one by which one does not force Nature, the one which is the least severe, the one that is the most proportional, I do not mean to the strengths, but to the weaknesses of the child."[10]

To understand the success of the *Natural History*, one must appreciate the mass of information that such a chapter as this presented to the reader without pedantry, mixed in with observations that provoke reflection on our customs and culture. Buffon here was a witness and an actor. A witness to the new interest in childhood, despite a traditional "disgust." A witness to the interest in what we call today "societal problems": breast-feeding, the use of nurses, public hygiene and the role of hospitals, and infant mortality—of which the first statistics revealed the enormity and which could no longer be considered inevitable. Buffon gave his opinion on all these problems and used, as needed, the examples of Negroes and Siamese to show the absurdity of our customs. With this chapter, the philosophy of the Enlightenment penetrated pediatrics.

In the chapter "On Puberty," "philosophy" again dominated, and Buffon felt the need to state it outright: "The age of puberty is the spring of Nature, the season of pleasures. Could we write the history of this age with enough circumspection to awaken only philosophical ideas in the imagination? Puberty, the circumstances that accompany it, circumcision, castration, virginity, impotence are nevertheless too essential to the history of man for us to suppress these relevant facts; we only attempt to enter into these details with that prudent self-restraint which characterizes a decent style and to present them as we have seen them ourselves, with that philosophical detachment which destroys all feeling in the expression and leaves only their simple meaning to words."[11] This was a useful warning, since in fact the chapter's subject was man's sex life, as we say today. The outline Buffon presented already showed that physiological observations would be perhaps less important than reflections on "habits" or customs, or even the superstitions associated with sexuality in different cultures.

What is totally absent from these pages is psychology. There is no mention of the vague "worries" of adolescence, of which the romantics would draw so many passionate paintings. Physiology itself was reduced to the minimum: the appearance of pubic hair of both sexes, of facial hair for men, of menstruation for women. Buffon did mention for each the variations that occurred according to the climate. Added to this were several fairly technical passages on sterility and its causes, on the signs of conception, and so forth. Buffon kept these passages short, however, since he felt he was encroaching on the area of medicine. One sole text offers real interest for the history of physiology: Buffon draws attention to the "singular relationships" that exist "between the parts of generation and those of the throat," as seen by the breaking of the voice in adolescence and the high voices of eunuchs.[12] Buffon had harsh criticism for classical physiology, which sought to understand only "the laws of gross mechanics that we have imagined, and to which we would like to reduce everything." There was an unquestionable "correspondence" between the parts of the human body: "Whether like the Ancients we call this singular correspondence sympathy . . . or like the Moderns we consider it to be an unknown interaction of the nerves, this sympathy or this interaction exists in all animals, and we could never apply ourselves too greatly in observing the effects of it if we want to perfect the theory of Medicine."[13] Buffon here intervened directly into contemporary medicine: he knew the early works of vitalist physicians from Montpellier, and this text would be constantly quoted by French vitalists up to the end of the eighteenth century.[14]

Once again, however, the main part of the text was devoted to customs associated with sexual life. Circumcision was described without comment, but infibulation and castration were presented as "barbarous and ridiculous operations," which "had been imagined by dark and fanatical minds, which through a base envy of humankind dictated painful and cruel laws that make privation a virtue and mutilation a merit."[15] Europeans should not be too proud of themselves either for ignoring these practices. Like other peoples, they are "jealous of preeminence of all kinds," and subject to "that kind of madness that has turned girls' virginity into a real entity." Thus, "virginity, which is a moral state, a virtue that consists in the heart's purity alone, has become a physical object with which all men have become concerned." Buffon "does not hope to succeed in destroying the ridiculous prejudices that have been formed on this topic." But he did try hard to prove that anatomists did not agree on the existence of a "membrane of the hymen" and that there was "nothing less certain than these imagined signs of the body's virginity." In addition, to make us even more ashamed, he linked virginity to the practice of infibulation of girls, carried out by "wild and barbarous nations."[16] After this point, he spoke of those peoples who prostituted their daughters to strangers, "to priests of their idols," even "to an iron idol." "This blind superstition of these peoples makes them commit these excesses in the guise of religion."[17] A word to the wise is sufficient! Let us note in passing that Buffon's remarks on virginity earned him severe criticism from the pious anatomist Haller, who reaffirmed the existence of the hymen and felt obligated to point out that "Nature never jests."[18]

Buffon all the same wisely defended monogamy and "marriage such as it is established by us" as "the natural state of men after puberty." "This law is that of Nature, as the number of females is just about equal to that of males." The "hateful seraglios, where the freedom and the heart of many women are sacrificed to the brutal or disdainful passion of a single man," were therefore contrary to "natural law."[19] Celibacy was not less abnormal, for men as for women, but "excesses are to be feared more than continence."[20] Let us end by noting that man had no power over his reproductive organs and that the use of the *congrès*,* to verify masculine potency or impotence, was a ridiculous practice.[21]

The image that Buffon gives us of sexuality, therefore, is nothing revolutionary. He simply removed it entirely from the religious sphere where it had been put. By comparing certain European customs to the "vulgar" and "barbarous" practices that "superstition" led to in "savages," Buffon indirectly discredited all the taboos that Christian morality had placed on sexual life. If he defended monogamy and a reasonable continence, it was in the name of nature. In this respect, he ushered in one of the important phenomena of the second half of the century: doctors would take the place of priests as the teachers of morals in this area.

The chapter "On Manhood" is the portrait of man in majesty: man and woman, of course, as "everything in both of them marks them as the

*Ed. note: Public sexual union. The practice was discontinued in the late eighteenth century.

masters of the earth." The title, however, betrayed old intellectual habits. Buffon specified that "man has the force and the majesty," while "the graces and beauty are the possession of the other sex," and what follows in the portrait of these "masters of the earth" shows that the painter was especially sensitive to the prestige of the male:

"Everything in man, even externally, marks his superiority over all living beings; he holds himself erect and tall, his attitude is one of command, his head looks to the sky and presents an august face on which is printed the character of his dignity; the soul's image is portrayed there by physiognomy, the excellence of his nature permeates his material organs and animates the features of his face with a divine fire; his majestic carriage, his firm and bold step indicate his nobility and his rank."[22] Nevertheless, Buffon further on denounced the servile state in which coarse savages maintained their wives and congratulated himself that "among nations civilized to the point of politeness," "women have obtained that equality of condition which, however, is so natural and so necessary to the gentleness of society."[23] An excessive optimist perhaps, but Buffon wrote this in the eighteenth century, and Napoleon had not yet dictated his civil code.*

Man's physical description took up only a few pages,[24] in which his comparison with animals was done very superficially and led to no general conclusion. On the other hand, Buffon lingered over human physiognomy and the expression of the emotions. Since the soul revealed itself in the head and the face, their descriptions were longer than that of the body.[25]

A few moral remarks on the importance that men attach to their appearance served to criticize their fashions and their manner of dress as "one of the most impractical . . . the one that takes the most time, the one that seems to me to be the least in harmony with Nature."[26] They also served to introduce a few strange customs of adornment, or deformation of the face. Ethnology was granted little room in these pages. Even though he remarked that "each nation has different preconceptions of beauty,"[27] Buffon referred to Greek statuary in order to define the canon of human beauty, "beautiful Nature."[28] This point must not be forgotten when we read his descriptions of "savage" peoples.

In the chapter "On Old Age and Death," physiology is apparently the concern. In reality, philosophy dominated.[29] Unlike animals, man knows he will die, and death's presence is in the center of his life. To speak about death is to speak about man himself. The eighteenth century witnessed the transformation of the classical perception of old age and death. Old age came to be valued differently: Molière's ridiculous old fogey progressively

*Ed. note: The Napoleonic Code was the legal monument of the First Empire (1804–1815) in which women's "equality" was severely limited.

became Greuze's noble old man.* Death changed even more. In the centuries dominated by Christian thought, death had been integrated with life. It was the necessary passage toward a great beyond where man would finally be himself. The eighteenth century, in its worried search for immediate happiness, was cautious of that great beyond, ceased to believe in it, or at least no longer wanted to wait for it. Death became an unbearable scandal, which must be forgotten or at least minimized, and above all put off as long as possible. The social prestige of medicine was higher than ever; the preoccupations of public hygiene haunted governments and individuals. It was during the eighteenth century that the mentality that is still ours today was born. As with pediatrics, Buffon was at the same time a witness and an actor in this transformation.

His physiological conception of aging was summary and relied on an image inherited from Aristotle, that of the drying up and progressive hardening of the solid parts of the organism: "The membranes become cartilaginous, the cartilages ossify, the bones become more solid, all the fibers grow harder, the skin dries out."[30] By insisting on the alteration of the solid parts of the organism rather than the alteration of the "humors," Buffon took part in the medical debates of his time, but his conception of aging was above all founded on an analogy with plants. When he described bone growth, following Duhamel du Monceau (whom he did not name), it was of his own research on the hardening of wood that he was thinking: bone ages like a tree, whose heart hardens and no longer allows the passage of sap.

From this fundamental image Buffon drew rash conclusions. As the parts of the body "are less solid and softer in women than in men" (an old Aristotelian idea), women age less quickly and live longer, as statistics prove.[31] In the same way, fish live longer than quadrupeds because their bones "never take on as much solidity as the bones of land animals."[32] The length of life was above all proportional to the length of the growing period: "Big animals live longer than small ones because they take longer to grow."[33]

The result of all this was that aging and death were natural phenomena, the necessary result of vital mechanisms. Far from being the opposite of life, death was only its consequence. The very notion of life was a relative one: as soon as the body ceases growing, it starts to die. The "too solid parts are already dead parts, because they cease being nourished; the body therefore dies little by little and in parts, its movement diminishes by degrees, life fades away by successive imperceptible changes, and death is only the last stage of this series of degrees, the last spark of life."[34]

---

*Ed. note: Jean-Baptiste Greuze (1724–1805) was a French painter whose painting *Father Explaining the Bible to His Children* gave a highly idealized picture of old age.

"The causes of our destruction are therefore necessary and death is inevitable; it is no more possible for us to push back the time of death than to change the laws of Nature."[35] It makes no difference whether one uses the body sparingly or "as much as it is capable of." No matter the food or the lifestyle. "The European, the Negro, the Chinese, the American, the civilized man, the savage man, the rich man, the poor man, the city dweller, the country dweller, all so different among themselves for everything else . . . each has only the same measure, the same interval of time to traverse from birth to death."[36] What Buffon meant was that a species' lifetime was inexorably fixed for man as for animals, and that no "fountain of youth" existed, despite what "a few visionaries" have said.* At most, it can be noted "that in high elevations, more old people are commonly found than in low places . . . but considering humankind in general . . . the man who does not die of accidental illnesses lives ninety or one hundred years everywhere."[37]

A delicate problem remained, that of the age of the patriarchs, who, according to the Bible, lived to the age of nine hundred years and more. In order to resolve this, Buffon went back to a device he had already used in his "Theory of the Earth," to which he referred. In the beginning, all materials were softer and therefore growth and hardening were slower. Capable of reproducing only at the age of one hundred and thirty years, or about ten times the current age, the patriarchs were able to live ten times longer than men "since the century of David."[38] We have the right to doubt Buffon's sincerity in this passage!

Let us return to serious matters. If "life starts to fade away long before it fades away entirely," if "we start to live by degrees and we finish dying as we started to live," why fear death? More specifically, "why therefore fear death, if one has lived well enough not to fear what comes after?"[39] Aside from this furtive allusion to eternal life, it was really of natural death, and of it alone, that Buffon wished to speak. It was this death that he tried hard to dedramatize by showing that most often it was without pain and in any case was never accompanied by "excessive pain." Even the illusory hope of healing was a benefit, when it supported the dying person until the last

---

*Among these "visionaries" was certainly Descartes, who had envisioned the possibility, thanks to the progress of medicine, of prolonging human life almost indefinitely. Réaumur can perhaps also be included. After having shown how it was possible to slow down the development of the chicken in the egg or the insect in the chrysalis, and after having shown the winterly torpor of marmots, Réaumur wondered "if there was absolutely no hope of prolonging the lifetime of the animal machines that interest us the most," that is, our own. He also wondered if a man could not enter into hibernation for one hundred years, live a few years in a new century, hibernate again until the following century, and in such a way "last for ten to twelve centuries." While playing with this idea, Réaumur presented it as a "pipe dream," but Buffon appeared to take these dreams seriously. See Réaumur 1734, II, pp. 44–7, and Buffon, "De la vieillesse et de la mort [On Old Age and Death]," *HN*, II, p. 570.

moment.[40] When we think of "Christian death," of what theologians called the "good death," of the lucidity it demands, of the preparation it requires, of the serious examination of oneself that it supposes, we see how far we have come.

An "abuse of philosophy" that Buffon wanted particularly to combat was the reasoning that supposed that "when the soul separated from the body," extreme pain must be felt, a pain that "may be very long lasting, since time has no other measure than the succession of our ideas; and an instant of very intense pain, during which these ideas succeed one another with a rapidity proportional to the violence of the pain, can appear to us as longer than a century."[41] Buffon accepted this quasi-Bergsonian definition of duration but defended a sort of psychological atomism, which required "a little time" between two ideas; this time remained the same in all circumstances "since it depends on the nature of our soul and the organization of our body." It was therefore impossible to imagine that a dying man could live a century in "one minute of pain."[42]

"I only elaborated on this subject to try to destroy a prejudice so contrary to man's happiness."[43] Buffon had in mind man's happiness on earth, of course, because this was the only important one. The "good death" of which Buffon spoke was not the "death of the sage," not the death of Socrates, a commonplace for non-Christian moralists. It was natural death, the death of Montaigne's peasants: "If you do not know how to die, do not worry; Nature will inform you immediately, completely, and sufficiently; she will do exactly that work for you; do not let that affect your life."[44] And it was perhaps in remembering Montaigne that Buffon remarked, "Common men, especially those from the country, see death without dread."[45]

The eighteenth century, which attempted to tame death, discovered a new worry. If death was only the ultimate step of a long process, how to be sure that a buried person was really dead? In that same year of 1749, two anatomists, Bruhier and Winslow, published a *Dissertation sur l'incertitude des signes de la mort et l'abus des enterrements et des embaumements précipités* [Essay on the Uncertainty of the Signs of Death and the Abuse of Hurried Burials and Embalmings]. An entire horrific literature of people buried alive who awoke in their tomb had already been in existence for many years, but it was now scholars who were pointing out the danger.[46] Buffon quoted them and asked, like them, that we wait a little longer before burying people. His train of thought here took an unexpected turn: "We will speak elsewhere of the customs of different peoples concerning funerals, burials, embalmings, etc. Even the majority of those who are savage pay more attention than we to these last moments; they consider the first duty what for us is only a ceremony; they respect their dead, they dress them, they speak to them, they recite their exploits, praise their virtues; and we, who like to think of ourselves as sensitive, are not even human; we flee them, we abandon them, we do not want to see them, we have neither the courage nor the will to speak of them."[47] What would Buffon say about our "funeral homes" or about our modern evasion of the deceased? He was denouncing today's mentality because it was the mentality of his time as well.

The statistical considerations ending the chapter are even more modern. Statistical research on the length of human life started in England in the seventeenth century, but this was the first time it was used with a purely "biological" point of view. The tables that Buffon published had been established by Dupré de Saint-Maur and concerned three Parisian parishes and twelve rural ones. They allow for the correction of distortions due to the floating character of the Parisian population and the great number of Parisian children put in the care of a nurse in the country, which explain why infant mortality was so much greater in the country than in Paris. From these lists, which he published in full, Buffon derived a general table of the "probabilities of the length of life," what we would call "life expectancy," according to age. This expectancy increased considerably in the first years, since it went from eight years at birth to forty-two years at the age of six, before decreasing regularly.

Buffon hoped to establish the life expectancy of man in general, independent of his conditions of existence and individual accidents. His concern was with the human species, with this abstract man whose life expectancy we have already seen to be fixed. Buffon did not think of any comparisons he might make between humans of diverse social classes or between the city and the country. His point of view was that of the naturalist.

The naturalist was, however, also a man. More specifically, Buffon was a man of forty-two years of age at the time he published these tables, which told him "that at the age of twenty-eight or twenty-nine years, a man has lived half his life" and at fifty had lived three-quarters of it.[48] To console himself, and to console his reader, he introduced "moral considerations": "A man must look at his first fifteen years as being nothing; everything that happened to him, everything that occurred in this long interval of time is erased from his memory. . . . We only start to live morally when we start to organize our thoughts, to turn them toward a specific future, and to give them some kind of consistency, a state relative to what we must be subsequently. By considering the length of life in this light, which is the most realistic, we will find that according to the table, at the age of twenty-five years one has lived only a quarter of one's life, that at the age of thirty-eight years one has lived only half of it, and that it is only at the age of fifty-six years that one has lived three-quarters of his life."[49] This meant, of course, that man only started to live from the time he became fully rational.

It also meant that in 1749 Buffon had still almost half of his life to finish his work. Yet we may also wonder, as we already have, why he so desperately tried to forget and even deny his childhood.

Variations in the Human Species

The last chapter of the *Natural History of Man* is the most famous and the one that has most attracted the attention of historians, not only because of its conclusion on the unity of human species but also, perhaps less legitimately, because of the worried concern that we have today for Western man's attitudes toward other peoples and other cultures. This question, although so important for us, must not lead us to hasty judgments on an older way of thinking. The historian here has to understand, not to judge; since otherwise we run the risk of not recognizing what is most original and most modern in this text.

Buffon led his reader on a vast ethnographic tour of the world, which passed successively across Europe and Asia, then Africa, and finally America. Between each continent, to retain our attention, he presented a few of the theoretical problems that he wanted to resolve.

His itinerary deserves study. Leaving Lapland, he went east, staying north of the Himalayas to China. From there, he descended south to Malaysia. After an excursion down to New Guinea and Australia he came back west, crossed India, Arabia, Egypt, and North Africa to the Atlantic Ocean. Then, using Kashmir to the north of India as a new starting point, he again headed west toward Western Europe. Passing to Africa, whose interior was almost unknown, he descended south by following the western coast and came back up north along the eastern coast. Finally, his itinerary in America went from north to south.

What controlled this itinerary was obviously the idea of "climate." Buffon first crossed the "cold climates" of the Old World and returned through the warm climates, between the twentieth and thirty-fifth parallels. Then he studied the populations of "temperate" climates. In Africa as in America, he followed the distribution of the climates in the same way, dividing the earth, so to speak, into horizontal climatic zones. It is clear that his theory of the climates had been entirely thought out when he wrote his chapter.

Buffon's documentation came from the immense travel literature already available, which he studied with care. Buffon cited his authors, compared their descriptions, corrected one by another, and did not hesitate to criticize them. He was very skeptical of "men with tails" that had been seen in the Phillipines and used this subject to remark that authors often copied one another.[50] He picked out accounts of the existence of white individuals in tropical countries: they formed a people, the Chacrelas, on the island of

Java, and were also found among the Papuans and in the isthmus of Panama. He concluded from these descriptions that these peoples suffered from a hereditary illness, "a kind of illness that they received from their fathers and mothers." They were therefore only individuals accidentally "degenerated from their race." He also noted the existence of a "white negro" studied by Maupertuis in his *Physical Venus* as well as his own observations on similar cases. These "white negroes" were actually albinos, and Buffon's only error was in seeing a "return to type" and in coming to the conclusion that "white therefore seems to be Nature's primary color."[51] We shall come back to this point.

The physical characteristic that Buffon pointed out first in all the peoples described was the color of the skin. Then came height, hair type and color—curly, straight, or wavy, black or blonde—and the shape of the lips, nose, face, and eyes. An extraordinary feature, like the famous "apron" of the Hottentots,* was of course mentioned.[52] As for "habits," whose description accompanied the physical portrait of the peoples, Buffon spoke of them quickly. He noted preferably and indiscriminately what was strangest, but he also carefully described the way they dressed, or did not dress, particularly the women. The greater or lesser freedom in sexual habits was also the object of a smug attention: excesses were criticized, but they were noted without displeasure.

Judgment almost always accompanied description, and the judgment was mainly aesthetic. For the Lapps, "the women are as ugly as the men." Among the Tartars, "the ugliest of all are the Kalmucks, whose appearance is something frightening." On the other hand, the inhabitants of Senegal are "well proportioned and of a very favorable size. . . . There are some, especially some women, who have very regular features." In addition, "they also have the same ideas of beauty as we, because they desire beautiful eyes, a small mouth, well-proportioned lips, and a well-made nose; it is only on the foundation of this picture that they think differently; the skin has to be very black and very shiny." Too bad, because "they also have very fine and very soft skin, and there are among them women as beautiful, apart from their color, as in any other country in the world; they are generally very well built, very gay, very vivacious, and very inclined toward love; they have a taste for all men, and in particular, white men." In short, they have almost all the benefits of Tahitian women, whom Bougainville had not yet discovered. Unfortunately, they smoked pipes! No one is perfect.

Nothing, however, was a match for Circassian women: "They have the most beautiful complexion and the most beautiful color of the world. . . . They have big eyes, soft and full of fire, well-made noses, ruby lips, and small laughing mouths, and the chin as it should be in order to make a

*Trans. note: Hottentot women stretched their labia majora to extraordinary lengths throughout childhood to increase their sexual attractiveness.

perfect oval; they have perfectly formed necks and throats, skin white as snow, tall and supple builds and hair of the most beautiful black." Add to this the fact that "in summer the common women only wear a simple chemise . . . and the chemise is open to the waist; they have perfectly well formed breasts, they are free enough with strangers and yet faithful to their husbands, who are not jealous of them."[53]

Were these indiscreet travelers well placed to give lessons in morality? They did not deprive themselves of it in any case. The Lapps were particularly badly treated: "rude, superstitious, stupid." "This despicable people has only enough morals for them to be looked down upon. They bathe naked and together, girls and boys, mothers and sons, brothers and sisters, and are not afraid to be seen in that state." The Tartars "have no religion, no moral restraint, no decency, they are all thieves." This list could be prolonged indefinitely. Very often, as a matter of fact, moral brutality went along with physical ugliness, and vice versa. The "Bengalis" "are beautiful and well built, they like commerce and have much moderation in their morals."[54] It was not an absolute rule, but for Buffon this fact was important.

It would be impossible here to examine the mix of errors, hasty generalizations, half-truths, and exact observations that Buffon echoed. Despite his critical intelligence, he was at the mercy of his informants, who often added to their European prejudices those of their station. A merchant does not have the same point of view as a missionary or a naturalist, and a missionary is not necessarily the severest judge. What was completely absent, in any case, was the idealization of primitive man. The physical and moral picture of all these peoples was generally somber, and human comprehension very limited. For Buffon that was not important. Beyond physical deformity and the extravagance of the customs, which he obligingly reported for the entertainment of his readers, a single question interested him: from where did this "variety in the human species" come, or, to ask the question in more direct terms, what was the origin of the human race?

This question was not new. It is the same one that the Mediterranean peoples had asked when they encountered black people. An ancient tradition said that the darkness of "Ethiopians" was the result of Noah's curse on his son Shem. This tradition, which has no justification in the biblical text,* was piously accepted by the Arabs, who, like the Greeks, the Romans, and many others, used black slaves. This tradition was still ritually put forward in the sixteenth century, but this did not prevent authors from advancing a physical cause, generally the intensity of the sun.[55] Black people were thus white people who had progressively "tanned." The idea of such a transformation, however, was repugnant to the fixist creationism at the end of the

---

*The list of Shem's children includes white people as well as Ethiopians, but Noah's curse doomed them all to servitude; the interpretation was very convenient (cf. Gen. 9:18–10:20).

seventeenth century, so little by little the idea became accepted that black people and then yellow people formed separate races or even different human species from white people from the very beginning.

This idea was untiringly defended by Voltaire, first because it contradicted the biblical idea of a single origin of humanity, and then because it corresponded to the unchanging character of created nature. "Never has a man with even a little education suggested that unmixed species degenerated," wrote Voltaire in his *Traité de métaphysique* [Treatise on Metaphysics]. "Pear trees, pine trees, oaks and apricot trees do not come from one same tree." Therefore, and no matter what "a man dressed in a long black cassock" might say about it, "bearded whites, wooly-haired negroes, yellow people with ponytails and men without beards do not come from the same man."[56] It was exactly this idea that Buffon attacked. It was not that he wanted to defend the Bible, but he did not think nature was completely immutable.

Let us first note that Buffon used the word "race" little, and employed it, as he himself specified later, only "in the broadest sense."[57] He occasionally, but very rarely, spoke of the "white race" and the "black race," and he divided blacks into two principal "races," Negroes and Kaffirs.[58] Most often he spoke of "peoples" or "nations," and noted that there was as much diversity among blacks as there was among whites. It is possible to find in Buffon the major traditional groups, whites, yellow peoples, blacks, and indigenous Americans, but it was not he who brought them to light.

If he rejected the major divisions and multiplied the minor ones, it was because in reality he was seeking to unite everything. There were not two skin colors, white and black: there were people more or less "brown," more or less "tanned," and many secondary shades—"copper" or "olive" skins: there is a continuity from white to black. Hair does not allow clear distinctions. "One finds, even in France, men who have [hair] as short and frizzy as Negroes."[59] Individual differences therefore blur racial differences.

The role of food, "customs," and "way of life" must not be forgotten either. "A civilized people that lives in a certain comfort," that is well governed and "sheltered to a certain extent from misery," "will be for that reason alone composed of stronger, more handsome, and better built men than a wild and independent nation," where the individuals would necessarily be "more swarthy, uglier, smaller, more wrinkled."[60] Moreover, many physical characteristics were attributed to systematic deformations or traditional practices. The Hottentots were black only because they rubbed themselves with fat and soot. Many Negroes have pug noses and large lips only because their parents methodically crushed them.[61] Physical characteristics therefore had only a relative value.

To that must be added the history of the peoples, in particular their migrations, which mixed them together. The Maldivians were "a people blended from all nations." The Persians were a mix of Georgian and Circassian blood, the Moors of black blood.[62] In order to know the origin of a people, it was sometimes safer to study its culture rather than its skin color. Thus the Ethiopians, "although extremely brown, get more from the white race than the black race," because "they have the same religion and the same customs as the Arabs."[63] This text is perhaps the only one in the chapter where Buffon spoke of the "black race" and "white race," and he did so to show that the division was deceptive.

Thus, even between the major ethnic groups, there were always "transitional peoples." Buffon did not believe any more in strict divisions in ethnography than he did in botany. The Ostiaks "appear to make the transition between the Lapp race and the Tartar race," which perhaps meant that the Lapps and all the peoples who lived around the Arctic Ocean and who resembled the Lapps so much were only "Tartars degenerated as much as possible." In the same way the Fulah (today the Peuls)* "seem to be a gray area between the Moors and the Negroes." In addition, they were "more civilized than Negroes" and were Muslims.[64]

The human species was not, therefore, a collection of isolated races in the beginning. It was an immense sheet which covered the globe, and one passed imperceptibly from one people to another when traveling across it. There was an extreme diversity everywhere among individuals or among groups; there was as much diversity among blacks and yellow people as among whites, and this diversity came from the very causes that Buffon attributed to "varieties in the human species."

These causes, as mentioned before, were climate, food, and "customs or way of life." Food "does a lot for the [physical] shape": "all peoples who live miserably are ugly and badly built," and that was true in the French countryside as well as elsewhere. Air and soil also played a large role. On "the hillsides and hilltops," the countrymen were "agile, full of energy, well built, spirited, and . . . the women there are generally pretty; while in the plains, where the earth is crude, the air thick, and the water less pure, the countrymen are coarse, heavy, badly built, stupid, and the countrywomen are almost all ugly."[65] This was valid for animals and plants as well as for men. As for the manner of life, it was enough to compare the Tartars with the Chinese in order to see its importance. The Tartars "are always exposed to air" and "live in a hard and wild way." The Chinese were whiter "because they lived in cities, because they are civilized, because they have all the means to protect themselves from the injuries of the air and the earth."[66]

It was the climate, however, that influenced skin color the most. The more "constant and always excessive" the heat, the blacker the skin. Buffon

---

*Ed. note: A Sudanese people of non-Negro extraction.

did not deny that skin color was a physical trait. What he claimed was that white passes imperceptibly to brown, brown to black, French to Spanish, Spanish to Moors, Moors to Negroes. "Apparently with time a white people transported from the North to the equator could become brown and even completely black, especially if this same people changed customs and used for food only products from the warm countries to which it had been transported."[67] Admittedly, color was a hereditary trait: "In any country where a Negro comes into the world, he will be black as if he had been born in his own country." But a hereditary trait can be modified in the long run, under the persistent action of another climate. What the climate had done, the climate could undo: "If one transported Negroes to a northern province, their descendants in the eighth, tenth, or twelfth generation would be much less black than their ancestors, and perhaps as white as the original peoples of the cold climate where they would live."[68]

The notion of "climate" was certainly not original. Without going all the way back to Hippocrates, Plato, or Aristotle, it is at least necessary to mention Jean Bodin's analyses in the sixteenth century, and to recall especially that Montesquieu had dedicated three books of his work *L'Esprit des lois* [The Spirit of the Laws] (1748) to the relationships among climate, habits, and legislation.[69] Buffon, who greatly admired Montesquieu, went even further by clearly stating that climate was the cause of differences between people. And it was for man that he introduced the notion of "degeneration," to which we shall have to return at greater length.

In any case, the climatic theory came up against a tremendous obstacle, of which Buffon was perfectly conscious: the American populations. If it was possible in the Old World to show that skin color was a direct function of the climate, in the New, the equivalent of Black Africans did not exist. Aside from the extreme north, whose inhabitants resembled the Lapps, in America there was "only one sole and same race of man," that was only "more or less tanned." Even the alleged giant Patagonians of the Tierra del Fuego, whose gigantic proportions were doubtful anyway, "resemble other men in size, and other Americans in color and hair."[70] Consequently, "the influence of the climate here is completely contradicted." Buffon wriggled out of this difficulty, first by showing that the tropical climates of the New World were not as hot as those of Africa, then by explaining that the civilizations found in Mexico or Peru were relatively recent and that the difference in their "way of life" compared with that of the Indians of North America had not yet had the time to create large differences between them. The Indians were therefore "new people"; there was no question that they came from Asia. The idea was not new, having been proposed in 1590 by José de Acosta. It holds true today. Buffon conjectured that the first wave of immigrants must have arrived in California and descended southward from there. As for the peoples of the far north, they came from Green-land.* Thus, "independent even of theological reasons," it was certain that the origin of the Americans "was the same as ours [the Europeans]."[71] The theory was saved, but it should be noted that the physical determinism of the climate theory had to be combined with the historical fact of migration.

The chapter's conclusion marked an important date in the history of anthropology and deserves to be quoted in its entirety: "Everything therefore comes together to prove that humankind is not made up of essentially different species, that to the contrary there was originally only one sole species, which, having multiplied and spread itself over the entire surface of the earth, underwent different changes, through the influence of the climate, differences in food, diversity in way of life, epidemic illnesses, and also the infinitely varied mix of more or less similar individuals; that at first these alterations were not so marked and only produced individual varieties; that they then became varieties of the species, because they became more general, more noticeable, and more constant through the continued action of these same causes; that they survived and continue to survive from generation to generation, like the deformities or the illnesses that their fathers and mothers pass to their children; and that, finally, since they were originally produced only through the concourse of exterior and accidental causes, and since they were preserved and rendered constant only by the passage of time and the continued action of these same causes, it is very probable that they would also disappear little by little, again over time, or even that they would become different from what they are today, if these same causes do not continue, or if they come to be varied in other circumstances and through other combinations."[72]

Buffon thus laid the foundation for what would later be called "monogenesis," and his conclusion was surprisingly close to those of modern anthropology, aside from the racial speculations of the nineteenth century.** There was obviously a major difference: Buffon did not imagine that the human species could be the result of an evolution from an animal species, a point to which we shall return. Moreover, the mechanisms that he put forward for the progressive variation of human types, in particular the prolonged action of the climate, did not lead to Darwin but rather to Lamarck and the inheritance of acquired characteristics. Here again we find the logic of his theory of reproduction. Otherwise Buffon's thoughts were very modern—the progressive evolution of forms within the human

---

*In 1778 Buffon held that the migration took place across the Bering Strait, which is the accepted idea today.

**Both the distance Buffon covered in his understanding and the astonishing convergence of his essential conclusions and those of modern anthropology can be measured by reading Langaney 1988.

species, the role of individual variations, the comparisons between new traits and hereditary illnesses, the effect of unions between different individuals. Perhaps this modernity came especially from this concept of species, which he had as yet only formulated in passing, and whose possibilities he had not yet exploited. It was because he did not identify a species as a collection of identical individuals but as a collection of individuals capable of reproducing among themselves that Buffon foresaw certain modern aspects of evolution, at least within one species, and it was for the human species that he described the results. Buffon did not yet possess the modern concept of "populations," but he was at least rid of the old logical categories of classification and creationism that they assumed, which underlay all naturalist thought at the beginning of the eighteenth century.

It is easy, having said all this, to show that Buffon remained a man of his time and that he naively shared the prejudices of his century: "The most temperate climate is from the fortieth degree to the fiftieth; it is also in that zone that the most handsome and the best built men are found; it is in this climate that the idea of man's true color must be taken; it is from there that the model or unity to which all other nuances of color and beauty must be compared, both extremes being equally far from truth and beauty."[73] We have, therefore, every reason to think that it was there that the human species was born, a hypothesis that had the tactical advantage of vaguely agreeing with the Bible. The same Eurocentrism manifested itself in his many judgments about the idleness of savages, the "laziness" of peoples who were satisfied to live from fishing and hunting, the stupidity of the inhabitants of Guinea, who lived "in wild places, and on sterile lands, when they could have lived in beautiful valleys, among pleasant, wooded hills, or in a verdant, fertile countryside crisscrossed with rivers and pleasant streams, but for them all this holds no pleasure."[74] We could multiply these examples, but we shall soon see that it was not just prejudice but in fact an entire vision of man that was at work. Buffon also used primitive customs to criticize European habits; his attitude was not one-way. Above all, though, affirming the unity of the human species was a gesture with considerable philosophical weight, all the more so since this affirmation ignored theological considerations and was exclusively founded on purely "scientific" proofs.

Buffon himself drew the first consequences of this position in a text that can serve as a conclusion to this chapter: "Even though Negroes have little intelligence, they do not fail to have a great deal of feeling; they are gay or melancholy, hard working or idle, friends or enemies according to the way they are treated; when they are well fed and they are not mistreated, they are happy, joyful, ready to do anything, and the satisfaction of their spirit is written on their faces; but when they are mistreated, they take the sorrow straight to heart and sometimes perish from sorrow; they are therefore very sensitive to kindnesses and insults, and they carry a mortal hatred against those who have mistreated them; when, on the contrary, they have a liking for a master, there is nothing they are not capable of doing to show him their zeal and devotion. They are naturally compassionate and even tender with their children, their friends, their compatriots; they willingly share the little they have with those they see as needy, without even knowing them except by their poverty. They have, therefore, as can be seen, an excellent heart. They have the seed of all virtues, and I cannot write their history without being moved by their state. Are they not unhappy enough at being reduced to servitude, at being obligated always to work without ever being able to acquire anything? Is it also necessary to exhaust them, strike them, and treat them like animals? Humanity revolts against these hateful treatments that a greed for profits has created. . . . We force them to work, we limit their food, even of the most ordinary kind; it is said they withstand hunger very well and that in order to live three days they only need a portion of one European's meal. However little they eat and sleep, they are always equally hardy, equally strong at work. How can men who still have some feeling of humanity adopt these maxims, make a prejudice from them, and using these reasons seek to legitimize the excesses that the thirst for gold drives them to commit?"[75]

Montesquieu had written about black slaves: "It is impossible for us to imagine that those people are men because, if we imagine them to be men, we start to believe that we ourselves are not Christians."[76] With Buffon, bitter irony gave way to a more sentimental plea, but also another syllogism; since these were men, their masters were not human.

Thus this *Natural History* of 1749 ended with man, as it had started. In all respects, man was in the center of it, and we have not finished with him. Everything that was strongest, newest, and least conventional in this work was about natural man, no less natural in his culture and his intellectual activity than in his physical characteristics. Man had been plunged into nature and was no longer the center of creation, above all because nature itself was barely a creation any more. Of course, God's creative intervention was not denied, but once stated, it receded behind the interplay of "secondary causes," which alone were the object of scientific research. Buffon thus cut science off from theology completely. In this sense, it could almost be said that Buffon's philosophical opinions on God or the human soul no longer have any importance; the science he wanted to construct needed to do without philosophy. This would be going too far, however, for at the time a certain philosophy was needed precisely to break the age-old but still existing ties between science and religion. Such a philosophy need not necessarily be atheistic although it would probably be materialistic, and certainly not Christian, at least according to the canons of Christian philosophy of the time.

In 1749 theologians and materialist philosophers were already battling each other and knew each other well. In a way, Buffon took everyone by surprise by standing on other ground. The very fact of separating science and religion had, at that time, such philosophical weight that no one could ignore it. In fact, Buffon's *Natural History* was the first really "natural" history since that of Pliny. It was inevitable that it would cause a scandal.

CHAPTER XIII

A Famous and
Scandalous Author

The first three volumes of the *Natural History* were an immediate and resounding success in sales. The first printing was sold out in six weeks,* and it was necessary to do a second one, which was soon followed by a duodecimo edition, much resembling our paperbacks in format. These editions appeared in April 1750 and sold as well as the first. English, Dutch, and German translations were started immediately abroad. Buffon was ecstatic, and his letters to his friends were reports of victory: the success was "prodigious."[1] This success continued during the entire time the work was published; we know that the *Natural History* was the most widespread work of the eighteenth century, beating the abbé Pluche's *Spectacle of Nature*, Diderot's and d'Alembert's *Encyclopédie*, and even the better-known works of Voltaire and Rousseau. Buffon had wanted to touch the general public; he had succeeded completely.

Thus it was that in September 1749 Buffon found that he had become famous overnight. In the salons the *Natural History* was the book to have read or to be able to talk about. The abbé Raynal, who did not like Buffon, expressed quite well his colleagues' frustration by noting that the work "did not succeed particularly well with educated people" but that "women, to the contrary, attach importance to it."[2] This criticism would be repeated hundreds of times. Unfortunately for the abbé, women's opinions counted

*The king paid for the publication of the work. We do not know the exact number of copies published in the first printing. Historians estimate it as between five hundred and one thousand; but of those the king must have taken more than two hundred fifty copies, which he gave to libraries, members of the royal family, several great lords, and high-ranking civil servants as well as foreign sovereigns. Buffon received eighty-six copies, of which fifty-six were bound. A quarto bound copy cost fifteen livres. The bookseller Panckoucke, who was responsible for the distribution, must have sold it for seventeen livres. At the time, the average wage for a common laborer was one livre per day. (Cf. Bourdier 1952a and Grinewald 1988.)

[184]

very much in the eighteenth century. Something fairly surprising, at least for us, was that Buffon immediately received the support of the Jesuits; their journal, the *Mémoirs de Trévoux,* reached a large audience. From October 1749 to May 1750, the journal dedicated no fewer than four articles, a total of about one hundred pages, to praising Buffon.[3] Another well-known journalist and Voltaire's great enemy, Fréron, also expressed his admiration for Buffon.[4] In its edition of October 1749 the *Journal des Savants* published a flattering analysis of the first volume then, curiously, remained silent on the subject for two years. The work's immediate success at first caught a growing opposition off guard.

The diversity of opinions on these first three volumes can be explained by at least three reasons. The first was the author's caution. Without hiding what he thought, Buffon knew how to take the necessary precautions so as not to run counter to the reigning orthodoxy. The Jesuits could therefore emphasize the spiritualism of the *Natural History of Man.* The second reason was that French intellectual opinion was very divided, but not along the lines that a reductive historiography has imposed for a long time. There was not at that time a "philosophes' party," "modern" and "progressive" by definition, or a "devout party," "traditionalist" and "reactionary." Christian intellectuals were too busy fighting one another to battle philosophes. The outlawed Jansenists were *intégriste,** Gallican, and Cartesian. They enjoyed public favor because they defied royal authority. The Jesuits were liberal, more open to new ideas. They preferred Locke and Bacon to Descartes because there they found Aristotelian ideas. Yet they were Ultramontane and therefore not very popular. "Reactionary" members of parlement and "progressive" philosophes joined forces to obtain their expulsion in 1762. As for the philosophes, they were not yet organized into a party; that would be left for d'Alembert, guided from afar by Voltaire. Even then, there would not be unanimity, neither as to which tactic to follow nor as to which ideas to defend. Anticlericalism, and the defense of the *Encyclopédie,* created only a superficial unity. Soon Voltaire would dedicate as much energy to battling d'Holbach's militant atheism as to "crushing the Despicable," that is, the Catholic Church. As for the royal government, it sometimes struck to the left, sometimes to the right, depending on the circumstances and changes in the ministry. Jansenists and philosophes took turns in the Bastille, or even found themselves there together. Mme. de Pompadour protected the philosophes when they were attacked by the "devout party." Malesherbes, overseer of the book trade** and therefore in charge of the censorship of books, personally helped the Encyclopedists to escape the police searches that he himself had ordered.

*Ed. note: Members of a religious party hostile to French royal absolutism.

**Ed. note: Malesherbe's title was Directeur de la Librairie, whose duties involved moral oversight of what was published.

Finally, Buffon's official situation and the nature of his work only added to the confusion of ideas. A high-ranking civil servant, friendly with the ministers, Buffon apparently had nothing to do with the obscure lampoonists, recorded in police files for their rotten morals and subversive writings. He was untouchable and even a priori above suspicion. No one had even thought that he could be sent to Vincennes like Diderot. Above all, he was a scholar and certified as such by his position at the Royal Botanical Gardens and his membership in the Academy of Sciences. Time would be needed in order to understand that a scientific work could convey ideas that were dangerous to orthodoxy, and time was therefore needed for the counterattack to organize. If this counterattack, when it came, was far from being unanimous, it was because Buffon's work had served to reveal the internal divisions of the important groups involved.

For the purposes of this analysis, I will examine the reactions of the various milieux—the religious, that of scholars, and finally of the philosophes. In reality, they were all linked, and nothing was more common for all of them than to denounce the scientific weaknesses of a work in order to ruin its philosophical prestige and to defend Christian orthodoxy as well as that of Voltaire.

The fact that the Jesuits praised the *Natural History* was perhaps reason enough for the Jansenists to attack it. Indeed, in the Jansenists' eyes, the Jesuits were mainly responsible for religion's decline in France, especially because of the culpable indulgence they showed for dangerous works. On February 6, 1750, the *Nouvelles ecclésiastiques* [Ecclesiastical News], the semi-clandestine publication of the party, which was officially forbidden but very regularly published, violently attacked the *Natural History;* it wanted to make "known the venom" of the work. The "First Discourse" and its analysis of the notion of truth were particularly criticized. By presenting mathematical truths as a product of the human mind, physical truths as only "probable," and moral truths as simple conventions, Buffon exhibited an absolute skepticism and expressed himself "as a perfect Pyrrhonist." He thus destroyed all human and moral values: two plus two no longer made four; it was no longer necessary to love God and one's fellow man; theft, adultery, and assassination were no longer anything more than convenience and probability. In addition, Buffon considered man to be an animal, he contradicted Genesis, and he believed that the world was eternal. If he were truly Christian as he claimed, he should follow Job's example and say, "I disapprove of my behavior and I repent, by covering myself with dust and ashes." If not, "should such a pernicious book go unstigmatized?" Besides the insult to God, "it dishonors the name of the king to whom it is dedicated." Let the sword of authority thus fall on this work, on its author, on the entire Academy of Sciences, on the *Journal des Savants,* which had published a laudatory account, and, at the same time, on the Academy of

Inscriptions, which had sponsored the translation of the *Essay on Man* by Alexander Pope, a deist, and on the Académie Française, which had just elected Voltaire.[5] It seemed as if the *Natural History* served as a detonator and revealed to what point the most respectable institutions were infected by the new spirit, with the complicity of an incapable government.

After this public denunciation, the Sorbonne, that is, the Faculty of Theology of Paris, had to react. In theory, as we have seen, the censorship of books depended on the royal administration, but the theologians could take charge of a contentious case, give their own decision, and ask the parlement to condemn a book they judged to be dangerous. These interventions up until then had been rare, but they became more frequent since the Faculty, like the parlement, was worried about the administration's laxity. It is probable, however, that the Sorbonne was very embarrassed. It was already involved in the business of *The Spirit of the Laws*, which Montesquieu had published the year before. For the College not to react was to expose itself to the vicious criticism of the Jansenists. To condemn a book published by the Royal Press, the work of a high-ranking civil servant well established at the Court, and which was already a commercial success, was to expose itself to ridicule. For his part, Buffon was quite annoyed to see his work thus denounced. "I hope . . . that it is out of the question to blacklist it," he wrote to the abbé Le Blanc in June 1750. "And in truth, I have done everything not to deserve it and to avoid theological harassment, which I fear much more than the criticisms of natural philosophers or geometers."[6] D'Argenson was perhaps exaggerating only slightly when he wrote in his *Journal*, "The Seigneur de Buffon . . . has been greatly affected by the grief that his book's success gives him. The devout are furious and want to have it burned by the executioner. Truly, he contradicts Genesis in every way."[7]

We do not know the details of the secret dealings which permitted Buffon and the Faculty to find an honorable solution for both parties. We know only that these dealings took place, certainly with Riballier, the Faculty's syndic, probably during the fall of 1750. On January 15, 1751, Buffon received the following letter:

"Sir,
We have been informed, by someone among us on your behalf, that when you learned that the *Natural History*, of which you are the author, was one of the works chosen by order of the Faculty of Theology to be examined and censured because it contained principles and maxims that are not in accordance with those of Religion, you declared to him that you did not have the intention of dissociating yourself from it and that you were prepared to satisfy the Faculty in regard to each of the articles it found reprehensible in your work; we cannot, Sir, praise

you enough for such a Christian resolution, and in order to put you in a position to carry it out, we are sending you the statements taken from your book that seemed to us to be contrary to the beliefs of the Church.

"We have the honor of being respectfully, Sir,

"Your very humble and obedient servants

"The Deputies and the Syndic of the Faculty of Theology of Paris

"In the House of the Faculty, January 15, 1751."

The tone, as we see, is very courteous. Attached to the letter was the list of the fourteen "reprehensible statements." The ideas that had bothered the deputies and the syndic were the cyclical, and therefore eternal, nature of the history of the earth, the theory of planet formation, and the discussion of the notion of truth and immaterialism following Berkeley, which was so visible in the analysis of the nature of man.

On March 12, Buffon sent his answer. As for the theory of the earth and planet formation, he stated that he believed "very firmly all that is told [in the Scriptures] about Creation, both as to the order of time and the circumstances of the facts," and had presented his theory "only as a pure philosophical supposition." His hypocrisy was enormous, but as Buffon himself said, "It is better to be humble than hung."[8] On the other hand, as for the idea of truth, the existence of the exterior world and the soul's nature, Buffon made only the necessary concessions and managed to save the essential parts of his principles. Finally, he offered to publish this entire correspondence in the next volume of his *Natural History*.

The proposal was clever, if it was true, for we have every reason to believe that Buffon's answer had been composed by the theologians themselves, and Buffon had only to sign it.[9] He therefore should have been sure of the success of his approach. The Faculty itself, however, was divided; it had its moderates and its conservatives. It was therefore with a sigh of relief that Buffon learned of the Sorbonne's decision, taken in its meeting of April 1, 1751. "I'm done with it, to my great satisfaction," he wrote to the abbé Le Blanc on April 24. "Of the one hundred twenty doctors assembled, I had one hundred fifteen [on my side], and their decision even contained praise which I was not expecting."[10] In fact, the College had "approved and praised" his deferential attitude. The Sorbonne's official letter was sent on May 4. It announced that the proposition for publication had been received "with extreme joy" by the College. "It hopes that you want to carry it out." Honor was saved, and the theologians must have been relieved also.

This entire correspondence was published at the beginning of the fourth volume of the *Natural History* in 1753[11] and was reproduced in all the subsequent volumes. It served for close to thirty years as a safeguard and protection against all official accusations of irreligion. During these thirty years, Buffon continued publishing these texts without changing a single word. His retraction was applauded by the Jesuits: "In such a precise and sincere statement, M. de Buffon shows the Doctors proof of his orthodoxy and the Philosophes an example of submission."[12] The Jansenists were less delicate: "What shame for the Faculty . . . to applaud such a declaration. . . . It amounts to hoping that everyone will be taken in, as the Sorbonnic carcass has been. . . . This academician has made fun of them, and they deserve it."[13] A more courteous theologian asked, "Does a Sorbonne exist?"[14] Later, the Sorbonne itself regretted its indulgence, as much as regards Montesquieu as Buffon.[15] In 1750 it had not yet realized the gravity of the situation.

Were the Jansenists more clear-sighted than the Jesuits and the College, or, because they were already in opposition, could they allow themselves to spare no one? Whatever the reason, they were not wrong to be on their guard. "Buffon comes from here," wrote the president, De Brosses, in 1753. "He gave me the key to his fourth volume, regarding the way that things said for the Sorbonne should be understood."[16] That an old friend like the president needed Buffon to give him a "key" at least proved that things were not simple and that mistakes could be made. Later it was Buffon himself who was criticized for taking part in this comedy.[17]

If the theologians were ill at ease, the naturalists were not any happier. Buffon's nomination to the Royal Botanical Garden had shocked many academicians who had judged Duhamel du Monceau as more qualified. Duhamel had been named to another important office, and the affair had apparently been peacefully settled. The botanists of the Academy, and primarily the two Jussieu brothers who worked at the Garden, must have noticed that Buffon was an active intendant who respected their freedom. The one who seemed to mistrust Buffon the earliest was Réaumur.

Born in 1683, almost one-quarter of a century before Buffon, René-Antoine Ferchault de Réaumur was one of the great figures in the Academy of Sciences.[18] He was interested in many things, including the production of steel, but he was foremost a famous naturalist, an authority in geology and particularly in entomology. His *Memoirs concerning the History of Insects,* which had begun to be published in 1734, had created numerous vocations. Charles Bonnet and Abraham Trembley, whose discoveries I have already mentioned, openly called themselves his disciples. For them, and for many others, he was "the oracle." He personified the spirit of rigorous and patient observation with a respect for facts, and suspicion of hypotheses. If he accepted the preexisting germ theory, it was primarily because of lack of a better one.[19] A devout Christian, he could be very hard on heretical materialists.* Finally, he was a model academician, respectful of the hierarchy, and left Paris only for the official vacation of the Academy, of which he was nine times the director between 1713 and 1752.

Réaumur's antipathy to Buffon went back to at least 1740 and was aimed more at the man and his ambition than the scholar.[20] When the journals announced the *Natural History*'s imminent publication in October 1748, Réaumur kept his distance. "I have no part in this work," he wrote to a correspondent. "In fact, I do not know it at all, even though of fifteen volumes that are promised, there are already three printed." Understand: Buffon and Daubenton had not respected Academic custom; they had not submitted their manuscript for the body's approval. And Réaumur added: "I do not know how they . . . will carry out [this enterprise], because I have seen nothing of this kind from either one of them. I know that they have taken many excerpts from naturalists and travelers, but I do not know that they have themselves observed."[21] From his pen, this was already a conviction without appeal.

As soon as the work appeared, Réaumur led a campaign against it, first in his letters to his friends, Trembley, Bonnet, and to Ludot, to whom he wrote, "Accustomed, Sir, as you are to analyzing ideas and only accepting exact ones, the imposing tone with which M. de Buffon advances such strange ones may not have been enough for you to find them acceptable. If you have read his first volume, you will not be happier with the second. The three together can only serve to hurt the progress of natural history and physics in general if the propositions that are advanced here are adopted. But I learn from everywhere that protests have been made against this work, which permits us to believe that the consequences are not to be feared. Indeed, one does not have a lot of confidence in facts that the author reports when he alone has seen them."[22]

No matter what Buffon had said with whom he had made his observations. The important point was that a "protest" had been raised against the work, and to be even more sure, Réaumur decided to raise one himself. In 1751 the *Lettres à un Amériquain sur l'Histoire naturelle, générale et particulière, de M. de Buffon* [Letters to an American on the General and Particular Natural History by M. de Buffon] appeared. The work was anonymous, officially published in Hamburg, but according to a contemporary note, printed at the Arsenal, "at the establishment of Mme. la Duchesse du Maine, Protector of M. de Réaumur."[23] Soon the author's name was known: an Oratorian, Father Lelarge de Lignac. It was also known that Réaumur

---

*Witness this extraordinary funeral oration for the materialist doctor La Mettrie, Frederic II's scandalous protégé: "Humankind has just been delivered from him, it was not in his bed that he should have perished, but at least he died there like a madman, crying, sobbing, yelling when he realized that he had killed himself by having himself bled twice to cure an indigestion" (letter to Seguier, December 9, 1751, in Réaumur 1886, p. 91).

Hippopotamus. From *Histoire naturelle,* p. 141.

sponsored the work. According to d'Argenson, a sarcastic witness of this entire business, "the real author is M. de Réaumur, from the same Academy of Sciences as M. de Buffon, a great enemy of the latter, envious and jealous of his works and rewards," who had only "joined with a little father of the Oratory, who wrote the work to avoid having the entire book deal with devotion and avenged religion."[24] It also became known that another academician, Bouguer, had collaborated on it. This did not stop Réaumur from recommending the *Letters to an American* as a "solid and ingenious critique" or from presenting the author as "a profound metaphysician" and "very much one of my friends."[25]

It is not very likely, however, that Réaumur had directly contributed to the writing of the first volumes of *Letters to an American* appearing in 1751.[26] The essential goal was to show that Buffon "contradicts Genesis in everything," ruins religion, and drives God out of natural history. "While other authors [read: Réaumur] know how to lift us to the Creator by entertaining us with the history of an insect, M. de Buffon barely lets us perceive Him in explaining the fabric of the universe."[27] Buffon fell at the same time "into the most insupportable Pyrrhonism" and surpassed the limits of "human pride" by making us the authors of mathematical truths. It is therefore understandable that "materialists consider his outrageous preface as . . . the re-establishment of epicureanism": "In his work, everything happens fortuitously."[28] His maxims were dangerous to morals and his descriptions offended decency. Father de Lignac had himself published a small work of natural history.[29] Scientifically speaking, however, his critique was very weak, and aside from a few justified remarks, was limited to quibbles or jokes: if we wanted to classify animals according to their proximity to man, we would have to start with the flea.

In short, Buffon was one of those "half-scholars . . . who must be accused of arresting the progress of the sciences and hastening its destruction; they have no other goal than to dry up all sources, because they feel that there are so many roads that lead to Christianity."[30] Lignac was careful to free the Academy of Sciences of any responsibility: Buffon "is from the Academy, but the frontispiece of his book does not indicate it: certain proof either that he did not dare communicate his work to the Academy or that, if he had presented it, he was unable to obtain approval."[31] This remark was quite correct, as we have seen,[32] and without a doubt inspired by Réaumur. Another academician, the abbé Nollet, who was Réaumur's friend but not Buffon's, made the same remark.[33]

Daubenton was no better treated than Buffon, and Lignac even tried to pick a nasty fight. When reporting on the Cabinet of the King and the order he wanted to put it in, Daubenton had cited a paper, giving the author's name and praising him highly, that Réaumur had read publicly at the Academy of Sciences in 1746 but that had not yet been published. Lignac saw there "a small literary theft." He was truly using all means possible. In fact, the last letters, which are about spontaneous generation and Needham's experiments, were the only ones having any scientific interest. Needham had been Réaumur's friend, and his religious feelings were above suspicion. Spontaneous generation, however, was inadmissible. Réaumur, during the academic vacation of 1751, repeated certain of

Needham's experiments in Lignac's presence. Lignac could only come up with one serious criticism: perhaps there were "animals dispersed in the air contained in the bottles" where Needham had seen life appear. The remark lost its force, drowned as it was in an interminable discussion of Needham's obscure metaphysics.[34] It was not until Spallanzani that Needham's experiments would be seriously studied again.

Obviously, the *Ecclesiastical News* applauded "an author who in truth seems infinitely better educated in the subjects he treats than the so-called scholar whom he has the talent of refuting with as much politeness as force."[35] According to the abbé Raynal, "the critic does not have as much standing as the author he attacks; but he is exact, he is clear, he is educated, and he has much wisdom. This work is making a stir and has a very large number of partisans."[36] The author's stand, his anonymity, and the role played in the wings by Réaumur and Bouguer, however, shocked a certain number of people. The journalist Pierre Clément accused the "triumvirate" of having "sometimes twisted ideas and abused Mr. de Buffon's terms . . . for having sworn that the only good thing in his work was the style." He denounced "the ill will, the *odium theologicum*," and exclaimed: "The most innocent supposition would not be permitted a Physicist! He is required to give a moral and constant commentary from Genesis!"[37] "Whence come these ridiculous fictions?" asked Deslandes in 1753. "Whence comes this disguise that fools no one; whence comes this deception in the title and the characters?"[38] Facing all this "clamor" (his word), Lignac did not renounce his crusade but understood that he had to change his tone.[39] In the end, the *Letters to an American* had helped Buffon in public opinion more than they harmed him.

As for Buffon, he refrained from reacting. On the subject of the *Ecclesiastical News,* he had written to a friend, "Everyone has his measure of pride; mine goes to the point of believing that certain people cannot even offend me."[40] As for the *Letters to an American,* he immediately attributed them "to a monk of the Oratory helped by a pedant from the Academy" and decided not to respond to the criticism "because it has not affected me and besides, I have much more indifference than one would think for the success of my opinions."[41] Was this indifference sincere or feigned? Buffon, in any case, did not miss a later opportunity to settle accounts with Réaumur.

In 1750 a German translation of the *Natural History* appeared in Hamburg and Leipzig, already attesting to the work's international effect. The second volume, which contained the theory of generation, began with a preface by Albrecht von Haller. A contemporary of Buffon, Haller was a professor at Göttingen, and he was already considered one of the great physiologists of his time. On the scientific plane,[42] Haller recognized that Buffon's theory had the merit of explaining the facts of heredity, but he was not able to understand the action of the "interior mold." Furthermore, he pointed out that the fetus possessed added parts that the parents did not have, the origin of which Buffon had difficulty explaining. He also underlined the fact that mutilations were not hereditary: the Hottentots and Swiss mountain men traditionally had one testicle amputated, yet their sons always had two at birth. The nontransmission of mutilations remained a classic objection to the inheritance of acquired traits until the end of the nineteenth century.

To this scientific criticism, careful yet courteous and without dogmatism, Haller added some religious thoughts. A convinced Christian, he thought that "the sciences belong to Religion by their very nature." But religion did not have to hold science by the reins, and it was not denying the Creator's omnipotence to admit a nature that is active and forces analogous to Newtonian gravity. "If matter has forces that allow it to create form, let us not believe that it has received them from blind destiny." But in fact Buffon had never said that Nature could create forms: "She only copies molds that were already created." Haller was less reassured with Needham's ideas and could not admit spontaneous generation. Unfamiliar with the intrigues of the Parisian academic milieu, however, as well as with the internal dissensions of French Catholicism, he took the texts as they were and did not see the need to suspect beforehand their authors' religious intentions. This could indicate a certain naiveté. Yet, through his own research on "irritability" (today called excitability) of muscular tissues, Haller was driven to admit an observable property in living matter that he could not explain mechanically. He thus escaped the intellectual block that artificially linked religious conviction and a traditional mechanical conception of nature. Réaumur and Buffon were victims of this block and showed it by taking antithetical, yet symmetrical, positions.

Subsequently, Haller became convinced of germ preexistence, and his criticism of Buffon and Needham grew harsher.[43] Meanwhile, his preface was translated into French and published in 1751.* Réaumur publicized the work,[44] without perhaps realizing that it showed the importance of Buffon's ideas.

Neither Réaumur nor Buffon, however, were conscious of the most trenchant criticism of the first three volumes of the *Natural History*, namely the *Observations* written, certainly very quickly, by a young magistrate fascinated by botany, Chrétien-Guillaume de Lamoignon de Malesherbes. Born in 1721 into a large family of parlementarians and early showing great promise as an administrator, Malesherbes had no traits of a bigot. As librarian to the king, he was as liberal as possible. "The principle of freedom of the press is the one for which I have been battling for nine years," he wrote in

*Ed. note: See Haller 1751 in Bibliography.

1759, "and for which I have made an enemy of the entire clergy, all devout people, even almost all those who are called statesmen."[45] He protected the *Encyclopédie* as best he could, but his liberalism did not prevent him from being guillotined at the age of seventy-three for having dared to serve voluntarily as the lawyer for Louis XVI before the Convention.

In 1750 Malesherbes became an honorary member of the Academy of Sciences and decided not to publish his *Observations* out of consideration for a colleague whom he esteemed. The *Observations* were published only in 1798, when Buffon's renown was being attacked from all sides. In this work, Malesherbes, a botanist by desire, first of all showed irritation with Buffon's attacks on classifiers.[46] He defended the honor of great botanists from Gesner to Tournefort. In particular he defended the existence of an order in nature. The idea that mathematical truths could be considered as a product of the human mind seemed inadmissible to him, and he could understand nothing of the discussion in the "First Discourse" on the notion of truth. Like many of his contemporaries, he reproached Buffon for scorning the facts because he believed, like Fontenelle, that the organization of facts must come from the facts themselves and that the human mind, far from having to "construct" a science, had only to "discover" the rationality of the natural order, which must reveal itself. At the same time, he did not understand how Buffon could hope to explain planet formation without first explaining the movement of the comet responsible for it. In addition, a "thousand events that we cannot imagine" could have been produced in the solar system, and it was perhaps better to have God intervene directly. As for the theory of generation, he was satisfied to underline its hypothetical character, but this area was less familiar to him and his criticism less sharp.

The interest of Malesherbes's *Observations* is that, free from all fanaticism, it expressed the perplexity of an enlightened contemporary faced with a work that contradicted the accepted conception of science in general and of natural history in particular. What had given the most offense in the whole work was the theory of knowledge in the "First Discourse." By uniting Locke's sensationalism with a Cartesianism free of metaphysical roots, Buffon contradicted everyone and was accused, at one and the same time, of embracing an absolute skepticism and of exalting an unlimited "human pride," of accumulating facts and then scorning them. Contradictory accusations, because everyone was caught off guard. After that, it was easy to single out the hypothetical character of the theories and to ignore the efforts Buffon made to find "models" that allowed him to organize the facts. Since these models were also substitutes for the Creator's direct intervention, the accusation of irreligion was inevitable and, in the logic of the time, not unjustified.

At least Buffon could hope for support from the philosophes. Their situation, however, was not clear, and divisions began to appear. The time of youthful friendships and of battles waged together for Newton's cause already belonged to the past. Maupertuis had left for Berlin to take over the direction of the Academy of Sciences renovated by Frederick II. His friendship for Buffon did not wane, and at his death in 1759 it was Buffon whom Maupertuis designated as sole heir. Yet the two men rarely saw each other. With Voltaire it was worse. As early as 1738 with the publication of the *Elements of Newton's Philosophy,* Buffon had to state that Voltaire retained everything in Newton that interested Buffon the least, including the topics Buffon would soon attack: creationism and God's omnipresence in nature. In the course of the debate that followed, Voltaire even specified that weight was a "quality . . . given by God" and was not "basic to matter," as Buffon believed.[47]

In 1746 Voltaire anonymously published a *Lettre italienne sur les change-ments arrivés à notre globe* [Italian Letter on the Changes Occurring to Our Globe]. His creationist philosophy prevented him from believing in these changes, and he violently attacked catastrophists like Whiston, Woodward, and Burnet. Yet he also rejected the existence of fossils, which required vast transformations in the earth's surface. Not knowing that the letter came from Voltaire, Buffon ironically commented: "Petrified fish are only, in his opinion, rare fish rejected from the Romans' tables because they were not fresh; and as for the shells, they were, he says, brought back from the seas of Levant by Syrian pilgrims in the time of the Crusades. . . . Why did he not add that it was monkeys who transported the shells to the top of the high mountains . . . it would not have spoiled anything and would have rendered his explanation even more plausible. How can it be that enlightened people who like to think they know a lot about philosophy still have such false ideas on the subject?"[48] In 1749 Voltaire had other worries than the *Natural History*. He had just seen Mme. de Châtelet die in childbirth in Lunéville for a child that was not even his. Completely disoriented, he was preparing to leave for Berlin, where Frederick II had wanted him to come for some time. There, he was to meet up with Maupertuis and to have a serious falling out with him. Buffon had foreseen it: "Those two men are not made to stay together in the same room."[49] The relationship between Buffon and Voltaire, however, did not break off, at least not officially.

Even Helvétius, Buffon's great friend who came to Montbard, saw Buffon less frequently. After his marriage in 1751, Helvétius more often stayed on his land with his wife. In 1758, when the publication of his book *De l'esprit* [On the Mind] stirred up a storm and the condemnations rained down, Buffon seems to have stayed away. Perhaps he thought, like d'Alembert and Grimm, that the book "had not been useful enough for men or for the progress of literature and philosophy to compensate for the blow it gave to freedom of thought and expression in France."[50] According to tradition,

Buffon had said that Helvétius, an old fermier général, would have been better off writing a new contract rather than a book—a remark that does not show much compassion.

In 1749 the philosophes' major occupation was the preparation of the *Encyclopédie*. Rather than Diderot—an already suspect philosopher whom the associated bookshops had much trouble getting out of Vincennes in order to save the business and the money they had invested—it was d'Alembert who served as the scientific guarantor of the work. It was also d'Alembert who, guided from afar by Voltaire, organized the party of philosophes, served as its recruiting sergeant, and assured the methodical conquest of the Academy of Sciences and especially the Académie Française.

The relationship between Buffon and d'Alembert was ambiguous, to say the least. At the time of the debate with Clairaut on the law of universal attraction, d'Alembert had been interested by some of Buffon's ideas but was not convinced by his "metaphysics."[51] When the *Natural History* appeared, he was shocked by the limits that the "First Discourse" claimed to impose on the power of mathematics, all the more so because Diderot had just supported analogous theses in his *Letters on the Blind*. In September 1749 d'Alembert wrote to Cramer, "You will find us badly treated in Buffon's new work. It is true that with a few calculations and a little more geometry he would not have risked so many notions about the formation of the earth." D'Alembert went on to compare Buffon's "system" to Descartes's *Traité du monde* [Treatise on the World], widely considered a "physics novel."[52]

Nevertheless, it was necessary to save appearances. After all, Buffon seemed to be on the correct side of the fence, he was Diderot's friend, and he was even close enough to the *Encyclopédie* enterprise to be sent proofs so that he could praise them to his correspondents.[53] The first volume of the *Encyclopédie* appeared in 1751. It opened with a long "Preliminary Discourse" signed by d'Alembert, who presented the analysis and the history of human knowledge. In discussing evidence, certitude, and probability, the author, with no reference, alluded to the definitions contained in Buffon's "First Discourse" without accepting them. For d'Alembert, probability always supposed chance and was especially applicable to history. As for certitude, it was not, as it was for Buffon, a very high degree of probability: it was derived from the evidence through the intermediary of logical deductions, seeming to indicate that the certitude of physical truths originated from mathematical evidence. This philosophy was quasi-Cartesian and was in opposition to Buffon's.[54] D'Alembert was willing to admit that algebra was sometimes abused in physics and that a "general and experimental Physics" existed alongside the "Physico-mathematical Sciences." For him, this former was strictly "only a systematic collection of experiments and observations." In conclusion he wrote, "The only true way to philosophize in Physics consists either in the application of mathematical analysis to experiments or in observation alone, enlightened by the spirit of method, sometimes helped by conjectures when they can furnish suggestions, but severely disengaged from all arbitrary hypotheses."[55] In short, the only serious "physicists" were d'Alembert and Réaumur!

Buffon did figure in the honor roll of contemporary glories; he was, however, curiously placed, right after Fontenelle, whom d'Alembert presented solely as a popularizer of great talent: "The Author of the *Natural History* followed a different route. A rival of Plato and Lucretius, he shows throughout his Work, whose reputation grows from day to day, that nobility and that elevation of style which are so proper to philosophical subjects and which in the writings of a Sage must be the picture of his soul." A singular way to compliment a scholar.

The worst had not yet been said. D'Alembert added immediately afterward: "Nevertheless, Philosophy, while hoping to please, seems not to have forgotten that it is principally made to instruct; it is for that reason that the taste for systems, more appropriate for flattering the imagination than enlightening reason, is today almost completely banished from good works. . . . A Writer among us today who writes in praise of systems has come too late. . . . The spirit of System is for Physics what Metaphysics is for Geometry. . . . Enlightened by the observation of Nature, he might glimpse the causes of phenomena: but it is up to calculation to assure, so to speak, the existence of these causes by determining exactly the effects that they can produce and by comparing these effects with those that experiments discover for us. Any hypothesis devoid of such help rarely reaches this degree of certitude, which must always be sought in the natural Sciences, and which nevertheless is so rarely found in those frivolous conjectures honored with the name of Systems."[56]

The target of this diatribe is seen clearly from the text's placement. It was Buffon. His theory of planet formation, a "frivolous conjecture" for which computation sufficed to show its inadequacies, particularly shocked d'Alembert. The text, moreover, killed two birds with one stone: it was Buffon who had introduced "Metaphysics into Geometry" in his controversy with Clairaut. And behind Buffon there was Jean Jacques Dortous de Mairan who, on November 13, 1748, at the public meeting of the reconvening of the Academy of Sciences ("among ourselves"), had praised systems by speaking of Newton, Kepler, Harvey, and Copernicus. Buffon was dangerous because he represented a contemporary reaction against the excessive devotion to the spirit of observation.[57] D'Alembert saved appearances, but deep down he was in agreement with Réaumur.

In order to keep peace with the Sorbonne, Buffon had signed a solemn declaration, which he later qualified as "foolish and absurd." He wrote a short but flattering letter about the "Preliminary Discourse" to d'Alembert:

"It is very elevated, very well written, and even better reasoned. It is the quintessence of human knowledge." After having let it be understood that the "Discourse" would fall victim to the same adversaries as the *Natural History*—same battle, therefore same enemies—he did not hesitate to add: "As for me, I am delighted and very flattered by the way you have treated me. If I did not have the honor of being one of your friends, I see by your elegy that I would be worthy of being so, and that is what touches me even more."[58] Did the two men really expect to fool each other? Be that as it may, d'Alembert's hostility would soon become more and more open, and Buffon would be less and less well armed to defend himself.

Only Diderot remained, with whom Buffon had been associated fairly intimately, it seems, for many years.[59] Diderot, imprisoned at Vincennes, trusted this relationship enough, in any case, to have the intendant of the Royal Botanical Garden figure on the list of people able to vouch for his good character. In fact, Diderot rose barely above being a literary bohemian, and these ties between a high-ranking civil servant and a suspected writer with a police file proved Buffon's great freedom of spirit and in particular a certain shared identity of thought. Buffon's public praise of the *Letter on the Blind* in the *Natural History* reinforced that impression. Diderot, we have seen, had had quarrels with Réaumur,[60] and traces of Buffon's influence can be detected in his *Letter*. In prison, he read and annotated the three volumes of the *Natural History*, with the intention of communicating his comments to the author. His notes, however, were confiscated and have disappeared. The article "Animal" in the *Encyclopédie*, where Diderot reproduced and commented on the first chapter of the *History of the Animals*, testified to this interest. In 1753 his *Pensées sur l'interprétation de la Nature* [Thoughts on the Interpretation of Nature] was very strongly inspired by the *Natural History*.[61] It is therefore possible to believe that it was at his request that Buffon promised to write the article "Nature" for the *Encyclopédie*, a promise that Diderot quickly announced in 1752, at the beginning of the second volume: "We hasten to announce that M. de Buffon has given us the article "Nature" for one of the volumes that will follow this one; an article whose importance is undeniable, since its subject is a fairly vague, often used, but poorly defined term that philosophers greatly abuse and that, in order to be raised and presented in all its different aspects, needs all the wisdom, accuracy, and elevation that appear in the subjects that M. de Buffon treats." Had Buffon actually "given" his article, or was it simply a promise? When the article "Nature" finally appeared, in volume XI, it carried d'Alembert's signature. In addition, it was Daubenton who, perhaps at Buffon's request, wrote a great number of articles on natural history for the *Encyclopédie*. This he apparently did with bad grace, if one can judge by the brevity and the dryness of his texts.

Buffon and Diderot would progressively drift apart, separated as much by their daily worries as by their personal intellectual development. They did continue, however, to see each other. In 1750 Diderot was perhaps the only philosophe who took the *Natural History* seriously, who had understood its importance for science and philosophy, and who openly supported it.

Aside from the required response given to the Sorbonne, Buffon answered no other critic. Was his silence due to self-confidence or wise politics? Perhaps both. "There have been only a few yelps from some people that I thought I should ignore," he wrote in February 1750. "I knew in advance that my work, containing new ideas, could not miss scaring the weak and outraging the proud; so, I was scarcely worried by their yappings."[62] The "Jansenist journalist" did not move him either: "I will not answer one single word."[63] He took the same attitude toward the *Letters to an American*, which disconcerted Father de Lignac. Moreover, it seems that Buffon imposed silence on those of his friends who wanted to defend him publicly. In 1752 the president de Ruffey had sent him a "document" that was supposed to be an analysis and defense of "systems" and "hypotheses." Buffon returned it to him saying, "I would find it good if I were not its subject, but I am praised in it much more than I deserve, and that is enough for me to beg you not to have it printed."[64] Ruffey yielded. At that date at least, Buffon did not display the vanity for which he has been ritually reproached.

Silence was certainly the best tactic. No criticism, in the long run, could harm the work's success. In 1753, when the worst of the storm had passed, two journalists summed up the situation. One, Fréron, was the sworn enemy of Voltaire and the philosophes; the other, Grimm, was on their side. In the former's opinion, "what is most sublime in Philosophy, the most curious in Physics, the most noble in Eloquence, and the most brilliant in Poetry are found gathered together in their [Buffon and Daubenton's] *Natural History*."[65] For the latter, the three volumes had been "received with a universal applause" and the "bad brochures that cabal and envy have forged against [this] immortal work" were no more than the price of success.[66] The fourth volume had just been published: Buffon had remained imperturbable and had continued his work. Whatever was said or done, he was from then on "the celebrated M. de Buffon."

Yet if the scholar was sure of being right, was the man so imperturbable? The fourth volume of the *Natural History* presented a psychological picture of the man where, unexpectedly, reflections on the difficulties of middle age are found. Buffon was forty-six years old when the text was published, and it is difficult not to hear an echo of his personal experience:

"It is at this age that cares emerge and that life is the most contentious; because a person has risen to a certain position, that is, he has entered by chance or by choice on a career in which it is always shameful not to do well, and often very dangerous to fulfill with brilliance. He therefore walks painfully between equally formidable pitfalls, scorn and hate; he is weakened by the efforts made to avoid them, and he becomes discouraged; because when at the limit of having lived and been recognized, and also having felt the injustice of men, he views discouragement as a necessary evil. When he has finally become accustomed to attaching less importance to the judgments of others than to his tranquility, and when his heart, hardened by the scars from blows that others have dealt it, has become more insensitive, he readily arrives at that state of indifference, at that idle peace, at which he would have blushed a few years before. Glory, that powerful motive of all great souls, once seen from afar as a shining goal that he endeavored to attain by brilliant actions and useful works, becomes no more than an object without appeal for those who have come close to it, and a vain and deceitful phantom for others who have remained at a distance from it. Laziness takes its place, and seems to offer easier paths and more solid benefits to everyone, but disgust precedes it and boredom follows it, boredom, that sad tyrant of all souls that think, against which wisdom can do less than madness."[67]

Did Buffon really feel this temptation to abandon everything, this lassitude when faced with hostility and lack of understanding, this sharp feeling of vanity and glory? Did he continue to work only in order to escape the "lethargy of boredom," as Voltaire said? Boredom terrorized this century, which has been said to be so carefree and which sought its salvation only in "entertainment." For Buffon, it was less "wisdom" that saved him than his phenomenal vital force, which would sustain him to the end. The athlete envied by Voltaire could not stop in the middle of his career. But, in spite of appearances, he was not invulnerable.

PART THREE

THE LONG GESTATION OF THE *NATURAL HISTORY*

CHAPTER XIV

The Life of an
Ordered Scholar

A Provincial Marriage

"Master Georges-Louis Leclerc, knight, lord of Buffon, Montbard, La Mairie, and other places, Intendant of the Royal Botanical Garden
in Paris, perpetual treasurer of the Royal Academy of Sciences, and [mem-
ber] of the Academies of London, of Berlin, . . . eldest son of my lord
Benjamin-François Leclerc of Buffon, honorary Adviser to the Burgundy
parlement, and of Lady Anne-Christine Marlin, the aforementioned lord of
Montbard usually residing in the city of Paris, in his mansion at the Royal
Botanical Garden, parish of Saint-Médard." Such were the titles and qualities registered by the royal notary who prepared the marriage contract. On
September 22, 1752, Buffon at the age of forty-five wed Marie-Françoise de
Saint-Belin-Malain, who was barely twenty years old. She came from a noble
family in Burgundy, so noble that the grandiose title of "Master George-
Louis Leclerc" paled before those of his future father-in-law: "High and
powerful and Lord and Master François-Henri de Saint-Belin." Yet the
powerful lord did not have a penny, and Marie-Françoise, as well as her
younger sister, was a boarder in the Ursulines convent of Montbard, whose
Mother Superior, Mother Saint-Paul, was Buffon's sister. It was there that
Buffon discovered Marie-Françoise while visiting his sister, who perhaps
had discreetly helped fate. Marie-Françoise was charming, gentle, pretty
rather than beautiful. She was stagnating in her convent, waiting for an
improbable marriage or a life as a nun without vocation. Even so, according to family legend, Buffon had to court her for two years, even causing a
scandal, before he convinced her to marry him. Perhaps for once he was
truly in love.

To leave bachelor life at the age of forty-five is not without its difficulties.

First there was Buffon's father, who opposed a marriage that upset his plans: he probably hoped that his son's fortune would go to his children from his second marriage. Then there were Buffon's friends, also bachelors, who could not fail to reproach him for his desertion; the abbé Le Blanc led this group. Buffon was so afraid of him that he did not dare tell him of his marriage: "You would have contradicted or blamed me," Buffon later wrote him, "and I wanted to avoid it, because I was decided, and whatever importance I may attach to my friends, there are things they should not be told; and among these things are those of which they disapprove and which one is determined to do."[1] Buffon perhaps was bragging when he wrote to Guéneau de Montbeillard, another old friend, but a married one, who was going to serve as his witness: "I will worry even less about criticisms of my marriage than of those of my book."

Escorted by Daubenton, who would be his second witness, Buffon thus left Paris the morning of September 19, stayed in Sens that night, left the next morning, stayed overnight in Cussy-les-Forges, and arrived on the twenty-first at the home of his future father-in-law for the signing of the contract. This was an important step as we know. The morning of the twenty-second, the religious ceremony was hastily performed in a church in the village, and the newlyweds immediately left for Montbard. Six weeks later, Buffon left for Paris, leaving his young wife to rule over her new empire.[2]

Buffon could have married in Paris, even entering into a powerful and well-established family who would have added to his wealth. He chose to marry a poor but noble and, above all, provincial girl. Despite his name, his reputation, and his comfortable income, in Paris he would have risked being bound to his wife's family and consequently to his wife. He would not have been the master in his own home. Also, in this way he avoided those young Parisian girls who so quickly became brilliant, flirtatious young women, who could not pass up a ball or the opera, and who would have refused outright to bury themselves in Montbard with their husband, and even more so to stay there alone while he returned to Paris. Provincial he was; provincial he stayed.

The gentle Marie-Françoise was just what he needed. She upheld her rank, she kept up the house, she gracefully accommodated all those Nadaults and Daubentons who haunted the "chateau," she received her husband's friends and their spouses well. Better yet, she admired her husband, and yes, she loved him. Probably not with a great romantic passion, but with a real affection mixed with gratitude for the man who had taken her out of her convent and made of her a peacefully happy woman. And Buffon? He surely loved her also in his way, with both a paternal and a sensual affection that her youth could inspire in a man of his age and temperament. Unfortunately, all this is no more than conjecture: we have none of the letters he wrote to her when he was in Paris and she in Montbard, and we possess no letters from her.

Nevertheless, Marie-Françoise did not always stay cloistered in Montbard and occasionally she came to Paris. There she maintained her place and her status perfectly as well. On May 25, 1758, she gave birth to a girl, Marie-Henriette. Buffon came back to Montbard, putting off another trip to the Royal Botanical Garden. The young woman, however, recovered slowly from childbirth; in July she was still not restored. All the same, she passed the winter in Paris. During the fall of 1759, she became seriously ill in Montbard, and the little Marie-Henriette died on October 14. She was buried in the crypt of the seignorial chapel that Buffon had just built next to the parish church. He had said to the workers who were digging the vault, "Make it immense and deep, as I will be there longer than here." That October 14, it was the little girl who was buried there.

"I lost a child who began to make herself understood, that is to say, loved," wrote Buffon, visibly grieved, he who had perhaps followed with more attention than any other the awakening of his child's intelligence. In order to distract his wife, he brought her to Paris, but he was in a dark mood, and the capital pleased him less than ever: "Everything here is expensive, everything here is sad." France was in the middle of the Seven Years War. The taxes were heavy, and the controller general of the Treasury, the famous Silhouette, had just created new ones. There was a general protest, and the state coffers were empty. Like his ancestor Louis XIV before him, the king sent his gold and silver dishes to the Treasury in order to pay the troops. The courtiers, the clergy, the nobility, and rich people felt obliged to do the same. Buffon complied in turn: "I just sent my dishes to your Treasury," he wrote to Ruffey. "It is still better to ask for money from well-off people than to overcharge the poor. You who are so honest and so good, do you not bemoan their misfortunes?"[3] For a long time in Montbard and at the Royal Botanical Garden, he and his wife were happy to eat off crockery or porcelain.

In 1764 Mme. de Buffon was pregnant again and she no longer left her house. On May 22, she gave birth to a boy, Georges-Louis-Marie, who was soon nicknamed Buffonet. He was immediately baptised and in accordance with the custom of the nobility of the time, his godfather and godmother were two poor people from the parish.[4] Buffonet grew up untroubled. When he was two years old, Guéneau de Montbeillard decided to inoculate his own son, who was seven, against smallpox.* This was a risky

---

*Ed. note: Unlike the later vaccination procedure invented by Dr. Jenner, which used the matter produced by cow pox (hence the term, from the Latin, *vacca*, for cow), inoculation used the matter from the pustules of someone recovering from smallpox. In modern terms, the virus was attenuated by the time it was used—except sometimes it wasn't and the result, then, was the full-blown disease and probable death, rather than immunity.

and very controversial operation. Guéneau, as his wife said, "had known filial love too well . . . not to carry paternal love to its highest degree." He came to Paris to learn from the "most famous inoculators" and, "so as not to alarm the public and principally in order not to harm anyone," chose "a country house isolated on all sides." There in the presence of a doctor, the loving father inoculated his son with a trembling hand. The operation succeeded. Guéneau made it the subject of a paper, which he read before the Academy of Dijon. Stating that "all the desires of caution were united in the cries of paternal love," he had wanted to give to his country "the example of a father inoculating his only son." This act won him local glory and even Diderot's congratulations for the heroic father: "The Ancients would have called you *bis pater* [twice a father]. . . . *Bis pater,* I embrace you with all my heart, and your son and his mother." It was a new sensibility that was emerging. Buffon, who had closely followed the business, decided not to follow his friend's example. Buffonet would not be inoculated and his father would be satisfied in being his father only once.[5]

True Happiness Is Tranquility

Life continued. There were always guests at Montbard, mostly intimate friends. For the first time in a long time, Buffon complained about his health: a bad cold that lasted several months, rheumatism in his right arm that prevented him from writing, tired eyes that prevented him from reading, violent "stomach colics." These bothers were all minor and not very frequent, but he no longer had his old rough vigor. Parisian visitors to Montbard were fewer. Even people from Dijon hardly came, and Buffon complained that he no longer saw either Ruffey or De Brosses. Ruffey sent him wine, acting as if he did not want to be paid, but Buffon insisted and paid. He advised Ruffey on the organization of a literary society that the president had created in Dijon in 1752. Buffon, of course, took part in the effort and helped it to gain official status. This put him in a slightly delicate position, since Ruffey intended to compete with the Academy of Dijon of which Buffon had also been a member since its creation in 1740. The Academy unexpectedly became renowned in 1750 by awarding first prize to the famous *Discours sur les sciences et les arts* [Discourse on the Sciences and the Arts] by Jean-Jacques Rousseau, but nothing indicates that Buffon had anything to do with it.[6] The two societies would merge in 1759 and everything would work out. Thanks to Ruffey again, Buffon settled many little financial affairs with the people in Dijon. To De Brosses, who in 1760 had just published in Holland his *Dissertations sur les dieux fétiches* [Dissertation on Fetish Gods], he gave cautionary advice: "The year is not good for metaphysics and philosophy. As if the persecutions of Helvétius, the *Encyclo-*

pédie, etc. were not enough, philosophers now appear in comedies, publicly acted out in the theater." Indeed, Palissot had just opened his comedy *Les Philosophes,* in which Voltaire, Rousseau, and others were vulgarly caricatured.

Buffon had no sympathy for "Palissot and the other little idiots of that clique." Not that he was delighted with the philosophes. He found that "since Voltaire has been living in Burgundy [read: at Ferney] he had become furiously prattling." Poor De Brosses knew about this (we will come to his experience with Voltaire later). Buffon also found that there was "a lot of constant harping" in *La Nouvelle Héloïse.* He was, however, well placed to know the high price of tranquility when one starts to write: "The basis of your ideas seems just and true to me," he said in a letter again to De Brosses. "It only seems to me that you were sometimes embarrassed to bring them completely to light, for the same reasons that embarrass us all when we want to speak the truth"—the fear of the censor. It was true that "often . . . things pass in a quarto volume that could not be said in a duodecimo one." But such subterfuges were despicable.[7]

Buffon was not a revolutionary. He was not one of those intellectuals who was ready to reform the state with a stroke of his pen. Yet faced with the incoherence of French politics, the parties confronting the Court, the ministers who followed one after the other, he expressed his worry for the environment in which he lived and which was truly his own: the provincial bourgeoisie with its direct contact with everyday realities. If the government created new taxes, he became indignant: "It will be necessary to put half of the provinces in prison and ruin all the poor people."[8] If, however, the clergy refused to declare its wealth and submit itself to taxation, if the Burgundian parlement was in opposition to royal authority, he also became indignant: "It is something quite remarkable that people get it in their heads that in acquiring a trust of twenty or thirty thousand livres, they are at the same time acquiring the quality of the tutors to kings."[9] Buffon saw the contradictions that would lead the regime to its downfall. But what to do? The temptation was great to withdraw into oneself, one's work, one's family. "True happiness is tranquility; the first way to procure it is to give it to others, and to let, as the monks say, *mundum ire quomodo vadit.* Instead, under the pretext and even with the purpose of doing good, a thousand times more bustle is necessarily made than would be normally; and it is this bustle that upsets and ruins everything."[10] Was Buffon cynical like Rameau's Nephew,* who cited this same maxim and angered Diderot so much? The Nephew added that the world "is going well, because the masses are satisfied with it." Buffon did not believe that the masses were satis-

---

*Trans. note: *Le Neveu de Rameau* [Rameau's Nephew], a work by Diderot, was presented as a dialogue between Diderot and the nephew of the composer Rameau.

fied with it. What guided him more was a cautious skepticism, like Montaigne's, and the fear of "novelties."

In addition, he needed tranquility to work. At Montbard, as usual, Buffon worked, primarily on the *Natural History*. In February 1750, still drunk with the success of the first volumes, he announced the fourth volume for that July and the entire history of the quadrupeds in two volumes for May of 1751.[11] That was the initial plan, which showed either a complete lack of realism or an ignorance of the subject. The disenchantment came very quickly. Volume IV did not appear until 1753 and, following two "Discourses" on the nature of animals and on domesticated animals, it contained the descriptions of only three species: the horse, the ass, and cattle. Which is not much. In fact, the quadrupeds would take up twelve volumes, the last of which only came out in 1767, and there would be nine volumes on birds.

It takes time to describe a species and before it can be described, to observe it. There was no menagerie at the Royal Botanical Garden, where Daubenton could only collect stuffed animals, dissect cadavers, or measure skeletons. It was for the Cabinet of the Garden and with the State's money that, in 1747, Buffon bought a monkey from Angola, for which he paid 1,200 livres.[12] In order to observe the behavior of living animals, Buffon installed a menagerie and an aviary at Montbard. It was there he kept the quadrupeds or birds given to him or which he was able to buy alive at his own expense. He left his boarders free as much as possible, and that gave the "chateau" and its surroundings a somewhat peculiar kind of animation. Since the foxes smelled bad, they were relegated to the stables and the cowshed. The problem was that after the age of six months, they started to chase the ducks and chickens. They therefore had to be chained up, which made them sad and even took away their desire to eat the chickens. On the other hand, the young wolves could be kept close to the house. As long as they were young, everything was fine: they were "docile enough" and "even affectionate." That state did not last: "At eighteen months or two years old, they reverted to their true nature," and without warning. One, "on his first attempt," killed all the chickens "in one night without eating any of them." Another fled after having killed a dog. The small animals wandered freely in the yard and even in the house. The badgers came to warm themselves by the fire and burned their paws. The stone marten returned to beg for food, just like cats and dogs. One day, "a hedgehog that slipped into the kitchen discovered a small pot, pulled the meat out of it, and did his business in it." One had to believe that that animal was being "mischievous, and in the same way as the monkey." As for a monkey, he had one, Jocko, who created a devil of a mess everywhere including the living room. The noble "chateau" must sometimes have smelled musky, and the servants probably cursed their master's extravagances.[13] While he, unruffled, observed his boarders or dictated a new text to his secretary, Mme. de Buffon took care of the birds.

Buffon was not only busy with the *Natural History:* in 1752 he experimented with electricity and lightning, and had a lightning rod installed, more for experiment's sake than protection against lightning.[14] He was already interested in the problem, but this rekindling of interest was not without its ulterior motives. At that time, the leader in the study of electricity in France was the abbé Nollet. A student of Du Fay's, Nollet had entered the Academy in 1739 under Du Fay's protection. After Du Fay's death, he became Réaumur's protégé. In 1746 he published an *Essai sur l'éléctricité des corps* [Essay on the Electricity of Bodies]. A great master of experimental physics, he shared Réaumur's antipathy for Buffon.[15]

Since 1746 and the invention of "the Leyden jar," electricity had become the latest fad. Acting like a condenser, the "bottle" produced electrical discharges much more powerful than any obtained before. The first ones to receive the shock believed they would die from it. Even better, the shock was transmitted if two people held hands; two or even more. Le Monier, Buffon's friend, thus electrified one hundred forty gentlemen in the presence of the king. Cut to the quick, Nollet made one hundred eighty French guardsmen leap into the air, again before the king, then more than two hundred Cistercian monks in their monastery in Paris. These collective leaps fascinated high society. Never had physics been so spectacular, never had it been talked about so much, never had Nollet been so famous.

The problem was that Nollet had formulated a theory to explain electrical phenomena, and the Leyden jar contradicted this theory and also all the others formulated up until then. It so happened that on the other side of the Atlantic in Philadelphia there was an author-printer-editor who was interested in the sciences and had leisure time: Benjamin Franklin. Franklin had an agent in London, Peter Collinson, a member of the Royal Society. Collinson came across an article in the *Gentleman's Magazine* about the marvels of the famous jar and sent it to Franklin with the material needed to reproduce the experiment.[16] Franklin set to work with his friends and invented a theory of his own, which he communicated to Collinson in 1750. Collinson published it in 1751 with the title *Experiments and Observations on Electricity Made at Philadelphia in America by Mr. Benjamin Franklin.* The book was an immediate success and was republished many times.[17]

Whether it was from Collinson himself, with whom he had stayed and would stay in contact, or by another member of the Royal Society, to which he himself belonged, Buffon received the book and had it translated immediately into French by Dalibard. Buffon was probably only too happy to do a bad turn to Nollet, whose thesis Franklin demolished. For good measure, Dalibard headed his translation with an "Abridged History of Electricity," so abridged that Nollet's name did not even figure in it. While Nollet wondered if Franklin really existed or if Buffon had invented him, Buffon and Dalibard were preparing a glorious feat.

Lightning had for a long time been compared to an electrical spark, but Franklin confirmed that a pointed metal strip could attract electricity from a storm cloud. All that remained was to attempt the experiment. It was done before the king in February 1752. Dalibard reproduced it in Marly on May 10, and Buffon at Montbard on the nineteenth. On June 22 Franklin did his famous kite experiment, of which Buffon learned only later. In July, Buffon was pursuing his experiments; he explained to Ruffey how to set about them and concluded with unconcealed satisfaction: "The abbé Nollet is dying of grief with all this."[18] Nevertheless, Buffon did not want only to upset the abbé Nollet. He wondered "if electricity might not be the chemists' phlogiston" and proposed an experiment on this question. This preoccupation with uniting physics and chemistry by identifying electricity with the "principle of combustion" of pre-Lavoisier chemistry already concerned Buffon, as well as Diderot.

At the Académie Française

Buffon liked to stay at Montbard and work there in peace. The majestic pendular rhythm of earlier times, however, seemed to have accelerated. Required to be at Montbard for certain family events, Buffon was also required to go to Paris, at least for short trips, even in the middle of summer.

This was the case on August 25, 1753. That day Buffon was inducted into the Académie Française. His election was made under unusual circumstances. His name as well as d'Alembert's had been advanced. According to certain reports, the two scholars withdrew before the candidature of Piron, who was supported by the king and Mme. de Pompadour. From Burgundy like Buffon, this elderly poet (born in 1689) had acquired a reputation as a playwright but especially as a light and pleasingly satirical poet. At the last moment, however, his candidature was torpedoed by one of his numerous enemies: a licentious poem he had written in his youth was brought to the king, and Louis XV made it known that he would refuse his election. According to d'Argenson, it was only at that moment that Buffon and d'Alembert discreetly withdrew from the competition "in order not to incur in their turn a libelous note of that sort, the former having contradicted Genesis."[19] The Académie was astir. Ten days later, Buffon was elected without having to make the traditional visits, of which he was particularly proud.[20] As for Piron, he received a pension of one thousand livres from the king's purse at Mme. de Pompadour's request, and consoled himself by writing his famous epitaph:

Here lies Piron who was nothing,
Not even an academician.

Fate willed that Buffon succeed Languet de Gergy, the archbishop of Sens, whom Buffon did not regard highly even though he was from Dijon. When the archbishop had drily been asked by the king to return to his diocese, Buffon had noted it with satisfaction. A little later he wrote, "The archbishop of Sens and the bishop of Auxerre treated each other abusively in their pastorals."[21] One condemned the Jansenists, the other defended them. As for Languet de Gergy's literary work, it was essentially a biography of a nun from Paray-le-Monial, Marguerite-Marie Alacoque (1647–1690), who began the devotion to the Sacré-Cœur and who would be canonized in 1864. The Jansenists refused to believe the revelations that she claimed to have received, and they condemned the new devotion. Praising the archbishop was a difficult task, and Buffon was perplexed: "I do not really know yet what I will say to them," he wrote in early July. "But perhaps a few inspirations will come to me, as to Marie Alacoque, but I will not speak of her for fear of jumping from one subject to another."[22]

Only just elected, Buffon left for Montbard in order to think about it more calmly. We know how he handled matters by giving his famous *Discours sur le style* [Discourse on Style]. Languet de Gergy was not even mentioned, and Buffon restricted himself to evoking in his last sentence "religion in tears which . . . seems to accuse me of delaying your regrets too long for a loss that we must all feel." The allusion was so obscure that it had to be explained with a footnote when the discourse was published. The bishops of the Académie were unhappy, and the public was enchanted. Grimm congratulated Buffon: "That famous man, disdaining the pale and heavy praise that is ordinarily the subject of this kind of discourse, judged it pertinent to treat a subject worthy of his pen and worthy of the Académie. These are ideas about style; and it has been said, on this topic, that the Académie has taken in a master of writing. It could be added, after having read M. de Moncrif's answer, that [the Académie] did well and needed such ideas. M. de Buffon's discourse . . . was interrupted during the Académie's session three or four times by applause from the public."[23] Ruffey publicly approved Buffon's "caution" on the subject of the archbishop: "Nothing other than mediocrity in style ever came out of his pen. . . . M. de Buffon could not praise his academic qualities without exposing himself to mockery and compromising his taste and his judgment."[24]

To be a member of the Académie Française did not bring only honors. There were also obligations. In 1760 Buffon was the director of the Académie: in that role he had to greet the newly elected members and praise them. This was easy in the case of La Condamine, received July 21, 1761. He was a scholar, an old friend, and his famous trip to Peru lent itself to eloquence. La Condamine had returned by descending the Amazon in a dugout canoe, and Buffon described his being "swept down the steep falls of those foaming torrents," then adding, "Nature, accustomed to the deepest silence, must have been astonished to hear herself questioned for the first time." There was a moment of silence, and the audience broke out in applause.

Buffon must have loved that sort of success, and Diderot admired him for it: "I received a visit from M. de Buffon this morning. I will go to spend a few hours with him one of these evenings. I like men who are very confident of their talents. He is the head of the Académie Française and, in that capacity, has the responsibility for three or four reception speeches, a cruel chore. What to say of a M. de Limoges? What to say of a M. Watelet? What to say of the dead and the living? Moreover, he is not allowed to offend them by scorn; he will have to praise them, and he said: 'So be it! I will praise them, I will praise them well, and I will be applauded. Does an eloquent man find any subject sterile? Is there anything about which he does not know how to speak?' It is with objectivity that I praise that assurance, for I do not have it."[25] Buffon perhaps showed more confidence than he had. On the subject of these same speeches, he wrote to De Brosses: "I sense that I will be obliged to take out the few good things that there are to say."[26] Was this not one more difficulty to overcome?

Once at the Académie, he also had to get his friends in. Buffon had two who were very dear to him. The first was the abbé Le Blanc, always a candidate and always rejected. In 1761 he almost succeeded, and Buffon did his best, but in vain. It was the abbé Trublet, another perpetual candidate, who was elected. "It is the time of the reign of the mediocre," Buffon wrote, expressing himself as "disgusted . . . with the Académie." Yet he needed to console the abbé and let him know that the guilty parties were the enemies of the philosophes, especially of Duclos* and Voltaire. For the abbé was not on bad terms with Voltaire, and Buffon congratulated him for it: "I am very happy that you are in touch with Voltaire," he wrote him in 1755. "He is indeed a very great man, and also a very kind man";[27] Le Blanc was talkative and it was necessary to be careful. Buffon's other efforts for his friends were not more successful. In reality, he was not very good at this kind of maneuver. First of all, he was absent from Paris too often; in the second place he was very busy; and finally he probably did not have the genius or the patience to flatter the "mediocre" whose votes he needed to win. He would always be beaten at this game, either by the philosophes' enemies, or on the other hand by d'Alembert, elected in 1754 under conditions that set tongues wagging but who obtained the entry of nine

*Ed. note: Charles Pinot Duclos (1704–1772), a minor author.

"philosophes," mediocre or not, into the Académie between 1760 and 1770.[28]

For the president De Brosses, it was worse. An old friend of Buffon, De Brosses had the misfortune of being at the same time a literary man, who hoped to make his career in Paris among the philosophes, and a great magistrate in Dijon, an influential member of the Burgundy parlement. In 1758 Voltaire burst into his life. He had fallen out with Frederick II, from whom he finally managed, not without difficulty, to escape. The poet was at that time living at the "Délices," a property just outside Geneva. Not being a Calvinist, he had to buy the house through the intermediary of a Genevan representative, and was looking to settle down more definitively. De Brosses possessed the domain of Tournay, in the area of Gex, a few kilometers north of Geneva. There was a fortified castle dating from the fifteenth century with the lands and the seignorial rights. In September of 1758, Voltaire proposed buying this land from De Brosses for his lifetime. This was the beginning of the president's misfortunes.

"Buy for his lifetime" meant that at Voltaire's death the land would revert to De Brosses. Voltaire thus obligingly engaged himself to die within two years and in that time spend twenty-five thousand livres for restoration of the castle and the land. In return for this he proposed such a complicated financial arrangement that it was not very easy to figure out what he would have to pay in cash. De Brosses was not enthusiastic, but how to say no to a famous man who, through the intermediary of his Parisian friends, could make or break literary reputations and elections to the Académie Française? Overwhelmed with more and more pressing messages from Voltaire, the president finished by giving in, not without reminding him that among the conditions of sale, Voltaire would not have the right to cut the trees on the property. Voltaire, who was in the process of buying Ferney in the name of his niece, Mme. Denis, pretended he had not understood, and started to cut down the trees. Nevertheless, De Brosses went, signed the contract, spent two weeks with Voltaire, and came back exhausted. Voltaire was, he said, "a man so flighty, who has neither order nor logic nor pause in his thoughts. I would rather fence with a flea."[29] At least now he hoped to have some peace.

Wrong. Voltaire did not stop bombarding him with letters and recriminations. First of all, as a French subject, he was required to pay rights that a Genevan would not have to pay. A monstrous injustice! De Brosses had to do something. Then other incidents arose, which all became catastrophes under Voltaire's pen. Each time, De Brosses was summoned to intervene. Suddenly Voltaire got it into his head to buy the land of Tournay outright. The financial arrangement that he proposed was so dizzying that the president understood nothing of it: "Your genius moves so rapidly that I lose myself trying to follow you," he wrote to the poet.[30] Then Voltaire had a grandiose plan: to buy back from the *ferme général* the rights of the salt tax for the area of Gex. A philosophical humanitarian plan if there ever was one: the unhappy peasant would stop groaning under the weight of the tax. If the president was a philosopher, as he claimed to be, he must make the plan succeed. This plan was also very profitable, as Voltaire explained at the same time to the Genevan capitalists with whom he wanted to create his company: the salt would be sold at an excellent profit. While De Brosses was wondering if he should sell or not, and started to feel at a loss, Voltaire changed his mind: he no longer wanted to buy Tournay, Ferney would be enough for him.

For all that, he did not stop writing to De Brosses: at Tournay, other affairs demanded the president's immediate intervention. "I can assure you that the divine poet does not have a single two-sous business deal that does not cost me at least twenty livres," moaned De Brosses in 1760. "He always ends by compensating himself for various other affairs with my woods, which he is cutting like turnips."[31] Exasperated, he had an inventory of the state of the forest taken, "such as it was transferred to you upon entering into possession."[32] A crime of lèse-poesy! Voltaire sent a bittersweet letter. Then he had wood delivered to him by a farmer and refused to pay for it, under the pretext that the farmer had bought the wood from De Brosses, that the latter had sold everything to Voltaire, and therefore the wood already belonged to him, Voltaire. The president discreetly advised the poet not to be ridiculous. The poet did not have a sense of humor. On October 21, 1761, he sent a furious letter in which he accused De Brosses of dishonesty.[33] In addition he denounced the president to the Burgundian parlement, to the Academy of Inscriptions, of which De Brosses was a member, and even to the Keeper of the Seals.[34] In short, he did everything to dishonor him.

De Brosses passed on the letters he had received from Voltaire to his friend Loppin de Gemeaux, with this disenchanted comment: "You will find, in the letters of this illustrious phenomenon of our century, sometimes witticisms, sometimes harping, and always an excessive self-interest and stinginess, along with a continuous bunch of lies advanced with extreme brazenness."[35] Buffon had taken the president's side, first because they were friends, and also because owners of forests do not like it when someone cuts their trees. "As I do not read any of Voltaire's foolish remarks," he wrote to De Brosses in 1768, "I knew only through my friend the ill he wanted to say of me. . . . He will be just as angry with you as soon as he sees you in the Académie."

De Brosses, however, would never be in the Académie.[36] Despite Buffon's efforts, Voltaire knew how to prevent his entry. More than ten years after the business of the Tournay trees, Voltaire again wrote to d'Alembert and the duke de Richelieu, wantonly slandering De Brosses in order to prevent his election. "You would make me die of grief earlier than necessary if you protect that man."[37] Whatever he thought of De Brosses, the "divine poet" had a logic in his ideas, especially when it concerned one of his grudges. More magnanimous, De Brosses would continue to admire Voltaire, at least the writer if not the man: "If I am not enthusiastic about Voltaire when he does, says, or writes weak and silly things, which happens at least seven times a day, I do not think less . . . that he will pass forever into posterity, and in the first class of noble minds. . . . I admire what he has produced as a mature man, just as posterity will admire him, not caring at all if the author had a vile soul and a contemptible character when the works are excellent."[38] The wiser of the two was not the "philosopher" poet.

The Enrichment of the Cabinet of the King

In Paris, Buffon had worries other than the Académie Française. As treasurer of the Academy of Sciences and intendant of the Royal Botanical Garden, he had to go to Versailles to confer with the ministers, or at least with the bureaus. Did he often go to the Court? It seems not, even though he was well received there. Mme. de Pompadour held him in great esteem but, according to popular myth, was upset with him for having written that in love "only the physical is good." "You are a nice boy," she said to him, a compliment difficult to interpret. Buffon had at least one thing in his favor: while going to Versailles in early 1750 he had an accident; the carriage overturned and he was half smothered by the weight of his two traveling companions.[39] He claimed that he was going to pay his respects to none other than the marquise, and the accident was counted as a mark of his respect. In 1764, a few months before her death, the marquise sent Buffon her pug dog, parrot, and sapajou,* who finished their days at Montbard.[40] As for the relationship between Buffon and the king, very little is known, but it seems to have been good: since Montbard's forests were famous for their roebucks, Louis XV asked Buffon to send him one. Buffon could send only half of one, and the king thanked him by sending half a pâté.

At Versailles, Buffon was particularly interested in the bureaus and the minister on whom he depended, that of the King's Household. Up until 1749, this office was held by Maurepas, who had always protected him. From 1749 to 1775, this office passed to Phélypeaux, count of Saint-Florentin and an honorary member of the Academy of Sciences. Apparently, Buffon had no complaints with him, for when the minister lost a hand in a hunting accident in 1765, Buffon declared himself as "very upset by the

*Ed. note: A tiny American monkey of the genus *Cebus*.

accident" and added, "He was, between us, the only one of our ministers in whom I consistently saw the desire for conservation."[41] Buffon had not perhaps looked very carefully, because Saint-Florentin had a reputation for signing *lettres de cachets* arbitrarily.

In the bureaus, Buffon knew how to conduct himself. Condoret noted this with perspicacity: "M. de Buffon always took care to acquire and keep the confidence of the Ministers and of those who, put in charge of the details by them, had an unavoidable influence on the decision and dispatching of affairs. He won some of them over by never permitting himself to advance opinions that could hurt them, by not seeming to pretend to judge them; he made sure of others by using a tone of equality with them which flattered them, and shedding the superiority that his glory and talents could give him. In such a way, none of the means of contributing to science's progress to which he was devoted was neglected."[42] We understand why in 1749 Buffon profusely thanked the minister who had had the goodness to send a crayfish claw to the Cabinet of the King; it was special, being completely covered with bristles instead of only underneath like the others. Something "singular enough for us to keep it with care in the Cabinet of the King."[43]

Even though Buffon was well regarded by the minister, he was nonetheless subjected to rigorous financial control. The letters he received from the ministry have been published. All the decisions for the Academy's finances were made at Versailles, and Buffon alone remained responsible for their execution, at least until 1758, when Tillet, an agronomist, entered the Academy and served as his assistant. Tillet, who became the inspector of the Mint in 1764, had to learn all about finances. Until then, if Buffon made an error, it was pointed out to him.[44] The instructions were particularly strict concerning the scientific instruments that belonged to the Academy, which the academicians were required to return after their use: Buffon was responsible for overseeing this practice from 1745 on. To do so, these instruments were deposited at the Royal Botanical Garden "so that you can more easily attend to putting and keeping them in good condition." First of all it was necessary "in order to prevent dispersal and deterioration, to make a new inventory of these things, and have you recover all those not currently in use that are in the hands of academicians." For the rest, they were asked to provide annual receipts.[45] Trust was not the rule, all the more so because upon the death of an academician "the heirs claimed ownership of the instruments found in their homes, even though they had been furnished at the expense of the king or the Academy."[46] The king, however, had presented as a gift to La Condamine a quadrant "that had been furnished upon his trip to Peru." Buffon could therefore nullify for La Condamine "the agreement that he made before his departure of bringing back the instrument or a sum of three thousand livres."[47] As for purchases, Buffon was reminded that the Academy could not make them for amounts greater than two hundred fifty livres without prior authorization.[48] The minister attended to the slightest details: when an individual who received a pension from the Academy neglected to settle a debt of one hundred livres, Buffon was called upon to retain that sum from the individual's pension and pay it to the creditor.[49] In all that, the treasurer of the Academy was only carrying out his orders. Because of a lack of documents, we do not know what influence Buffon could have had on the minister's decisions.

At the Royal Botanical Garden, Buffon had more power. He still could not hire a gardener without the minister's authorization, and he was courteously brought back into line if someone was not named to fill a vacant position quickly enough.[50] When a porter from the Garden, fired by Buffon, wrote to the minister to complain, the minister asked Buffon to "write down the reasons that had led him to remove the man from his position."[51] The intendant was not all-powerful.

The big responsibility was the enrichment of the Cabinet of the King. The minister did not show his interest only by sending crayfish claws. If an individual or a foreign prince made a gift of remarkable objects, the intendant and the minister consulted about what acknowledgment to send. Very often it was necessary to make purchases, but Buffon could do nothing without authorization. When he recommended the purchase of collections gathered in Senegal by Adanson, the minister did not say no, but France was involved in the Seven Years War. "The current circumstances not being favorable to propose acquisitions to His Majesty," Saint-Florentin suggested selling duplicate objects in the Cabinet, agreeing upon a price and in this way only asking the agreement of the comptroller général of the Treasury.[52] In the end, the acquisition was made in return for a life annuity, which burdened the Treasury with less debt for the time. In the same way, a pension paid for the "collection of agate minerals from the late king of Poland."[53] When a purchase was to be paid in cash, it had to be paid quickly. If Buffon was late in paying for "a sperm whale's jaw," the minister called him back into line.[54] And when the monks of Pont-à-Mousson offered to give the Cabinet of the King a fossil ammonite shell "that they are conserving preciously," Saint-Florentin asked if the Cabinet needed it and what could be the object's value, "though it does not seem as if the Monks are requiring any price," but they would have to be thanked.[55] Apparently, Buffon did nothing without the minister, but the minister did nothing without Buffon.

On October 17, 1757, Réaumur was suddenly taken ill during a horseback ride in the country, and died. He was seventy-four years old. After having read volume IV of the *Natural History*, where he was personally attacked, though he was not named, he felt hatred toward "M. de Buffon and his entire clique."[56] In his will, he left his collections of natural history and his papers to the Academy of Sciences, to which he especially recommended his friend, the abbé Nollet. This was the beginning of a somber business.[57] In order to take care of this inheritance, the Academy first proposed Nollet and Buffon as curators: strange bedfellows! Then it added Guettard, a Jansenist who hardly liked Buffon, and also Bernard de Jussieu, who would have perhaps wanted to stay neutral but who nonetheless was attached to the Royal Botanical Garden. Then the king, or rather Saint-Florentin, named Clairaut, Buffon, and Nollet, which did not help anything. Finally on January 2, 1758, a royal order entrusted to Buffon, who was perhaps secretly pushing himself forward, the responsibility of taking care of the entire legacy. The minister took advantage of this order to reorganize all the collections. Everything concerning the arts, the technical arts, that is, had to be displayed to the academicians who had already been working for quite some time on a *Histoire et description des arts* [History and Description of the Arts]. There were in particular numerous engraved plates, ordered by the Academy and paid for by the king, which had been greatly plundered, it was said, by Diderot and d'Alembert for the *Encyclopédie*.[58] The collection of machines, deposited at the Royal Botanical Garden, was to be brought back to the Louvre, the seat of the Academy.

Everything relating to natural history was to be transported to the Cabinet of the King. There were many minerals "which had been collected at the king's expense," as the ordinance stated! Also, there were many birds "prepared and dried, the majority of which are in a decrepit state." It was, so it was said, the arrival of that collection which determined Buffon to write the natural history of birds immediately after that of the quadrupeds.[59] And then, of course, there were many insects, "prepared and dried," the remains of all kinds of animals, and even "many weapons and clothes of savages." Finally there were the papers—"several manuscripts on insects and other subjects of natural history"—and the correspondence.

That this correspondence should fall into Buffon's hands must have worried those scholars who had not missed any opportunity of telling Réaumur how little they esteemed the *Natural History*. The one who worried most was Charles Bonnet. He soon learned that Nollet had been given the responsibility of examining the papers and that according to Nollet himself, Réaumur's "entire collection of letters [from] correspondents" had been misappropriated. Brisson, Réaumur's former assistant, even affirmed that copies were being sold in Paris. Duhamel du Monceau, all too happy to harm Buffon, confirmed Bonnet's fears. As soon as Buffon had had the manuscripts in hand, "everything had been plundered." And Bonnet was in turn very happy to confide to Haller: "If Buffon had taken the trouble to read those letters, he would have seen things with which he would not have been satisfied."[60] Buffon, forewarned by Lalande of Bonnet's suspicions, defended himself: Before he had intervened by order of the king, and the theft had been committed, the seals had been broken by those close to the deceased. This could very well mean Brisson. Suddenly Bonnet found much merit in Buffon: "I owe him much gratitude for having prevented by his care the loss of a cabinet which had cost so much and which was the epitome of nature. This great writer has an admirable sense of color. It is too bad that the drawing is not always correct." Now it was Duhamel who was furious: Bonnet had warned Lalande of his accusations against Buffon, and Lalande had repeated them to Buffon. Bonnet justified himself: had not Duhamel publicly accused Buffon of being responsible for the loss of these manuscripts? Finally everyone was reconciled, at least officially, and Bonnet wrote to Lalande: "M. de Buffon could not have a more excellent apologist than I, and I assure you truthfully that I am very pleased to know him innocent. . . . I would like him to be pleased enough to defend Brisson." In fact, most of Bonnet's letters to Réaumur are today found in the archives of the Academy of Sciences. We just do not know the route they took to get there.

Space would be needed for all these new collections. The improvements of 1745 were not enough. Specimens arrived from everywhere in boxes and barrels. "M. d'Aubenton must inform you of what he will find appropriate for the Royal Botanical Garden in a great number of boxes that have been sent from Santo Domingo and delivered by M. le Marquis de Marigny," Saint-Florentin wrote to Buffon in 1763. He also told of his intention to visit the Cabinet of the King himself. "I can well imagine that you are too cramped by the location, but circumstances do not yet permit a request for money for the enlargement of the Cabinet. It is necessary to wait until M. the Comptroller-Général has a bit more affluence."[61] The eternal promises of ministers. For once, as we shall see, these would end up by coming true.

Saint-Florentin appreciated Buffon's work. In 1769, after having given his permission for the purchase of a collection of minerals which had belonged to the king of Poland, he answered Buffon's thanks: "I am very happy, Sir, to have been able to contribute to increasing the collections of the Royal Botanical Garden's Cabinet of Natural History, which, thanks to your care, has become the finest in Europe."[62] All ministers did not show this courtesy to the state's servants. It was true that Buffon had difficulty in obtaining a new pension, or what we would call today a "promotion" or simply a raise. In 1759 he was able to obtain for himself and Daubenton only an exceptional "gratification": 3,000 livres for himself, 1,500 for Daubenton. The "fixed pension" came later.[63] It was also true that Buffon must not have been satisfied just with putting his work and his fame into the Garden's service: he must have made large cash advances to it, and have done so for some time. In 1762 he received a sum of 26,863 livres, 2 sous, 10 deniers from M. Salvan, first assistant to the comptroller général of the

Treasury, as a reimbursement for money spent for the Garden and which he could only have obtained, it seems, by borrowing.[64] This system hardly corresponds to modern methods of public accounting, but they were normal at the time: Buffon resorted to it more and more often.

At the Peak of Glory

What did Buffon do in Paris when he was not working? First of all, in 1760 or 1761, he had his portrait and that of his wife painted by Drouais. This famous portrait of Buffon has been copied and engraved time and time again.[65] The naturalist is shown here in all his glory, at the age of 53. In his rich embroidered clothes, he breathes dignity, opulence, self-confidence, and a certain good-heartedness all at the same time. To see him, it is understandable that his contemporaries had spoken of the "imposing" air of the naturalist, and it is easy to forget that this athlete stood barely five feet five.[66] Diderot greatly admired this portrait, "where the nobility and the vigor of the truly picturesque head of this philosopher can be seen." Drouais here had depicted "the male traits of genius." All traces of Mme. de Buffon's portrait have been lost for some time. It must be, like that of her husband, in a private collection. It is known that she was very pretty and richly clothed in it, "her face young and fresh . . . her figure elegant and svelte."[67] This description does not quite correspond to what Diderot wrote of her to his friend Sophie Volland in 1760: "M. and Mme. de Buffon have arrived. I saw madame. She no longer has any neck; her chin has made half the trip, guess what has made the other half? In return for this, her three chins rest on two quite fat pillows."[68] Either Drouais knew how to please his clients or Diderot was being a bit malicious.

Having been immortalized in painting, Buffon began to figure in poetry. Weary of the traditional mythology, little tempted by epigrams and satire, the young poets were seeking modern subjects for a new heroic poetry. Voltaire in France and Pope in England had already sung the praises of Newton, the genius who had discovered the order of the heavens. In 1760 a young poet, Ponce-Denis Ecouchard-Lebrun, wrote an *Ode à M. de Buffon sur ses détracteurs* [Ode to M. de Buffon on His Detractors]. The style was noble and still "in the Antique style":

> Buffon, let it rumble, Envy;
> It's the tribute of its terror;
> What can its dark low fury
> Do against your life's glamor?
> Olympus, which a storm did beleaguer,
> Disdained the impotent anger

> Of the north wind's turbulence;
> Whilst the Tempest an uproar led
> At his feet, his noble head
> Enjoyed a majestic tranquility.

There were fifteen stanzas in this vein. Buffon, who was compared successively to Hercules and Jason the Argonaut, saw that he was promised an uncontested glory but, like Montesquieu, only after his death.[69] We do not know if this posthumous revenge seemed a bit late to him, but he thanked the poet by having the good taste to praise a stanza in which he was not even directly mentioned.[70] Ecouchard-Lebrun was a forerunner; there would be many imitators.

Did M. and Mme. de Buffon frequent the Parisian world? We can imagine so, but, in truth, we really do not know. Buffon's letters spoke of many of his Parisian acquaintances: they were almost all intellectual or professional. They were colleagues or fellow members of the two Academies, or people in high places who might look favorably upon Buffon's projects or discuss ideas with him, even if he happened to maintain more personal friendly relationships with them. He frequented *fermiers généraux* like La Popelinière or Boulongne, great magistrates or high-ranking civil servants like Trudaine, the father of the famous minister; Marc-René de Voyer d'Argenson, the director of the Stables and a member of an illustrious family; sophisticated people linked to the scientific world, like the Marquis de L'Hôpital, the son of the mathematician; or Dupré de Saint-Maur, Montesquieu's friend, an economist and statistician, who had communicated to Buffon the tables of mortality mentioned earlier. There were also intellectuals, like Duclos, the king's historiographer, and the grammarian d'Olivet. The names that stand out by their absence are those of the philosophes and the women who entertained in their salons, Mme. du Deffand, Mme. Geoffrin, Mlle. de Lespinasse.[71]

As for the last one, the abbé Morellet told a famous anecdote, which paints a rather unexpected image of Buffon: "Mlle. de Lespinasse . . . loving intellectual men with passion and neglecting nothing in order to meet them and pull them into her company, had deeply desired to see M. de Buffon. Mme. Geoffrin, putting herself in charge of procuring this joy for her, had bound Buffon to come spend the evening at her home. With that, Mlle. de Lespinasse was in heaven, promising herself to observe this famous man, and to lose nothing that came out of his mouth.

"The conversation having started, on Mlle. de Lespinasse's part, with flattering and fine compliments as she knew how to do, they came to speak of the art of writing, and someone remarked with praise how M. de Buffon had known how to unite clarity with an elevation of style, a difficult and rare reunion. 'Oh, the devil!' said M. de Buffon, head high, eyes half closed, and with a half-simple, half-inspired air; 'Oh, the devil! When it is a question of clarifying one's style, that is another kettle of fish!'

"At this exclamation, at this uncouth comparison, Mlle. de Lespinasse was quite perturbed: her facial expression altered, she fell back in her armchair, repeating between her teeth: 'Another kettle of fish! Clarify one's style!' She would not recover from it the entire evening.

"It was, however, necessary to allow Buffon these trivial and popular forms, with which he seeded conversations, especially in the beginning. He made up for it once he was allowed to expound on the subjects of his work, about which he liked to speak, as if he were testing public opinion ahead of time. I have heard him thus present two of the most brilliant developments of his work: one on the power of man over nature; the other on the painting of uncultivated nature; and in truth, it was equally beautiful as his book, aside from a few trivial expressions that he employed not only without scruples but with a sort of satisfaction, and which, although not in keeping with the rest and not being in the style of the thing, nonetheless served to make it understood."[72]

To this Buffon who spoke like a book, except when he used "trivial and popular forms," and who seemed "half-simple, half-inspired," let us add, still according to Morellet, a Buffon who spoke "with his Burgundian accent."[73] This hardly corresponds with the traditional image that we have of the great man. Who would dare to read aloud the *Discours sur le style* with an accent from Burgundy? Is it true that d'Alembert with "his little shrill voice," as Voltaire said,[74] was not better off. And perhaps people were less sensitive to provincial accents then than they are today. In particular, the philosophical salons of the second half of the century no longer had the brilliant lightness of the salons before, during the good old days of Mme. de Lambert, Fontenelle, Marivaux, and Montesquieu. With Diderot, d'Holbach, and their colleagues, it was necessary to tolerate "inspired" monologues, or slightly weighty essays. Still out of Morellet's account came the fact that Buffon was neither a regular at Mlle. de Lespinasse's salon, where d'Alembert reigned, nor even one at the salon of Mme. Geoffrin, who had to invite him especially for the occasion. In the philosophical salons, where he did not have many friends, he could not find either the pleasures of a familiar conversation or contacts useful for his work.

And his work absorbed him. From 1753 to 1767 he published, with Daubenton, twelve quarto volumes on quadrupeds, or about six thousand pages. What was worse, he was always late with his plans and his promises to the public. The unrealistic expectations of 1750 had vanished. Volume IV appeared only in 1753, volume V two years later, and at the end of volume VI, which appeared in 1756, our naturalists were still only at the rabbit, which meant they had described thirteen species out of two hundred! To defend himself, Buffon spoke of "obstacles that others have put under our feet" and that "have multiplied despite the public's voice and the silence of the authors, who, having undertaken their work only in order to satisfy more fully the duty of their positions, and do not claim to receive any other glory from it, have stayed calm, and have expected everything from the effect of time and from the protection with which the king is happy to honor them."[75] In short, if Buffon was late, the fault lay with the Sorbonne and Réaumur.

There were more serious things than tardiness and the immensity of the work: there was a slump in sales. In 1764 the bookseller Durand, to whom Buffon had given the edition of the *Natural History*, had just died, and his business was put up for public sale. Now, as Buffon explained at length to Daubenton, the two authors had "with the contract, given up ownership and the follow-up of the work and even the reprintings and new editions." In order to keep it from falling into the hands of a bookseller whom he would not have chosen, Buffon decided to buy back the entire work. He managed this, not without difficulty, for the enormous sum of 179,000 livres: "I am losing more than 20,000 ecus with this transaction, but there was no way to do it otherwise. . . . I was obliged to borrow around 50,000 francs in order to give the cash that was necessary; I gave 100,000 francs in nine letters of exchange [today called drafts] at different rates." He could find only 30,000 francs in his own treasury.

What is more, the *Natural History*'s sales were not enough to offset publication costs. Each run of quarto volumes cost approximately 24,000 livres (1 livre = 1 franc). Durand sold them for twelve francs a copy, less a forty-sous discount "for all peddlers and booksellers in the provinces." Thus, "it was necessary to sell 2,400 copies before covering the expenses." This number had not even been reached for the first three volumes of 1749, available for fifteen years, of which 3,000 copies had certainly been printed. For the following volumes, it was even worse: "More than ten years are needed to sell more than 2,000." "This reduction in sales comes in large part because the volumes were too weak in substance and subjects that one can read."[76] The conclusion: it was necessary to enrich the volumes and put in more "subjects that one can read." The two "Views" *De la nature* [On Nature], placed at the beginning of volumes XII and XIII, were added precisely for this reason.

Buffon needed to re-establish his finances: in 1767 he sold a duodecimo edition of the *Natural History* to the bookseller Panckoucke which, in order to include more "subjects that one can read," left out the anatomical descriptions, Daubenton's "guts." Daubenton, it seems, was rather vexed. In any case, his reaction to the announcement of the sale was reserved: "I do not know if I must congratulate you for the sale that you have achieved for your two editions. It seems to me that you have given copies at a low price in comparison to what the public pays for them; but I do not understand business affairs." In any case, it has perhaps been said too quickly that the two men quarreled, for in the same letter Daubenton did not envision ceasing his collaboration with Buffon: "As for the new outline that you have made for the following volumes, I find the cetaceans very well placed after the quadrupeds, and I will be very happy to work on their descriptions such as you have proposed it to me."[77]

The quadrupeds done, Buffon now threw himself into other projects. First of all came a *Histoire des cétacés* [History of the Cetaceans], which would never see the light of day: that was one of the reasons Daubenton no longer collaborated on the *Natural History*. Then he had to write the history "of cartilaginous fish; then that of oviparous quadrupeds, and reptiles, and finally general subjects on plants and minerals."[78] In fact, in 1771 Buffon and Guéneau de Montbeillard would ask Panckoucke to obtain Jean-Jacques Rousseau's collaboration for a *Histoire du règne végétal* [History of the Plant Kingdom].[79] This project was also aborted. The only one to be finished would be the *Histoire des minéraux* [History of Minerals]. In 1760, however, Buffon had started to prepare a *Natural History of Birds*. It seemed that at first he had envisioned it as a series of commentaries for in-folio plates, engraved and "illuminated," that is, colored by hand, an undertaking for which he recruited his old friend Guéneau de Montbeillard. In 1762 they were up to the "third section."[80] In 1766 four sections had already appeared, including twenty-four plates each, and Buffon sold each section for one louis (ten francs). He pressed his friends to buy them quickly, as "I have only twenty-six remaining." There were already close to two hundred engraved plates, and the total should have made four or five volumes.[81] In fact, there was much more text than originally thought, and in the end the edition included nine quarto volumes with uncolored engravings, the first volume appearing in 1770 and the last in 1783.

Buffon had wanted to progress quickly, as a common friend, M. de Montmirail, explained to Daubenton in January, 1764: "In the very attractive and very broad new outlines, he [Buffon] has aspired to give the history of birds in three years, and for that reason all anatomy is left out." This is why Daubenton did not participate in the undertaking. "But he also wanted to omit most of the birds of the country by including, for example, only four species of pigeons, of titmouse, of sparrow, etc." This actually meant treating the birds by "genus" and not by "species." "I fought hard against this project," Montmirail added, but "all my reasons were rejected, and the desire to accelerate and to tighten [the text] prevailed over everything." Finally it was decided to treat foreign birds first and to speak only of the "birds of the country" where space was available. "This arrangement . . . was adopted. It is followed today with enthusiasm."[82]

Buffon therefore stayed as unrealistic as ever but was still as full of "enthusiasm" and plans. These plans, however, were all in the same line as what he was already doing. In 1767, when the last volume, number fifteen, of the *History of Quadrupeds* was published, he was at the peak of his glory. He was sixty years old and was watching old age approach without any particular fear, so long as it left him the opportunity to work. In February 1766 he had written to Ruffey: "I count much, my dear friend, despite the fact that I am fifty-eight years old as of last September, on finishing the entire *Natural History* before I am sixty, that is, before I start rambling."[83] He did not know that he still had twenty years of work in front of him, and that he had reached a turning point in his life and his work.

CHAPTER XV

Man in Nature

Buffon placed man in the center of the *Natural History*. Animal species appeared successively in the order of their proximity to man: domestic animals, then wild animals. Since it was man who was constructing science, he had the right to impose his order on nature. Buffon, like Diderot, was convinced of this.* So, at first, Western man could believe to be at the center of the world. Later on, as we shall see, Buffon had to distance himself from this anthropocentric position, which caused a scandal when he announced it in 1749.

Man is not only a knowledgeable subject, he is also an animal species like the others. An exceptional species, certainly, but one that lives among the others and is subject like them to the laws of nature. What relationships does he maintain with the living world? What does he have in common with the other species, and what is unique? Has his empire over nature always existed? Is it legitimate and definitive? These fundamental questions run throughout the *Natural History of Quadrupeds* and come up again later. No one before Buffon had asked these questions in such precise terms. No one would dare say today that these questions are not fundamental.

Man in the "Economy of Nature"

"Nature is the system of laws established by the Creator for the existence of things and for the succession of beings." She is "a perpetually living

---

*"It is man's presence that makes the existence of things interesting," wrote Diderot, "and what better suggestion can be made than for us to consider things from this point of view? Why should we not introduce man . . . as he is placed in the universe? Why should we not make him the common center?" (*Encyclopédie*, art. "Encyclopédie" [Volume V, p. 642, recto B]; Diderot 1975, VII, p. 212).

work, a constantly active worker. . . . The springs that she uses are living forces, which space and time can only measure and limit without ever destroying: forces that balance one another, merge, and oppose each other without ever destroying one another."[1] Nothing could be less static, less frozen than nature, an immense combination of forces and motions rather than a collection of entities created once and for all and definitively fixed in an unchanging structure.

This dynamic universe was also a stable universe. Before Laplace, Buffon proclaimed the stability of the solar system: "It is from the very essence of motion that the equilibrium of worlds and the Universe's repose is born."[2] The same dynamic stability reigns in the world of the living. In order to understand it, it is necessary to look at it not from the individual's point of view, but from that of the species. In this way we see that each species "holds its place, survives, defends itself from others, and all of them together make up and represent living Nature, which maintains and will maintain itself as it has maintained itself. . . . They are all equally dear to [Nature], since to each one of them she has given the means to exist and to last just as long as she does."[3]

Just as antagonistic forces maintain the equilibrium of the universe, reproduction and destruction maintain the balance of life. "It is the order of Nature that death serves life, that reproduction is born from destruction." In such a way "everything is well, because in the physical universe the bad contributes to the good." "The violent death of animals is a legitimate, innocent happening, since it is based in Nature, and they are born only under that condition." Just think of herring and their "immense multiplication": "They alone would cover the entire surface of the ocean . . . if they were not destroyed by others in great numbers."[4]

Nature was therefore "the war of everyone against everyone," according to Hobbes's famous saying. Buffon admitted that this situation was inhumane: "the motive for doubting it honors humanity: animals, at least those that have senses, flesh and blood, are sensitive creatures like us; they are capable of pleasure and subject to pain. There is therefore a sort of cruel insensitivity in the needless sacrifice of beings who are close to us, who live with us, and whose sensitivity is revealed to us by signs of pain."[5]

Man's humanity here utterly opposes nature, or at least it should. Buffon, unlike so many biologists and political thinkers of the nineteenth century, was never tempted to propose a "natural model" for human society. On the contrary, he preferred to think that Nature was less inhumane than believed and that she "knew how to compensate for everything": "Uniquely attentive to the preservation of species," she made sure that the large predators produced few offspring, and that the smaller animals, the "inferior species, were in a state to resist and to endure because of their numbers." She had even made multiple similar species, such that "if one of them disappeared, the gap in this genus would be barely perceptible."[6]

Although nature maintained the balance between species, it was still not a perfectly stable balance. At certain times certain species seemed to threaten taking over the earth: grasshoppers or ants in warm climates, or rats in the northern countries suddenly multiplied prodigiously. Yet "these animals in innumerable multitudes, which appear suddenly, disappear in the same way," and "the numbers of these species are not increased."[7]

Was it the same for the human species? Buffon hesitated. At times, in seeing man drive animals that he could not domesticate back to the wilderness, he thought "that with time, the human species has expanded, multiplied, and scattered."[8] On the other hand, he thought that the great invasions, "these floods of the human species, the Norse, the Alans, the Huns, the Goths, the peoples, or rather the tribes of animals with human faces," that were seen "marching in wild bands [and] oppressing everything without any other strength than numbers," were only "minor tribulations in the ordinary course of living nature." For "the quantity of men . . . depends on the balance of physical causes . . . that men's efforts, despite all moral considerations, cannot break; these considerations themselves depend on the physical causes of which they are no more than the specific effects."

Buffon here went further than Montesquieu: "When a portion of the earth is overcrowded with people, they scatter, they spread out, they destroy one another, and at the same time, they establish laws and customs that often serve well to prevent this excess of multiplication. In extremely fertile climates, as in China, Egypt, and Guinea, children are abandoned, mutilated, drowned; here they are condemned to a perpetual period of celibacy. Those who exist claim rights over those who do not exist; as existing beings, they annihilate the contingent ones, they suppress future generations for their comfort, for their convenience. What happens to humans, without us noticing it, is what happens to animals." Thus, "as all these moral effects depend on physical causes that . . . are fixed and in permanent equilibrium, it seems that for man as for animals, the number of individuals in the species can only be constant."[9] Malthus would need only to draw the conclusions from this state of things.

This model of natural equilibrium was accepted by many authors of the second half of the eighteenth century. Buffon studied its functioning in living nature but linked it explicitly to the equilibrium of the solar system, which Laplace later demonstrated at the end of the century. It was also the model adopted by Adam Smith in political economy: the fluctuations in price oscillate constantly around a perfect equilibrium as if maintained by an "invisible hand."[10] Indeed for Adam Smith, the laws of economics were natural laws. In this way, God did not need to intervene, as Newton believed, in order to maintain the universe in balance. Nature was now a self-regulating system that corrected its own fluctuations.

Although the law of nature seems inhumane to us, wild animals did not necessarily suffer from it: "Nature has given them means and resources to use against other animals; they are their equals, they know their force and their skill, they judge their intentions, their habits, and if they cannot avoid them, at least they can defend themselves physically."[11] They live among one another, and they are free. In them, Nature shows herself "naked, adorned only in her simplicity, but more striking for her innocent beauty, her light step, her free air, and by other attributes of nobility and independence." Among them, Nature was seen "to share her domain . . . allot to each one its element, its climate, its food . . . distribute her gifts with equity, balance good and evil," and give to all "freedom . . . desires and love always easily satisfied and always followed by a fortunate fertility." "Love and freedom, what bounty! Do these animals that we call wild because they are not submissive need more in order to be happy? They still have equality, they are neither slaves nor oppressors of their kindred; the individual need not, like man, fear the rest of his species, and war only comes to them from other beasts or from us."[12] "The majority ask only for tranquility, peace, and the moderate and innocent use of the air and the earth; they are even encouraged by Nature to live together, to join in families, to form sorts of societies."[13]

Later, in talking about birds, Buffon became almost sentimental. For them, there was "more tenderness, more attachment, more morals in love" than for quadrupeds. They offered "examples of conjugal chastity," "of the care of fathers for their offspring." The male is very attentive to the female. When she nests, he "brings her her sustenance; he even replaces her at times, or joins with her, in order to increase the nest's heat, and share the problems of the situation." In short, "birds represent to us . . . everything that happens in an honest household; of love followed by an attachment without division, which then is shared by the whole family."[14] In 1753 Buffon celebrated the beauty of wild horses, "proud of their independence," the nobility of the deer, "one of those innocent, gentle, and peaceful creatures, which seems to be created only in order to beautify and animate the forests' solitude, and to occupy, far from us, the peaceful retreats of these gardens of Nature," the grace, vivacity, and elegance of the roe buck, the lion's pride.[15] "Nature is more beautiful than art, and the freedom of movement of living beings makes Nature beautiful."[16] Nature is innocent, as long as she is protected by man.

Man's Tyrannical Power

It was man who introduced terror and slavery into nature. Animals were defenseless against him: "What can they do against beings who know how

Large mongoose. From *Histoire naturelle,* p. 106.

to find them without seeing them, and to subdue them without approaching them?" It is "man who disturbs them, who drives them away, who
scatters them, and who makes them a thousand times wilder than they
ordinarily would be." "In countries . . . where men settle, terror seems to
live with them."[17] Man is the greatest predator in nature: "If to harm is to
destroy living animals, is not man, considered to be part of the general

environment of these beings . . . the most harmful species of all? He alone slays, annihilates more living individuals than all carnivorous animals devour. They are only harmful therefore because they are man's rivals . . . and because, in order to provide for their fundamental needs, they sometimes dispute over a prey that he was reserving for his excesses; because we sacrifice even more for our excesses than we do for our needs. Born destroyers of beings who are below us, we would exhaust Nature if she were not inexhaustible."[18] Granted, man is not a vegetarian: the structure of his digestive system proves this. If he were content "to enjoy moderately the goods which are offered him . . . to share with equity . . . to repair in proportion to what he destroys . . . to renew when he annihilates," there would be nothing to add. Yet he destroys "more from evil custom than from necessity,"[19] and nothing justifies his destructions.

If he terrorizes wild animals, man does not treat domestic animals any better: "A domestic animal is a slave with which one entertains oneself, which is used . . . abused . . . altered . . . disoriented and . . . denatured."[20] Too often, it is scorned and mistreated. If the ass "did not have such a great fund of good qualities, it would lose them by the manner in which it is treated: it is the toy, the target, the butt of jokes of boors who drive it with stick in hand, who hit it, overload it, exhaust it without mercy and without care."[21] Horses are not better off; they "always carry the marks of servitude, and often the cruel imprints of work and suffering, the mouth is deformed by folds that the bit has made, the flanks are cut by wounds, or furrowed by scars made by the spurs. . . . Even those whose slavery is the gentlest, who are only fed, who are only kept up for show and magnificence, and whose golden chains are used less for their harness than for their master's vanity, are even more dishonored by the elegance of their forelocks, by the braids in their manes, by the gold and the silk with which they are covered, than by the irons on their feet."[22]

The violent tone is remarkable, and this passage of the *Histoire Naturelle du cheval* [Natural History of the Horse] deserves to be cited as an example of the author's eloquence, rather than the text's famous introduction: "The most noble conquest that man has ever made is that of this proud and fiery animal," a sentence eternally linked to Buffon's name, in which it is not understood that the beast's grandeur, and not that of its master, is being celebrated.

This theme of man's tyrannical power over domestic animals is found throughout the *Natural History*. In his appetite for domination, man blinds himself to his own interests. For example, in order to maintain the beauty and vitality of domestic animals, he should let the females have the freedom to choose the male with which they would like to mate. But man keeps only a few reproductive males: "With this practice men preferred their convenience to other advantages; we have not sought to maintain, to beau-

tify Nature, but to make her submissive to us and to enjoy her more despoti-
cally; the males represent the species' glory; they are more courageous,
prouder, always less submissive; a great number of males in our troops
would render them less docile, more difficult to lead, to keep: it has even
been necessary in these slaves of the lowest rank to force down all the heads
that could rise up."[23] Yet were there not other slaves above "these slaves of
the lowest rank"?

The indictment of man immediately became an indictment of society. Of
animals, "whose flesh flatters his taste, [man] has made domestic slaves . . .
and through the care that he takes to have them multiply, he seems to have
acquired the right to sacrifice them to himself." All men did not share in
the feast: "The rich man glories in consuming, shows his grandeur by
wasting on his table in one day more goods than would be needed to allow
several families to live; he abuses equally animals and men, the rest of
whom live in hunger, languish in misery, and work only in order to satisfy
the inordinate appetite and the even more insatiable vanity of this man,
who, destroying the others by dearth, destroys himself through excess."

It is the very ones who produce these riches who are deprived of them:
"Poor families are today reduced to depending on their cow! These same
men who every day groan in work and are bent over the plow from morning til night, drawing from the earth only black bread, and are forced to
give to others the flower, the substance of their seed, it is through them and
not for them that the harvests are abundant; these same men who raise,
who increase the livestock, who tend it and are constantly taking care of it,
do not dare enjoy the fruit of their labors. The meat of these animals is a
food that they are forced to deny themselves, forced by the poverty of their
condition, that is, by the domination of other men, to live like horses from
barley and oats, coarse vegetables, and sour milk."[24]

What relationship, one might ask, exists between man's harshness to
animals and our society's injustice? For Buffon, the relationship would
soon be evident. In France, in our "civilized" society, we mistreat "our oxen,
who only know us by our ill-treatment: the goads, clubs, and famine make
them stupid and weak: in short, we do not know enough, even about our
own interests, for it would be advantageous to treat better what depends on
us." It is not the same in less civilized or less unequal societies: "Men of the
inferior state, and peoples who are least governed, seem to feel the laws of
equality and the nuances of natural inequality better than others; the
handyman of a farmer is, so to speak, on equal footing with his master; the
Arabs' horses, the Hottentots' oxen are cherished servants, companions in
practice, helpers in work, with whom are shared residence, bed, and table;
man, by this sharing, degrades himself less than the beast rises and human-
izes itself: it becomes affectionate, sensitive, intelligent; it does, then,
through love everything it does here through fear; it does much more;

because its nature has risen owing to the gentleness of its education and the continuity of care, it becomes capable of almost human things."[25] It is not man himself but society that is responsible for the manner in which we treat domestic animals, as it is responsible for the unhappiness of the majority of men.

Nature, however, is not perfect, and we must not look to her for lessons in morality: "If Nature is disturbed by the unjust division that society makes of happiness among men; she herself, in her rapid change, seems to have neglected certain animals that, through an imperfection of organs, are condemned to endure suffering and destined for penury: deprived off-spring, born into destitution, forced to live in need. . . . The heron presents us the image of this life of suffering, of anxiety, of indigence."[26] For man, things are clear: "Evil in truth comes more from us than from Nature. A man who is born weak, impotent, or deformed is unlucky by Nature, while millions of men are that way because of the harshness of their fellows."[27]

It is not necessary to mention here how close these themes are to those Rousseau passionately developed in his *Discourse on the Sciences and the Arts* and *Discours de l'origine de l'inégalité parmi les hommes* [Discourse on the Origin of Inequality among Men]. The meeting of minds of the naturalist and the political philosopher is that much more remarkable since the two men were so different, especially in their personalities and lives. We shall come back to their relationship. It is enough for us to note here this haunting presence of Rousseauist themes throughout the *Natural History*.

Man, "Vassal of Heaven" and "King of the Earth"

This division forces itself on us, because I have gathered together texts scattered among Buffon's works that other texts immediately following them seem to contradict. Buffon affirmed from the very beginning that "man's dominion over animals is a legitimate dominion that no revolution can destroy." "It is not only a right of Nature, a power founded on unalterable laws, but it is also a gift of God."[28] Are we going to see Buffon return to the providentialism of the abbé Pluche and proclaim that God created bees in order to give us honey and sheep for us to obtain wool?

His point was probably something else entirely. First of all, man's domination has limits. Many animals escape it "by the rapidity of their flight, by the faintness of their paths, by the obscurity of their lairs, by the distance that the element in which they live puts between them and man"; others "by their small size alone." Many even "far from recognizing their sovereign, attack him with open force, not to mention those insects that seem to insult him with their stings, those snakes whose bite carries poison and death, and so many other foul, disagreeable, useless creatures, which seem to exist only in order to form the distinction between good and bad, and to make man feel how little he is respected since his fall."[29] In short, far from being a sovereign by divine right, man is a contested master. Let us temporarily leave aside the opportune allusion to original sin and try to define, in the current state of things, man's place in this nature that he dominates without completely mastering it.

It is necessary, Buffon tells us, "to differentiate God's dominion from man's domain." Man can do nothing about the general order of nature, "about the motions of celestial bodies . . . about animals, plants, minerals in general." In nature, where "everything that happens, that is in order, and in which things succeed one another, [everything] renews itself, and is moved by an irresistible power." Man is "himself carried by the torrent of time" and "can do nothing about his own life span." "He is forced to obey universal law: he obeys the same force, and like all the rest, he is born, grows, and perishes."[30] Nevertheless, his power is considerable. He knows how to overcome the resistance of inert bodies, "he is the master of plants, for through his industry he can increase, diminish, renew, denature, destroy, or multiply them to infinity."[31] "Wheat, for example, is a plant that man has changed to the point where it no longer exists anywhere in a state of nature," and "transforming a sterile herb into wheat" was "a kind of creation."[32]

Man "can do nothing about species," Buffon wrote in 1753.[33] Two years later, he had changed his mind: man can, "with time, change, modify, and perfect species; it is even the greatest power that he has over Nature."[34] From then on, "instead of looking at the ass as a degenerate horse" and the goat as a more rustic ewe, "it would be more correct to say that the horse is a perfected ass, that the ewe is only a more delicate kind of goat, which we have cared for, perfected, and propagated for our use, and that in general the most perfect species, especially among domesticated animals, originated from the least perfect species of wild animals that resemble them the most; Nature alone is not able to do as much as Nature and man together."[35] Buffon, we shall see, did not hold to this idea. It is, however, significant that he had had it.

But what can nature do alone? As the *Natural History* progressively distanced itself from Europe and its countryside, marked for centuries by man's work, and as it discovered those lands of tropical America where man, far from reigning, barely survived, nature became ferocious, hostile, hideous. In 1764, in a famous text, Buffon opposed nature left to herself to nature cultivated by man: "Look at those deserted beaches, those sad regions where man has never lived: covered or rather bristling in all the higher parts with thick and dark woods, trees without bark and without tops, curved, broken, falling with age[;] many more, lying at their feet, rotting on the already rotted heap, suffocating, burying seeds ready to sprout. . . . The earth . . . offers . . . only an overcrowded space, filled with old trees loaded with parasitic plants, lichens, fungi, the impure fruits of corruption: in all the lower parts, dead and stagnant waters . . . muddy ground . . . marshes covered with aquatic and fetid plants, which nourish only venomous insects and serve as a den for foul animals." What can man do "in these savage places" where "no vestige of intelligence" appears? He could simply flee. But no: "He turns around and says: raw Nature is hideous and dying, it is *I, I* alone who can make it pleasant and living: let us dry up these swamps, let us vivify these dead waters by making them run . . . ; let us use this active and devouring element that was hidden from us and that we owe only to ourselves; let us set fire to this superfluous waste, these old forests already half dead; let us finish destroying with iron what the fire was not able to burn: soon, instead of rushes, and water lilies, from which the toad made his venom, we shall see buttercups, clover, sweet and healthy grasses; herds of gamboling animals will crowd this earth, hitherto unusable . . . ; they will multiply and multiply again. . . .

"How beautiful is this cultivated Nature! Through the care of man she shines and is sumptuously adorned! . . . He brings to light through his art all that she concealed in her bosom; so many unknown treasures, so many new riches! . . . In the valleys of the laughing prairies, in the plains of rich pastures or even richer harvests; the hills laden with vines and fruits, their summits crowned by useful trees and young forests; the wilderness has become cities inhabited by an immense population . . . ; open and frequented roads . . . a thousand other monuments of power and glory prove clearly that man, master of the earth, has brought changes to it, renewed its entire surface, and from time immemorial has shared Nature's rule."[36]

Buffon of course specified beforehand, to reassure pious souls, that it was from God himself that man had received this power: "Made in order to adore the Creator, he commands all creatures." But if he is "the vassal of Heaven," he is also "king of the Earth, he ennobles it, peoples it, and enriches it; he establishes order, subordination and harmony among living beings."[37] Had God not done this work himself? "Nature is the visible throne of the divine magnificence," Buffon wrote. But this is so only when she has been "ennobled" by man. Buffon could doubtless hide behind the authority of Genesis: had not God wanted man "to rule over the fish in the ocean, over the birds in the sky, over domestic animals, and over the entire earth," and had He not placed him in the Garden of Eden "to cultivate it and to watch over it"?[38] Yet the very precise allusion to fire, "which was hidden from us and which we owe only to ourselves," reveals the Promethean character of this human undertaking. If man has become "like a master and possessor of Nature," according to Descartes's famous statement, which Buffon adapted practically word for word, it was only from the moment that man "could walk in strength in order to conquer the Universe."[39]

Moreover, this empire of man is never secure or definitive. Nature "never fails to reassert her rights as soon as she is left to act freely: wheat sown on uncultivated earth degenerates during the first year."[40] Man "reigns only by right of conquest; he rejoices rather than possesses, he conserves only by ever constant care; if [the care] stops, everything languishes, everything alters itself, everything changes, everything falls back under Nature's hand: she takes back her rights, erases man's works, covers with dust and moss his most sumptuous monuments, destroys them with time, and leaves him only with the regret of having lost by his mistakes what his ancestors had conquered by their works."[41]

So, was man a legitimate king or a tyrant of nature? Both, probably. Buffon participated in the double movement of his century, which was more and more aware of the power that science and technology offered man but which nevertheless, did not find happiness in human society alone. If "the wilderness [has] become cities inhabited by an immense population," at least a part of this population detested these cities, where vice and unhappiness ruled, and aspired to the simple pleasures of nature, but a nature made smiling and orderly by the work of man. The abbé Pluche's characters fled the city to find themselves in their chateaux, in their parks, among tilled fields and hills covered with vines, in short, in "Nature's garden," as Buffon said. A garden duly cultivated by the neighboring peasants. The "awful" spectacles of lofty mountains, of New World wildernesses, of tropical forests still found few enthusiasts.

Buffon was not, however, the abbé Pluche. The idea that he had of nature and of man protected him from the temptation of seeing only an immediate harmony in their relationships. Although he began with the Burgundian countryside, where the dog accompanied the horse, the description of exotic animals led him very far from this peaceful countryside. And so, faced with the spectacle of raw and savage nature, his first reaction was one of horror and recoil: he knew that nature was not made for man, and that she could be grandiose when she is most inhuman. The naturalist hesitated, but how could the writer resist the temptation of describing the indescribable, of depicting, for example, that wasteland most hostile to man, the desert of Arabia?

"Imagine a landscape without greenery and without water, a burning sun, a forever rainless sky, sandy plains, even more arid mountains, on which the eye rests and sight loses itself in the distance without being able to discern any living object; a dead earth, so to speak, scorched by the winds, and which offers only bones, scattered pebbles, upright or overturned rocks, a desert entirely bare in which travelers have never breathed in the shade, where nothing accompanies them, nothing recalls living Nature: absolute solitude, a thousand times more ghostly than that of forests; for trees are still beings for the man who finds himself alone; more isolated, more abandoned, more lost in these empty and limitless places, he sees space everywhere as his tomb: the light of day, more depressing than the shadow of night, returns only to light his isolation, his impotence, and presents to him the horror of his situation, by pushing back the barriers of emptiness before his eyes, by spreading around him the immensity of the abyss that separates him from the inhabited earth: an immensity he would try in vain to cross because hunger, thirst, and burning heat all foreshadow the moments remaining for him between despair and death."[42]

Man, nevertheless, has not recoiled before these deadly solitudes: "The Arab, with the camel's help, knows how to surmount and even to adapt to these extremes of Nature, they serve him as a refuge, they assure his rest and maintain him in his independence."

Buffon would sixteen years later contrast to "this picture of absolute dryness" "one of vast miry plains of the drowned savannas of the new continent," where "earth and water seem to dispute these boundless areas," "Nature's sewers," "unworkable lands, still unformed": "Enormous serpents trace wide furrows on this swampy earth: crocodiles, toads, lizards, and a thousand other reptiles with broad feet knead the mire; millions of insects multiplied by the humid heat lift up the sludge from it, and this entire corrupt population slithers in the silt or hums in the air that it obscures; all this vermin with which the earth swarms attracts flocks of voracious birds whose raucous cries, multiplied by and mixed with the croakings of the reptiles, trouble the silence of these awful wastes and seem to add fear to the horror in order to repel man and forbid the entry of other sentient beings."[43]

The overabundance of the stagnant water's swarming life is almost more horrible than the deathly dryness of rock and sun. For Buffon there was almost an instinctive horror of stagnant waters and swamps, where a "foul population" of animals multiplied. He always dreamed of "making the water run."[44] It was the first civilizing gesture of man: "Let us dry these marshes, let us vivify these dead waters by making them run."[45]

The naturalist, however, had to surmount this instinctive horror, because the power of nature is discovered at the farthest point from man: "It is not by walking in our cultivated fields, nor even by traveling all of man's domains on the earth that the great effects of Nature's variations can be known: it is by going from the burning sands of the Torrid regions to the glaciers of the Poles, it is by descending from the summits of mountains to the bottom of the oceans, it is by comparing wastelands with wastelands that we will judge her better and admire her more."[46] If nature shows her power best where man can barely penetrate and survive, what, then, is man's place in nature? First of all, what is man? We might believe that the *Natural History of Man* had already answered this question. Apparently this was not so, since Buffon comes back to it at length in the beginning of volume IV of his *Natural History*.

Man and Animal

Let us recall the two main statements of 1749. On the one hand, man "must put himself in the class of the animals that he resembles physically."[47] On the other hand, "man is a reasonable being, and animals are creatures without reason."[48] The analysis of the operation of the senses in man had allowed, not without ambiguity, the deepening of the understanding of the mechanisms of the human mind. Buffon took up this entire problem again, with the pretext of using a new method. "After having considered man by himself," he now wanted to compare him to animals, because "only through comparison do we know anything."[49] Thus we will arrive at the same time at a better knowledge of man, a better understanding of the "principal effects of the vital mechanism," and a clearer demonstration of man's superiority over animals. We come back, thus, to man in a text titled *Discours sur la nature des animaux* [Discourse on the Nature of Animals]. The work seems to be nothing more than a deeper look at an earlier analysis in order to make it more precise.

In reality, it was a polemical text. Buffon confronted the defenders of animals' souls, those who humanized animals by giving them an intelligence and human feelings that they do not have. Buffon did not attack Condillac, who had not yet appeared on the scene. It was Condillac who would attack Buffon in 1755, in his *Traité des animaux* [Treatise on Animals].[50] It was with Réaumur that Buffon engaged. In this sense, the *Discourse on the Nature of Animals* was a response to *Letters to an American*.

By studying bees, Réaumur had, like others before him, shown the admirable geometry revealed in the construction of their hexagonal cells, which allowed them to use the least wax possible to obtain the maximum capacity without losing any space.* "The more one studies the construction of these cells," wrote Réaumur, "the more one admires it. It is necessary to be as skilled in geometry as [man] has become since the new methods that have been discovered** in order to know the perfection of the rules that the bees follow during their work."[51] In particular, Réaumur insisted that these insects' behavior was not mechanical and automatic, but that they were capable of inventing and facing unexpected situations. In short, far from being pure machines, the bees had a soul. Réaumur did not go so far as to claim that bees knew geometry, but it was necessary to recognize that God

---

*Given as a constant the thickness of the inner wall, the relationship between the capacity of a container and the quantity of matter necessary to make it is equal to the relationship between the volume and the surface. The best solution is the cylinder, like our cans. But the cylinders cannot be exactly juxtaposed: they leave spaces between them. Of all the juxtapositional volumes, it is the hexagonal cell that gives the best proportion of volume to surface. Moreover, the rows of wax are made up of two rows of cells facing in opposite directions, joined by the bottom and which are fitted together with an even more complicated geometry.

**Ed. note: Réaumur here refers to analytical geometry and the calculus, both products of the seventeenth century.

had given them a certain "genius" and even, like other insects, a certain intelligence."

This statement was exactly what Buffon rejected. With a surprising malice, he caricatured "our observers" who "admire bees' intelligence and talents to the point of envy." "The more this basket of bees is observed, the more one discovers marvels." This is hardly surprising: "We always admire more what we observe more and reason about less." "It is not curiosity that I blame here, it is the reasoning and the exclamations . . . it is the morality, it is the theology of insects that I cannot hear preached; it is the marvels that observers place there and about which they exclaim as if they were indeed there." "Finally a bee must take up more space in the head of a Naturalist than it does in Nature; and this marvelous republic will never be in the eyes of reason anything more than a crowd of little creatures that have no other relationship to us than that of furnishing us with wax and honey."

Behind this admiration for bees, there was something worse than "the enthusiasm someone feels for his subject." There was a deep error of metaphysics, a misunderstanding of the true grandeur of God's work: "Is not Nature astonishing enough in and of herself, without our seeking to surprise ourselves by further stunning ourselves with marvels that are not there and that we put there? Is not the Creator great enough through his works, and do we believe to make him greater by our imbecility? This would be the way if it were possible to lower Him. Who, indeed, has the grander view of the Supreme Being, he who sees Him as the creator of the Universe, the orderer of all things, founder of Nature on invariable and eternal laws, or he who seeks Him and wants to find Him occupied with leading a republic of flies and very busy with the way a beetle's wing should fold?"[52]

We can imagine Réaumur's fury in reading these texts. "All the fuss about bees and other insects is because I like them and dare admire them. So much so that it makes M. de Buffon and his crowd speak of them with scorn. They should not however be embarrassed by having them react with intelligence; once one admits the existence of material souls conscious of their own existence, everything can be credited to matter. But what a metaphysics that is."[53]

Buffon was indeed at a loss. He did not accept a material soul, although he admitted that it was necessary to recognize "a kind of foresight" in many animals. But prior to any attempt at explanation, he posed a principle: nature is not understood by imagining that for each particular fact there is a particular "little law." To explain is, on the contrary, to bring a "specific effect" back to a "general effect," to a law of nature. To make up as many "particular statutes" as there are living species was both "hardly philosophical and little deserving of the idea that we must have of the Creator." If God exists, he acts only through "general wills," which are so many universal laws of nature. He is not a meticulous craftsman. He is a supreme legislator. If we are not able to relate all the observed facts to general laws, it is better "to put these facts aside and abstain from wanting to explain them until, through new facts and new analogies, we can know the causes of them."[54]

Let us look, therefore, for a "general effect" that can take into account these "specific effects," and let us start, right here, with the hexagonal cells of bees. Notice first that according to these observers, "solitary bees do not have any mind as compared with bees who live together." Individually, these bees "have less intelligence than the dog, the monkey, and the majority of animals." Notice also that the hexagon is a common shape in nature: it is found in certain crystals, "in seeds, in their hulls, in certain flowers, etc." And let us do a very simple experiment: "Fill a receptacle with peas, or rather with some other cylindrical seed, and close it after having poured in exactly as much water as the spaces remaining between these seeds can hold; when the water is boiled all these cylinders become columns with six sides. The reason, which one sees clearly, is purely mechanical: each seed, whose shape is cylindrical, tends to occupy the most space possible within a given space by its swelling; they therefore necessarily become hexagonal by reciprocal compression." In the same way, "the little scales of dogfish [are] hexagonal, because all the scales, growing at the same time, become obstacles to one another and tend to occupy the most space possible in a given space."

If animal morphogenesis is submitted to mechanical constraints, as we would say today, would not the same apply to beehive cells? "Each bee is seeking to occupy in the same way the most space possible in a given space, and it is therefore also necessary . . . that their cells be hexagonal." "The more numerous they are . . . the more mechanical constraint, forced regularity, and apparent perfection in their productions there is."[55]

It was therefore necessary to put insects back in their place, which is at the bottom of the animal hierarchy, at the lowest part of this "scale" that can be established "by taking the material part of man for the first level" and descending from those that resemble him the most "in exterior form [and] interior organization" to those that resemble him the least. In order came quadrupeds, crustaceans, birds. Only oysters and polyps would be lower than insects.[56]

Animals Do Not Think

This closed the case of Réaumur and his beloved insects. But what about higher animals? The answer was clear: animals in general are "material being[s], which neither think nor reflect." Their behavior was therefore "a purely mechanical effect, and absolutely dependent on [their] organization." Nevertheless, animals "react and seem to make up their minds."[57] In order to understand this, the "animal machine" must be carefully studied, starting with the observation of animal behavior: "Let us see the effects of this mechanics in general, and without wanting to reason on the causes initially, let us be content to notice the effects."[58]

In interpreting this behavior, let us be wary of all anthropomorphism. If we see a starving dog that wants a piece of meat beg for it from its master rather than throw itself upon it, the "common interpretation" would be that the animal reasoned "much like a man who would like very much to seize others' property, and who, although violently tempted, is restrained by the fear of punishment." "The analogy, it is said, is well founded, since the organization and the conformation of the senses . . . are similar in animals and in man." If that were true, animals could do, and sometimes would do, all that man does, which is not the case: animals "neither invent [nor] perfect anything. [They] just do the same things in the same way." It was therefore necessary to seek the mechanism by which they acted.[59] In short, it was necessary to take up the Cartesian theory of the "animal-machine," which had stirred up storms, but by giving it a more precise and more persuasive form.

Let us start at the simplest level. There are two parts of the "animal economy." The essential part is organic life or, as Buffon called it, "vegetative life." An animal that is asleep breathes and its heart beats: these functions never stop. This form of life is common to all living things, from plants to man. In order to rise to "animal life," more is necessary: senses and limbs. The more these "exterior parts" are developed, that is, the more they resemble those of humans, "the more the animal's life seems complete to us and the more perfect the animal is."[60] Among these "exterior parts," the most important are "the brain and the senses." The principle of "animal life" is very simple: external objects act on the sensory organs, these transmit a "modified impression" to the brain, which there becomes a "sensation." The brain transmits "the disturbance" that it just received to the nerves, which the nerves transmit to the muscles, putting the animal into motion. A small sensation can provoke a violent reaction, just as a spark can make "a citadel blow up": none of this is contrary to the laws of physics.

Sensation is enough to put the animal into motion, but not just any motion. Above all, the same sensation does not always provoke the same movement. We are getting away from the simple mechanics of brute matter. The "disturbances" that make up the sensation "are pleasant or unpleasant, relative or contrary to the animal's nature, and bring forth appetite or disgust, depending on the state or the present disposition of the animal."[61] Between the action of the external object and the animal's reaction, there is an internal mechanism that determines the reaction, as a function of the "state" or the "disposition" of the animal. It is not enough to say that the sensation gives rise to the desire to approach or flee, or that an internally felt natural need, such as hunger, has the same result. It is also necessary to explain why the animal approaches *or* flees, why it eats one food and not another, in short why it seems "to deliberate," to decide, to choose. For that, it is first necessary to have an organ where the sensations can be assembled in order to be compared, and also where the animal's "present disposition" is sensed in a certain manner so that the useful movement, and it alone, will be activated. This organ is the brain, which is therefore "a general and common internal sense." It is a purely material organ that has a specific property, namely that the disturbance transmitted by the organs of the senses "survives and continues a long time after the action of the exterior object," in the same way that the impression of a bright light stays in the eye. This property of the "internal sense," which constitutes a sort of sensory memory, plays a fundamental role in the explanation of animals' faculties.

Let us now return to our dog. The dog would like to throw itself on the meat, but its "interior sense" retains "the prior impressions of pain with which this action had been previously accompanied." The two "impressions" cancel each other out. At the same time, "it renews in [its] brain . . . a third disturbance, which had often accompanied the first two," "caused by its master's action" when he gave it a coveted morsel. Nothing counterbalances this "disturbance." "The dog will thus be determined to move toward its master."[62] This is what had to be proved.

Buffon defended his reduction of animals "to being only simple machines, only insensitive automatons." He "accorded them everything, with the exception of thought and reflection; they have feelings, they even have them in a higher degree than we have; they also have consciousness of their current existence . . . ; they have sensations."[63] But they do not think, and this was what an even more precise analysis intended to prove.*

Animals Do Not Have Memories

Let us first note that the senses in animals and in man are not developed in the same way. Animals possess a highly developed sense of "smell, which is the sense of the appetite." They also move more easily: the newborn animal is capable of finding the teats of its mother, whereas the newborn human cannot. On the other hand, infants already have open eyes, which is not the case of newborn animals. Man obviously possesses a highly developed sense of sight, "which is a sense more relative to knowledge than to appetite."[64] The sense of touch is also much more highly developed in man.

*In speaking about animals, Descartes said the same: "I have never denied them what one commonly calls life, corporeal soul, and organic sense." But he did not bother to explain exactly what these common notions meant (*Méditations*, Answers to the sixth objections, § 3, in Descartes 1963, II, p. 866 [Adam-Tannery, IX, p. 229]).

Possessing senses, especially that of "appetite," animals have "feeling." What is "feeling"? It is "sensation become pleasant or unpleasant."[65] Animals, therefore, feel "pain and pleasure like us, they do not know good and bad, but they feel it: what is agreeable to them is good, what is disagreeable to them is bad . . . in a word, pleasure is physically good and pain is physically bad." Without this simple mechanism, animals could not survive.[66] "It is on feeling that all exterior motion and the activity of all the animal's forces depend."[67] It is enough to explain their behavior. Let us not therefore go any further, and let us not give animals faculties that belong to man only.

Buffon thus reviewed these supposed faculties of animals. The first statement was purposely provoking: no matter what was said, animals have no memory. Buffon recognized that everything seemed to prove the opposite, starting with the dog. Do not animals "remember the punishments they suffered, the caresses . . . the lessons given to them"? Buffon was careful when he spoke of "memory" for the dog; what animals possessed was "reminiscence," that is, "renewal [of] sensations or rather the disturbances that had caused them," which made possible the particular property of the "interior sense" of which we have spoken. The present sensation "wakes up" the past sensation, which again becomes present. The two sensations can then be combined by the interior sense to determine the animal's reaction. Thus, animals "see the present and the past together, but without distinguishing between the two, without comparing them, and consequently without recognizing them." True memory, human memory, requires more than "the renewal of anterior sensations: it assumes the comparison of ideas, the reconstruction of an order of succession, that only our soul is capable of realizing." It is this memory that "weaves the web of our existences with a continuous thread of ideas." "The child who babbles and the elderly man who rambles on" have no memory, as their reason is still too weak or too deteriorated. Lacking a memory, in the true sense of the word,* animals have no idea of time, no knowledge of the past, no notion of the future.[68] Because he has a notion of the future, man is a being who plans ahead. Animals live for the moment.

But, someone will say, animals dream, which means they have a memo-

---

*Descartes likewise distinguished a corporeal memory, similar "to the creases that remain in a piece of paper, once it has been folded," which resides in the entire body as well as in the brain (letter to Meysonnier, January 29, 1640, in Descartes 1963, II, pp. 157–8), and a memory that is "totally intellectual, which does not depend on the soul alone" (letter to Mersenne, June 11, 1640, in Descartes 1963, II, p. 247). The corporeal memory, however, registers only habits, as in the fingers of a lute player. It was therefore very different from the "reminiscence" of which Buffon spoke. Descartes said almost nothing about the "intellectual" memory.

ry.* Yes, but neither in animals nor in man does a dream suppose an activity of the soul. Dreams "are only driven by sensations and not at all by ideas. The idea of time, for example, never enters into them. . . . In dreams one sees a lot, one rarely hears, one does not reason, one feels vividly, sensations succeed one another without the soul either comparing or reuniting them; one, therefore, has only sensations and no ideas. . . . Thus dreams only reside in the material interior sense" and come, not from memory, but from "material recollection." From this it followed that man, because he thinks and possesses true memory, was capable of marking the difference between dream and reality, which animals cannot do. Buffon took advantage of this analysis of dreams to distinguish two phases of sleep: deep sleep, when the "interior sense" was asleep, and lighter sleep, when the interior sense was awake and spontaneously revived past sensations.[69]

Just as there were two memories, so there were two imaginations.** One is "the power we have to compare images with ideas . . . to see clearly the distant relationships of the objects we are considering." This was "the most brilliant quality" of our soul, our "genius." It belonged only to man. The other, which depended uniquely on the corporeal organs was only the vivid and deep impression" that objects produced on our interior senses, and which there provoked violent and irrational reactions. This one, of course, we have in common with animals.[70]

In the same way, fear and all the "primary passions"—compassion, horror, "natural pity,"[71] love, jealousy—are found in animals, since these passions depend on the experience of "natural sentiment" and need no thought. For domestic animals, one could even add passions that are born "from education, example, imitation, and habit": a "sort of friendship," like a dog's "attachment" to its master, a "sort of pride," a "kind of ambition." All this did not presuppose any "power to reflect." Likewise, imitation, which allows the monkey to mimic human actions, or the canary to learn musical melodies and the parrot to speak, was only the product of a mechanical apprenticeship. "Monkeys are at the very most creatures of talent

---

*"Various theologians and philosophers" had already objected to Descartes's ideas on dreams by animals, and in particular dogs, who "know . . . in dreaming, that they bark" (*Méditations*, Sixth objections, "3d scruple," in Descartes 1963, II, p. 851 [Adam-Tannery, IX, p. 219]). Descartes answered that it was up to those who presented such assertions about animals "as if they were in secret understanding with them and see everything that passes in their hearts" to offer proofs of them (Descartes 1963, II, p. 886 [Adam-Tannery, IX, pp. 228–9]).

**Descartes as well admitted a "fantasy or corporeal imagination" (*Méditations*, Answers to the third objections, in Descartes 1963, II, pp. 611–2) whose mechanism he analyzed in article 21 of *Passions de l'Ame* (Passions of the Soul) (ibid., III, pp. 968–9 [Adam-Tannery, XI, pp. 344–5]). The description of the "awakened dream" that Descartes gave in this text perhaps suggested to Buffon an analogous description ("Sur la nature des animaux," *HN*, IV, pp. 66–7).

that we take for intelligent people," and if they imitate us so well, it is only because their conformation is so similar to ours.[72]

To conclude our discussion of the soul of beasts, let us examine some famous examples of "foresight" that can be observed, it is said, in animals. Bees gather honey and wax for winter, field mice and ants make provision for the harsh season, birds build nests in which to lay their eggs, owls "know how to use sparingly their supply of mice by cutting off their paws to prevent them from fleeing." How not to see here "a comparison of weather, a notion of the future, a reasoned anxiety"? Agreed. But first of all, many of these marvels were only popular tales that had never been verified by "judicious people." Let us admit, however, that all of them were true. What results from this? That it is necessary to give animals an infallible knowledge of the future, thus a knowledge superior to ours, which always remains "conjectural." "I wonder if that consequence is not as repugnant to religion as it is to reason." It is rather ironic to see Buffon call on religion for help.

Religion aside, let us look closer: reason supposes doubt and deliberation; animals act with "assurance": thus "movements that reveal only decisiveness and certainty prove mechanism and stupidity at the same time." In fact, animals "that have a fixed home" bring everything they can there. Bees make a lot of honey and wax, and if, when the flowers are wilted around their hive, bees are taken to "other regions where there are still flowers," they "start to work again, they continue to gather, to hoard." "It is, therefore, not from the product of their intelligence but from the effects of their stupidity that we profit." As for the field mice, the quantity of their provisions does not depend on the length of winter but on the size of their hole. And if birds making their nests are observed, it is discovered that "all these maneuvers are relative to their organization and dependent on feeling." "That intuitive foresight, that certain knowledge of the future that is supposed in them," is not found.[73]

Buffon proposed in principle that animals do not think, that is, they do not have ideas. All his analyses tended to show that animal behavior could be explained by "feeling" alone: by combinations of sensations operating mechanically and by the intervention of the "interior sense." To convince his reader, he adopted the method that Darwin later used, which consisted of examining all the objections that could be made, one by one, and showing that the mechanism of "feeling" sufficed to explain all the appearances of reasoning that can be seen in animals. In fact, almost all the objections that he had posed to himself had already been made to Descartes, and it is clear that Buffon had read Descartes very closely.[74] The two theories, nevertheless, are clearly different.

Without going into a detailed comparison, let us notice that the two

theories differ on a central point. In both cases, the theories start from a
mechanical phenomenon caused in the sensory organs by exterior objects:
motion of animal spirits according to Descartes, "disturbance" of the
nerves according to Buffon. For Descartes, however, the motion of spirits
was transmitted directly to the "pineal gland" situated in the middle of the
brain, which was the seat of the soul.* It was, therefore, in the soul that a
"perception" was born, which necessarily included a judgment.[75] For Buffon, the "disturbance" was transmitted to the "interior sense," where it gave
rise to a "sensation." The difference in vocabulary is significant, all the
more so because the two ideas seem to exclude each other: Descartes never
spoke of "sensation," nor Buffon of "perception." On the whole, Buffon's
animal machine had much superior powers than Descartes' animal machine, even if the latter accepted animal "feelings"** and a "corporeal
memory." For Descartes, even the internal sensations like hunger and thirst
in man required the soul's intervention: they could not, therefore, exist in
animals.*** For Buffon, they clearly came from the "appetite," common to
man and animals, and were purely "material." What did not exist for Descartes was this "internal sense" which, for Buffon, received the sensations
and was capable of combining them and commanding the appropriate
actions for the organs of movement. This organ was completely material,
but its function assured the animal a behavior adapted to all kinds of
situations, so adapted that it could even seem "rational."

Buffon, like Descartes, was ignorant of all the physiological mechanisms
which underlie the behavior of living beings. Descartes described with
confidence the path of "animal spirits" with a wealth of images taken from
hydraulics. Buffon, more careful, spoke of "disturbances" with more conviction than precision. Buffon, perhaps even more than Descartes, would

*Descartes admitted however that a simple mechanical movement of animal spirits can
cause a muscular movement at the same time, as it is perceived by the soul, but automatically
and independent of it (*Les Passions de l'âme* in Descartes 1963, III [Adam-Tannery, XI], art.
38).

**"I see no reason which proves that animals think, if it is not that, having eyes, ears, a
tongue, and the other sensory organs as we do, it is likely that they have feelings like us, and
that, since thought is enclosed in the feelings that we have, it is necessary to attribute to theirs
a similar thought" (letter to Morus, February 5, 1649, in Descartes 1953, pp. 124–5). Yet we
do not know if a thought is "enclosed" in beasts' feelings as in ours. In addition, the French
statement "*elles ont du sentiment*" is a translation of the Latin "*sentire*," which can mean only the
operation of the "sensory organs."

***Descartes distinguished "three degrees" of the "certitude of the senses." The first was
simply the "motion of particles" in the "corporeal organ." "The second contains everything
that results immediately in the soul, from which it is united to the bodily organ . . . and such
are the feelings of pain, tickling, hunger, thirst, colors," etc. "The third makes up all the
judgments that we are accustomed to make" (*Méditations*, Answers to the sixth objections, § 9,
in Descartes 1963, II, pp. 878–9 [Adam-Tannery, IX, pp. 236–7]). The precise difference
between Descartes and Buffon concerned the "second degree of the senses."

have been delighted by modern knowledge of the brain's functioning.*
Nonetheless, deprived of this knowledge and under the influence of vitalist
physiology, Buffon distanced himself even more from Descartes and decided in 1758 that it was not the brain but the diaphragm that was the seat
of the "interior sense." The brain itself was unfeeling, and a part of it can be
removed without the animal's behavior being modified. Even though sensations arrived at the brain, and even though they combined there, it was in
the diaphragm that the impression of a "lively feeling" was felt. The brain
therefore served only to nourish the nerves.[76] This idea had the unexpected result of attracting Buffon's attention to the role of the spinal cord.
It also prevented him from differentiating reflex motion from other neuro-
physiological reactions.[77]

Deprived of the necessary physiological knowledge, on what did Buffon
base his analyses? On the observation of animal behavior, but also on
analysis through introspection of human feelings. At every moment, this
analysis provided a means to distinguish that which, for man, came from
the "interior feeling" and that which came from the "soul." This method,
perhaps uncertain and perilous, raised another question, that of the animal presence in man.

---

*The current position, especially for neurophysiologists, tends to remove all distinctions
between the "body" and the "soul" or, to use a more modern vocabulary, the brain and the
mind. This has given rise to debates and an immense literature, in which linguistics and the
"cognitive sciences," not to mention psychoanalysis, play more and more important roles. See
in particular Changeux 1983 and Vincent 1986.

CHAPTER XVI

From Animals to Man

I was by studying man that Buffon believed himself able to reconstruct the functioning of the animal machine, for the animal is in man. Both possess the same "material interior sense." The difference is that in man this material principle is "infinitely subordinate" to the "spiritual principle."[1] It is, or at least it should have been, but it was clear that it was never subordinated completely or without difficulty. If man ruled nature only by right of conquest, then the soul ruled the senses only by an unceasing effort. Man has a twofold nature, and the analysis of this double nature must not only lead to the comprehension of all the powers of the "material principle" in animals, as we have seen, but also provide a better grasp of the nature of the "spiritual principle" in man and, perhaps, raise the question of its origin and development. Before even confronting these major metaphysical problems, however, the naturalist has to participate in a more pressing debate, one that was at the core of all eighteenth-century thought: in what does man's happiness consist?[2]

"Homo Duplex"

The idea that man has two natures goes back at least to Plato.[3] Matter and mind, soul and body: man was torn between contradictory desires. This theme became a common ground in moral and religious literature, which meant that, most often, the conclusion was of a moral order. Christian or pagan virtue consisted of assuring the triumph of the spiritual principle over the material, of the soul, or the mind, over the body. Buffon's intention was different: he did not want to teach us to be virtuous but simply to be happy.

For animals are happy. "Guided by feeling alone, they are never wrong in their choices, their desires are always proportional to their powers of enjoyment, they feel as much as they enjoy and only enjoy as much as they feel."[4] They "feel neither internal conflicts nor opposition nor troubled; they have neither our regrets nor our remorse, our hopes, or our fears." This is because their "nature is simply and purely material."[5] In the same way, man "is happy during his childhood, because the material principle dominates and acts almost continually. . . . If he were left entirely to himself, [the child] would be perfectly happy."[6] The grown man sometimes knows this happiness: "He abandons himself fervently to dissipation, to his own tastes and passions, and hardly reflects for an instant upon those very objects that constantly occupy and concern us." In this state, "we are happy. . . . We do not feel any internal conflict. Our *self* seems simple to us, because we feel only a simple impulse, and it is in that unity of action that our happiness consists."

Unfortunately, this state does not last, since the "superior principle" does not delay in manifesting itself. "Through reflection we come to blame our pleasures . . . we lose the unity of our existence, which is the basis of our tranquility."[7] Then why not obey the "superior principle"? "So long as the rational faculty dominates, we quietly take care of ourselves, of our friends, of our affairs," and are happy. But this state does not last either. "We perceive, if only through involuntary distractions, the presence of the other principle." "When . . . through the violence of our passions, we seek to hate reason," "the internal conflict renews itself . . . and the two principles make themselves felt and manifest themselves through doubts, worries, and remorse."

The worst moment is when the two powers oppose each other violently and with equal force: "Here is the point of deepest boredom and of that horrible self-disgust, which leaves us with no other desire than death, and allows us only enough power as is necessary to destroy ourselves." Without going that far, how many "times of trouble, irresolution, and unhappiness"[8] there are. Unhappiness is frequent in middle age: "We still run . . . after the pleasures of youth, and seek them out of habit and not need; and as we age, with increasing frequency we feel less the pleasure than the inability to enjoy it; we find ourselves contradicted and humiliated by our own weakness."[9] This is also the age, as we know, when we discover the vanity of glory and the temptation of laziness.[10]

Was Buffon about to moralize? It is possible to believe so: "We therefore prepare pains for ourselves every time we seek pleasure: we are unhappy as soon as we desire to become happier. Happiness is inside ourselves, it has been given to us; unhappiness is outside, and we go looking for it. Why are we not convinced that the peaceful enjoyment of our soul is our only and true good . . . that the less we desire, the more we possess; that, finally, all that we want beyond what Nature can give us is painful, and that nothing is pleasurable except what she offers us."[11] A rather ambiguous lesson. What "Nature offers us" is physical pleasure, and "there is infinitely more good than bad in this." "We have, as well, another source of pleasure, that of exercising our mind, whose appetite is to know." This is an "abundant and pure"[12] source of pleasure, but not the only one.

Passions, however, blind the soul and turn it away from "contemplation," which is its true function. The soul "was given to us only to know; we should not want to use it only to feel." The passions, true "fits of madness," these "illnesses of the soul," are the cause of all our misfortunes: "Most of those who say they are unhappy are passionate men, that is, crazy people."[13] At the base of the passions lies the corporeal imagination that "forces us to act like animals." "Enemy of our soul, it is the source of illusion, the mother of the passions that master us."[14] The soul cannot be responsible for its own passions.

This idea was repugnant to Buffon: "I do not know, but it seems to me that everything commanding the soul is outside it; and it seems to me that the principle of knowledge is not that of feeling, that the seed of our passions is in our appetites, that illusions come from our senses and reside in our material internal sense. At first the soul exists there silently, but when it joins in, it is subjugated, and perverted when it enjoys."[15]

It is, nevertheless, in the soul that the "morality" of passion was to be found and that passion accomplished its ravages. Love is the best example: "Love! innate desire! Nature's soul! inexhaustible principle of existence! sovereign power that can do everything, and against which nothing prevails, by which everything acts, everything breathes and everything renews itself! divine flame! unique and fertile source of all pleasure, of all sensual delight! love! why do you produce happiness for all beings and unhappiness for man!

"It is because only the physical side of this passion is good; it is because, despite what lovers may say, passion has nothing to do with morality."

This brutal conclusion hardly corresponds to the lyricism of the first paragraph, but perhaps Buffon amused himself greatly by introducing it this way. He knew he would shock "sensitive souls," the readers of Richardson and the abbé Prévost, and Diderot himself, who had praised passions in his *Pensées philosophiques* [Philosophical Thoughts]. Even, it was said, Mme. de Pompadour was scandalized.

Buffon continued his analysis unperturbed: "Indeed what is the morality of love? Vanity: vanity in the pleasure of conquest, an error that comes from making too much of it; vanity in the desire to enjoy it exclusively, an unhappy state that always accompanies jealousy, a petty passion, so base that everyone would like to hide it . . . vanity even in the very way of losing it, for each wants to be the first to break away; for if one is jilted, what humiliation! And this humiliation becomes despair when one recognizes that one has been fooled and wronged for a long time." Conclusion: "Animals are not subject to all these miseries, they do not seek pleasures where there are not any."[16] It is for that reason that they are happy.*

Can there be a happy man? Yes, the "wise man": "master of himself, he is master of events; content with his state . . . sufficient unto himself, he only has a slight need of others, he can not be a burden to them; continually occupied in exercising the faculties of his soul, he perfects his understanding, he cultivates his mind, he acquires new knowledge." A pure mind? Certainly not: "he joins to the pleasures of his body, which he has in common with animals, the joys of the mind, which belong only to him: he has two ways of being happy, which help and fortify each other mutually." "Such a man is without a doubt the happiest being in Nature."[17] He enjoys his body and his intelligence, and is wary of his heart. A single "attachment" is worthy of man, that of friendship, "the only one that does not degrade him" because it "comes only from reason."[18]

M. de Buffon, it could be said, was a dreary friend. Was this a self-portrait? Up to a certain point, it probably was. At least it was Buffon as he would have liked to be or as he would have liked to have been seen, and this very ideal betrays a sensual, intellectual, and minimally sentimental temperament. Unfortunately, rather than insisting on the wise man "continually occupied in exercising the powers of his soul," biographers have preferred to emphasize Buffon's taste for quick and undemanding loves: "He looked only at young girls," wrote Hérault de Séchelles, "not wanting to have women who would waste his time."[19] It is true that on that occasion Hérault was quoting exactly the famous text on love that we just read. The work makes the writer.

There were other passions than love. The wise man Buffon described, independent of other men and of public opinion, "content with his state" and without ambition, could not have written the *Natural History*, directed the Royal Garden, or undertaken all the projects that Buffon would never tire of launching. If we want to find a portrait of the author in his work, it is necessary to look elsewhere, or at least to complete it. "Young girls" aside, the portrait of the wise man is not really that of Buffon. It is a stereotype, that of the wise epicurean who knows how to use pleasures with moderation and seeks above all for "tranquility," that state of which Buffon spoke too much to have truly found.

Thus, the important thing is wisdom and not virtue, happiness and not morals, especially not Christian morals. This leads us to ask ourselves an-

*In the *Discours sur l'origine de l'inégalité* [Discourse on the Origin of Inequality], published in 1755, Rousseau also stated that "the morality of love is an artificial sentiment." It was, according to him, "born from the custom of society," and had been imposed by women! (Cf. Rousseau 1964, pp. 157–8.)

other question: what therefore was this "soul" of which Buffon spoke so much and which created man's grandeur?

On Man and His Soul

The soul was "thought, reflection," "understanding," "the mind, memory," "reason." It "was given to us to know," not "to feel." It constituted the "principle of knowledge," what we would call today the whole of our "cognitive faculties." Essentially it was "the power that produces ideas," that is capable of "comparing" sensations and therefore of "judging."[20] It was not a "spiritual principle" but an intellectual one: nothing in Buffon's texts evokes the Christian idea of the soul.

This principle exists only in man, but we already knew that man does not have the same sensory organization as animals: the sense of sight, known as the "organ of knowledge," predominates in man. This principle exists in all men but does not act the same in all of them. A somewhat emblematic figure returns periodically in Buffon's writings, that of the "imbecile." Buffon found it in Locke, in a passage where the author was striving to mark the difference between man and animals.[21] The imbecile was human, and therefore had a soul, but the soul was not active in him. This inaction was all the more mysterious, since Buffon had explicitly stated that the soul's activity did not depend on the activity of the senses: "It does not seem that the people who have dulled senses, are short sighted or hard of hearing . . . have less intellect than others."[22] For the imbecile, it is therefore necessary to suppose an absence of communication between the soul and the "bodily organs." Left to themselves, the latter act alone, and the imbecile is similar to animals.[23] Later it was the "organization" of the imbecile that was called into question: "The defect is certainly in the material organs." The soul was quite "independent of matter," but "as it pleased the divine Author to unite it with the body, its acts depend on the constitution of the material organs," as could be seen in the imbecile and also in the "sick man in delirium," the sleeper, "the newborn child who does not yet think," or "the old decrepit man who no longer thinks."[24]

There is more: all men are not equally "spiritual." Some "never receive two sensations without comparing them and without consequently forming an idea from them"; for others, stronger sensations were necessary; "finally others, the majority, have so little life in their soul and are so apathetic in their thinking that they compare and combine nothing. . . . These men are more or less stupid and seem to differ from animals only by that small number of ideas which their soul has so much trouble in producing."[25] Many men are capable only of an animal "attachment": the attachment "of a woman for her canary, a child for his toy": "These puerile habits only last

Small gibbon. From *Histoire naturelle,* p. 39.

through idleness and have force only because of the head's emptiness; and is not the taste for Chinese porcelain figures (magots) and the cult of idols—the attachment, in a word, to inanimate things—is that not the last degree of stupidity? How many creators are there, however, of idols and magots in this world! How many people who adore the clay that they have kneaded! How many others are in love with the sod that they have tilled!"[26]

Of course, nothing prevented a clever reader from including all the believers of his time in the list of "creators of idols."

In fact, "we should distinguish two different operations in the understanding." The first consists in "comparing sensations" and "forming ideas from them." "The second is to compare the ideas themselves and to form reasonable arguments from them." The majority of men stop at the first. They are therefore "reduced to a servile imitation" and "only do what they see done. . . . Formulas, methods, crafts fill the entire capacity of their understanding, and prevent thinking so much that they cannot create."[27] Granted, but how does it happen that the soul, this "superior principle," acts this way in some but not in others? And how does it happen that men, who possess this principle, react almost as mechanically as animals, who do not?

Finally, what can be said about the child? It is happy, we have seen, because it obeys only the "material principle," like animals. How does it become a man? "The spiritual principle manifests itself later," Buffon explained. "It develops, it perfects itself through education; it is through the communication of the ideas of others that the child acquires it and becomes thinking and rational; without that communication it would be either stupid or odd, according to the degree of inactivity or activity of its interior material sense."[28] Deprived of communication with adults, a child would therefore remain an animal. Would the "spiritual principle" be absent? That was impossible, since it had been given to all men. So why would it stay inactive? And, if one dares ask the question, how did the first man become a man? Buffon, of course, did not ask this question in this manner, but it is one that he would answer.

How Did Man Become Man?

The eighteenth century was fascinated by the problem of origins: the origin of ideas, the origin of knowledge, the origin of society. In order to solve these problems, however, it used an analytical rather than historical approach, proceeding logically from the explicit or implicit principle that the simple existed before the complex. Buffon himself, we know, reconstituted the origin of our ideas by referring them back to combinations of elementary sensations. Condillac and Charles Bonnet used the same method. They did not claim, for all that, to have reconstructed the history of a "primitive man."

The case of Jean-Jacques Rousseau is more ambiguous. In the *Discourse on the Origin of Inequality*, Rousseau evoked a "man of Nature" anterior to all society and all history. This man, at ground zero of humanity, solitary, "wandering in the forests, without industry, without words, without a home, without war and without ties, without any need of his fellows and without any desire to harm them," joining with members of the opposite sex only for rapid couplings, distinguishable from animals only by his perfectibility. Exceptional events such as natural catastrophes, droughts, and floods were needed to force him out of his solitude and into society.[29] It was then that history began, a history in which man perfected his intellectual capacities and lost his primitive innocence. Whatever the status of man in nature may be, it is at least certain that modern man is the product of a history.

Buffon engaged directly with Rousseau. He rejected any possible existence of Rousseau's "man of Nature," in which Rousseau saw only a modernized version of the Golden Age, "a moral allegory, a fable, in which man is treated like an animal in order to give us lessons or examples." The disagreement was twofold: it bears on what it is that we dignify with the name man—we shall come back to this—and it bears as well on the physical possibility of this "man of Nature." "Let us first distance ourselves from all facts, as they do nothing for the problem," Rousseau had written. By this he meant that since the man of nature had disappeared without leaving any historical trace, and since all observable men live in society, no available document can help to reconstruct the original man. Buffon could not accept this method: "When someone wants to reason from the facts, it is necessary to distance oneself from suppositions." Now, we do not find man living alone in nature. We do not find, "in traveling over all the isolated places of the globe, human animals lacking words, deaf to voices as well as signs, dispersed males and females, abandoned children." "The most solitary, the most independent of men do not let the opportunity go by to form families and to be submissive to their fathers." The reason for this is simple and comes from physiology: "Children would die if they were not helped and cared for over several years; newborn animals, on the other hand need their mother only for a few months." Therefore, "it is not possible to maintain that man has ever existed without forming families," and Rousseau's natural man was no more than a myth.

This meant that the distance between the natural man and the modern savage, such as Rousseau had established it, did not exist. Rousseau's man of nature must have crossed this distance when natural catastrophes had forced him to leave the solitude of the forests and to combine with his fellows. In these earliest historical times, he must therefore have learned to communicate with other men, to start a family, and to organize a society, simple as it may have been. Between the state of nature and the "age of huts," a historic process of humanization had transformed an animal that was still only a potential man into a true man.

It was this historical process of humanization that Buffon apparently rejected. He did not want to believe "that there is a greater distance between man in pure Nature and the savage than between the savage and us."

For "the state of pure Nature is a known state; it is the Savage living in the wild, with his family, recognizing his children, known by them, using words, and making himself understood."[30] At this simplest degree of man, man was already a man, for he possessed language and therefore thought.

This situation, the most "primitive" that can be imagined, is nonetheless the result of a natural process. It is the slowness of the growth of the human infant that required the family's existence. Therefore "the union of fathers and mothers to children is Natural because it is necessary. This union cannot fail to produce a mutual and lasting attachment between the parents and the child, and that alone was enough for them to become accustomed among themselves to gestures, signs, and sounds—in a word, to all the expressions of feeling and need."[31]

We are here at the origin of language and society. Are we not also at the origin of man? Let us be careful with the vocabulary: between parents and children, there was an "attachment" that made them live together and brought them to communicate their "feelings" and their "needs." All these words, we know, can be applied to the operations of the "material principle." Was this primitive family already human? What separated it from certain animal families, other than that it lasts longer? In certain animals and especially in birds, there is also a lasting "attachment" between the male, the female, and their babies. Was the true primitive human family nothing other than a particular animal family, and if so, did it not have some distance to travel before becoming a family of savages, who already used words? In short, was it not the animal that in Buffon plays the role of Rousseau's "man of Nature?"

Buffon never posed the question in these terms, and yet it does need to be asked, especially in reference to Rousseau. For Buffon knew well the innocent happiness that Rousseau gave to his "man of Nature": it was the happiness that Buffon himself attributed to wild animals, specifically leaving solitude out. There was, therefore, for both authors a process of humanization. For Rousseau, it took place between the "man of Nature" and the man of society; for Buffon, between animals and man. Rousseau clearly gave this process a historical character. Buffon analyzed its mechanism but refused to situate it in a history.

On several occasions, Buffon repeated his argument: in 1753 in the *Discourse on the Nature of Animals*, in 1758 in *Les Animaux carnassiers* [Carnivorous Animals], in 1766 in the *Nomenclature des singes* [Nomenclature of Monkeys]. In all these texts, he wanted to show that man is not an animal, even when the physical resemblance is striking, as with the orangutan. And he always repeated the same necessary sequence: slow growth, therefore long education; durable families, therefore language and thought. In addition, he always maintained his two major assertions: the soul is a "living, immortal force," a "divine light,"[32] and although the Creator chose to give man an "animal form similar to that of the ape, He suffused this animal body with His divine breath."[33] Without a long education, however, the child would not become a man: "If he only had as a teacher his Hottentot mother, and at two months of age his body was developed enough to go without her care and separate himself from her forever, would not this child be worse than an imbecile, and *as for the exterior* quite on the level of animals?" Let us go further: "If one should like to suppose *falsely* that this mother in the state of Nature possesses nothing, not even words, would not this long association with her child be enough to give birth to a language?"[34] In these two quotations, let us take out the words I have emphasized: the child without education became truly an animal; language and therefore thought become purely natural phenomena.

It is therefore possible to believe that Buffon cautiously added a number of reassuring statements about the divine character of the human soul, and these statements became perhaps even more numerous after the Sorbonne's criticism. It is equally sure that he attempted to establish an essential difference between man and animals. Buffon presented a most repulsive portrait of the most hideous of the savages, the Hottentot: "A stupid or ferocious look; hairy ears, body, and limbs; skin as hard as black or tanned leather; long, thick, and hooked nails . . . for sexual attributes, long and limp breasts, the skin of the stomach hanging to the knees; the children wallowing and crawling in ordure; the father and mother squatting on their heels, both hideous, both covered with stinking filth." Let us add all the resemblances we want between the worst savage and the ape: despite everything, "the distance separating them is immense, since in the inside the savage is filled with thoughts and on the outside with words." If an ape or any other animal species had possessed these powers belonging only to man, "this species would have soon become man's rival; enlivened by the mind, it would have prevailed over the others, it would have thought, it would have spoken."[35] Nothing like this has ever occurred. Even though his origin was purely natural, man's superiority is absolute. For the naturalist, that alone counted; the problem of the origin, divine or not, of this superiority does not fall within his jurisdiction. Within his competence are only those mechanisms that, according to some, allow the soul to express itself or, according to others, give birth to thought. Although Buffon probably belonged to the second group and adopted the language of the first only out of caution, his description of man was the important thing.

Society Makes Man

Let us now return to our "Savage in the family." "No matter how little he prospers, he will soon be the head of a more numerous family, of which all the members will have the same manners, will follow the same customs and will speak the same language." With time "new families . . . always united by the common ties of custom and language, will form a small nation." If they have vast space and a mild climate at their disposal, these nations will remain savage. Constrained by a lack of space or a severe climate, they will confront one another or mix with other peoples. A society will be created everywhere: "It is a constant effect of a necessary cause, as it derives from the very essence of the species, that is, its propagation."[36] Thus, "the necessity of the parents' prolonged caring for the child produces society in the middle of the wilderness."[37] It also produces man, as "all actions that can be called *human* are relative to society."[38]

In this way, Buffon dismissed the problem Rousseau attempted to solve, that of the origin of society: there was no man without society, and it was even society which created man. To establish man's absolute superiority, Buffon was therefore obliged to deny the existence of true animal societies, based on real communication, and consequently on a language, primitive though it might be. Certain observers of animal life such as Le Roy asserted that such languages do exist.[39] If that is so, it would be possible to take the step that Buffon refused to take, and to see human society as the product of an evolution of animal society. This is what Lamarck would do in the famous pages of his *Philosophie zoologique* [Zoological Philosophy].[40] It was Buffon, however, who first made the consideration of this idea possible by indissolubly linking humanization and society. Despite his animallike appearance, Rousseau's natural man was further from the animal than Buffon's "spiritual" man: his "perfectibility" had no natural explanation.

With the extension of the primitive family, we should enter into history, a history that would be at the same time that of the species and that of society. Buffon indicated the means for this history; it was once again education, but a human education, of another nature than animal education. "A young animal . . . learns in a few weeks to do everything its father and mother do." With man, on the contrary, "the education of the child is no longer a purely individual education, since his parents communicate to him not only what they know about Nature but also what they have received from their ancestors and the society of which they are a part; it is no longer a communication made by isolated individuals, who, as with animals, do no more than transmit their simple practices; it is an institution in which the entire species participates, and whose product forms the basis and the connections of the society."[41] The instrument of this transmission is obviously language. Thus was born what modern ethnologists call a "culture," which became from then on a part of man's heritage, and therefore of his very being, just as much as his biological inheritance.

Buffon was, however, too much a man of his century to ignore all the fashionable theories of the origin of society. Without fear of contradicting himself, he did not hesitate to affirm that "among men, society depends less on physical fitness than moral relationships." He presented society as the product of a conscious decision of man, who understood "that solitude was no more for him than a state of danger and war" and that it was necessary to seek "security and peace in society" and even "to renounce the unlimited use of his will to acquire a claim on the will of others."[42] This is a surprising picture, for it supposed the existence of rational man before society appeared. This owed much to Hobbes and all the social contract theorists before Rousseau. Buffon returned to this picture in the following paragraph by explaining the necessity of the family for the human child's survival, according to the scheme that we already know. This text appeared before the *Discourse on the Origin of Inequality;* it was the dialogue with Rousseau that forced Buffon to be more precise in his ideas. Here the important issue involved contrasting human society with the alleged animal societies and showing that human society "presupposed the rational faculty in man."[43]

It also required showing that without society man could do nothing: "He is calm, strong, he is great, and controls the Universe only because he knows how to control himself, to tame himself, to humble himself and to impose laws on himself: man, in a word, is only man because he knows how to join himself to others."[44] In a space of twenty lines, the contradictions pile up. One is tempted to ask Buffon how a man who was not yet a man could have decided that society was the only reasonable solution for becoming one. It is more interesting to notice that the contradiction is not accidental, that it was at the very heart of Buffon's thoughts on man, always split between the absolute superiority of reason and the philosophico-scientific requirement to explain that superiority by natural causes.

The main point remained: it was in society and through society that man became himself and took the place in nature that he currently occupies: "It was necessary that he himself be civilized in order to know how to instruct and command others. His domination over animals, like all his other dominions, had been founded only after society. It was from society that man received his power, it was in society that he perfected his reason, exercised his mind and focused his forces; before society, man was perhaps the most savage and most terrified animal of all: naked, without weapons and without shelter, the earth for him was only a vast wilderness occupied by monsters, whose prey he often became."* It was "only from the gifts of the arts and society" that "man could move forward and conquer the Universe."[45] Every society presupposes authority. If one "descends by insensible degrees from the most enlightened, the most polite societies" to "coarser" people,

*To protect himself from theological criticisms, Buffon specified that he was only speaking about men who had "become criminal and ferocious," that is, fallen through original sin. In Genesis, indeed, God had given man all power over animals.

Chimpanzee. From *Histoire naturelle*, p. 40.

these are "still submissive to Kings and to laws." "Chiefs" are still found among the savages, and at the very least "customs." "The most isolated, the most independent do not fail . . . to submit to their fathers. An Empire, a Monarchy, a family, a father, these are the two extremes of society."[46] Buffon could imagine society only as a monarchy.

Although all men live in society, all societies are not equal. It should therefore be possible to arrange them in ascending order, from the primitive family to the "most enlightened nations," and thus to reconstruct the history of humanity. We have just seen that Buffon followed the opposite order, descending from the enlightened nations to "the most isolated" families, exactly as he had descended from man to the polyp when putting the animal species in order. He made only furtive allusions to a possible history of humanity, which did not assume a continuous and regular rise toward civilization. He did not doubt "that the sciences were cultivated and perfected in very ancient times, perhaps beyond what they are today . . . during centuries of enlightenment" that were followed by a "cycle of darkness."[47] He evoked, "a time when man, still half wild, was, like animals, subject to all the laws, and even to the excesses of Nature," "those overflowings of the human species, the Normans, Alans, Huns, Goths, peoples or rather tribes of animals with human faces" who came "marching in unrestrained hordes . . . to ravage cities, overturn empires, and after having destroyed nations and devastated the Earth, repopulated it with men as new and more barbarous than they."[48] The progress of civilization is never definitive.

It is in this nonhistorical perspective that it is necessary to place the savage, whom we are tempted today, as in the eighteenth century, to describe as "primitive." We have seen that the ethnological journey that Buffon presented to us in the *Natural History of Man* assumed historical events, migrations, the mixing of populations, but did not seek to classify people by an ascending order of "civilization." We have also seen that the physical agent responsible for these "varieties in the human species" was the climate. But although there is no history of humanity, there are different degrees of civilization, and there is a direct correlation between the degree of civilization that a given people attains and the mastery that they exercise over nature.

Buffon expressed this idea most strongly with regard to the American Indians. Persuaded that American animal species were smaller than those of the Old World, he found an analogous inferiority among the savages of the New World, not in size but in activity. Buffon's analysis is interesting for the manner in which it linked physical with social aspects. It begins with the climate, which gave rise to a very specific physical inferiority: "The Savage has weak and small organs of generation; he has neither hair nor beard and no ardor for his female." His other inferiorities follow from this. The Indians "are lacking ardor for their females, and consequently love for their fellows. . . . The most intimate society of all, that of the same family, only contains weak links for them; the connection of one family to another does not exist; thus, no fellowship, no republic, no social unit. The physical side of love defines their morals; their hearts are icy, their society cold and their rule hard. They look at their women only as servants at best or beasts of burden. . . . They have only a few children; and take little care of them

. . . and that indifference to sex, [by] destroying the seeds of life, cuts at the same time at the root of society."[49] In fact, it was the very status of American man that Buffon challenged.

"Under this miserly sky and in this empty land," man, "in small numbers," "scattered, wandering," did not know how "to master this territory as his domain." He had "no rule" there, he had never "dominated either animals or the elements," never "either tamed the seas or guided the rivers or worked the land." He was barely a man: he was "in himself only a being of the lowest order," "a being without consequence, a sort of impotent automaton," incapable of "improving" or "assisting" nature.

Abandoned to herself, being "neither caressed nor cultivated by man," nature, "hidden under her old garments," had never "bared her beneficent breast." The "stagnant waters," which no human hand had made flow, the "coarse, thick, bushy grasses," which no fire has come to purify, holds moisture in, and this "cold mass . . . will produce only moist things, plants, reptiles, and insects, and will be able to support only cold people and feeble animals."[50] We recognize here the depiction of that "raw, hideous and dying Nature" that Buffon would give several years later in his "First View" in *De la Nature* [On Nature], inspired by descriptions of tropical America.

Nature is what it is only because of man, and man is man only if he masters nature. It is through action as much as through knowledge that he realizes himself completely. The chain of causes—slowness of growth and of education, family, language, thought, society—takes on its meaning and reaches its goal only in the domination of nature. Man was not born man; he became it through a double labor on himself and on the world, and he must not slacken his effort. Man is an imperial animal: where he does not exercise his rule, he is not truly a man.

To the extent that this portrait of man reveals Buffon's intellectual temperament, we can correct, or at least complete, the idealized portrait of the sage, in which we wanted to see his self-portrait. Buffon would not consider himself a complete man if he had been imprisoned in reflection or contemplative knowledge. His obsession with marshes that needed draining reminds us of the tree nursery of Montbard, established on carefully drained lands. The wild forests that needed to be aerated and tended evoke his activity as a woodlands owner. If he shared the horror of his century for "miasmas" that poisoned the atmosphere, he believed as well that man had the duty to purify nature in order to allow himself to bloom. It was because he himself had contributed to this task that Buffon could consider himself a man. Against all the moral tales that celebrate, along with Rousseau, the frugality of the man of nature, or that affirm, like Epicurus or the Stoics, that "the lack of pains is equal to the enjoyment of pleasures" and that "in order to be happy" it is enough "to desire nothing,"

Buffon preached a morality of enjoyment and action, both physical and intellectual. For if all these so-called sages were right, "then let us say at the same time that it is sweeter to vegetate than to live, to want nothing rather than satisfy one's appetite, to sleep a listless sleep rather than open one's eyes to see and to sense; let us consent to leave our soul in numbness, our mind in darkness, never to use either the one or the other, to put ourselves below the animals, and finally to be only masses of brute matter attached to the earth."[51] If we want to draw a portrait of Buffon from the *Natural History*, we will discover there not a sage like that of antiquity but rather an "entrepreneur."[52]

Man Rules Only through Right of Conquest

Although man has made himself man by mastering nature, it was at the expense of other living creatures. He did not content himself with pursuing them, decimating them, and reducing their numbers. He has had a much more harmful and profound influence on them: "As much as man has raised himself above the state of Nature, animals have lowered themselves below it; submissive and reduced to servitude, or treated like rebels and dispersed by force, their societies have vanished, their industry has become sterile, their feeble arts have disappeared. . . . What views, what designs, what projects can slaves without souls, or the banished without power, have? Crawling or fleeing, and always existing in a solitary way, building nothing, producing nothing, transmitting nothing, and always languishing in misfortune, demeaning themselves, perpetuating themselves without multiplying, losing, in short, by endurance as much as and more than they had acquired through time."[53]

Throughout the *Natural History*, we shall see that Buffon showed himself more and more conscious of the existence of animal societies. So why would not life in society have the same effects for animals as for man? Let us not speak of bee societies, which remained forever in Buffon's eyes a purely mechanical phenomenon.[54] But what about beavers? "Their society, not being a forced union, was formed, on the contrary, by a sort of choice, and suggests at least general support and common views in those that make it up. It also indicates at least a glimmer of intelligence, which, although very different from that of man in principle, produces nonetheless fairly similar effects so that one can compare them, not to a complete and powerful society such as exists among peoples long used to order, but to a nascent society like that of savage men, which is the only one that can fairly be compared to that of animals."[55]

Buffon sketched this comparison, from which it rather curiously emerged that he was unaware of the cultural and therefore social character of the fabrication of primitive tools by man.* The main point for him was to contrast the solitary beaver, whose "genius, faded by fear, no longer blossoms," to the extraordinary accomplishments of beavers living in societies. Despite a critical willingness to reject "all that seems too difficult [for him] to believe," Buffon described at length "a type of republic" that knows how to build dikes, to raise "villages made up of twenty or twenty-five huts," where "everyone had first worked as a group to construct the large public works, and then as a society built individual dwellings." An astonishing society, since "stable peace is there maintained." A human vocabulary as well as a comparison to men, seemed natural: beavers "enjoy all the benefits that man can only desire."[56]

In order to find "groups of beavers forming societies," it was necessary to go "to those uninhabited lands, which men in society only entered much later." Everywhere else, beavers were not themselves. We can, therefore, no longer know what animals are and what they really can be, because we do not know what they were when "each species could manifest its Natural talents in freedom and perfect them in peace by uniting into a lasting society."[57] Those that we observe have lost their "abilities" and their "talents." The same history that made man has degraded animals. "The weight of an empire as terrible as it is absolute . . . removes all means of liberty from them, all idea of society, and destroys their intelligence to the core. What they have become, what they will become again, does not tell us enough of what they were, or what they could be. Who knows to which one of them the scepter of the earth would belong if the human species were annihilated!"[58] The actual and "Natural" distance between man and animals is perhaps much less than what we believe. It would not be impossible for the earth to have another king. This is just another way of saying that man is very much a natural creature, that his power is only the result of history, and that another history could put everything back into question. Once again, man rules only through right of conquest.

Therein lies the problem. If man had received reason from God, that "divine light," as Buffon says repeatedly, his rule would be legitimate, and he would have exercised it from the beginning. In reality, however, he only owed this dominance to a series of natural causes; he has developed it through history and he maintains his power only through force. At the same time, by exercising it, he plays his role in the "economy of Nature."

---

*Buffon, like Rousseau, saw the action of society only in enterprises that required the cooperation of several individuals: "putting up a hut, making a dug-out canoe." All that man can do alone, such as polish a stone axe, make bows and arrows, "are purely individual acts" that "depend only on adaptation, as they only suppose the use of the hand" ("Le castor," *HN*, VIII [1760], pp. 285–6). One might ask Buffon why an ape was not capable of this. By putting these activities on the same level as those of animals, he obviously wanted to underline society's role, but at the same time suppress all the distance between animals and solitary man.

Up until the end, Buffon hesitated between two images of man or, perhaps we should say, tried to hold "both ends of the chain." The contradiction was not in the naturalist's thought: it is in man himself, when he attempts to understand himself and to situate himself in that nature from which he has received everything and of which he has become the master, or the tyrant. It is necessary to legitimize this tyranny, to smother, if possible, the mute reproach addressed to us by all other living creatures to which we are in reality so close; perhaps our successor to the mastery of the earth is already there among these creatures.

At the same time drunk and terrified with his power, understanding that he has become the master of the earth without being sure of being the master of himself, modern man understands this contradiction too well to reproach Buffon for not resolving it. But if Buffon knew how to recognize it as well as he did, it was because, despite his linguistic prudence, he studied man as a naturalist.

CHAPTER XVII

To Identify, Name, and Describe

Buffon had general ideas to spare: all his large treatises were full of them. For this philosopher naturalist, the drafting of these discourses was a moment of happiness and intellectual excitement. For the historian, these texts are an ideal object for analysis and commentary, a temptation that I have not resisted. After having stated and restated that "the exact description and the faithful history of every thing is . . . the only goal that one must first propose,"[1] the naturalist must get to the point, that is, to the animals themselves. Without even speaking of minerals, which we shall consider later, Buffon's *Natural History* comprises a description of more than four hundred species or genera of quadrupeds and birds. That represents thousands of pages: it is understandable that Buffon defined genius as "a greater aptitude for patience."

In reading these pages, we discover that the profession of naturalist in the eighteenth century was not an easy one. The "observers," those who had decided to speak only about what they had in front of their eyes or in their hand, had in one sense chosen the easy road. The difficulty consisted in presenting as complete and precise a tableau as possible of all the animal species in a specific category for all quadrupeds and all birds. Even the most common, the most familiar species could cause unexpected problems. What could be more familiar than the swallow? And yet, where do they go in the fall when they suddenly disappear from our sky? They went to Africa, some said. Not at all, affirmed others, "They swoop down into the swamps and . . . stay torpid there until the return of spring." Who to believe? The testimonies were contradictory and almost all seemed worthy of credence.[2]

And what to say of those exotic animals that were known only through uncertain stories of travelers, where direct observation and fanciful accounts collected from the natives were mixed together? Should we believe Regnard when he noted that the "petit gris," a squirrel from Lapland, crossed lakes by boarding a piece of pine or birch and used its tail as a sail? "This detail could pass for a fable if I had not seen it myself," wrote Regnard. What a charming vision, that of squirrel flotillas, but should we take the word of a writer "who has given us excellent theater plays [but] has never been much concerned with natural history"[3]? Buffon preferred to cite him without commentary.

The question was often even more fundamental: about which animal were we speaking? The descriptions were uncertain, the names were different, faulty, or approximate. How to figure it out? What name finally to adopt? And, in the description itself, what characteristics to retain? For describing simply to describe was not very interesting. The knowledge of animal species only made sense if it led to general conclusions, which meant that it was necessary to ask good questions. Buffon's early thoughts, as we have seen, dealt with the major questions of biology. He discovered problems specific to natural history only little by little, from one day to the next, and so to speak, in the field. He was often the first to consider them, which without doubt accounts for his very great originality and the importance of his work in the history of the natural sciences.

The articles of the *Natural History* were not written according to a standard plan. Each animal raised a problem of its own, unlike the one before, and it was with this problem that Buffon started. Nevertheless, the same questions, more or less developed, are found more or less throughout. There is of course the problem of the animal's identification and that of its name: the name that it had or, very often, the name it was necessary to give it. Then there was the actual description, starting with the anatomical one, that of the "interior parts." That was Daubenton's area, whose text and drawings accompanied Buffon's text. That description, however, was not always enough to identify the animal. At most it yielded the definition of a few large categories, for example, that of ruminants. Between many related animals there exists a great "conformity of the interior parts," and Buffon and Daubenton drew important conclusions from this. They carefully observed morphological exterior characteristics as well as physiological characteristics: the period of heat, duration of gestation, number of litters per year, offspring per litter, and duration of growth, which was sometimes the only means of evaluating the length of life. Then came the description of the animal's "nature," its "temperament," and finally its geographical distribution.

All these characteristics were described less for themselves than to serve for the identification of species, for the separation or the reunion of apparently neighboring species. They were also described in order to answer a few general questions that Buffon asked about living species and that he treated in long theoretical discussions. In fact, the elements of a synthetic view of living nature appeared very early, but Buffon would present it at the end of his history of the quadrupeds.

Buffon did not forget that his readers would not all be naturalists. He addressed himself to a cultivated public, one that was interested in science and philosophy, but he also wrote, more or less consciously, for an aristocratic elite, which he imagined to be like the English model of large landowners, passionate about horses and hunting and concerned to make their lands profitable. The articles dedicated to domesticated animals sometimes contain long discussions on husbandry, on illness and the care to be given to livestock. It is significant that these discussions are much longer and more detailed for the horse, which directly concerned the lord, than for cattle or sheep, which were more easily left to the steward or the farmer. Often, Buffon ended an article by briefly indicating the use of certain parts of the animal, for example the horns, all the while apologizing for leaving natural history to enter into the "history of the arts," that is to say, of techniques. He also often indicated how an animal was hunted. Discussing the deer, he highly praised hunting on horseback with dogs, which scandalized the philosopher Grimm* but which probably charmed the great lords, whom Buffon did not wish to displease; among other things he was not averse to showing them his familiarity with this noble pastime.

The philosophes could also find nourishment in these texts. Like Diderot in the *Encyclopédie,* Buffon in passing denounced many fables. It is false that the shrew is poisonous or that the porcupine shoots its quills and that they penetrate flesh by themselves: "Marvels, which are but falsehoods that we like to believe, increase and grow as they pass through more and more heads."[4] As for the buffalo, he indicated how the Indians had made of the bull "an idol, a benevolent and powerful divinity," their consideration for this useful animal having "degenerated into superstition, the last term of blind respect."[5] The explanation was not perhaps good, but it could please Grimm. Once again in passing, Buffon explained the belief in metempsychosis as an obscure survival of an ancient belief that organic molecules passed from one animal to another. At the same time, he explained the origin of the idea of the soul: one imagined "that what had been alive in the animal was apparently something indestructible, which separated from the body after death. This ideal was called a soul, which would soon be looked upon as a truly existing being."

It then became this "fantastic being" that was supposed to pass from body

*Grimm used the article "Hunt" in the *Encyclopédie* to oppose Buffon's position. Here Diderot stated that the taste for hunting "almost always degenerates into a passion, that it then takes up precious time, damages health, and gives rise to expenditures that upset the wealth of men in high places and ruins some individuals" (*Correspondance littéraire,* November 1756, III, p. 303).

to body. Buffon continued with a long discussion showing how a philosophical truth, delivered to the multitude, degenerates into prejudice, then into religion, and finally into superstition; he then thanked God for having given us the true religion, "which, not dependent on our opinions, is inalterable."[6] But were the Brahmins alone in believing in the soul's existence? The procedure was classic for the period. It is only surprising to find these subjects approached in a natural history.

Identifying

The first question was therefore one of knowing which animal was under consideration. For the large animals of Europe, there was no difficulty, at least for most of them. But what is a *buffle,* or buffalo? The animal was common in Greece and Italy. The Ancients, however, had not known it. Aristotle had spoken of a *bonasus;* no one today knows what it was. The very name buffalo is neither Greek nor Latin.[*] Julius Caesar had mentioned the *urus:* Pliny, the bison. These two animals were subsumed under the name *bubalus.* "The confusion only increased with time: to *bonasus, bubalus, urus, bison* were added *catoblepas, thur,* the *bubalus* of Belon, the Scottish bison. . . . All our Naturalists have created as many different species as they have found names. The truth is here enveloped in so many clouds, surrounded by so many mistakes, that I will perhaps be thanked for having undertaken the clarification of this part of the *Natural History,* which the contradiction in the accounts, the variety of the descriptions, the multiplicity of names, the diversity of the places, the difference of languages, and the obscurity of the times seem to have condemned to eternal darkness."[7] So Buffon tried to bring a little order into this confusion.

Not unexpectedly, the small European animals presented even more surprises. Two species of bats had been identified. Daubenton showed that there were in fact seven, which "were very distinct, quite different from one another, and never even live in the same place."[8] Buffon described them, backed up by engravings. What was worse, unknown species like the "surmulot" (gray rat) appeared. "It is only in the past nine or ten years that this species has spread in the surroundings of Paris: we do not know where these animals came from. . . . The places where they appeared for the first time, and where they soon made themselves noticed by their damage were Chantilly, Marly-la-Ville, and Versailles."[9]

When it came to exotic animals, the situation was even more complicated. Buffon started by creating a bibliography, citing and comparing a large number of authors from whom his secretaries had made "extracts."

---

[*]Ed. note: The *Oxford English Dictionary* suggests that the word comes from the Portuguese *bufalo,* a corruption of the vulgar Latin, *bufalus,* referring to an ox.

Puffin and two penguins. From *Oeuvres complètes*, XXVIII, p. 378.

Of course the authors did not agree in either the names, the descriptions, or the regions where the animals were found. A close comparison of the texts allowed a certain number of claimed "testimonies" to be put aside: Buffon concluded a long discussion of the possum by stating: "The truth of all this is that Valentine, who reassures [us] that nothing is so common as these animals in the East Indies, especially in Solor, had probably never seen any there. All that he says about them, including his most obvious mistakes, was copied from Pison and Marcgrave, who were, in this respect, only copiers of Ximenes themselves, and who were mistaken in everything that they added of their own. . . . What we wonder now is whether it would be reasonable to have faith in the testimony of three men of whom the first had made a mistake in seeing, the second had amplified the first's errors, and the last had copied the other two."[10] Here is someone who was starting to clear the ground. It was also possible to spot the authors of whom it was necessary to be wary: "As for Kolbe, we count his testimony for nothing, since a man who at the Cape of Good Hope has seen elk and lynx very similar to those in Prussia could just as well have seen tamanduas [Brazilian anteaters] there."[11]

First, then, the authors must be compared. If possible, verification was made by direct observation. Each time Buffon was unable to see an exotic animal, he noted it. Often he saw only the stuffed animal, but he had also seen a great number of live animals and, when he could, he kept them in captivity for observation before Daubenton dissected them and gave a very precise description.

Observing a live exotic animal was not easy. To see a rhinoceros, it was necessary for him to go to the Saint-Germain fair, where one was shown in 1749. Buffon had it drawn, which allowed him to state that Dürer's famous drawing, so often reproduced, was full of inaccuracies. The animal attracted the curious, and the animal painter Oudry painted it. Then the rhinoceros left for London, where it met with the same success. He saw another curiosity: an elk, "but as it stayed only a few days in Paris, we were unable to finish the drawing."[12] Many observations introduced in the *Supplement* had been made at the Saint-Germain fair. At Versailles, there was the King's Menagerie. Buffon was able to see an elephant, a zebra, a zebu,* and several other exotic beasts there. At the Royal Botanical Garden, where there was no menagerie, Buffon still kept a few small animals, including a beaver who lived there in freedom and left one fine day to explore the quarries that were under the garden. In Montbard, where he had built a menagerie and even a pit in which to put bears and lions, Buffon kept many live animals, including a peccary that he kept two years and that he tried in vain to cross with a domestic sow. On the other hand, the only

*Ed. note: A small, humped ox domesticated in the East and parts of Africa (*bos indicus*).

jaguar that he could see had been conserved in alcohol. The Cabinet of the King also contained many skeletons, various bones, and horns of uncertain origin. But how to identify an unknown species of gazelle that no one had mentioned and that was represented by only one horn? Buffon managed it through a stroke of luck. It is difficult today to imagine the work that the compilation of the *Natural History* required for the identification of the animals alone.

The work was thankless, difficult, and without glory. Buffon made this clear in his writing on gazelles: "It is now easy to see how difficult it was to arrange all these animals, which number more than thirty—ten goats, twelve or thirteen gazelles, as many musk deer and prong-horned deer, all different from one another, many absolutely unknown, others presented pell mell by Naturalists, and mistaken for one another by Travelers. It is also for the third time today that I am rewriting their history, and I admit that the work here is much greater than the product; but at least I will have done what was possible with the given materials and the acquired knowledge, which I have had more trouble in gathering than in using."[13] Very often, dozens of pages were necessary to cite and compare accounts and descriptions, before being able to distinguish two different animals or to show that the same animal was hidden behind two or more names. Buffon did not recoil faced with these minute details and laborious debates: he was not just an unrepentant theorist.

In order to guide himself through this maze, Buffon used several standards. The first and most obvious was morphology. Thanks to Daubenton's precise descriptions, to which he often referred, Buffon could compare sizes and shapes in detail. He knew the standards used by the "nomenclators" and "classifiers," and used them himself: number of teeth, nature of the horns, structure of the digestive system. These characteristics distinguish the large groups, not the species. The nature and the coloring of the coat, taking into account seasonal or individual variations, the differences between the male and the female could also all be useful: the marten and the sable, the weasel and the ermine do not all have the same fur. All this was very classic but not always enough.

Every time he could, Buffon used what he called the "nature," the "instinct," or sometimes the "temperament." Though the stone marten and the marten, the skunk and the ferret, the vole and the field mouse resemble each other greatly, they are easily distinguishable by their "nature." He used the same method to differentiate the ferret from the Greeks' *ictis:* according to Aristotle, the *ictis* loved honey. At Buffon's request, Le Roy, the king's inspector of the hunt at Versailles, experimented with a ferret. Negative result: the ferret did not like honey. Buffon himself tried the experiment with an ermine by offering it only honey to eat: "It died after a few days."[14] Conclusion: Aristotle's *ictis* was neither a ferret nor an ermine.

Finally, we come to the last criterion to which Buffon attached great importance: geographical distribution. In contrast to man, animals can live only in certain climates. It was therefore not possible to confuse under the same name as "muskrat" three animals as different as the *ondatra,* the *desman,* and the *pilori:* "The *ondatra* is found in Canada, the *desman* in Lapland and Muscovy, and the *pilori* in Martinique and the other islands of the Antilles."[15] Although the hippopotamus was sometimes designated under the names of "sea horse" or "sea ox," a little common sense was enough not to confuse it with the "sea cow," "which lives only in the North seas," while Buffon was "very close to believing with M. Adanson that the hippopotamus is only found, at least today, in the large rivers of Africa."[16] The question of the geographical distribution of species would have a crucial importance, as we shall see, in the problem of American fauna.

Naming

To all these identified species, a name, and only one, had to be given. Starting with the "Natural History of the Deer" (1756), Buffon indicated in a footnote at the beginning of every article the scientific name that the "nomenclators" had given that particular species. Also cited were ancient naturalists, Aristotle, Pliny, Oppian; then the authors of the sixteenth century, Gesner, Aldrovandi, Belon; and finally the modern ones, John Ray, Linnaeus, Klein, Brisson, who most often are the only ones cited. From time to time, Buffon criticized this scientific nomenclature, which he always gave, of course in Latin, and which allowed his readers to find their way among names that very often differed.

A more original idea was that Buffon, again starting with and including the "Natural History of the Deer," indicated the "common" name of each animal in a number of languages: Greek and Latin, but also Italian, Spanish, Portuguese, English, German, Polish, Danish, Swedish, Dutch, "Slavonic," Russian, Turkish, Persian, Arabic, and even "Savoyard," Old French, and "Grison" (Romanche). For the animals of the New World he gave the Indian, "Mexican," or "Brazilian" name, or simply the one used by the "French of the New World," which was most often a Frenchified indigenous name. Buffon even made up a list of Mexican names where the "pelon ichiatl oquitli" (the llama) appears with the "macatlchichiltic" or "temamaçama" (an animal resembling the gazelle) and the "quauhtla caymat" or "quapizotl," strange names for which Buffon remarked with a double-edged irony that they "are almost all so difficult to pronounce that it is amazing that Europeans took the trouble to write them."[17]

Was this only a display of scholarship? Surely not. If an animal had a name in a language, that meant first of all it existed there where the language was spoken. The buffalo "was known neither by the Greeks nor the Romans, since it had never had a name in the language of these peoples: the very word *buffalo* indicates a foreign origin." Indeed, it "was brought and naturalized to Italy only around the seventh century."[18] The aurochs and the bisons were animals from the North: "The ox without a hump was called *urochs* and *turochs* in the German language, and the wild ox with a hump was called *visen* in the same language." The Romans made "urus" and "bison" out of it.[19] We therefore know where these animals lived.

This same logic held for the elk. Caesar, the first Latin author to mention it, called it *alce:* this name "seems to have been taken from the Celtic language, in which the elk was called *elch* or *elk.*" Caesar also described reindeer, "which seems to prove that it then existed in the forests of Germania." Better yet, Gaston Phebus (Gaston III de Foix, 1331–1391, author of a famous treatise on hunting) described under the name of *rangier* an animal that was probably a reindeer; the animal therefore existed in France at that time, at least in the high mountains of the Pyrenees from which it has now disappeared either because it had been destroyed by man or because the climate of France was "much wetter and colder because of the extent of forests and swamps than it is today."

The philological discussion, which continued for a dozen pages, thus ended with conclusions on the geographical distribution of species and the modification of climates.[20] These conclusions rendered impossible the use of the scientific Latin nomenclature, made up entirely by classifiers. Another advantage of common names was that they sometimes allowed one to recognize the same animal under two different names. Thus the Cayenne "cariacou" is perhaps the "*cuguacu* or *cougouacou-apara* of Brazil," "because of the similarity of the names."

When the animal had a common name in French, it was therefore necessary to keep it. If it did not have one, it had to be given one. This must be a simple name. Buffon gleefully wrote in a footnote that for Brisson the fox was a "Canis fulvus, pileis cinereis intermixtis" and that according to Linnaeus, the ferret was a "Mustela flavescente nigricans, ore albo, collari flavescente putorius," but he wanted to call a cat a cat: "Why make up jargon and phrases when one can speak clearly, by pronouncing only one simple name!"[21] For the new bats discovered by Daubenton, Buffon proposed "oreillar," "noctule," "sérotine," "pipistrelle," "barbastrelle," and "fer à cheval," sometimes inspired by a morphological peculiarity of the animal, sometimes by Italian names. It was from Italy again that he took the name "campagnol" for the "little field rat." To the new species of rat which had just invaded the Parisian region and which Brisson called "wood rat," he gave the name "surmulot" (gray rat): "since it differs as much from the rat as the *mulot* (field mouse) from the mouse, which have their own names, it should also have a specific name."22

As for exotic animals, the necessity of giving each species a "particular name" raised another problem. European travelers had given animals of the New World names that they knew, which brought on great confusion. Buffon discovered the problem with the tiger: "In Europe the first mistake was to call all animals with tigerlike skin from Asia and Africa *tigers:* this mistake, brought to America, was compounded; animals in this new land whose skin was marked with rounded and separated spots, were given the name *tigers,* even though they were neither of the true tiger species nor even of any of the animals with tigerlike skin from Asia or Africa."

Thus, "names confused things."23 It was in order to introduce a little clarity in this confusion that Buffon decided to write three "discourses" on "animals of the Old Continent," "animals of the New World," and "animals common to the two Continents," discourses of considerable importance to which we shall return later. Buffon also apologized for having let it be believed that cats existed in America before the arrival of the Europeans: "I was not then as aware as I am today of all the abuses made of names, and I admit that I did not yet know the animals well enough to distinguish clearly in travelers' accounts between names that had been poorly adapted, borrowed, badly applied, or invented."24 Travelers took the easy way out by using European names, and the nomenclators tangled everything up by proposing their own inventions. It was especially important to avoid names such as "camel of Peru" for the llama, or "Strasbourg marmot" for the hamster, "since it does not sleep like the marmot, and it is not found in Strasbourg."25

Meanwhile, how to name all these false tigers and false camels? How to name all those animals that had been given European names? Buffon's answer was simple: it was necessary to give them the name they have in the country where they live. Thus the llama was neither a camel nor a sheep, but simply a llama. The "muskrat of Canada" was called *ondatra* by "the Savages of northern America." The "muskrat of Muscovy" was called *desman* in Sweden. Why not call them this, since they are different animals, and neither one of them is a rat? The caracal resembles the lynx, but certainly belongs to another species: let us call it a "caracal," a name "that we have from the Turkish language, *karrah-kulak* or *karacoulac*." Among false American tigers, let us distinguish between the "jaguar," from the Indian *jaguara* or *janowara,* and the "cougar," "a contraction of its Brazilian name *cuguacu-ara,* pronounced *cougouacou-ara,* which the French have again inappropriately called a red tiger." And in order to separate two species of "sloths," let us call them "unau" and "ai," which are "the names they have in their native country." Sometimes, however, Buffon shrank before the indigenous name. How to bring into French the *tlacoosclotl,* the "Mexican" (that is, Indian) name of another of those "false tigers"? Buffon proposed "*chat-pard*" (leopard cat). For the countless gazelles that he succeeded in identifying, he adopted the African names, which he generally learned through Adanson. And if he adopted "vampire" for an American bat, it was only because "the American name has not been sent to us."

Thus Buffon created or adopted an entire series of foreign names that he Frenchified, usually artfully. Many of these have entered the language: the "tamanoir" (ant bear), the "cabiai" (water-cavy), the "coati" and the "agouti," the "gibbon," the "jaguar," the "cougar," and many others are found in French dictionaries at the end of the nineteenth century. Others have disappeared, some are still used. In the same way, these exotic names entered other European languages, for the same reasons. Buffon thus enriched the French language. He almost invented "raton-laveur" (raccoon): he proposed "raton" to translate the English *rattoon* or *rackoon,* and noticed that the animal, which he had observed alive, had the habit of dipping everything it ate in water.[26] The *raton-laveur* would get its complete name only when it was necessary to distinguish it from a neighboring species, the *raton-crabier* (crab-eating raccoon).

Describing

Once the animal had been identified and named, it needed to be described, an essential task. Since Buffon almost always took the trouble to say so, we know when he was describing from nature and when he was reporting the descriptions of others. Very often, he preferred to cite word for word the descriptions given by others, choosing those he considered to be the most exact and citing his references. He was quite proud to have discovered the first good description of the hippopotamus, given in 1603 by Frederico Zerenghi, a surgeon from Narni in Italy: he translated it and quoted it almost in its entirety, before showing that "all Naturalists for one hundred sixty years" had used an author, Columna, who had only copied Zerenghi without being "in this article, either original, or exact, or even honest."[27] As for the hamster, which he had also observed himself, he quoted not only the great authors at length but also a paper written by "M. de Waitz, State Minister of the Prince Landgrave of Hasse-Cassel," which had been sent to him with two live animals by M. le Marquis de Montmirail.[28] A good occasion to thank his high-ranking correspondents and to encourage others. M. de Montmirail was actually cited several times for having communicated texts from German naturalists to Buffon. Thus, the vanity of being original was surpassed by the concern to be precise and procure new information.

Description was an art in which Buffon excelled, and he knew it. As a naturalist, he had to give all the necessary information. As a "philosopher naturalist," he had to choose useful information. Finally, as a writer, he had to avoid monotony and above all have a familiar or exotic animal, known or unknown, come alive for his readers. Thus those dry descriptions in Linnaeus are not found in Buffon's writings; Buffon cited Linnaeus's descriptions and sometimes criticized them when he judged them to be inexact. Beyond the fact that they were written in Latin and described only the animal's morphology, often leaving out all that Buffon judged essential, they did not make the animal "come alive" under the readers' eyes.

Curiously, it was Daubenton who, in volume IV of the *Natural History*, discussed the theory of an animal's description at length. It was as necessary, he said, to choose the facts as it was to choose the words. It was especially necessary to know how to compose the "tableau," to give a view of the whole, "to start by describing an animal in the state of rest," then show it "in the state of movement": thus the "portrait" became a "historical tableau." The difficulty lay in "grasping the physiognomy" of the animal. Even the best painters might not succeed.[29] It is clear that here Daubenton was not speaking for himself but was expressing Buffon's ideal. He became more personal when he explained the difference between anatomy as practiced by naturalists and by the professional anatomist.[30] This is a matter of defining the boundary between two neighboring but different scientific disciplines. With this as his starting point, Daubenton proceeded to define comparative anatomy.

For morphology, Buffon often used a comparison with a known animal. The "polatouche" (or flying squirrel) resembled the squirrel "in the size of its eyes and the shape of its tail," but even more so the dormouse "in the shape of its body," the ears and the hair on the tail. The tapir "is the size of a small cow" but "without horns and without a tail; short legs, a curved body, like that of a pig, with certain characteristics when young resembling the deer." It has "a large and long head with a sort of horn, like the rhinoceros." All this was useful only in providing a verbal picture of the animal, to prepare the way for the inspection of the engraving that followed the article. More often, the morphological description served to distinguish between two similar animals, and Buffon did not hesitate to enter into detail.

For example, here are the details that differentiate two neighboring species of exotic squirrels, which Buffon named the "barbaresque" and the "palmiste": "The *barbaresque* has a more curved head and nose, bigger ears, a tail with bushier and longer fur than the *palmiste*. [It] has four white stripes, while the *palmiste* only has three: the white stripe is on the spine, while at the same spot of the *barbaresque* there is a black stripe mixed with red, etc." These are, therefore, two different animals, which Buffon compared with precision because "we have both of them in the Cabinet of the King."[31]

Nevertheless, the description of shapes rarely remained neutral and purely scientific: esthetic considerations were often present. Buffon showed himself sensitive to the beauty of animals. Thus the deer: "Its elegant and light shape, its size, as svelte as it is slender, its flexible and sinuous members, its head decorated rather than armed with living branches . . . its size, its lightness, its force, distinguish it well from the other inhabitants of the forest."[32] The roe deer was no less pleasant to look at: "If it has less nobility, less strength, and much less height than the deer, it has more grace, more vivacity, and even more courage; it is livelier, nimbler, more aware; its form is rounder, more elegant and its face more pleasant; its eyes in particular are more beautiful, more brilliant, and seem to be animated by a livelier sentiment; its limbs are suppler, its movements readier, and it leaps, without effort, with as much strength as lightness."[33] Sometimes we wonder if Buffon is speaking about an animal or a beautiful woman!

The lion, on the other hand, inspired respect and admiration in him: "It has an imposing face, an assured look, a proud walk, a terrible voice; its size is not excessive . . . but is, on the contrary, so slender and well proportioned that the lion's body seems to be the model of strength joined with agility."[34] On the other hand, "the tiger, too long of body, too low on its legs, with naked head, wild eyes, its tongue the color of blood always outside of its muzzle, has only the characteristics of base viciousness and insatiable cruelty." With animals as with man, "the shape of the body is normally in accordance with its nature."[35] Buffon would go so far as to find "physical and moral affinities" between the eagle and the lion and between the vulture and the tiger, pursuing the parallel beyond what was necessary or even likely.[36]

Buffon was very conscious of what was subjective in these judgments. In principle, each creature was "equally perfect in and of itself, as all came from the hands of the Creator. There are, however, relative to us, finished beings, and others that seem to be imperfect and deformed." In fact those "that seem hideous to us are those whose qualities are harmful to us, those whose nature distances itself from the common nature, and whose shape is too different from the ordinary forms from which we have received our first sensations and conceived the ideas that serve as models by which we judge." Thus we consider the bat a monster, as it is neither quadruped nor bird: "A quadruped should have four feet; a bird feathers and wings; for the bat, its front feet are neither feet nor wings [but] deformed extremities, whose bones are monstrously elongated . . . in a word, the parts have more an air of a whim than of a regular production." Add to this "the deformities of the head," and we have a perfectly monstrous and repulsive animal.[37]

The repulsion that bats inspire did not stop Buffon from going to visit the caves of Arcy-sur-Cure, where they lived in large numbers, or from examining their excrement, which showed that they fed on insects. It did not stop him either from noting that like birds, "they have much stronger pectoral muscles than any of the quadrupeds." For, although anatomy was in principle the responsibility of Daubenton, who described and measured everything he observed, Buffon borrowed from him everything he judged important. For example, it was important to note that in the seal, the "oval hole" that separates the two ventricles of the heart stays open, which allows the animal to stay in the water a long time without breathing. In the possum, which was one of the first known "marsupials," Buffon noted not only the existence of the "pouch" (marsupium) where the young stay to finish their development, but also the particular muscles and bones that support this pouch and allow it to open and close. For the "polatouche," or flying squirrel, Buffon carefully described the skin's conformation, which allows the animal to glide when jumping from one tree to another. I could go on with examples of this type, even without speaking about the anatomy of the digestive system, which gives rise to the distinction between ruminants and other animals. Thus, Buffon could state that the rabbit was not a ruminant, no matter what had been said before, but he had not observed the particular behavior that had given rise to this error.[38]

This information was given only when it was necessary, and especially when it was linked to the animal's gait and speed. Here is where the "historical tableau" of which Daubenton spoke, appeared, and here is where Buffon excelled. The description could be very precise, even technical: thus the long analysis of the horse's gaits, in particular the gallop, compared with those of the other quadrupeds.[39] After having read these pages, it is clear that the posture of Géricault's horses, in his famous *Derby d'Epsom,* used poetic license to its advantage. Every time that Buffon had been able to observe an animal himself, or collect the necessary evidence, he made the animal come alive under the reader's eyes:

"See these horses that have multiplied in the regions of Spanish America and that live there as free horses, their gait, their running, their jumping, are neither hampered nor measured. . . . They wander, they leap in freedom in the immense prairies. . . . They also travel in herds and reunite for the sole pleasure of being together, for they have no fear, but they become attached to one another. . . . All this can be remarked in the young horses, which are raised together and which live in herds; they have gentle habits and social qualities, their strength and their passion are ordinarily only marked by signs of emulation; they seek to overtake one another in running; they inure themselves to peril and even become livelier when faced with it, daring one another to cross a river or jump a ditch."[40]

Buffon had never seen herds of horses in the pampas, but he imagined them in accordance with those he knew. He had never seen an elephant outside of Versailles, and it was from accounts by travelers and other naturalists that he sang its praises: "It is necessary to give it at least the beaver's intelligence, the monkey's dexterity, the dog's feelings, and then add [to this] the particular, unique advantages of strength, size and long life." Above all, it must be added "that to that incredible force, it also adds courage, caution, calm, and strict obedience; that it remains moderate even in its strongest passions; that it is more steady than impetuous in love; that in anger it does not ignore its friends; that it only attacks those who have offended it; that it remembers good turns just as long as insults; . . . that finally, it is loved by all, as all respect it and have no reason to fear it."

Buffon does not go so far as to follow the Ancients, who "had no fear of giving these animals a rational morality, a natural and innate religion, the observance of a cult, the daily adoration of the Sun and the Moon, the custom of washing before adoration, the spirit of divination, piety toward Heaven and toward their fellow creatures, whom they support in death and whom, after death, they water with their tears and cover with earth, etc." But he includes all the enlightened descriptions of the moral virtues and intellectual capacities of the elephant, its skillfulness and its cleverness in the execution of tasks assigned it. Like many others, he emphasizes most of all its modesty, which led it to hide its love affairs carefully: "One never sees them mate in public; they are especially afraid of the stares of their fellow creatures and know perhaps better than we that pure sensual delight of enjoyment in silence, and of concern only for the loved one."

The domesticated elephant refuses to reproduce: "Here the isolated individual is a slave, the species remains independent and constantly refuses to increase for the tyrant's profit. That alone supposes feelings in the elephant raised above those of the common nature of beasts: to feel the strongest passions and refuse at the same time to satisfy them, to enter into the tempest of love and remain modest are perhaps the ultimate test of human virtue but in this majestic animal are only ordinary acts, which he never fails to do."[41] "We would like, if it were possible," adds Buffon, "to doubt this fact," but the authors were unanimous. We know today that these authors were wrong, but Buffon took visible pleasure in singing the praises of this "animal" that was so far above "beasts" (the change in vocabulary is significant), and we understand why Le Roy, unlike Buffon a firm believer in the souls of animals, did not fail to tell his reader to read "the interesting story that he tells about the elephant."[42]

It is indeed clear that in his descriptions of animal behavior Buffon seems to have forgotten his grand theory of the "animal-machine." His vocabulary attests to this evolution. In the beginning he spoke only of animals' "feelings," which we know was a purely physical phenomenon. From 1755 on, he spoke of "instinct" and "nature," vaguer terms, which he eventually defined: instinct "is only the result" of "feelings" or "the faculty of feeling," and "natural [behavior] . . . is simply the habitual exercise of instinct, as guided and even produced by feeling."[43] Thus the theory was saved.

Buffon, however, forgot it often, used a human vocabulary, spoke of certain animals' "intelligence," and did not hesitate to write that the fox was "skillful as well as circumspect, clever and careful, even patient." Although he was wary of the shrewdness attributed to the otter, he praised the mole, in which he observed "a lively and reciprocal attachment of the male and female . . . the gentle habits of rest and solitude, the skill in finding a secure place, of quickly creating a safe asylum, a domicile for itself, the ease of enlarging it and keeping in it an abundant subsistence without having to leave it. That is its nature, its habits, and its talents, which are probably preferable to more brilliant qualities that are more incompatible with happiness than the darkest obscurity."[44] Who would have expected to see the mole's happiness lauded in the century of the Enlightenment?

There was, in all these descriptions, a great deal of anthropocentrism. Buffon acquired a bad reputation with cat lovers, by denouncing in this "unfaithful domestic" "an innate malice, a false character, a perverse nature, which age increases even more and which education only masks." Having distanced himself from the beginning from the opinion of those who "raise cats only to amuse themselves," which he considered an "abuse," he developed its portrait at leisure:

"Determined thieves; only when they have been well-raised do they become flexible and flattering like rogues; they have the same deftness, the same taste for wrongdoing, the same bent for plundering; like them they know how to cover their tracks, conceal their designs, be on the look out for opportunities, wait, choose, seize the moment to carry out their deeds, then shy away from punishment. . . . They only give the appearance of attachment; it is seen by their shifty movements, by their ambiguous looks. . . . The cat seems to have feelings only for itself, to love only conditionally, to lend itself to relations only in order to abuse them; and because of this natural disposition it resembles man more than does the dog, which is entirely sincere."[45]

The portrait of the cat was obviously the antithesis of the dog, which "comes crawling to offer its courage, its strength, its talents at its master's feet." "Without having, like man, the light of reason, it has all the warmth of feeling; it has in addition loyalty, steadiness in its affections, no ambition, no self-interest, no fear except that of displeasing; it is all zeal, all ardor, and all obedience."[46] One ends up wondering if man deserves such devotion.

When he wrote of the "nature" of the numerous animals he had directly observed, Buffon was especially attentive to their relationships with man. In the case of the deer and the wolf, it was of course the ruses of the hunted animal he described with precision. Whenever he could, Buffon tried to experiment; an example is the migratory behavior of swallows. If it were true, as certain people claimed, that they spent their winters in a torpor in swamps, it was because cold weather produced in them the same numbing effect as for certain quadrupeds. "I have therefore done research to find out which species are subject to numbing, and to discover if the swallow is one of them; I had a few of them closed up in an icebox, where I kept them for a certain time. They did not become torpid; the majority died there and not one started to move again in the sun's rays."[47] Added to the precise observations that Adanson sent to Buffon about swallows he had seen in Senegal and Buffon's own observations of the behavior of these birds before their departure from France, these experiments settled the question: swallows migrate to Africa in the winter, or else we have to believe that we are dealing with two different species.*

In order to study the "nature" of wild animals, Buffon tried first to tame them and often let them run freely in the house or garden. To describe these relationships of animals to man he used a precise vocabulary. There were those that were "deprived," those that were "tamed," those that "come when called," those that "become attached," those that come to be fed without showing any "attachment." Two questions underlie this classification: how had man been able to domesticate animals? Which wild animals might he now domesticate?

Buffon did not forget that these animals did not behave normally in captivity. A female hedgehog put in a barrel with her offspring devoured them rather than eat all the food put at her disposal. Buffon commented: "One could not have imagined that such a slow, lazy animal, which lacked only freedom, would be in such a bad mood and so angry to be in prison."[48] In captivity, the tiger does not lose any of its ferocity, but wastes away physically. In fact "animals of all species raised and fed in menageries . . . never attain their natural size."[49] The chained fox is no longer interested in chickens, even when it is hungry. In fact, "it languishes when it is no longer free, and dies of boredom when one tries to keep it too long in captivity."[50]

On the contrary, animals keep their "nature" in semi-freedom. Young foxes run spontaneously after chickens and ducks, even if they are not hungry. The "raccoon" still washes its food. Animals taken very young let

*Buffon here refers to a "discourse" dedicated to bird migration, but he never published it. He described elsewhere the agitation of the caged quail at the time of its annual migrations, disturbances that he attributed to "an affection of instinct." Guéneau de Montbeillard later wrote at length on the subject of swallows and refuted the fable of their hibernation, which he said was upheld by Linnaeus among others. In passing, he cited an experiment that could very well be one of the first examples of the "banding" of migratory birds. See "La caille," *HNO*, II (1771), pp. 452–65, and "Les hirondelles," *HNO*, VI (1779), pp. 556–88.

themselves be tamed, but with age they "return to their nature." Raised in freedom, young wolves are "docile enough" and "even affectionate." Around two years old, they become ferocious and it is necessary to chain them.

Buffon noted all that he could learn of animal behavior, in particular sexual behavior and everything concerning it: the relationship of the male and female before and after mating, the preparation of a nest or a "bed" by the female who was going to give birth, precautions of certain females like the cat, who hides herself and her young to prevent the father from eating them, yet who sometimes devours them herself. What most interested him, however, was the attitude of one species toward another. Thus the ferret "is naturally the mortal enemy of the rabbit: when a rabbit, even a dead one, is presented to a young ferret, which has never seen one, it throws itself upon it and bites it with furor; if it is alive, it takes it by the neck or by the nose and sucks its blood."[51] Even more interesting were the relationships between two neighboring species. As soon as they "return to their nature," wolves throttle dogs.[52] The nature of the animal is part of the species' definition.

For animals running free in their natural environment, available observations were much rarer. Buffon nevertheless observed the difference between animals that have a "domicile" and those that do not: "This difference, which makes itself felt even among men, has much greater effects and implies much larger causes among animals. The very idea of a domicile presupposes a singular attention to oneself; then the choice of place, the art of making one's dwelling, making it comfortable, concealing the entrance, are all indications of a higher feeling."[53] This, in short, is where civilization started, for animals as well as man. It also started with society, even for animals. Here Buffon described the social behavior of fallow deer:

"When they exist in large numbers in parks, they ordinarily form two herds, which are quite distinct, well separated, and which soon become enemies because they both want to occupy the same part of the park: each of these herds has its leader who walks at the head and is the strongest and the oldest; the others follow, and all place themselves to fight and chase the other herd from the region. These combats are unique in the order that seems to rule in them; they attack one another in order, battle each other with courage [and members of each herd] support one another. . . . The battle is renewed daily, until the stronger chases away the weaker and drives them to a poorer area."[54]

The notions of "territory," of animal societies, and of "male dominance" thus appear, although only in passing. Later, basing his ideas on a memoir sent by one of his correspondents, "M. Sanchez, former first physician of the armies of Russia," Buffon proposed with regard to the horses of the

Ukraine that "men are not the only ones who live in society, and who together obey the commands of a few": "Each one of these horse herds has a lead horse that commands it, leads it, turns and organizes it when it is necessary to walk or stop. . . . This leader is very vigilant and always alert. . . . Moreover, the lead horse occupies this important place, which is most fatiguing, for four or five years; and when it starts to become weaker and less active, another horse, one with the ambition to command and with the strength to do so, comes from the herd, attacks the old leader, who keeps its command if it is not vanquished but retreats in shame into the body of the herd if it is beaten, and the victorious horse places itself at the head of all the others and makes itself obeyed."[55]

Buffon also noted how birds warn each other of a fox's arrival: "They have for it such a strong antipathy that as soon as they see it they give a short warning call: in particular, jays and especially blackbirds accompany it from the tops of trees, often repeating the short warning call, and sometimes follow it at a distance of more than two or three hundred paces."[56] This behavior arises from hatred rather than fear. The proof "is that they follow it from far enough away, and they cry out against all voracious and carnivorous animals . . . and never against a deer, a goat, or a hare, etc."[57]

By thus describing the "nature" of animals, Buffon went against the current of his time, and he was often criticized for having renewed the anthropomorphism of the old naturalists. We have seen, however, that all his observations are not anthropomorphic. The birth of an entirely different scientific discipline is a complex phenomenon, and we are not arguing that Buffon created ethology. But we can say that with Buffon the description of animal behavior became a necessary element in the description of a species, and even of its definition. Buffon was conscious of the insufficiency of the facts he possessed and of the experiments that he tried to do on animals that were not in their natural habitat. These experiments testified nevertheless to a scientific drive, as witness his critical attitude toward the marvels reported by the Ancients concerning the lion or the elephant. His stance was probably not critical enough in our eyes, but a naturalist in Buffon's position could hardly do better.

Behind this scientific drive there was doubtless a philosophical plan. The description of animal behavior tended to blur more and more the sharp line that Buffon had drawn between animals and man, and we have seen all the difficulties it created. Throughout the volumes of the *Natural History*, as the grand declarations of principle with which the work began recede more and more, the reader cannot but admire the intelligence of these beasts, their ingenuity, their ruses, their precautions, all that they know about surviving as best they can in a hostile world. The reader is all the more tempted to admire them because Buffon carefully abstained from all philosophical conclusions in these pages and in the volumes that follow one another. The facts accumulate: it is left to the reader, if he wishes, to reflect upon them.

Beyond a scientific or philosophical plan, however, there was a historical situation and a literary project, or perhaps even a literary imperative. The relationship of man to animals is not only a relationship of knowledge or domination. Between these beings and us there are complicities, hostilities, friendships, repulsions, even terrors, be they reasonable or not. For the man of the eighteenth century, the animal world was still very near. There were still wolves in the forests, foxes that got into henhouses, horses in the streets of the cities. Certainly, pure scientific knowledge should be based on a perfect objectivity, yet few naturalists were capable of practicing it. And then, at that time, and at least for Buffon, natural history was part of literature; he was often criticized for this.

The description of animal behavior, sometimes so close to human behavior, renders natural history "interesting," in the very particular sense of the adjective during the century that used it so much, that is, human and touching. If Buffon had felt nothing toward animals, the "painter of nature" would have been only a cold draftsman. If he himself had not known admiration, respect, tenderness, and fear, he would not have known how to make the animals he described "come alive." It was not by chance that Buffon was a zoologist, that those plans of writing a botany never came to fruition, that he left fish and reptiles to Lacépède. He was fascinated by animals because they spoke to his sensitivity as well as his intelligence, and finally because they spoke to him of man. For minerals, his motives were different. He also would not have been able to interest so many readers in natural history had he displayed that inhuman objectivity that is theoretically the scientist's ideal. That is well known by all those who, in their writings or images, fascinate large audiences with the "life of animals" today.

CHAPTER XVIII

Unity and Diversity
of Living Forms

All That Can Be Is

From the first pages of the "First Discourse," Buffon underlined the infinite variety of living forms. Faced with the "prodigious" number of "Nature's productions," "one is as surprised by the variety in design as by the multiplicity of the means of execution." He nevertheless introduced a first classification in "that multitude of Quadrupeds, Birds, Fish, Insects, Plants, Minerals."[1] In fact, it was to quadrupeds alone that he devoted the twelve volumes of descriptions which, starting in 1753, make up the first series of the *Natural History.*

But was it justifiable thus to isolate the quadrupeds? For as long as he worked on animals with four legs, Buffon did not ask himself that question. But what about monkeys? The question was first asked about them, and the immediate response was that many of the alleged quadrupeds were not such: "The name *quadruped* implies that the animal has *four feet;* if it is missing two like the manatee, it is no longer a quadruped; if it has arms and hands, it is no longer a quadruped; if it has wings like the bat, it is no longer a quadruped, and this general designation is abused when applied to these animals."[2] Indeed man is bimanual and bipedal, the manatee is bimanual, the bat is bipedal, and the monkey is quadrumanous. If all the animals that resemble them and all those that "use the front feet like hands" are counted, we realize that half the quadrupeds do not deserve that name.

Linnaeus had already encountered the same difficulty, and it forced him to modify considerably the very principle of his animal classification. In 1758, in the tenth edition of the *Systema Naturae,* he had abandoned, at least in part, classing animals by morphological characteristics alone and decided that "the natural division of animals is indicated by their internal structure."[3] Suddenly, the class of quadrupeds disappeared and was replaced by that of *Mammalia*, which today we call mammals,* and which includes everything from primates to cetaceans. This solution was very reasonable, since it allowed the grouping of closely related animals by a few essential characteristics of their physiology, but it also offered a rich future, because it attracted attention to a structural unity that the diversity of exterior forms did not immediately reveal. Buffon must have known of this solution, for he cited this tenth edition of the *Systema Naturae*.

Buffon refused nevertheless to follow Linnaeus because he was looking for, or saw, other things in nature. Replacing the quadruped class with that of mammals perhaps solved a technical problem, but it also created a class that was arbitrarily isolated from other living animals. Quadrupeds are "neither superior in everything nor separated by constant attributes or unique characteristics from all other beings." Lizards, frogs, etc. also have four feet: "A separate class of them has been quite rightly made." Cetaceans are viviparous, like quadrupeds. And certain quadrupeds do not have hair: armadillos have a "crust" or "shell" like crustaceans, pangolins (scaly anteaters) have scales like fish, porcupines a "sort of feather that is sharp and without barbs but whose shaft is the same as that of birds' feathers." Thus in quadrupeds themselves, "Nature varies by bringing together three other very different classes that remind us of birds, scaled fish, and crustaceans." To join cetaceans to quadrupeds in order to create the class of mammals unduly privileged one characteristic, viviparity. It was, however, "only through the union of all the attributes and through the enumeration of all the characteristics that one can judge the essential shape of each of Nature's productions."[4] Translated into Aristotelian language, which Linnaeus understood very well, this was the same as saying that a being's essence was the sum of its accidents, which Linnaeus would have obviously rejected.

Buffon, therefore, did not abandon the conviction he expressed in 1749, that "Nature works by unknown degrees" and can pass "from one species to another species, and often from one genus to another genus through imperceptible gradations."[5] Throughout the *Natural History*, he insisted on the fact that certain species "bridge the gap" between one species and another. Thus the shrew came between the rat and the mole,[6] and even more so the pig, which resembles nothing, and of which "the [only] thing that one could say is that it bridges the gap, in certain respects, between the *solipeds* [whole-hoofed animals] and the *cloven-hoofed* animals, and in other respects between *cloven-hoofed* animals and *fissipeds* [animals with divided paws]."[7] Yet we can go further:

*In French, these were first called "mammaires" or "mammaux." In 1802 Gilibert judged it still useful to specify that "mammaires" included quadrupeds and cetaceans. See Gilibert 1802, title page.

Mouse shrew. From *Histoire naturelle,* p. 9.

"Let us bring all quadruped animals together for a moment; let us then make a group of them, or rather let us form a troop of them in which the intervals and the rows more or less represent the proximity or the distance found between each species: let us place in the center the most numerous genera, and on the sides, on the wings, those that are least numerous; let us squeeze them all into the smallest space in order to see them better; and we will find that it is not possible to draw a circle around this enclosure. For, although all quadrupeds are as close as they are to other beings among themselves, there are nevertheless some that push toward the outside and seem to branch out in order to create other classes of Nature; monkeys tend to approach man . . . ; bats are like monkeys among the birds . . . ; porcupines and hedgehogs, by the spines that cover them, seem to indicate to us that feathers could belong to others than birds; armadillos with their scaly shells come close to turtles and crustaceans; beavers with the scales on their tails resemble fish . . . ; finally seals, walruses, and manatees are a little separate group that form the most prominent branch leading to the cetaceans."[8]

Among all beings, therefore, a network of "relationships" existed that a classification, whatever it may be, could only obscure. This text marked a clear evolution in Buffon's thinking. In 1749, we recall, man could, by looking below himself, see "with astonishment that one can descend by almost imperceptible degrees from the most perfect creature to unformed matter, from the most organized animals to the crudest mineral."[9] For this classic image of a linear chain of beings, Buffon now substituted the image of a network in which each species maintained, through each of its characteristics, different "relationships" with a diversity of species.[10]

This passage from one image to another obviously had a philosophical significance. The image of the "great chain of beings," a linear chain that ascended by imperceptible degrees from mineral to man, described a creation ordered according to increasing perfection. The intermediary beings that "bridge the gap" between minerals and plants, between plants and animals, between animals and man, testified to the infinite power of God, who had not wished to leave an empty space in His creation. Such was the meaning of the image of Leibnitz or for Charles Bonnet, who stretched the "chain of beings" beyond man to angels and other celestial creatures.[11] In contrast, the image of a network is a horizontal one, which erased hierarchies, which tangled a universal order, which observed so many "relationships" between so many different beings that it gave rise to the feeling that the forms had appeared by chance without any overall plan, combining in absolute disorder elements of structure that had no other merit than being possible and compatible. Such an image of the living world inevitably evoked the description that Lucretius had given of its birth.

To establish these "connections," these "relations," these "affinities" among living forms, Buffon took into account all the characteristics that served to define a species, which he took from morphology, anatomy and "internal organization," and the environment or behavior. He refused to establish a hierarchy among these various characteristics that would permit a classification. In this respect, he was close to Adanson, with whom he was in personal contact and whom he often cited when he mentioned African animals. Adanson, however, wanted to establish a natural classification, which was not yet the case for Buffon. From then on, it was easy for him to show, as we have seen, that the "class" of quadrupeds "is connected to" other classes: fish, birds, or crustaceans. There were even two ways to go from quadrupeds to birds: it is possible to go from one to the other through the intermediary of "quadrupeds" that fly, like the bat, or one can return from birds to quadrupeds through the intermediary of birds that cannot fly, like the ostrich.[12] But quadrupeds and birds are also connected to fish, the first through the intermediary of walruses and seals, the second through the intermediary of penguins.

Between species, these "connections" are countless, and Buffon was always discovering new ones. In certain cases, he used them to describe an unknown animal by referring to known forms. Take, for example, the case of *cayopollins* (small possums of the forests of South America), with "their split muzzles like a pike, their batlike ears, their tails resembling grass snakes, and monkeylike feet." They are no less "a bizarre form," even though their "internal conformation" and the physiology of their reproduction are very close to those of the marmoset and the possum. They are, therefore, "marsupials," as Linnaeus would say.[13] In other cases, on the contrary, we have the feeling that these "connections" are more basic: such as when it is said that the hartebeest "resembles deer, gazelles, and the ox by a few obvious enough connections."[14] Sometimes, finally, certain species are situated at the intersection of several groups. Thus the cariama, the secretary bird, and the kamichi all belong to the group of shore birds. But the cariama and the secretary bird are birds of prey, whereas the kamichi belongs to the gallinaceous order.[15]

What should one conclude? That nature has left a tangled trail, that she created beings that are bizarre, even monstrous, or which at least seem so to us, that she had "produced a world of related and unrelated creatures all at the same time, an infinity of harmonious and contrary combinations," in short "that all that can be is."* It was not by chance that the statement of 1749 appeared in the article on the pig: "It is necessary to see nothing as impossible, to expect anything, and to suppose that all that can be is. Ambiguous species, irregular productions, anomalous beings will from now on cease to astonish us."[16] The same statement was repeated in 1780 apropos of the toucan's beak, which is both enormous and fragile. This beak, "far from serving a purpose, only harms the bird," whose flight it unbalances, without it even being useful for grinding or crushing food. "The true characteristics of Nature's mistakes are disproportion joined to inutility: all the parts in animals that are excessive, overabundant, placed the wrong way, and that are at the same time more harmful than useful, must not be put in the great plan of Nature's direct intent, but in the little list of her whims or, if one prefers, her mistakes." Yet "these same extraordinary productions indicate to us that all that can be is." These "less careful productions" prove to us that Nature did not limit her power "to those ideas of proportion and regularity to which we would like to refer everything."[17]

Let us therefore not seek to make everything reasonable and, above all, let us not try to find a "final cause" for all the peculiarities of these bizarre beings. The pig "obviously has useless parts, or rather parts of which it cannot make use, digits in which all the bones are perfectly formed and which nevertheless do not serve any purpose. Nature is therefore very far from submitting to final causes in the composition of beings; why should she not add superabundant parts sometimes, since she so often misses putting essential parts there? . . . Why does one want each part to be useful to the others and necessary to the whole in each individual? Is it not enough that, placed together, they do not harm each other? . . . All that

---

*The text does not mention nature but rather the "hand of the Creator"; but this Creator is apparently so inattentive that Buffon would rather have nature be responsible for the mistakes of creation.

can subsist together survives; and perhaps there are, in the majority of beings, fewer related, useful, or necessary parts than indifferent, useless, or superabundant ones."[18] There was, therefore, neither final cause, nor moral rule, nor perfection in design in nature. In short, there is nothing divine in nature, nor is there anything human.

Thus, all our attempts to reduce nature's complexity to a simple order are destined in advance to fail. It is not just that our words do not correspond to things: things that our words are intended to designate simply do not exist. If the word "quadruped" is inadequate, if all "those general terms, which seem to be the masterpieces of thought," are in the end defective, it is because the path of our intelligence is not that of nature. Human intelligence "travels on a line to arrive at a point; and if it wants to grasp another point, it can only reach it by another line. . . . Nature, on the contrary, does not take a single step except to go in all directions: in marching forward, she extends to the sides and above."[19] This text from 1766 apparently repeats some statements from 1749; in fact it was based on fifteen years of work, study, and description of animal species. So was it necessary to despair of introducing some order into this chaos? Was it necessary to be resigned to listing indefinitely the individual descriptions of species that, because of the "connections" they have with all the others, turn out in the end to be isolated monsters, senseless oddities, or, as in a felicitous statement that I am borrowing, "collages" of heteroclitic forms?[20] Was there still something to understand in the infinite diversity of nature's productions?

The "Primitive and General Design"

As early as 1753, Buffon and Daubenton proposed a principle of extreme importance for the history of biology and the natural sciences, which later would be called the principle of the unity of the plan of composition. In short, this meant that animals, or at least a large number of them, are built according to the same "plan." The idea was not new. In his *Histoire naturelle des oiseaux* [Natural History of Birds], published in 1555, the French naturalist Pierre Belon had already had the idea of presenting a human skeleton and a bird skeleton side by side, placed in the same position to show a resemblance, which he emphasized by using the same letters to designate the same bones. The engraving had been reproduced by Aldrovandi, and Buffon could not have been ignorant of it. He perhaps did not know, however, about the very precise comparison that Leonardo da Vinci had established between the leg and foot bones of man and those of the hind leg of a horse.[21]

It was precisely for a horse and an ass that Buffon and Daubenton

asserted this "unity of the plan of composition." In his description of the
horse, Daubenton had meticulously shown the exact resemblance in structure between the skeleton of the horse and that of man, comparing each
bone of one to a bone of the other, thus establishing the list of what would
later be called the "homologies" between the two structures. In the article
on the ass, Buffon took up this theme giving it a wider generality: "Let one
consider, as M. Daubenton has remarked, that the foot of a horse, in
appearance so different from the hand of man, is nevertheless composed
of the same bones, and that we have at the extremity of each of our fingers
the same knuckle bone in the form of a horse shoe that is at the end of the
foot of that animal, and one may judge if that hidden resemblance is not
more marvelous than the apparent differences.

Let us go further: it is the entire skeleton of man that we can transform
into a horse skeleton simply by modifying the dimensions of the bones and
the angles of their articulation without changing anything either in the
order or the general organization of the whole: the same bones are found
in the same place. We can go even further: "If a few parts essential to the
shape are considered separately, the ribs, for example, they will be found in
all quadrupeds, in birds, in fish, and their vestiges [can] be followed up to
the turtle, where they seem still to be designated by furrows under its shell."
From this, it was necessary to conclude that even if all the organized beings
"existed alone," even if they "varied by differences graduated to infinity, a
primitive and general design exists that can be traced very widely, whose
degradations are much less noticeable than those of shapes and other
apparent relationships."

An orthodox and even edifying shape can be given to this statement, and
Buffon did not miss the opportunity: "In creating animals, the supreme
Being wanted to use only one idea, and at the same time vary it in all
possible ways, so that man could admire equally both the magnificence of
the execution and the simplicity of the design."[22] Yet, independent of an
apologetic interpretation, obviously welcome to someone who must constantly prove his orthodoxy, "this constant conformity and this design,
followed from man to quadrupeds, from quadrupeds to cetaceans, from
cetaceans to birds, from birds to reptiles, from reptiles to fish, etc.," opened
a new field of research. We see here the birth of comparative anatomy,
which, beginning with Daubenton's works and continuing through the
intermediary of Vicq d'Azyr, arrived at Cuvier, with whom it attained the
stature of an autonomous and recognized scientific discipline. But we are
also witnessing the birth of a different discipline, morphology, which received its name and scientific dignity from Goethe and which became, with
Étienne Geoffroy Saint-Hilaire, Owen, and the German morphologists, one
of the essential elements of the evolutionary debates in the first half of the
nineteenth century.

Here Buffon perhaps played a more important role than Daubenton. Whereas comparative anatomy, especially with Cuvier, became an auxiliary of general physiology and placed emphasis on the different plans of organization of the major organic functions such as digestion, circulation, the nervous system, and so on, morphology was more concerned with the problem of forms, apart from physiological functions, and therefore focused particularly on the structure of the skeleton. Buffon seems to have clearly made the distinction when he wrote: "Without speaking of the organs of digestion, circulation and generation, which are part of all animals, and without which the animal would stop being an animal and could neither survive nor reproduce, there is an incredible similarity in the very parts that contribute most to the variety of the exterior shape, which necessarily reminds us of the idea of an original design according to which all seems to have been conceived."[23] The major physiological systems that allow animal life were not the object of particular attention: the internal organization is pretty much the same everywhere, and its variations did not particularly interest Buffon, at least here. For philosophical reasons that he expressed clearly in the article on the pig, as we have seen, Buffon did not establish the close ties between form and function that Cuvier later established; such links practically implied a teleology, an argument by final causes that Buffon rejected. It was, therefore, on the form that he concentrated, and it was the comparison of a human and a horse skeleton, in particular the demonstration of how it was possible to pass from one to the other by successive deformations of the bony parts, that illustrated this "original design" and its transformations. Neither Camper nor Goethe, the fathers of morphology, would forget this text of Buffon's.

Buffon, however, did not have the precise ideas that the morphologists of the end of the century would have. Returning to the question in 1766, he did not hesitate to extend this unity of the plan of vertebrates to insects, zoophytes, and even plants, and the notion of "plan" lost its morphological worth by being submerged, in a sense, in the essential functions of life, nutrition, development, and reproduction, "general and common features of all organized substances, eternal and divine features that time, far from erasing or destroying, only renews and renders more obvious."[24]

The intervention of time was remarkable here, because it seemed that only those essential functions of life escaped its action. Was it the same for the morphological plan, for the "original design?" Whence did it come? Where can it be observed in its original purity? And what was the cause of that "degradation" which made it modify itself slowly from species to species before disappearing from view? These questions arise in the reading of the texts themselves. It was not certain that Buffon wanted to, or could, answer them.

If we asked where this "original design" came from, Buffon had a ready

Mouse shrew bat. From *Histoire naturelle*, p. 19.

response, with which he was satisfied: it was the work of God the creator. In fact, if we want to substitute nature for God, nothing about the problem changed. The original design, without being an immediate given of observation, was nevertheless built on the observation of existing forms, and the question of its origin has no answer. If we now ask how this plan "is degraded" progressively from species to species, the answer is not any easier.

Buffon, like Aristotle, took the human skeleton as a starting and comparative point, both for practical reasons (it was the best known) and esthetic ones: it was nature's masterpiece. And if here it is very difficult for us to enter into Buffon's way of thinking, it is because we think "spontaneously" of evolution, and neither Buffon nor Goethe later on thought in these terms. Buffon's "original design," like Goethe's "Urtypus," did not exist anywhere in observable nature, and its transformations were not the result of an evolution in the modern sense of the word.

From then on, the only question that was asked and that Buffon could answer was to know what happened within each species. It was again in the *Natural History of the Horse* that Buffon raised the question: "There is in Nature a general prototype of each species on which every individual is modeled, but which seems, in becoming a reality, to alter itself or perfect itself through circumstances. . . . The first animal, the first horse, for example, was the exterior model and the interior mold on which all horses that are born, all those that exist, and all those that will be born are formed; but this model, of which we only know the copies, was able to alter or perfect itself in communicating its form and multiplying itself."[25] We are tempted to ask, where does the "prototype" come from? This is a scientifically unanswerable question, Buffon would answer, just as unanswerable as that of the origin of "organic molecules" or of the "original design." He would later attempt to answer it. Meanwhile, it was better to let it be understood that the "first horse" had been directly created by God. There were other questions to ask, which could be answered.

We can nonetheless draw from this text an unexpected consequence: although in each species "the original impression survives whole in each individual," "none of these individuals, however, is entirely similar to any other individual, or consequently to the model whose imprint it carries." This was an important statement, since it destroyed all definitions of a species as a collection of absolutely similar beings. It is therefore tempting to see here the origin of the "idea of populations" in the sense used by Ernst Mayr,[26] that is, a conception of the species as a population composed of individuals all differing among themselves. This conception is generally attributed to Darwin.

The Search for the "Lands of Origin"

It was with the horse that Buffon started his investigations into what he later called the "land of origin" of an animal. In fact, the question had already been implicitly posed for man, and the chapter "Variations of the Human Species" suggested that it was necessary to seek the species' "prototype" in the "most temperate climate," where "the most beautiful and well-made men" were found with "the true natural color of man."[27] The same purely esthetic standards guided Buffon with the horse. Because "Arabian horses have always been and still are the finest horses in the world, as much for beauty as for excellence," he thought that "the climate of Arabia was perhaps the true climate for horses," the only one where it was not necessary to "cross breeds with foreign breeds." Nevertheless, the Arabs themselves had contributed greatly to their horses' perfection "through the particular care that they have always taken to ennoble the breeds, by only breeding together animals that were the best made and of the finest quality." Thus, "they could perfect the species beyond what Nature would have done in the best climate," and one can only conclude "that warmer rather than cooler climes, and especially dry regions, are best suited to the nature of these animals."[28] We can see from this passage that Buffon was not unaware of the merits of the systematic selection by breeders, yet the Arabian breeders did work in a favorable climate.

For all the wild animals he studied from 1756 on, Buffon sought to fix a "land of origin." Most often, it sufficed to define their "area of geographical distribution," as we would say today, since "wild animals live constantly in the same way; one does not see them wandering from climate to climate; the woods in which they are born is a homeland to which they are faithfully attached."[29] For species that are more dispersed, it was the climate where all the varieties of the species could be found that must be considered the place of origin. Buffon thus introduced the notion of what would later be called "centers of creation." That meant that all the living species were not born in the same place, but that they must have appeared in various points of the globe, and that, in short, the biblical account of creation cannot be accepted.

In order to see just how original Buffon was, it is enough to compare Linnaeus's text that I have already cited to Buffon's in which Linnaeus argued that "in the infancy of the world" there existed only a "unique island, in the middle [of an] immense sea, where all the Animals lived comfortably and the Plants sprang up abundantly." This island was big enough to shelter them all, since God had only created a couple of each species: it was the Garden of Eden, where "Adam named each animal that the [Holy] Spirit produced before him."[30] The idea of a unique center of creation was firmly entrenched in people's minds. The debates on this point would last a long time.

In a few privileged cases, it was the animals' perfect adaptation to the "climate" that proved its place of origin. So it was for the camel: "It appears to be original to Arabia; for not only is this the region where it is most numerous, but it is also the one to which it is best adapted; Arabia is the most arid country of the world, where water is the rarest; the camel is the most abstemious of animals and can pass several days without drinking; the terrain is dry and sandy almost everywhere; the camel has feet made to walk in sand, and cannot on the contrary maintain itself on wet and slippery terrains. . . . One can hardly be mistaken about the natural country of animals when one judges by these relationships of conformity; their true homeland is the earth they resemble, that is, to which their nature seems to be totally adapted."[31]

As early as 1761, Buffon established a true connection between the climate and animals: "Each one has its region, its natural homeland in which each is retained by physical necessity, each one is the son of the earth it inhabits, and it is in this sense that one must say that this or that animal is original to this or that climate."[32] The geographical distribution of the species thus became an indispensable element of their very definition, which would lead Buffon to a discovery he was particularly proud of, the originality of South American fauna.

"In the new continent animals from the southern provinces are all very small in comparison to the animals from the warm regions of the old continent." The elephant, the rhinoceros, the hippopotamus, the giraffe, the tiger were all much larger than the tapir, the llama, or the jaguar.[33] This was because "the heat was in general much less in this part of the world, and the humidity much greater,"[34] which Buffon explained by the arrangement of the topography and the system of the winds. For the same reason, insects "are nowhere as large as in the new world," the same for snakes, "toads, frogs, and other beasts of this type."[35] If it was added that since men were few in number, not very active, and had "not restricted the torrents, or directed the rivers, or drained the swamps," or cultivated the earth, it was understandable that this earth could produce "only moist and humid beings, plants, reptiles, insects" and could "nourish only cold men and weak animals."[36] Regardless of this fantasy vision of tropical America, we will here retain only the conclusion: "Thus one must cease to be amazed by this general fact, which at first seems to be very singular, and that no one before us had even suspected, that is, that no animals of the torrid zone of either continent are found on the other."[37]

Domestication and Degeneration

The diversity of climates and of "lands of origin" was a first explanation for the diversity of forms. The study of domestic animals immediately suggested another possibility: the modification of already existing forms. Here Buffon opened an immense field of research, of which no one before him had had any idea. Like Darwin later, he started necessarily with the study of domestic animals, not just because he knew them better, but especially because it was well known that breeders had the power to modify forms and it was easier to study their actions. After starting from the same point, however, the two naturalists went in two separate directions. Where Darwin would see selection and the possibility of evolution, Buffon saw "degeneration." The two naturalists probably did not refer to the same techniques of breeding. The action of the breeder is primarily the action of man on nature: that is perhaps where the essential problem lay.

Buffon tackled it again using the horse. The French breeder for whom he wrote did not find himself with the same favorable conditions as the Arabian breeder we saw above. France was not the horse's "natural home-land," and breeders had to fight at first against the influence of the climate by importing breeding stock from other countries: "Without that, seeds, flowers, and animals degenerate, or rather they are so strongly influenced by the climate that matter dominates over form and seems to bastardize it: the imprint rests, but it is disfigured by all the traits that are not essential to it."[38] Did this opposition between "form" and "matter," surprising in Buffon, mark a return to an "essentialism" or even an "idealism" of a Platonic rather than Aristotelian nature? In Plato's *Republic,* matter can only alter the purity of form and push it toward degeneration. We are perhaps here at one of the confluences in Buffon's thinking, a point where his theory of reproduction and his equivocal vision of man's role in nature were joined in an unexpected manner.

Let us look at his theory of reproduction first, which was in fact only a theory of "re-production" and has nothing to say about the appearance of the original, which plays the role of an "interior mold" and therefore of the guardian of the original form. This theory presented a very simple mechanism of "degeneration": even when born of foreign parents, the young animal is subject "at a tender and weak age," to the influences of climate and food, which affect its development and modify its nature. It will transmit to its descendants the modifications thus induced and, after a few generations, "horses from Spain or Barbary . . . become French horses."[39] Buffon here described what later would be called the inheritance of acquired characteristics, which we have seen was a necessary consequence of his theory of reproduction.

It was probably not by chance that the question was discussed for the horse, "the most noble conquest that man has ever made." The best that a French breeder could hope for was to preserve the nobility of the "proto-type" of the horse. But what about sheep? They were weak and defenseless: "It is only through our help and our care that this species has endured, endures, and will continue to endure: it seems that it would not survive by itself."[40] Here there was no wild "prototype." Where did the sheep come from then? In comparing it with the goat, Buffon had a new idea: along with horses and sheep, there are asses and goats, "auxiliary species, which could, in many respects, replace the first," and which "could do better without man's help."

From then on, "instead of imagining that these subordinate species were produced only by the degeneration of the primary species and instead of seeing the ass as a degenerated horse, it makes more sense to say that the horse is a perfected ass and that the sheep is only a more delicate species of goat that we have cared for, perfected, and propagated for our use, and that in general the most perfect species, especially domestic animals, get their origin from the less perfect species of wild animals that come closest to it, nature alone not being able to do as much as nature and man combined."[41] The idea could have led very far, perhaps even to the very idea of evolution: if man can perfect Nature, cannot Nature perfect herself? That would be Darwin's reasoning. But Buffon saw the matter completely differently, for reasons that deserve to be analyzed.

The first reason was purely scientific. Whether the modifications came from "degeneration" or "perfecting," they could act only within the species, defined by interbreeding. Such was at least Buffon's conviction in 1755. To say that sheep were "perfected" goats, it was first necessary to prove that interbreeding took place between the two species. But even though a billy goat could mate with a ewe, "no intermediary species was formed, . . . no new races of intermediate animals."[42] We must therefore consider the two species to be distinct. In fact, we know too little about the interbreeding of neighboring species, and it would be necessary to do a fair number of experiments to prove anything on this subject. Buffon continued to work on this problem, and his ideas later changed deeply.

The second obstacle was one of a completely different order, which concerned man's very capacity to "perfect" nature. We have seen the ambiguity of Buffon's positions on this subject. Certainly, nature when cultivated was more beautiful than nature when wild, "hideous," and "dying." Yet the wild animal was free, and the domestic animal was a slave, and the best-trained horse did not have the nobility of the wild horse. How then could domestication be a "perfection"? Buffon would never admit this, just as he would never admit that the action of climate could perfect the "prototype."

In both cases, the image of the first form, taken directly from the hands of nature, in all its primitive beauty, nobility, and freedom, imposed itself as an ideal beyond all possible "perfection." Buffon has been criticized for writing that certain species were "nobler" than others and for thus introducing social divisions of his time into nature. But it has not been noticed sufficiently that these "noble" species were all wild species and that neither man nor climate could make them degenerate. They were "noble" because they had stayed what they were as they emerged from the hands of nature. "All is good, coming from the hands of the author of things: everything

degenerates in man's hands." Buffon could subscribe to this famous statement by Jean-Jacques Rousseau.[43] Here is the heart of a debate that divided
the entire thought of the Enlightenment, and this agreement between two
of the greatest minds of the century was truly remarkable. Now we understand why Buffon could never conceive of a theory of evolution.

In 1764 he came back to sheep, in an article whose very title was a sort of
provocation: "The Mouflon and Other Sheep." Buffon started by reviewing
all the various races of sheep, all different because of the different climates
in which they lived, which allowed him to show that certain characteristics,
like the shape of the horns or nature of the wool, could vary and were
therefore not essential to the species. Yet all these sheep were domestic:
"None is strong enough, agile enough, lively enough to escape from carnivorous animals . . . all have an equal need for shelter, care, protection; all
must therefore be considered degenerate races, formed by the hand of
man, and propagated by him for his use." A "wild, stronger, less manage-
able, and, in consequence, more inconvenient and less useful race" must
exist. Pursued by man, "it will therefore be found only in small numbers in
a few sparsely inhabited places, where it can maintain itself." This race was
the mouflon, "which seems to us to be the primitive source of all sheep; it
exists in the state of nature, it subsists and multiplies without man's help; it
resembles all domestic sheep more than any other wild animal; it is livelier,
stronger, and more agile than any among them . . . ; finally it reproduces
with the domestic sheep, which alone would suffice to show that it is the
same species and that it is the founder of it."[44]

Who would dare say that the stupid and defenseless sheep was a "perfected" mouflon? The difference between the mouflon and the sheep became the paradigm of domestication. Buffon came back to this point with
even more force in 1766, having clarified his thoughts: sheep, "relative to
us, have been perfected in certain respects and corrupted in others; but
perfection or corruption is the same thing relative to Nature. They have, in
any case, become denatured."[45] Thus the "perfection" of a domestic animal was as equivocal an idea for Buffon as the perfection of man by civilization was for Rousseau.

Domestication could therefore be only a "degeneration," and domestic
animals had most often been disfigured by man. The camel's callosities,
even its hump, were "imprints of servitude and stigmata of pain." "These
deformities, which at first were only accidental and individual, became
general and permanent throughout the entire species." It is impossible to
be sure of this, since the species does not exist anywhere in the wild state,
but Buffon was guided in his reasoning by a very clear principle: "We must
suppose that all [that these animals] have in the way of goodness and
beauty they receive from Nature and that what they have [that is] defective
and deformed comes to them from the rule of man and the results of

slavery."[46] Following this same principle, the "bison's hump, like that of the camel, is less a product of Nature than an effect of work, a stigma of slavery."[47] Why there are wild and humped bisons remained to be explained. Buffon got out of this as well as he could by supposing that "humped and enslaved cattle might have escaped or been abandoned in the woods and created a posterity [that was] wild and affected with the same deformity."[48]

Not a very likely explanation, but one that conforms to the stated principle. Buffon even contradicted himself. He had started by identifying "two primitive races, both of them formerly subsisting in the state of nature"[49]: the aurochs, which he considered with just cause to be the ancestor of the common bull, and the bison, ancestor of all cattle with humps. For the aurochs, no difficulty: it "is only larger, stronger" than the bull, "the way all animals that enjoy their freedom will win in size and strength over those who have been reduced to slavery for a long time."[50] The bison's hump, however, shocked the naturalist too much for him to see nature's work in it.*

If man made the wild species "degenerate," it was not only because he fed them badly, mistreated them often, and required them to work hard. It was also because he kept only a few males to assure reproduction. Even if this breeding stock was chosen "among the most beautiful," the result could not be good: "The first products of this chosen male will be, if you will, strong and vigorous: but by dint of pulling copies from this sole and only mold, the imprint becomes deformed, or at least does not render Nature in all her perfection: the race must consequently weaken itself, shrink, and degenerate." Since all the individuals in the same species are different, the use of a single breeder diminished this diversity and therefore the vitality of the species. Furthermore, the females have not been given due consideration:

"When there is only one male for a large number of females, the females do not have the freedom to be guided by their taste; gaiety, idle pleasures, gentle emotions are taken away from them; there is nothing stimulating in their love affairs, they suffer from their ardor, they languish in waiting for the cold approaches of a male whom they have not chosen, who often does not suit them, and who always flatters them less than another they might have preferred; from these sad love affairs, these loveless couplings, equally sad products must be born, insipid beings who will never have either the

---

*Buffon returned to the callosities of the camel and the llama in 1766, in his discourse "On the Degeneration of Animals" (*HN*, XIV, pp. 325–6). He encountered the same difficulty as with the bison's hump: certain monkeys, who live for the most part seated, have similar calluses "above the area of the buttocks." Yet these monkeys were free. Buffon got out of it by specifying that, for them, the callosities were "dry and healthy" because they were "the effects of the animals' natural habits," whereas for the camel they were "filled with pus and contaminated blood." We know also that the camel's callosities, already visible in the fetus before birth, were a classical argument in favor of the inheritance of acquired characteristics.

courage or the strength that Nature can propagate in each species only by leaving these faculties complete in all individuals, especially their freedom of choice and even of their chance meetings."[51]

Pure anthropomorphism, one might say. Certainly, and Buffon did not hide it, which on this occasion suggested the practice of Oriental harems. But on this matter of "feeling," why refuse to the animal what is noticed in man? It was the same reasoning that later drove Charles Darwin, following his grandfather Erasmus, to emphasize the role of sexual selection, specifically the free choice of partners by females, in the survival and the evolution of species. Buffon, however, used this point mainly to repeat that man's blind despotism prevented him from "maintaining" and "beautifying" Nature."[52]

The fragility of varieties and artificial races of plants and animals created by man thanks to an intensive selection, the impoverishment of their collective genetic patrimony, and the threat caused by significant variations of the environment are well understood today. Buffon knew nothing about all this. He was guided by his deep and primarily philosophical feeling that by transforming wild species for his profit man impoverished and "denatured" them. What was good for man was not good for the species. Buffon seems to contradict himself endlessly, sometimes admiring man for having "cultivated" nature, at other times accusing him of making it "degenerate." He admires man for having created wheat, but he blames him for having reduced the proud mouflon to the sad state of sheep. This is because he was not always talking about the same nature. To drain marshes and get rid of snakes and other foul beasts was useful and profitable work, which helped nature to correct her mistakes. To degrade a beautiful wild animal was to act like a tyrant. To all the ambiguities that he saw in man's status in nature Buffon added all the ambiguities of moral and esthetic feeling. However unscientific it may be, this feeling was not always a bad guide. Who would dare to say that Buffon has been the only naturalist to feel these divided sentiments?

Migrations, Climate, and Natural "Degeneration"

Although domestication was almost always a "degeneration," domestic animals were not the only ones to have degenerated. If they have degenerated more than wild animals, it was first of all because man had transported them to different climates, and this cause alone "must have produced more varieties than all the others combined."[53] This cause, however, also came into play for wild animals.

It was again for animals of the New World that Buffon took up the question. While it was true that many of these were indigenous and essentially different from species of the Old World, it was also true that in the north of America animals existed that were so similar to those of Asia or Europe that one was tempted to believe that they belonged to the same species. This was the case for bears, for roe deer, for the Canadian caribou, which was the "same animal" as the reindeer of Lapland, for beavers, for wolves and foxes, and for several other small animals. They resembled one another, "and even though there is always some fairly marked difference, one can nevertheless not refuse to consider them the same, and believe that in the past they passed from one continent to the other through lands of the North, perhaps currently unknown, or rather, which sank beneath the sea many years ago."[54] The difficulty was that the American animals resembled those of Europe more than those of Asia, while "it is with Asia that America communicates by contiguous lands, and on the contrary it seems that the north of Europe is and always has been separated by seas large enough that no quadruped could have crossed them." Buffon re- solved this difficulty for the time being by remarking that "the eastern parts of the north of Asia" were not well enough explored to know "if the animals from the north of Europe are found there or not."[55]

There was, however, another question to be asked: why did the animals of the Old World emigrate to the New? Buffon saw only two possible reasons. First, perhaps, "because of revolutions of the globe."[56] But more important, "since the beginning of time [animals] have been hunted, ban- ished by man, or at least by those among them who were the strongest and meanest." They were "forced to flee, to abandon their native country, and to accustom themselves to less felicitous lands." Human or natural, it always concerned violence. "Those whose nature was flexible enough to adapt to that new situation spread out fairly far, while the others had no other resort than to confine themselves to the neighboring wilderness in their region." There was therefore a "degeneration" in nature. The nature of wild ani- mals "seems to vary according to the different climates." The only differ- ence from domestic animals was that although their nature had varied, "no part of it was degraded."[57] Only man could "degrade" nature.

Here again it was the animals common to the Old and the New Worlds that served as examples. The new climate had, in general, made them smaller. Just as indigenous American animals were smaller than their equiv- alents in the Old World, so domestic animals transported to America "be- came smaller there." Similarly, those that went there by themselves, those, in short, that are common to the two worlds . . . are also considerably smaller in America than in Europe, and this is so without any exception."[58] Later Buffon would discover an exception: roe deer and deer were larger in the New World than in the Old.[59] Buffon was convinced that in the New World "there are obstacles to the development and perhaps the formation of the seeds for large size; those very ones that, through the gentle influence of another climate, acquired their full form and their entire size, shrink and grow smaller under this spare sky on this empty earth."[60] Such a general affirmation would, we shall see, incite the wrath of North Americans.[61]

Thus the climate was the factor responsible for the diversity of living forms, whether it controlled the "formation of seeds" or modified already existing forms. The way this "degeneration" functioned for the human species had already been described. Buffon's description, which I cited earlier,[62] is remarkable not only because it added the influence of food and the "way of life" to that of the "climate" but also and because it underlined the fact that the "alterations" were only "individual variations" that "perpetuate themselves from generation to generation, like the deformities or the illnesses of fathers and mothers that pass to their children." From then on, the variations within a species "depend on the combination of the number in individuals, as much from those that produce them as from those that are produced." "Where females often change males . . . fairly numerous variations are found; and as *in all Nature there is not one single individual that resembles another perfectly,* there are thus many more variations in animals when the number of their progeny is larger and more frequent."[63]

Attentive to the problems of heredity, as his theory of reproduction proves, and in particular to the inheritance of accidental malformations and illnesses, Buffon, like several of his contemporaries, considered the variations to be malformations.* That they were an effect of the climate did not prevent them from being "originally produced . . . by the combination of exterior and accidental causes,"[64] and their hereditary transmission depended on the conditions of reproduction within the species, the species itself being clearly conceived as a population of individuals all different from one another. It was difficult at this time to distance oneself more from a typological conception of the species or to go much further in this line of thought, which would be taken up by Darwin and modern population genetics.

"The Great Worker of Nature Is Time"

In order for individual variations induced by the action of a new climate to spread out in a population, it was necessary that they be "established and rendered constant . . . by time and the continuous action" of the same

*Réaumur had attempted a few experiments, whose results he did not give. Maupertuis had studied a case of *sexdigitism* (an extra digit of the hand or foot) in a Berlin family and showed that the anomaly was hereditary. As for physicians, they had already been concerned for quite some time with the inheritance of pathological illnesses or dispositions. See Réaumur 1749, volume II, 4th thesis; Maupertuis 1756, II, pp. 275–82; *Encyclopédie*, article "Héréditaire (Médecine)."

causes.[65] Here Buffon was both very close to and very far from modern science. Very close because he admitted that living forms change and are changed in the course of the ages; very far because, persuaded of the inheritance of acquired characteristics, he thought that the continuous action of the climate must reinforce the same variation from generation to generation. Nothing was more foreign to him than the idea of a purely accidental variation,* which, once it appeared in a population, could maintain itself and spread or, on the contrary, be eliminated.

The continuous action of the climate was necessary for the reinforcement and the propagation of the variation, and this is what took time: "These changes are only made slowly, imperceptibly; the great worker of Nature is Time; as it always moves with an equal, uniform, and regulated pace, it does everything; and these changes, at first imperceptible, become noticeable little by little, and finally leave results about which one cannot be mistaken."[66] Just as he would not admit catastrophes in geology, Buffon would not admit an abrupt modification of forms. Living forms continue to be transformed, much as the history of the earth continues before our eyes. "Nature, I admit, is in a continuous movement of flux."[67]

Moreover, not only were certain animals "often changed to the point of being unrecognizable," but some had purely and simply disappeared. The idea that the history of nature could erase creation's work seemed such a sacrilege that naturalists preferred to think that fossil species of which "living analogues" had not yet been found still existed somewhere and would eventually be discovered. This could easily be accepted for marine mollusks, which could hide themselves at the bottom of the sea. But what about the "prodigious mammoth, a quadruped, whose bones we have often considered with astonishment"? Such a beast could not pass unnoticed. Thomas Jefferson, we shall see, tried vainly to discover it in the American wilderness. Buffon considered it extinct: "This species was certainly the foremost, the largest, the strongest of all quadrupeds: since it has disappeared, how many others—smaller, weaker, and less noticeable—must have perished without having left us either evidence or information about their past existence?"

It was clear that "the least perfect, most delicate, heaviest, least active, least armed, etc. species have already disappeared or will disappear."[68] The best examples were the unau and the ai, both better known under the name of "sloth." Buffon gives a pitiable description of them. They survive only because they live in "the wilds." If not, "these species would not have come down to us, they would have been destroyed by others, as they will be

*When Buffon said that variations were due to "exterior and accidental causes," he meant that these causes were exterior to the normal life of the species. In this sense, it can be said that they intervene "by chance," but this chance was not the kind that would intervene in the Darwinian theory of evolution. On this point, see Lenay 1989.

one day." Rather than develop the theme of the "struggle for life," Buffon preferred to return to "all that can be is" and to a Lucretian vision of the world: "Everything reminds us of these monsters by their failure, these imperfect rough drafts, a thousand times tried and executed by Nature, which, having barely the ability to exist, must only have survived for a time, and have since been erased from the list of creatures."

Instead of committing himself to a path that might have brought him closer if not actually to Darwin, Buffon thus returned to two of his favorite themes: the condemnation of final causes and a comparison, typically Rousseauistic, between nature and society. As was normal, he remained a prisoner of the problems of his century. Yet it is clear that Buffon owed the possibility of unhesitatingly admitting the existence of extinct species to Lucretius's philosophy, to the idea of a universe born from chance in which an apparent order was merely the result of the elimination of nonviable entities. This is precisely the philosophy that Diderot had taken up in the *Letter on the Blind.*

Thus the diversity of living, or at least surviving, forms, has multiple causes. In the beginning, it was the diversity of "prototypes," in principle inexplicable but known to correspond very precisely to the diversity of climates in the "lands of origin." Then came "degeneration," which affected wild animals only when they had to emigrate and which hit domestic animals hard and fundamentally. Overall, this was a deeply pessimistic view, in which nature had been completely desacralized and abandoned to history. Buffon, we shall see, would sink even deeper into pessimism.

The naturalist, however, could not be satisfied with the philosopher's conclusions. He could not give up the task of putting nature in order. "Nature, I admit, is in a continuous movement of flux; but it is enough for man to grasp her in the moment of his century." This did not prevent the naturalist in him from "glancing behind and ahead, to try to catch a glimpse of what she could have been before, and what she may become in the future."[69] The reconstruction of the past would thus be put at the service of the knowledge of the present. Likewise, the observation of the present would be put at the service of the reconstruction of the past. And this reconstruction of the past, this effort to reconstitute and describe the history of nature, would eventually become Buffon's sole obsession, invading his thoughts and monopolizing his research.

CHAPTER XIX

Species, Genus, Family:
Toward a New Classification

The eighteenth century had a passion for classification, which it called, revealingly, "methods." It wanted to classify everything: sciences, books, illnesses, and of course, stones, plants, and animals. This "project of a general science of order"[1] showed a will to master things in order to manage them better and submit them to human reason, the reflection of the divine reason. The order of the classification must be the very order of creation.

In natural history, botany showed the way: Mayol, John Ray, and Tournefort had laid down the principles, and Linnaeus, who came on the scene in 1735, was foremost a botanist. The zoologists were not long in following their example. John Ray in 1675 and then Theodor Klein in 1730 undertook the classification of the animal "kingdom." The first edition of Linnaeus's *Systema Naturae*, which dates from 1735, already presented a brief classification of animals: two pages in-folio, which became an octavo volume of 824 pages in the tenth edition of 1758.[2] Later, Brisson, Réaumur's former secretary, presented his own animal taxonomy. The classification of plants responded, in part, to a practical need and served medicine. The classification of animals sought to meet a purely intellectual demand.

What did this amount to, in a practical sense? "To gather the similar beings and separate the dissimilar beings," according to Linnaeus's statement.[3] The principle was simple, the application difficult. Resemblances and dissimilarities between beings varied infinitely. Yet the ground was already well marked, perhaps too much so. Ancient philosophers and medieval scholastics in particular had reflected a lot on these questions, which were primarily questions of logic, and imposed two notions, those of "genus" and "species." Genus was a "universal notion formed by the abstraction of qualities that are the same in certain species, just as the idea of

[309]

species is formed by the abstraction of things that are found similar in all individuals." Such was Formey's definition in the *Encyclopédie*.[4] The process of abstraction could continue, and "inferior genera" brought together into a "superior genus" and so on. As Formey explained, if the hound were the species, the inferior genus would be the dog, the superior genus the quadruped, and the next superior genus the animal. At the end of the series, one arrives "at the supreme genus, at Being."

In order to clarify the ideas, the classifiers had introduced an entire vocabulary that distinguished among the different degrees of superior "genera": between "kingdom" and "genus" they introduced "class" and then an entire series of intermediary terms inspired by a military or political hierarchy: orders, tribes, legions, cohorts, families.[5] It is understandable that Linnaeus wanted to put order into this chaos by accepting only five categories in a kingdom: class, order, genus, species, and variety.

Two fundamental problems remained. First of all, in order to classify beings as complex in their structure as plants and animals, how to choose the significant resemblances or differences? Was it necessary to retain the *habitus*, that is, the general look of the plant, or to choose one single characteristic? Though theoretically easier, the second method presented no fewer difficult problems. The chosen characteristic must be present in all the species of the same genus, and absent, or different, in all those of the neighboring genera. Could the same characteristic serve to unite genera into orders and orders into classes, or was it permissible to use different characteristics in each step of the classification? And could one hope to establish thereby a "natural" classification, that is, to discover the real order of creation?

The problem was already well known by botanists, who had long ago established the existence of "natural families" of plants united by a great number of characteristics: examples are *Graminae* or *Umbelliferae*.* To classify all plants in such families would be to establish a "natural method," which everyone wanted but everyone considered impossible. Meanwhile, by choosing this or that particular characteristic, an "artificial method" or "system"[6] could be constructed. Officially, Tournefort and Linnaeus did not claim to do more than that. At the same time, however, they endeavored to show that the characteristic they had chosen was "essential" and defined the plant's very "nature." Their method was therefore not as artificial as they said.[7]

Classification brought an old medieval debate back to life, which has remained in history under the name of the "quarrel of the universals." The question is a fundamental one. In reality, we are acquainted only with individuals—Tom, Dick, or Harry—and we say they are "men." But "man"

*Ed. note: Grassy, or plants with umbrella-like flowers.

is an abstract notion, a "universal." Did "man" have a real existence, as "realists" said, or was it only a name that allowed one to group together individuals who were in reality different, as the "nominalists" maintained? From the fourteenth century* until today, the question has been and is still being debated, as it poses the entire problem of knowledge. If we know only individuals through direct experience, if we cannot generalize our experience, then science is not possible.

If the problem arose for species, it was even more pressing for the category immediately above it, the genus. There is no a priori reason for the genus to be a "natural" category, with a real existence in nature: this was, however, what Tournefort[8] and later Linnaeus stated.[9] As Malesherbes said, "The establishment of *genera* is completely independent of systems of botany."[10] Not only does the genus truly exist in nature, but it constitutes the basic unit. When Linnaeus proposed his binomial nomenclature in 1758, the noun indicated the genus, and the species was indicated by an adjective. Linnaeus only simplified a practice that was already common: "The name that one gives a species," wrote Malesherbes in 1749, "is the name of the genus with some epithets."[11] The species was defined by the genus, and not the genus as a grouping of species. It was therefore the genus that was the primary unit, the species being only a subdivision. As Locke had already noticed, this manner of naming only introduced into nature a logician's rule: "A definition must consist of genus and differentia."[12] This subordinate position of the species must not be forgotten if Buffon's criticisms are to be understood.

The Criticism of Classifications

Starting with the "First Discourse," as we have seen, Buffon violently attacked the classifiers, beginning with Linnaeus. Remember Malesherbes's scandalized reaction: "It is not in such terms that one criticizes a generally respected author, especially when the one who criticizes him (whatever talents he may have) is still a man new to science." Which let it be understood that Buffon was criticizing Linnaeus without having taken the trouble to read him, and that if he had not understood him, it was because "he had badly understood the elementary terms that botanists use, like those of *genus, species, system,* etc. These are, however, the first notions that one ordinarily learns when one devotes oneself to botany."[13] Following Malesherbes, historians have for a long time been content to accuse Buffon of ignorance, superficiality, and presumption, or at least of not having understood anything of his adversaries' thoughts.[14] Thanks especially to the binomial nomenclature, Linnaeus's authority has for the most part

*Ed. note: Actually the debate began in the late eleventh century with Peter Abelard.

survived his classification, since by international convention it is the tenth edition of the *Systema Naturae* that is used as a starting point for modern zoological nomenclature.[15] To attack Linnaeus is still a bit sacrilegious, and it is only recently that historians have undertaken the task of examining more objectively the criticisms that Buffon addressed to him.[16]

To attack Linnaeus in 1749 showed a certain courage. The *Systema Naturae* was already in its seventh edition, and Linnaeus's renown was well established in France and even in all of Europe.[17] Not content with his first attack, Buffon did not miss the opportunity in later volumes of the *Natural History* to attack "nomenclators" in general and Linnaeus in particular. He redoubled these attacks in the last volumes, from 1761 to 1766.[18] It cannot therefore be said that Buffon, in "becoming a naturalist" and in "learning natural history" through practice, came to understand the classifiers better. Nor can Buffon's "levity" be contrasted to Daubenton's "seriousness": in 1753 the latter had given a detailed critique of classifiers in which he agreed with all of Buffon's positions.[19] Linnaeus never deigned to answer these criticisms: even before Buffon publicly attacked him, he had explained that as long as his methods "did not displease the Divine Master of the true method," he was ready to tolerate "with a tranquil spirit the barking of dogs." Was it then by chance that in 1748 he had given the name of *Buffonia* to a plant with a particularly unpleasant odor? In 1768, in a long section of the *Systema Naturae* dedicated to the classification of quadrupeds, he did not even condescend to mention Buffon and Daubenton's work.[20] Usually it was Buffon who was attacked and did not respond. One wonders why in this case, and in this case alone, he was so relentless in his attacks— attacks directed, it is true, not so much against an individual as against a way of thinking.

We have already seen the major criticisms that Buffon had leveled at "methods":[21] they overload the memory without teaching anything; they assume, at least for Linnaeus's "sexual system," that flowers are examined with a magnifying glass; they are founded on a unique characteristic, whereas one should take into account all characteristics; finally they claim to introduce clean divisions into nature, whereas "Nature works by insensible degrees." In order to illustrate his ideas, Buffon took a wicked pleasure in showing the strange associations to which the Linnean method led: the sloth and the scaly lizard were put in the same class as man, the hippopotamus was placed with the shrew, and so on. Later he would amuse himself by underlining the strange modifications that Linnaeus imposed on his classification, changing even the number of orders, which went from five to seven: "One can judge, by these essential and very general changes, of everything that is found in the genera, how much the species, which are the only real things, are tossed about, moved around, and badly placed together." Buffon also criticized a new species of man that Linnaeus had just invented, the "nocturnal troglodytic man": "Is that not adding fables to absurdities?"[22]

In these attacks against classification, his contemporaries, Malesherbes and Lelarge de Lignac, saw an attack against all possibility of science and called it an "unbearable Pyrrhonism." They immediately denounced its origin: Locke's philosophy. Locke had in fact restudied the problem of "universals" in his own way and devoted an entire chapter of his *Essay on Human Understanding* to the question of "general terms." His conclusion was frankly nominalist: "real species" can be reduced in the end to "nominal species," and both of them are creations of the understanding. Abstract ideas are useful but without reality in nature.[23] Locke was perfectly informed of the problems that the classification of animals posed. His analysis was perhaps of use to John Ray, whom Buffon knew well.[24] We find too many ideas in Buffon that could only have come from Locke to have doubts about the origin of his criticism of classifications. Lord Monboddo, a friend of Linnaeus, drew from this what he felt was the legitimate conclusion that Buffon seemed not to know that "a distribution in genera and species is the foundation of all human knowledge." Buffon was therefore not only Linnaeus's enemy, "he is at the same time the enemy of all philosophy and all human knowledge."[25] This conclusion was rather hasty, as we shall see. In fact, Buffon refused to follow Locke to the conclusion of his argument.

The Definition of Species

As a philosopher, Locke constructed his discussion according to the problem he intended to combat, that of Aristotelian logic. The first "universal" whose existence in nature he claimed to reject was the species. The very category of *man* or *horse* seemed to him to be an abstraction, created by the human mind. He did not deny the existence of living species, but he argued that "the boundaries of the species . . . are made by men."[26] "I would fain know," he asked, "why a shock [poodle] and a hound are not as distinct species as a spaniel and an elephant."[27] "The frequent productions of monsters, in all the species of animals, and of changelings, and other strange issues of human birth" constantly posed the problem of the limits of the species,[28] and thus showed that despite appearances, even the notion of animal or plant species did not correspond to anything specific in nature.

Reading Locke led the English naturalist John Ray to revise the principles of his classification and to disengage himself from the method founded on the search for the "essential characteristic." In doing so, he became involved in a controversy with Tournefort and with the German naturalist Rivinus.[29] Ray had already sought a way to distinguish species from varieties and he had concluded: "no surer criterion for determining species has occurred to me than the distinguishing features that perpetuate themselves in propagation from seed."[30] From what was for Ray simply a way of distinguishing "good species," Buffon would make a definition of species, period.

In 1749, in the *History of Animals,* the primary concern was finding a means of "recognizing species" and "distinguishing them from one another": "One ought to consider as the same species those that perpetuate and conserve the similarity of the species by means of copulation, and as different those that through the same means can produce nothing together."[31] This statement permitted Buffon, as we have seen, to affirm immediately the unity of the human species. He had already made a passing reference to "that chain of successive existence of individuals that constitutes the real existence of the species."[32]

In 1753 Buffon went further and truly responded to Locke's analysis: "It is neither the number nor the collectivity of similar individuals that make the species; it is the constant succession and the uninterrupted renewal of these individuals that constitute it. . . . Species is, therefore, an abstract and general word, which describes something that exists only by considering Nature in the succession of time and the constant destruction and the equally constant renewal of beings."

Locke had asked why a spaniel and an elephant did not belong to the same species, unlike the poodle and the hound. Buffon could answer him: "The ass resembles the horse more than the water spaniel the hound, but nonetheless the water spaniel and the hound are only one species, since they together produce individuals that can themselves produce others, whereas the horse and the ass are certainly from different species, since they together produce only defective and barren individuals."[33] Species was perhaps "an abstract and general word," but the thing truly existed in nature. It was moreover something defined by its very limits, which were natural limits, contrary to what Locke thought: the criterion of interfertility, as we would say today, allows the recognition of which animals belong to the same species and which ones belong to another. No theoretical skepticism can prevail over experience.

The concept of species as presented by Buffon has nothing "nominalist" in it. In fact, beyond Locke and John Ray, it goes back to Aristotle, who considered the universe eternal and saw that living beings were mortal; he had therefore, thanks to "generation," transferred to the species the eternity denied to individuals. The difficulty was that Aristotle used the same vocabulary, but with different meanings, in his treatises on logic and in his biological works. *Species* could thus, depending on the text, be either the logical category inferior to *genus* or the living *form* that perpetuated itself through generation. The classifiers, criticized by Locke, emphasized the

OISEAUX.

Pl.14

Three owls. From *Oeuvres complètes*, XX, p. 327.

logical sense. Buffon escaped Locke's criticism because, probably in following William Harvey here, he emphasized the biological meaning, which rested on an empirical reality.[34]

His concept of species was much less nominalist because it was based on his theory of reproduction. The guardian of the species was the "interior mold," which thus played the same role as the Aristotelian "form." This interior mold, however, was not an abstract entity, it was the very body of the reproducing animal, and this was how "generation" became "reproduction." If the body were subject to individual modifications due to food or climate, these modifications, at least within certain limits, would be transmitted to the descendants. Individual varieties, and even "constant varieties," no longer posed any problems and no longer threatened to ruin the definition of a species. There was room for that vast phenomenon of "degeneration" that we studied in the previous chapter. At every moment, at least theoretically, we can know where we are: the experience of hybridization ought to allow us to know if we are dealing with two varieties of the same species or with two different species.

Remember the questions Buffon asked himself about the relationship between goats and sheep: is the sheep only a "perfected" or a "degenerate" goat? In 1755 he simply noted that fertile matings were possible between the two species. Nevertheless, "no intermediate species between the goat and the sheep has been formed." It was necessary, therefore, to consider them distinct species. It was a true research program that Buffon established at that time:

"We wander in shadows or we walk with perplexity between prejudices and probabilities, not even knowing the extent of the possibility of things and confusing men's opinions with Nature's acts at all times. Examples present themselves in multitudes, yet without considering any from anywhere but our subject here, we know that the goat and the sheep mate and reproduce together, but no one has yet told us if a sterile mule, or a fertile animal that can found a line for new generations or ones similar to the first results from this. . . . We believe that mules in general, that is, animals that come from the mix of two different species, are sterile. . . . This opinion, however, may be unfounded. . . . It is, therefore, necessary to deny or confirm these facts, which arise from obscurity about the real distinction between animals and about the theory of generation. . . . We do not know if the zebra can reproduce with the horse or the ass . . . ; if the dog can reproduce with the fox and the wolf; if the deer reproduces with the cow, the hind with the fallow deer, etc. . . . I have spent a few years making attempts of this sort, . . . but I will confess right now that they have furnished me with only a few hints, and that the majority of these tests where without success.

"On this, however, depends our entire knowledge of animals, the exact division of their species, the perfect understanding of their history; on this also depends our manner of writing it and the art of treating it: but since we are deprived of this knowledge so necessary to our object, . . . we can do no better than to go step by step, to consider each animal individually, to regard all those that do not breed together before our eyes as different species, and to write their histories in separate articles while retaining the possibility of joining them or combining them, as soon as we are better informed through our own experience or that of others."[35]

Buffon was certainly not the first to have experimented with hybridization: Leeuwenhoek and Réaumur, to mention only two, had done so before him. For them, however, it involved knowing if the "preexistent germ" was contained in the sperm or the egg. Buffon's problem was completely different and original.

Buffon did several of these same experiments himself with the peccary and the pig, the dog and the wolf, the dog and the fox, the hare and the rabbit—all without positive results.[36] He also carefully noted the experiments done by others, whether they had been intentional or not. He learned thus that the peccary did not reproduce with either European or African pigs brought to America; on the other hand, the European and Siamese pigs belonged to the same species, like the wild cat and the domestic cat. M. de La Nux had successfully crossed the "cattle* with a hump from India" with the "cattle of Europe." The inhabitants of Louisiana ought therefore to attempt to "mix the bison of America with the cow of Europe." In case of success, "one would be assured that the cattle of Europe, the humped cattle of the Isle of Bourbon, the bull of the East Indies, and the bison of America are all one single species."[37] Only the buffalo was left out: it often lived in the company of cattle, but they "have always refused to be joined."[38]

Certain experiments had given ambiguous results that Buffon did not know how to interpret: "When a billy goat mates with a domestic sheep, the offspring is a sort of mouflon, for it is a lamb covered with hair, not an infertile mule but a crossbreed that returns to the original species, and which seems to indicate that our goats and our domestic sheep have something in common in their origin." However, "the ram does not breed with the nanny goat." Therefore the goat was the "dominant species" and the sheep the "subordinate species," "a species much more degenerate than that of the goat." By mating the goat and the mouflon, perhaps "kids that would return to the goat species" could be obtained.[39]

All this was not very clear, and things soon became even more complicated. In studying the ibex and the chamois, Buffon persuaded himself

*Ed. note: Buffon uses the word *boeuf* which, taken literally, means ox, but also is the generic name for cattle. Oxen and steers, in English, are castrated bulls and, therefore, cannot reproduce. There is no general name in English for the singular of cattle.

that they made up a single species with the goat. Certainly, no one knew if they could "reproduce with our goats." The analysis of morphological characteristics suggested this possibility, but since the females resemble one another more than the males, it was necessary to conclude that the females "are of a constant nature," while the males "undergo variations that render them different from one another." Hence arises the idea that "the ibex would be the male in the original race of goats, and the chamois the female." And thus "it is very likely that the domestic nanny goat would breed equally with [the] three different males." "Thus there can be sometimes two races in the same species, one masculine and one feminine, which . . . seem to make up two species," and then it becomes impossible "to fix the line between what Naturalists call *species* and *variety.*"

Buffon gathered together all the information at his disposal, including what he found in Linnaeus, in order to show that the ibex, the chamois, and the goat belonged to a single species and that the "nine or ten species of which the Nomenclators speak" are instead "ten different races produced by the influence of the climate." Pliny had already said this, but he was wrong to include "under the *generic name of goats,* not only those that we just mentioned but also the roe deer, gazelles, the antelope, etc.," which "are nonetheless all different species." Buffon concluded: "In the entire history of the quadrupeds, I have found nothing more difficult to explain, more confusing to understand, and more historically uncertain than this story of goats, gazelles, and other species that resemble one another; I have made efforts and focused my entire attention to bring some light to it, and I will not regret my time if what I write about it today can serve in the future to prevent errors, clarify ideas, and lead to truth, by extending the views of those who wish to study Nature."[40] What was certain in any case was that things were not simple and that the definition of species by interfertility did not resolve all the problems.

For wild animals, experiments were not possible, and one had to be satisfied to establish that certain apparently neighboring species coexisted in the same place without ever mixing or producing an "intermediate race." Such was the case for the deer and the fallow deer.[41] When experiments are impossible, "inductions" from certain physiological facts could be made, in particular the length of gestation. Does "carry more than eight months," while "the roe deer carry five and one-half months": "This difference alone would suffice to prove that these animals come from species separated enough never to be able to come closer, or mix, or produce together an intermediate race." On the other hand, "the goat carries about the same time, and the roe deer can be considered a wild goat."[42]

There were perhaps other obstacles to the crossing of species. For example, foxes were not interested in bitches. Buffon had done the experiment: "Even if they had never seen a female of their species and they seem urgently to need to [have sexual relations], they cannot decide, they constantly reject all female dogs; yet as soon as a female of their own kind was presented to them, they covered her, even though chained, and she produced four offspring."[43]

The same difficulty with the zebra: "Female asses in heat were presented to the [zebra] that was in the Menagerie at Versailles in 1761. It disdained them, or rather it was not moved at all, at least no visible sign of emotion appeared; nonetheless it played with them and mounted them, but without erection or whinny, and one can hardly attribute this coldness to any other cause than the disparity of nature: for this zebra, four years old, was very strong and very nimble in all other activities."[44] This "disparity of nature" anticipated what we call today an "ethological barrier" between two neighboring species. It was a notion that Buffon would use more often and which, we shall see, would lead him to modify his definition of species.

The entire natural history of the quadrupeds nonetheless proceeded according to the outline announced in 1749: domestic animals, then wild animals of the Old World, then animals of the New World. This outline already implied a certain number of regroupings, and from the beginning exotic varieties were joined with European species: the goat from Angora or the pig from Siam were described in the same article as the European goat and pig. They were merely variations of the same species. Other tacit regroupings brought together neighboring species. It was not the dog that followed the horse, no matter what Buffon had said in the "First Discourse": it was the ass. From the badger to the ermine, from the rat to the vole, two series of small carnivores or small rodents parade through volume VII. "Distinct and separate species, differing so little that they were able, in some way, to replace each other and assure that if one of them were to be missing, the gap in this genus would be barely perceptible."[45] The large felines succeed one another in the same way in volume IX: Lion, tiger, panther, snow leopard, and leopard were immediately followed by their relatives from the New World, the jaguar and the cougar. Three marsupials followed one another in volume X, then an entire series of animals similar to squirrels, another made up of anteaters and similar animals, a third including the armadillo and its close relatives. Volume XII brought together a large group ranging from elk and reindeer to gazelles, and volume XIII included a new group of American cats, then, following the jackal, a group of animals that resembled wolves, foxes, and dogs.

Progressively, more and more often the articles brought together several neighboring species, even larger and larger groups, such as seals, walruses, and manatees. Certainly none of this was systematic. The hamster was placed between the *coquallin*, which resembles a squirrel, and the *bobak*, a kind of marmot. Buffon did not hesitate to admit: "If we did not give its history with that of the other rats, it was because we had not seen it, and because we had only procured one recently."[46] Yet we have the feeling that Buffon tried, without saying so, to reconstitute "natural families," analogous to those that botanists had recognized for a long time in plants. Theoretical reflections soon followed the practice.

Genus and Family

We have seen the importance of the notion of "genus" for the classifiers, for Tournefort as well as Linnaeus. In 1749, at the very moment when he attacked the classifiers' methods, Buffon had shown no particular hostility to "genera." He admitted that individuals were arranged in species, genera, and classes, as a function of the relative importance of their similarities and their differences. He even fully accepted Linnaeus's principle "of putting together things that resemble one another, and to separate those that differ from one another."[47] What he rejected was founding these classifications on a unique characteristic and being willing "to decide about the resemblance and the differences of animals, by using only the number of fingers or dew-claws, teeth or teats."[48] Above all, he insisted that methods were only convenient artifices "and that genera, orders, and classes exist only in our imagination."[49] Only the species, we have seen, truly existed in nature.

What shocked Buffon the most about the genera established by Linneaus was that they seemed to mix up the species. Linnaeus, indeed, named the genus after one of the species it contained. The species itself was only defined by a supplemental specification, a "phrase" of a few words, which would be reduced to an adjective in the binomial nomenclature of 1758. Thus the horse was "a horse with a tail entirely full of hair" (*Equus cauda undique setosa*), and the ass was "a horse with a tail full of hair at the extremity" (*Equus cauda extremo setosa*). For Buffon, that meant that the ass was "a kind of horse," and according to the same principle "the lynx is only a kind of cat, the fox and the wolf a kind of dog . . . the rhinoceros a kind of elephant." "Would it not be simpler, more natural, and truer to say that the ass is an ass, and a cat a cat, than to believe, without knowing why, that an ass is a horse and a cat a lynx."[50] Obviously, this was not what Linnaeus intended. Nevertheless, his descriptions most often started with a "primary species," which often gave its name to the genus itself. From then on, the primary species was defined only by the "phrase" or the adjective.[51] This was a logical position if the genus was the essential reality. Brisson, it is true, went further: he had "made a genus of a dozen animals . . . under the name of rabbit-like genus," but the rabbit was not included in it: "It is necessary to look for it in the genus of the hare."[52]

Buffon's misunderstanding had numerous consequences. In 1753, again for the ass, he challenged not only the notion of "genus" but also, what is very instructive, that of "family." Noting the Linnean nomenclature, where the ass and the horse were both called *equus*, he asked a question that Linnaeus had certainly not thought of: was the ass a degenerate horse? And from that question, which was possible only in his personal agenda of problems, he asked a second question: "Did the ass and the horse therefore originally come from the same ancestor? Are they, as the nomenclators say, from the same *family*? Or are they not, and have they not always been, different animals?" This is a question "whose generality, difficulty, and consequences the physicists* will appreciate," and which "pertains to the reproduction of beings more closely related than any other."[53]

Certainly a "primitive and general design" exists, one that is found in all animals: we have already spoken of it. In this sense, all animals are part of the same family. "But should one conclude from this that in this large and numerous *family*, which God alone conceived and created from the void, there are other little *families* planned by Nature and produced by time? If these *families* indeed exist, they could only have been formed by the crossing, the successive variation, and the degeneration of the original species; and if one once admits that there are *families* in plants and animals, that the ass is of the horse *family* and that it differs only because it has degenerated, one can likewise say that the monkey is of the *family* of man, that it is a degenerated man, that man and monkeys have a common origin like the horse and the ass, that each *family*, of animals no less than of plants, had only a single ancestor, and even that all animals came from one single animal, which over time produced all the races of the other animals by perfecting itself and by degenerating. . . . If it were true that the ass is only a degenerate horse, there would be no more limits to the power of Nature, and one would not be wrong to suppose that she in time was able to produce all the other organized beings from one single being." "The Naturalists who so lightly establish families in animals and in plants do not seem to have fully realized the entire extent of these consequences."[54]

And for good reason, for these naturalists did not claim at all that plants from the same "family" had true links of kinship. The meaning that Buffon gave to the word was purely personal,[55] and it was in his own thinking that the question arose. More precisely, it came from the strictly biological concept of the species that he held at that time. Even in this passage, where he seemed to envision the theoretical possibilities of a generalized transformism, Buffon did not say that one single animal could produce all the living *species*, but "all the *races* of the other animals." For him, a "race" was only a "constant variety" within the species. A generalized transformation meant that all the animals were from the same species. The family was

---

*Ed. note: Not to be confused with modern English *physicist*. The 1799 edition of the *Dictionnaire de l'Academie Françoise* gives "Physique—science whose object is natural things," which clearly encompasses living creatures.

implicitly assimilated to the biological species, defined by interfertility. If the animals "came from the same ancestor, if they were indeed from the same *family*, one could bring them closer, blend them once again, and undo with time what time had done."[56] We already knew that this was impossible.

When he published this text in 1753, Buffon knew of the hypothesis of the unique origin of all living forms, which Maupertuis had proposed in 1751 in his *System of Nature*. Instead of calling upon a "degeneration" under the influence of the "climate" as Buffon had, Maupertuis noted the sudden and apparently accidental appearances of abnormal forms within a species, which were transmitted to descendants. He attributed these appearances to an accident in the course of the process of reproduction. This phenomenon, which later became the Darwinian "variation," served as the point of departure for Maupertuis's hypothesis: "Could one not explain by this how from two individuals the multiplication of the most dissimilar species could have resulted? They would have owed their initial origin only to a few fortuitous reproductions, in which the elementary parts would not have retained the order that they held in the father and mother animals: each degree of error would have made a new species, and with the multiplication of repeated deviations would have come the infinite diversity of animals that we see today, which will perhaps increase with time, but to which the following centuries will perhaps bring only imperceptible increases."[57]

This is a seductive hypothesis, especially to us, but Buffon could not agree with it. It is easy to see why: it supposed a plasticity of the "interior mold," which Buffon was not ready to accept, that is, a very unfocused if not a meaningless definition of a species. Buffon could allow not imposing "limits to the power of Nature," but he was not about to abandon his definition of species.

Since he had no hypothesis to offer at that time for the origin of species and the primeval formation of the "interior mold," he was quick, as he was so often from 1753 on, to underline his religious orthodoxy: "It is certain, through revelation, that all the animals participated equally in the grace of creation, that the first two members of each species and of all species emerged fully formed from the hands of the Creator, and we must believe that they were such then, more or less, as they are represented to us today by their descendants."[58]

Some have wrongly concluded that Buffon did not, through fear of the censor, dare to show his adherence to an evolutionary theory, when in reality his conception of the species prevented him from accepting one. He gave several other scientific arguments against evolution. First of all, no new species had appeared since Aristotle. In addition, and this was in direct response to Maupertuis, who was in fact cited in a note, the "modifications of variation" that have been noticed produced only "individual variations that did not separate these individuals from their species." And again, in order for a new species to appear, it would be necessary to suppose that at least two individuals, a "male and female, of a certain species, not only have degenerated so much that they no longer belong to this species, that is, they no longer are able to reproduce with those to which they were similar, but also have both degenerated precisely to the same point to be able to breed together . . . and how could it be that a tainted origin, a deprivation, a negation, could form a rootstock?" Finally, "if the species of the ass comes from the species of the horse, that could only have been done successively and by small degrees, and there would have been a large number of intermediate animals between the horse and the ass. . . . Why should we not see today the representatives, the descendants of these intermediate species? Why do only the two extremes remain?"[59]

Here we recognize many of the difficulties that the theory of evolution would have to surmount in the nineteenth century and which would only finally be overcome in the twentieth. It was first of all his definition of species that prevented Buffon from conceiving a generalized evolution. Therefore, "although one cannot prove that the production of a species by degeneration is impossible for Nature, the number of contrary probabilities is so enormous that even philosophically one can hardly doubt them." The conclusion was obvious: "The ass is therefore an ass, and is not a degenerate horse, a horse with a naked tail."[60]

The end of the sentence, that Parthian shot directed toward Linnaeus, brings us back to where we started: classification and the problem of the "genus." It was not unintentionally that Buffon spoke of "family" as opposed to "genus": the word lent itself much better to the long critique that we just saw. He refused, however, to distinguish the two. The only reality that he recognized was the species: "We will use *families*, genera, orders, and classes no more than Nature uses them."[61] The very most he would admit was that the idea of "family" could be convenient when speaking about species of small animals: "The small species . . . are very numerous, and at the same time border more closely on each other, in such a way that one is much more tempted to mix them together in the same *family* because they embarrass us and fatigue us by their multitude and their small differences, with which we are required to load our memory: but it must not be forgotten that these *families* are our work, that we only created them for the comfort of our mind, that if they cannot comprise the real series of all creatures, this is our fault and not that of Nature, which does not know these alleged *families* and indeed only contains individuals."[62]

Here we are forewarned: when Buffon speaks of families or genera, it is only for convenience.

Not only does the genus not exist in nature, but the use of this idea constituted an epistomological regression. It could be used "when one sees the objects only broadly," but it "fades away as soon as one applies it to reality, and one comes to consider Nature in detail." Such indeed was the very history of the human mind: "For primitive peoples and all nascent languages, there are almost only general names, that is to say, vague and undefined expressions of things of the same order. Particular names only came after a detailed examination was made of each sort of thing: the number of these names grew as Nature was studied more and known better. . . . Therefore when it is presented to us today in general terms, that is, by genera, it sends us back to the *abc*'s of all knowledge and recalls the darkness of man's childhood: Ignorance made genera, Science made and will make proper nouns."[63]

No one could have rejected the very foundations of classification more completely. This did not stop Buffon from pursuing classifiers on their own turf and, for example, from reproaching Linnaeus for having put the flying fox in the same genus as the European bat, since they do not have the same number of teeth in the lower jaw: "Thus they cannot be from the same genus according to a method that, like this Author's, is based on the number and the order of the teeth."[64] Besides that, the flying fox is American, and Linnaeus fell into the error "of forming generic names, which confound together a large number of species, not only different, but often very removed from one another";[65] here one must understand: geographically very distant.

The true satisfaction came from showing that the classifiers could not agree among themselves: "The Nomenclators who never want a creature to be only what it is, that is, alone in its genus, have differed greatly on the subject of the mongoose. M. Linnaeus made it first a badger, then he made it a ferret; Hasselquist, in accordance with the primary lessons of his master, also made it a badger; Messieurs Klein and Brisson placed it in the genus of weasels, others made it an otter, and others a rat; I cite these ideas only to show how little consistency they have even in the minds of those who imagine them, and also to put people on guard against these appellations that they call generic, and which are almost all false, or at least arbitrary, vague, and equivocal."[66]

The year was 1765, and we can see that Buffon's aggressiveness had not diminished. Among Linnaeus's disciples, Hasselquist was the only one to enjoy the dubious honor of being named. First of all, he had insulted France by saying at the end of "his long and dry description of the mongoose": "Frenchmen who live in Egypt, and who impose their own imaginary names on all things they have no knowledge of, have called this animal Pharaoh's rat." If he had really read the authors he attacked, he would have seen that it was the Egyptians who called it this, "and he would have abstained from taking an occasion here to speak badly of our nation; but we must not be surprised to find the accusation of a pedant in the work of a schoolboy." In addition he did not even know that what the Arabs call *nems* is a ferret and not a mongoose.[67] Above all, in his interminable descriptions, he forgot the essential part: in his "very long, yet very dry" description of the giraffe, he omitted mentioning the nature of its horns. Now "if its horns fall every year, it is of the genus of deer, and if on the contrary its horns are permanent, it is of that of cattle or goats." Hasselquist had amassed "methodically, that is, like a schoolboy, a hundred minor, useless characteristics." This was what was meant by using methods "that do away with reasoning, and permit a person to believe himself to be that much more scholarly the less intelligent he is."[68] Had Buffon, however, not fallen into a trap? Here he was no longer discussing the existence of genera but whether the giraffe belonged to the "genus of deer" or to "that of cattle or goats."

The Two Meanings of the Word "Species"

In fact, Buffon had already accepted the idea of genus, but by giving the word a very personal meaning and by restricting its use to the case of domestic animals alone. Why this difference? "Because in Nature, only individuals and the descendants of individuals exist, that is, species; . . . on the contrary, we have altered, modified, and changed the individuals of domestic animals: we have therefore created physical and real genera, very different from those metaphysical and arbitrary genera that have existed only as an idea; these physical genera are truly composed of all the species that we have manipulated, modified, and changed; and since all these species altered differently by man's hand nevertheless have a unique and common origin in Nature, the entire genus must form only one species."[69]

The difficulty with this text is that within the space of a few lines the word "species" is used with two different meanings. A "genus" grouped together different animals, but ones that can be crossed and that produce fertile young. All these animals, therefore, belong to the same biological "species," made up of "varieties"—or "races" when the varieties are constant. These domestic races all issued from the same wild "prototype," and they diversified through the effect of a "degeneration" for which man was responsible, as we saw in the preceding chapter. These races Buffon now decided to call "species," while the biological species to which they all belong became a "genus."

Before coming back to the meaning and consequences of this change in vocabulary, let us simply note that "genus" in Buffon's sense had nothing in common with the "genus" of the "Nomenclators," whom Buffon immediately attacked, once again, accusing them of wanting to "judge Nature" "through small particular characteristics," and to obscure natural history "by increasing the names, and the species as much as the names, without any necessity; by inventing arbitrary genera that Nature does not know, by perpetually confusing the true creatures with beings of reason; by giving us false ideas about the essence of species." "It is for this reason that the Nomenclators err constantly and write almost as many errors as lines."[70] Whatever else has been said about him, Buffon never accepted traditional classifications.

As we have also seen, however, he had discovered that the process of "degeneration" not only affected domestic animals, but was also found in nature, when animals had been forced to change climates. From then on, might it not have been legitimate to use the idea of "genus," such as Buffon defined it, for wild animals as well as for domestic animals? In fact, it seems that from 1761 on, in his work *Animals Common to the Two Continents*, Buffon's vocabulary started to become a bit uncertain.

"Hares, squirrels, hedgehogs, muskrats, otters, marmots, rats, shrews, bats, moles are all species that could be considered common to the two continents, even though in all these *genera* there is no species which is perfectly similar in America to those in Europe; and one feels that it is very difficult, not to say impossible, to decide whether they are truly different species or simply varieties within the same species, which became stable only through the influence of the climate."[71]

Was Buffon speaking of "genus" for simple convenience and because he was discussing these small species for which the use of the word was inconsequential? Or was he using the word with a new meaning? What is sure is that the word "species" was already in the process of changing sense. Living forms were so profoundly transformed by the influence of climate that we were "surprised at the rapidity with which the species vary and at the ease with which they alter themselves by taking on new forms." "It would therefore not be impossible that even without inverting Nature's order all these animals of the New World are basically the same as those of the Old, from which they had earlier drawn their origin. . . . That must not, however, prevent us from regarding them today as animals of different species: from whatever cause this difference came, whether it was produced by time, the climate, and the land, or whether it dates from the creation, it is not less real."[72]

How could these animals of the New World be at the same time "the same" and "different species"? And what made up this "real difference" between the species? The word "species" cannot be understood here as the biological species. If it were, the animals of the two worlds, having a common origin, would always be of the same species.

Buffon was required here to rethink his definition of a species as based on interfertility. This was, first, because he was less and less sure that hybrids, which he generally called "mules" or "mongrels," were sterile. The "mule" type was the offspring of the ass and the mare. In 1753 Buffon was persuaded that this mule was sterile. The ass and the horse, therefore, were two distinct species. Later, he accumulated evidence that he would publish in 1776 in a special article he had announced as early as 1755,[73] but whose results he gave only in 1766.[74] These results indicated that "the male mule can . . . beget and the female mule can conceive, carry, and give birth in certain circumstances."[75] The mating of the male and female mules probably has no result: this is why an intermediate "race" was not created between the horse and the ass. But it was no longer possible to say that these two species were separate. What was true for this "mule" type was even more true for "mules" or "mongrels" in general, that is, for hybrids. We have seen that the mating of the male goat and female sheep could produce fertile young. Buffon suspected that this must be the same for matings between dogs and wolves, or dogs and foxes: "Each one of these species is truly so close to the others" and the "individuals resemble each other so much . . . that one has difficulty in conceiving why these animals cannot reproduce together."

We have seen that Buffon tried the experiment. He therefore understood that the obstacle did not arise either from physiology or anatomy but from a "repugnance" that came "from the wolf and the fox rather than from the dog," and that probably resulted from the state of captivity in which the wild animals were kept. "Yet I am persuaded that in the state of freedom and celibacy . . . the dog can indeed unite with the wolf and the fox, especially if, becoming wild, it loses its odor of domesticity, and has at the same time come closer to the customs and natural habits of these animals."[76] Buffon would not be surprised to learn in 1773 that a Belgian lord, the Marquis de Spontin-Beaufort, had succeeded in mating a male dog and a female wolf, which gave birth to three offspring. Not able to make the dog wild, he had tamed the young wolf and thus removed the obstacle of "repugnance." Very excited, Buffon asked that the experiment be done with a male wolf and a female dog, and between a dog and a fox. He apologized for "requiring too much," but his excuse was beautiful: "I admit that the discovery of a new fact in Nature has always transported me with delight."[77] Especially when the fact confirmed one of his suppositions!

What was true for quadrupeds was even more so with birds. Not only are there three varieties of canaries from the Canary Islands, which produce among themselves "fertile cross-breeds whose races propagate themselves," but "there is likewise a mix of canaries with siskins, goldfinches, linnets, buntings, chaffinches, and perhaps even sparrows."[78] Of course, these mixes had been obtained in captivity by taking certain measures. The point is that "in animals as in man, and even in our small birds, the lack of character, or if one prefers the difference of moral qualities often affects the fitness of physical qualities."[79] Even here the conditions for love were more than physical in nature.

What should be concluded from all this? If the dog reproduces with the wolf and should reproduce with the fox, the three animals belong to the same biological species. Let us note in passing that Buffon did not believe in the fertile mating of the wolf and the fox, which never occurs in nature. The dog serves as a bridge, so to speak, between the two animals. And why not include the jackal and the arctic fox, especially if "the jackal and the dog easily breed together?"[80] But now the biological species included such a large number of forms that employing the word "species" became impossible. In all these texts, the dog, wolf, fox, and so forth became different "species," the collectivity of which constituted a "genus."

Meanwhile, Buffon had to admit new characteristics into his definition of species. Starting in 1761, he asserted that the only law "that can guide us in the knowledge of Animals" was to "judge them as much by the climate and by their nature as by their shape and structure."[81] An animal's "nature" here included the "repugnance" that prevented one animal from mating with another. The species thus defined was much more restricted than the biological species. Interfertility still ruled but was no longer enough to define it. The geographical habitat and the "affinities" of the "nature" had to be added. Suddenly, the limits between these restricted species were much hazier: "In reality, it is the number of agreements or disagreements that constitute or distinguish species."[82] It was, however, a question of more or less: "There are fewer agreements, fewer similarities of nature between the horse and the female ass . . . than between the male and the female ass," but there are some all the same, since they reproduce together.

Basically, things were simple. More sensitive to the diversity of the living world, more a "naturalist" and less a "biologist," Buffon simply changed his vocabulary. The biological species became the genus, and the variety or race became the species. The difficulty was that Buffon did not warn the reader about the change in the meaning of the words. What was worse, he continued to use the word "species" in both ways, the broad and biological meaning and the restricted meaning. This drove him to contradictions, apparent but bothersome, for they related to an important point in his thinking, the problem of the fixity or the variability of a species. In 1765, in the "Second View of Nature," he wrote: "The imprint of each species is a type whose principal traits are engraved in indelible and forever permanent characters, yet all the subordinate appearances vary, no individual perfectly resembles another, no species exists without a great number of varieties."[83] The following year, in the discourse "De la dégénération des animaux," [On the Degeneration of Animals], after having spoken of the varieties encountered in many species, he wrote, to the contrary:

"After the fleeting look, however, that we just took at these varieties,

which indicate to us the specific differences of each species, a more important consideration presents itself whose perspective is broader, that of the
change of the species themselves. It is degeneration, that older [process]
existing from time immemorial, which seems to have occurred in each
family, or if one prefers in each of the genera within which one can include
similar species that differ little from one another: we have in all terrestrial
animals only a few isolated species that, like man, make up the species and
the genus at the same time: the elephant, the rhinoceros, the hippopotamus, the giraffe form genera or simple species that propagate only
themselves in a direct line and have no parallel branches; all the others
seem to make up families in which one ordinarily notices a principal and
common ancestor, from which different stems seem to have come and
which are much more numerous when the individuals in each species are
small and more fertile."[84]

Not only does the second text contradict the first, but it seems to show
that Buffon was ready to accept a natural process of "speciation," a natural
formation of new species, and therefore almost a theory of evolution. In
reality, it was nothing of the kind. Here again the word "species" does not
have the same meaning in the two texts: in the first, the species in question
is a biological one, and the "imprint of each species" was what transmitted
the "interior mold"; in the second text, species is used in a restricted sense,
corresponding to the "variety" in the first text. There was therefore no
contradiction. In fact, the two texts are complementary, and it is their
comparison that gives meaning to Buffon's works.

The Order of Nature: Toward a New Classification

The work of the naturalist could not be limited to the detailed description of observable reality. If Buffon, arriving at the end of his *Natural
History of Quadrupeds,* felt the need to put his two broad "Views of Nature"
at the beginning of volumes XII and XIII, it was not just, as he claimed,
because "the details of the *Natural History* are interesting only for those who
apply themselves uniquely to this science" and to allow "a return to our
details with more courage."[85] Nor were they even intended to accelerate
the sale of the volumes. It was because he felt himself at a turning point in
his intellectual life, and he wanted to sum up the general conclusions
drawn from twenty-five years of fierce work by offering his views on nature—"philosophical" views in the sense he used the word, syntheses of all
that his scientific research had taught him.

For Buffon, nature was both order and activity: "Nature is the system of
laws established by the Creator, for the existence of things and the succession of beings. Nature is not a thing, for this thing would be everything.

Nature is not a being, for that being would be God; but one can consider her an immense living power, which embraces everything, which animates everything. . . . Nature is herself a perpetually living finished product, a worker ceaselessly active, who knows how to employ everything, who in working by herself always on the same resources, far from exhausting them, renders them inexhaustible."[86] "Guardian . . . of the immutable decrees" of God, Nature is above all an order of laws: she "never distances herself from the laws that have been stipulated for her," and it was because of this that a science was possible.

Nature is also, however, an arrangement of structures: "She alters nothing in the plans that are laid out for her, and in all her works she presents the seal of the Eternal One: this divine imprint, inalterable prototype of existences, is the model on which she operates, a model on which all the properties are expressed in indelible characters enduring forever; an always new model, that the number of molds or copies, however infinite they may be, only renews."[87] It was in this sense that the biological species was immutable: it was an element of the order of the world, the basis for the organization of the living world. A real and physical being, it nonetheless plays the rational role of a logical entity, and expressed again the double value of the Aristotelian "form": a principle of organization of living beings and a logical foundation of knowledge.[88] The species in the restricted sense of the word, on the contrary, was the product of nature's activity. It was an empirical reality, the daughter of time and of "circumstances." Dependant on "external and accidental causes," it was not an element of the order of the world. This hierarchy between the biological species and species in the restricted sense was parallel, if not similar, to the hierarchy that Tournefort and Linnaeus established between the genus and the species. From then on, the biological species, promoted to the rank of "genus" or "family," would play the role for Buffon that the "genus" had played for them. It also would serve to put the diversity of the species in order; it would serve as a principle of classification.

The discourse "On the Degeneration of Animals" was not primarily a reflection on the possible fertility of "mules" or even on the possible extent of artificial or natural degeneration: it was above all a sketch of a new classification of quadrupeds: a sketch that brought together and summed up fifteen years of research and reflection, and which was founded on a principle that Buffon expressed as early as 1761: "The true work of a Nomenclator does not consist . . . of doing research to make his list longer, but [in making] reasonable comparisons to shorten it."[89] And the result of Buffon's work would indeed be to reduce to thirty-eight families (that is, to this number of biological species) the two hundred species (in the restricted sense) of quadrupeds that he described.

Some of these families were represented by a unique species, which had never "degenerated" or which had degenerated so little that its "varieties" remained varieties in the new nomenclature. At the head of these "isolated species" was the human species. It was unique because it was found in all climates and it had nevertheless undergone only "superficial alterations," which could even be erased by time and a return to the original climate. Buffon had not changed his ideas on this point since 1749. Among the quadrupeds, isolated species had two common characteristics: they were large in size (except for one) and had always lived in the same climate. Certain ones contained varieties, but they remained only varieties. In the Old World, these species were the elephant, rhinoceros, hippopotamus, camel, lion, and tiger. Those that were common to the two continents were the bear and the mole. Those belonging only to the New World were the tapir, water cavy, llama, and peccary. The majority were there, it seems, because Buffon did not know to which family they should be attributed. We have seen that he could not decide whether to put the giraffe in the "genus of deer" or in that of oxen or goats. Finally he placed them separately. He still wondered if he could not put the vicuna with the llama and the guinea pig with the water cavy. The biggest surprise in this list was perhaps to find the lion and the tiger in it: Buffon must have been impressed by their size, by the lion's morphology and by the geographical isolation of the tiger. Note that in this text, the isolated species were never called "noble species."

The list of the twenty-five genera can be found in the accompanying table. It is, however, important to note that Buffon did not attempt to group his genera into higher categories, into an "order" of animals with cloven hoofs or *fissiped* (divided paws), for example. Such a regrouping did not interest him, as he did not see the possibility of attributing a common origin to these different animals. On the other hand, he fully respected, as a first approximation, the divisions between animals of the Old World, animals of the New World, and animals common to both.[90]

Only as a first approximation, however, because the "ten genera" and the "four isolated species . . . which are specific to the new world . . . nevertheless have distant relationships that seem to indicate something common in their formation and that lead us to go back to larger and perhaps older causes of degeneration than all the others."[91] The tapir, "despite the unimportant relationships [that it] is found to have with the rhinoceros, hippopotamus, and ass," is nevertheless "of a singular and different genus from all the others." But "the llama and the vicuna seem to have more significant signs of their former kinship, the first with the camel and the second with the ewe." Other cases were difficult, like that of monkeys. For the peccary, there was no doubt: it was "from the same genus" as the pig. In addition, "the tigers of America, which we have indicated by the names of jaguars, cougars, ocelots, and margays, although different species from the panther, the leopard, the snow leopard, the cheetah, and the serval, are nonetheless

Classification of quadrupeds

ISOLATED SPECIES, WHICH MAKE UP A GENUS BY THEMSELVES

Specific to the Old World: elephant, rhinoceros, hippopotamus, giraffe, camel, lion, and tiger.
Common to both continents: bear and mole.
Specific to the New World: tapir, water cavy, llama, and peccary.

GENERA BRINGING TOGETHER SEVERAL SPECIES

Specific to the Old World or common to both continents:

1. *Actual solipeds [whole-hoofed animals]:* horse, zebra, ass, and mule, fertile or infertile (specific to the Old World).
2. *Large cloven-hoofed animals with hollow horns:* cattle, buffalo, bison (common to both continents).
3. *Small cloven-hoofed animals with hollow horns:* sheep, goat, gazelle, musk deer (specific to the Old World).
4. *Cloven-hoofed animals with solid antlers:* elk and moose, reindeer and caribou, deer, fallow deer, axis, roe deer (common to both continents).
5. *Equivocal cloven-hoofed animals:* boar, pig and its varieties (specific to the Old World, but the peccary should probably be included).
6. *Fissiped [divided-pawed] carnivores with hooked and retractable claws:* panther, leopard, cheetah, snow leopard, serval, cat, and their varieties (specific to the Old World, but ultimately connected to their American analogues).
7. *Fissiped carnivores with nonretractable claws:* wolf, fox, jackal, artic fox, dog, and their varieties (common to both continents).
8. *Fissiped carnivores with nonretractible claws, with a pocket under the tail:* hyena, civet, zibeth (Indian civet), genette, badger, etc. (specific to the Old World).
9. *Fissiped carnivores with a very long body, with five digits on each paw:* stone martens, martens, polecat, ferret, mongoose, weasel, etc. (common to both continents).
10. *Fissipeds with two large incisor teeth on each jaw and no quills on the body:* hare, rabbit, and all species of squirrels, dormice, marmots, and rats (common to both continents).
11. *Fissipeds whose bodies are covered with quills:* porcupines, hedgehogs (specific to the Old World, despite the proximity of the tree porcupine and the Canada porcupine).
12. *Fissipeds covered with scales:* pangolin and phatagin (specific to the Old World: American anteaters, "to which one can compare them," are covered with hair and "differ too much")
13. *Amphibious fissipeds:* otter, beaver, desman, walruses, and seals (common to both continents).
14. *Quadrumanes:* monkeys, baboons, guenons, lemur, loris, etc. (specific to the Old World: the American quadrumanes are different).
15. *Winged fissipeds:* flying foxes and bats (common to both continents).

Specific to the New World:

1. *Sapajous:* eight species.
2. *Squirrel monkeys:* six species.
3. *Possums:* marmosets, *kaïopollons* (a small possum from the forests of South America), phalanger, tarsiers, etc.
4. *Jaguars:* cougars, ocelots, margays, etc. (ultimately linked to "Carnivores with retractible claws of the Old World").
5. *Coatis:* three or four species.
6. *Skunks:* four or five species.
7. *The genus of agouti:* acouchi, paca, aperea, and tapeti.
8. *Armadillos:* seven or eight species.
9. *Anteaters:* two or three species.
10. *Sloths:* the unau and the ai.

N.B. Notice that for the New World "genera," Buffon often uses the name of a "principle species," after having reproached Linnaeus for doing so. The principal species was probably the founding species for him. Notice as well that this classification suffers from at least two major flaws: Buffon attributed the same importance to all the characteristics, and he dares not give sufficient power to "degeneration" to reduce all the quadrumanes of the two worlds to two genera.

most certainly from the same genus." "Finally four genera and two isolated species remain, the possums, the coatis, the armadillos, the sloths, the tapirs and the water cavies, which can neither be related nor even compared to any of the genera or species of the old continent." How to explain both these resemblances and these differences: the "most reasonable" way was "to think that in other times these two continents were contiguous or continuous" and that "the species that had been confined to these regions of the new world . . . were enclosed and separated from the others there by the flooding of the waters when they divided Africa from America."[92]

Many obscurities and uncertainties remained in all this. Buffon at least believed that the "genera" he had created were "real and physical." In our eyes, his merit is to have been the first to try to establish a phylogenetic classification by grouping species according to their community of origin. Certainly, when creating genera he used morphological analogies and comparisons much more often than the irrefutable but also impossible experimental test of interfertility. To our eyes, moreover, the failure of this experiment would prove nothing. What was important here was less the detail, necessarily erroneous, of this classification than the new problems that it created. Within each genus, species did not have the same status and did not have among themselves the same degree of relationship. In each genus, there were "principle stems" and "accessory branches." The first were present at the origin of the genus, the second were more recent. "One must regard the wolf and the fox as major stems," and that meant that "in order to reduce these two species to one, it is necessary . . . to go back to an older state of nature."[93] As we would say today, the two lines "diverged" earlier.

From the notions of "kinship among species" and "agreements" or "disagreements" among species, Buffon was able to ask some very modern questions: "In general, the kinship among species is one of those deep mysteries of Nature that man can probe only with experiments that are as repetitive as they are long and difficult. How could one learn the degrees of relationship among species other than through the results of thousands and thousands of attempts to breed animals of different species? Is the ass a closer relation to the horse than to the zebra? Is the wolf closer to the dog than the fox or the jackal? At what distance from man should we put the great apes, whose bodily structure is so similar to his? . . . What relationships can we establish between this kinship among species and another, better-known kinship, that of different races within the same species? Does not a race, like a mixed species, generally result from a divergence from the pure species in individuals who were the founders of the race? . . . How many other questions we have about this material alone, of which only a few can we answer!"[94]

Even if we resist the temptation to modernize Buffon's thoughts unduly, how can we not see here the questions that modern geneticists and taxonomists endeavor to answer with knowledge and concepts that Buffon could not even imagine? At the very least it was in terms of filiation that he approached the problem of "kinship" among species. Does not the modern practice of comparing genotypes measure precisely these "divergences," which were still for him only an intuition?

In the *Natural History of Birds,* Buffon and Guéneau de Montbeillard renounced the practice of "treating birds one by one, that is to say, by distinct and separate species." After announcing this Buffon added, "I will unite several together under one same genus," except for "domestic birds and a few major or particularly remarkable species." This decision was especially motivated, we have seen, by the desire to advance more quickly, which a friend of Buffon's, M. de Montmirail, vigorously combatted, using Buffon's very own principles: "Despite the fact that I had said that in nature there is indeed only one species of sparrow, of titmouse, etc., that this same name has been given to other birds only because of some more or less distant relationship with these first species, and that for example the black-headed titmouse is no more a titmouse than the ass is a horse, all my reasons were rejected, and the desire to accelerate and tie things up carried the day."[95] Montmirail was ten years behind, and Buffon, in order to legitimize his decision, emphasized the fact that there were many "neighboring" species, especially among small birds, which were "similar enough to warrant being looked upon as parallel branches of the same stem, or of a stem so close to another that one could suppose a common ancestor."

Buffon certainly did not dare to guarantee "that each of our articles will really and exclusively contain only the species that have the degree of kinship of which we speak. It would be necessary to be more learned than we are, and than we can be."[96] As Guéneau de Montbeillard explained apropos of "foreign cuckoos," he had only been able to group together species that "have more relation to the genus of the cuckoo than to any other." In practice, the method hardly differed from that of traditional classifiers. It was the ambition that was different: "to compose a genus that is not strict, rigorous, and in consequence imaginary, but a real and true genus,"[97] which brings together beings of the same origin. Rather than the results of completed research, it was, apart from an introductory "historical sketch," an ambition and a program of research that the *Natural History of Birds* claimed to offer posterity.[98]

The program of research is one that Buffon knew would be fertile, and not only for the natural history of birds: "Far from becoming discouraged, the Philosopher must praise Nature, even when she seems miserly or too mysterious to him, and he must be satisfied with knowing that as he lifts a part of her veil, she lets him glimpse an immense number of new objects all worthy of his research. For what we already know must make us judge what we could know. The human mind has no limits; it expands as the Universe reveals itself. Man therefore can and must attempt everything; he needs only time in order to know everything . . . and what more pardonable or even nobler enthusiasm is there than that of believing man capable of recognizing all the powers and discovering through her works all the secrets of Nature!"[99] Buffon was correct in thinking that it was through the study of "mules," that is to say hybrids, that we would be able to answer all the questions he asked. He himself, however, turned to other goals. Since the living world was ever more visibly the product of time and nature's activity, it was the history of that nature that he wanted to study from then on:

"It is therefore necessary, to explain the origin of these animals, to go back to the time when the two continents were not yet separated; it is necessary to recall the first changes that happened on the surface of the globe; it is necessary at the same time to imagine the two hundred species of quadruped animals reduced to thirty-eight families: and even though the state of Nature back then is not the one that has reached us and that we have represented, even though it is, on the contrary, a much older state, which we can hardly imagine except by means of inductions and relationships almost as fleeting as the time that seems to have erased all traces of them, we shall nevertheless try to go back in time using facts and the still-existing vestiges of these first ages of Nature, and to represent only those epochs that seem clearly delineated to us."[100]

This return to nature's past was the logical although distant consequence of the very definition of species that Buffon had sketched beginning in 1749. From the simple reconstruction of forms, Buffon passed to their "degeneration," their diversification, and their multiplication. "The great worker of Nature is time." As Buffon already indicated here, however, and as we shall see in more detail, the history of the earth now entered powerfully into his thoughts and transformed them profoundly.

The grandeur and the effect of the works that would follow must not make us underestimate the importance of the *Natural History of Quadrupeds,* whose discourse "On the Degeneration of Animals" is both a high point and a synthesis. In its way of conceiving the "history" of animals, in its definition of species, and in the new mode of classification it offered, this natural history broke completely with the intellectual habits of its time. The finished work probably did not always fulfill its aims, and Buffon was conscious of this. He had, however, asked new questions and imposed new directions of research, which no one would pursue exactly in the same way but which no one after him could ignore either. Probably no naturalist since Aristotle had so deeply transformed his science.

CHAPTER XX

Chronicle of Ordinary Glory

After the shock of the three volumes that appeared in 1749, after the success, the fuss in the journals, and the controversies, and in the calm that was gradually reestablishing itself, the fourth volume of the *Natural History* was awaited. It did not appear until 1753, just two days after Buffon's entry to the Académie Française, when all literary Paris was talking about his discourse. It was, in its way, an event. Many journals commented on it, and the *Journal des Savants* even gave two excerpts from it. And then, progressively, as the following volumes appeared, reviews became rare, and the *Journal des Savants* remained the only French journal to publish some regularly.

When volume IV appeared, Grimm, considering that "it is only after a careful reading that one can give an account of such an important work, which so honors the author, his nation, and his century," first limited himself to a scornful remark about the two letters from the Sorbonne that opened the volume: "Aside from the woes that are their object, these two documents are very remarkable for the barbarism of style that prevails there."[1] He came back to the work one month later, with dazzling praise: "This book, which belongs to the small number of works that will pass to posterity and deserve to do so, has been universally acclaimed from the beginning." The history of the horse, the ass, and cattle were "models" for "those who would like to learn how to write." They "will learn how one speaks with dignity of the most common things" and "the nature of genius and talent." It was, however, the "the admirable discourse on the nature of animals" that was the object of the longest analysis. Grimm did not always agree with it, defending the passions and the "morality" of love in opposition to Buffon, but admired—who would have believed it?—the "modesty and the fairness with which [Buffon] took care in qualifying his arguments." Only a "true philosophe," a "superior man" could do it. In short, he was enthusiastic.[2]

When volume V appeared two years later, there was still admiration for the style, but less enthusiasm for the ideas: too much poetry in the history of the dog, too much assurance in questionable conclusions. On the other hand, Daubenton's texts "have the merit of exactitude and doctrine."[3] One year later, for volume VI, Grimm was even harsher. "People do not talk about [M. Daubenton's] work in Paris. Since it is a more useful than brilliant piece of research, it hardly interests people who seek only to amuse themselves and not at all to learn." People, he said, talked too much about Buffon, who wrote "with a pomp, a harmony, and a magnificence of style that cannot help but turn our heads. Indeed, the case that one makes of style in Paris is a very singular thing; there is nothing that cannot be made to succeed by this means." But "M. de Buffon's merit will lose some of its brilliance with posterity as well as with foreigners," because posterity "neglects form [and] will be able to judge only by the ideas and the content. On the contrary, M. Daubenton's reputation will only increase from this. His merit is durable and solid; it does not appeal to the idle people of Paris." Besides that, Buffon was "infatuated with his systems," exactly like Rousseau, and Grimm set off full tilt against "system builders."

Grimm thus began a pattern of criticism that would become stereotypical: Buffon was a great writer but a bad philosopher. The true scholar was Daubenton. In fact, Grimm was irritated by the praise of game hunting that the article on deer contained. He said nothing in 1753 on the discourse "Domestic Animals." He now said nothing about the discourse on wild animals or the thoughts on the large infestations or on the infanticide contained in the article dedicated to the hare. He did not, however, like hunting, for he considered "that there is nothing more barbarous . . . than to seek one's entertainment in the agonies and the slow tortures of a living creature," and here Buffon was condemned. "I do not want to hint that he wanted to pay court to great men and to flatter their dominating taste at the expense of truth and its sacred rights; this would be an unpardonable baseness." In fact, of course, the hint was there, and even the innermost certitude: Buffon had betrayed "philosophy" and truth.[4]

In 1759, after having waited "several months" to speak about volume VII, Grimm directly attacked the discourse on carnivorous animals, "that which is the weakest part of the *Natural History,*" and which he rewrote in his own way. The questions he raised were serious ones, Grimm observes, since they concerned the war among the species and the possibility of their transformation during the course of "millions of centuries." It was Diderot, obviously, who inspired him. Buffon preferred to speak of other things; of the "organization" and the "primitive state of nature." He was wrong to do so, these were not the subject. It was from here that a somewhat surprising overall judgment came: "This writer does not abound in ideas." True, the *Natural History* "was presented in the middle of the persecution incited against philosophy. It was not achieved without frequently sacrificing the liberty and the boldness that speaking the truth demands." Buffon had had to censor his work and put in "several corrections before daring to have it appear in public." He was obviously too clever or too timid.[5]

Two weeks later, Grimm returned to the charge, again denouncing Buffon's dogmatism but approving the fact that Buffon attacked the supporters of "final causes" and refuted "this chimeric state of nature that M. Rousseau has wanted to establish."[6] Could Diderot have had anything to do with this? Two months later, the *Literary Correspondence* published "Observations sur quelques auteurs d'Histoire naturelle" [Observations on a Few Authors of Natural History] by Charles Bonnet. After a long and subtle praise of Réaumur, Bonnet devoted several lines to Buffon and launched the famous statement, traditionally attributed to Voltaire: "Is this Natural History natural enough?" A great writer, "a sublime and bold genius" but "too carried away by the spirit of system," Buffon was on the verge of a "philosophical novel."[7] Elsewhere, we shall see, Bonnet would be more subtle, but the classic image of Buffon was already established.

In March 1762, after a three-month delay, volumes VIII and IX were examined, and the same themes were stressed, this time for the section "Animals Common to the Two Continents": too many assertions founded on too few facts. The mysteries of nature are impenetrable and it is necessary to be modest. Nothing precise was known about animals, and after having read Buffon, nothing more was known. In fact, "science was not made for man." "It is useful to him only insofar as it lessens the pains of life, serves as a diversion . . . and contributes to making us more human."[8] To a certain extent, this impression was corrected two years later with volumes X and XI. Buffon clearly spent too much time criticizing other authors, and he himself, in treating exotic animals, "shed no light on their nature, their species, their instinct, their habits, etc." The only exceptions: the history of the elephant and the camel, and a "fine enough and broad view" of the animals that passed from the Old to the New World. "Yet one admires in all M. de Buffon's articles that philosophical view, that sound and wise intellect, that noble, elevated, and wise style which, so to speak, enchants and broadens the reader." In short, "M. de Buffon will always have the reputation of a distinguished philosopher," and the *Natural History* that he has produced with Daubenton "will be counted among the works that have illuminated the century of Louis XV."[9]

To understand Grimm's waverings, we must not forget that he had to give an account every two weeks of a pile of new publications, books, poems, or ephemeral brochures, all the gossip at the Opéra and the Comédie Française, the news of politics, in short, everything concerning Parisian life. In this din of current events, the volumes of the *Natural History* contrasted strongly with their intellectual seriousness and their style. Buffon obviously dominated his time. Grimm might very well criticize him, sometimes violently; but he could not prevent himself from admiring him. When he panned d'Alembert, primarily because of his style, it was through comparison with the greatest stylists—Voltaire, Montesquieu, Diderot, Rousseau, and, naturally, Buffon.[10]

The entire party of philosophes, Diderot in the lead, who sometimes collaborated on the *Literary Correspondence,* was behind Grimm.[11] The party was quickly torn apart by internal quarrels: a rupture between d'Alembert and Rousseau after the article "Genève" [Geneva] in the *Encyclopédie* in 1757; a partial rupture between Diderot and d'Alembert when the latter left their enterprise in 1758; a shattering break between Diderot and Rousseau the same year. Grimm was plunged into this milieu and felt the backlash of these dissensions. It was inevitable that his attitude to Buffon would show the effects of this. In many respects, Buffon was close to the philosophers, and they, especially Diderot, knew it. At the same time he was not one of them, holding himself apart from all their quarrels and carefully protecting himself from all the attacks with which they overwhelmed him. His prudence and his apparent detachment were like an insult to their difficulties. They could neither love nor hate him.

The only thing they probably could do was envy him. For, throughout all these years, Buffon had been spared by the theologians. Aside from ironic or furious reflections that the correspondence between Buffon and the Sorbonne in 1753 occasioned, the theologians paid no attention to the *Natural History.* It is true that they were busy elsewhere and that Helvétius, the *Encyclopédie,* or Marmontel's *Bélisaire* gave them enough work. It is also true that Buffon, always very careful, as we have seen, multiplied his orthodox declarations. What his true thoughts were behind these declarations were probably less Catholic, but it would have been necessary to be very attentive to detect any possible dangers in the discourse "On the Degeneration of Animals." In the sciences of nature, only geology and the contradictions that it could bring to the story of creation according to Moses attracted the theologians' attention. Thus it would be necessary to wait until 1778 and the *Epoques de la Nature* [Epochs of Nature] for the Sorbonne to become interested in Buffon again.

Diderot's case was different. I have already spoken of their relationship in 1749. The article "Animal" in the *Encyclopédie,* signed by Diderot, was in fact no more than a long annotated extract from the first chapter of the "History of Animals," even if the commentary proved that Diderot had already gone further than Buffon and tended to remove all barriers between living and inorganic matter, as he did between thought and sensitivity.[12] At the end of 1753, Diderot published his *Pensées sur l'interprétation de la Nature* [Thoughts on the Interpretation of Nature], in which Buffon was immediately cited.[13] It still mostly concerned the Buffon of 1749. From the "History of the Ass," Diderot retained primarily the idea of the unity of the plan of composition, and he just as quickly also used it as an argument to show the plausibility of a generalized evolution, proposed the same year by Maupertuis under the name of Doctor Baumann. "Who would not be inclined to believe that there has been only one prototypical being for all beings? Whether one admits this philosophical conjecture, along with doctor Baumann, as true or rejects it as false, along with M. de Buffon, it is undeniable that it is necessary to embrace it as a hypothesis essential to the progress of experimental physics, to that of rational philosophy, to discovery, and to the explanation of the phenomena that depend on organization."[14]

Diderot was not satisfied to combine Buffon's vocabulary and Maupertuis's thought. He pursued his personal philosophy, which already led him not only to lend to "organic molecules" a "dull sensitivity,"[15] which Buffon could not help but deny, but also to a vision of the living world where "entire species" were like an individual who "is born, so to speak, grows, endures, wastes away, and passes on."[16] While Buffon's position slowly evolved from an absolute creationism toward a limited conception of the transformation of forms—a transformation limited by the imprint of the "prototype," which assured the permanence of order in nature—Diderot leaned toward the vision of the world that he later offered in the *Rêve de d'Alembert* [D'Alembert's Dream] in 1769: a universe where everything changed incessantly, where everything was born and died. "Who knows what races of animals have preceded us? Who knows what breeds of animals will succeed ours? Everything changes, everything passes, only the whole remains."[17] This was a universe in perpetual "fermentation," where chance ruled as master, where no enduring order can be detected; a universe very far from the one in which Buffon would never renounce finding or preserving order. From 1753 on, the two trains of thought diverged from each other. The two men remained on cordial terms, but it was no longer Buffon who would serve as a guide to the philosophe in his reflections on life and living nature.

The *Encyclopédie*, nevertheless, continued to take up certain themes of the *Natural History* and spread its influence. The article "Espèce (Histoire naturelle)" [Species] reproduced the definition Buffon gave in the "History of the Ass" purely and simply between quotation marks and with a reference.[18] The articles "Homme (Histoire naturelle)" [Man] and "Humaine, Espèce (Histoire naturelle)" [Human, Species], both signed by Diderot, were explicitly presented as taken verbatim from the *Natural History of Man*. The articles written by Daubenton, despite their dryness, remained faithful to the main lines of the *Natural History*. Buffon was

constantly present, with his "organic molecules" and his theory of reproduction in articles by vitalist physicians, and in particular by Ménuret de
Chambaud.[19] Yet here again, it was primarily the Buffon of 1749 who was
cited, used, or discussed. Since the *Encyclopédie* and the *Natural History*
appeared at the same time, the dictionary borrowed only a few things from
the *History of Quadrupeds,* and Jaucourt, who wrote so many articles and
often filled in the blanks left by Daubenton, very often ignored Buffon's
texts.[20]

It was once again with the Buffon of 1749 that Condillac wanted to settle
a score when he published his *Treatise on Animals* in 1755. Here, however,
the matter was a little more complicated. In 1754 Condillac had published
a *Traité des sensations* [Treatise on the Sensations], in which he developed
Locke's sensualist philosophy and imagined a statue that he progressively
endowed with sense organs, starting with smell, to show how sensations
combined progressively. Grimm had already given a fairly favorable account of the work, but he had immediately compared it to Diderot's *Letter
on the Blind* and to the awakening of a man as described by Buffon. The
comparison was not flattering to Condillac: "This precise philosopher does
not compare well to the philosopher of genius."[21] As for Raynal, he had
absolutely demolished the work,[22] and it was said that "the abbé de Condillac had dumped the statue of M. de Buffon into a barrel of cold water.
This criticism and the meager success of the work have embittered our
author and wounded his pride; he has just written an entire work against
M. de Buffon."[23] This work was the new *Treatise.*

Without going into a detailed analysis of the book,[24] we can say that
Condillac thought he was catching Buffon flagrantly contradicting himself
by simultaneously stating that matter was insensitive and that animals, purely material beings according to him, were capable of "feeling." He also
accused Buffon of materialism. The criticism was not groundless, but it
ignored the importance of organization and the true meaning that Buffon
gave to the word "feeling." Condillac himself preferred to give a soul to
beasts, but he found himself faced with a corresponding problem: how to
distinguish man from animals? In reality, the two philosophers were fairly
close to each other in their analyses. The paradox was that Buffon seems to
have been more dualistic than Condillac. The latter attributed language to
animals (referring to cries of alarm made by birds, evidence that Buffon
also described), a capacity for invention (Condillac attributed it to the
individual, Buffon to the species), and finally an aptitude for life in society.

Grimm was very hard on Condillac, whom he accused of being inexcusa-
bly impolite and lacking genius: "M. de Buffon puts more into one discourse than our abbé will put into all his works in his lifetime."[25] It is true
that the abbé was embroiled with Diderot at this time. On the other hand,
Grimm had much more praise for the *Lettres sur les animaux* [Letters on

Animals] that Le Roy, lieutenant of the hunt for the Park of Versailles, published anonymously in 1768.[26] This work in fact contained a collection of articles that had appeared in the *Journal encyclopédique* [Encyclopedic Journal] in 1762 and in the *Gazette littéraire de l'Europe* [Literary Gazette of Europe] in 1764–1765. Le Roy was one of the collaborators in the *Encyclopédie,* for which he had written the articles "Homme (Morale)" [Man (Morality)] and "Instinct," among others. As early as 1756, he maintained personal relationships with Buffon, Condillac, and Rousseau[27] and discussed animal behavior with Rousseau.[28] In short, he was one of the tribe, and Grimm had no reason to mistreat him.

Le Roy had in fact supported just about the same theses as Condillac, "that excellent philosopher," but he based them on his personal observations of animal behavior. For Le Roy, "beasts think, compare, judge, reflect, conclude, etc. They therefore have, in their trains of ideas, all that they need to speak,"[29] and indeed they speak among themselves in their own way. He even said that "beasts count, that is certain."[30] The *Journal des Savants* in January 1765 had attacked these views, considering them materialistic and calling in particular on Buffon's authority and on his distinction between human and animal memory. Le Roy reacted fairly violently on this point: "I do not recall which are M. de Buffon's ideas. . . . I ask . . . what is the *kind* of animal memory, and if he knows of two *kinds* of memory. Up until now, I admit, I have thought that there was only one and that it consisted uniquely of remembering felt sensations."[31] Which was the same thing as attributing to man the type of memory Buffon reserved for animals. In fact, for Le Roy, the only difference between man and animals was that man had more needs, and therefore more intelligence.[32] Buffon was thereby formally contradicted, or rather ignored. Nevertheless, Le Roy argued that certain articles of the *Natural History,* in particular the "History of the Elephant," supported his position, and the two men continued to have an apparently cordial relationship. This was because Buffon, as we have seen, had slowly changed. Without ever retreating from his original positions, he increasingly recognized a sort of "intelligence," a mode of communication, and a social life in animals. It is not very likely that Condillac had anything to do with this.

Let us remember the reactions of Haller and Charles Bonnet to the theory of organic molecules and reproduction presented in 1749. These reactions were much more measured for Haller, who had provisionally abandoned the theory of the preexistence of germs. In 1762 Bonnet published his *Considérations sur les corps organisés* [Considerations on Organized Bodies), in which he assembled all his ideas and his research on generation. The book was in fact a defense of the theory of preexisting germs and a formal refutation of the theories of Needham and Buffon, presented with even more force since Bonnet felt himself supported by the authority of

Haller, who on the basis of new experiments had become a convinced partisan of preexistence.

Bonnet, Haller, and soon Spallanzani formed a common front against Buffon and his supporters.[33] Bonnet criticized Needham even more than Buffon; it was organic molecules in particular that he could not accept. His criticism remained polite: "I respect this great Writer; but I respect truth even more."[34] He was sincere, and we have the proof of his capacity for dialogue in the regular correspondence that he exchanged with Needham, where the divergence of viewpoints prevented neither friendship nor esteem. It is true that Needham and Bonnet agreed strongly on the defense of the Christian faith against the impious; Bonnet was much less sure of Buffon's stand and made a few sharp statements in this respect: "Many Readers will probably reproach me for having expanded too much on M. de Buffon's System. They will claim that Dreams that are not in the least philosophical do not deserve consideration. I do not seek to justify myself for this reproach, but I will admit that I believe I owe something to the fame of the Dreamer and to the singularity of his Dreams."[35] And he urged Buffon to "rework his own observations" and "go deeper into this interesting subject. He has so much sagacity that it would be very strange if the truth should escape him. But surely it will not escape him should he forget, at least for a time, his organic Molecules, his Molds, and all the paraphernalia of a system that his fertile genius has been pleased to invent and that his reason in becoming strict will perhaps abandon some day."[36] Although polite, the criticism was nonetheless firm.

To Charles Bonnet's stupefaction, the sale of the *Considerations* was forbidden in France. He learned this fact from Malesherbes himself, to whom he had sent a copy. "The sensitivity of the subjects treated in a work of metaphysics can render the reading dangerous to the public," explained Malesherbes. Bonnet was indignant: "Here, I have only been a simple naturalist, I have always brought my reader back to the Being of beings." There was obviously a misunderstanding, or there was a plot: the ban came "from M. de Buffon and his cabal."[37] Finally the name of the nervous censor was known: it was Guettard, a well-known naturalist. And Lalande, after having deplored his "stupid timidity," explained to Bonnet: "It was not out of friendship for M. de Buffon that Guettard opposed it, for he was a disciple and intimate friend of M. de Réaumur; it was out of [religious] devotion." The ban was lifted, and Bonnet, very happy, hastened to write to Haller to clear Buffon of all suspicion "on the subject of this ridiculous business."[38]

The criticisms of Bonnet, Haller, and shortly Spallanzani, whose works appeared in Italian in 1765 and in French in 1769, were still concerned with the theories of 1749. Neither Buffon nor Needham, it is true, wanted to budge an inch. Haller and Bonnet, however, read the volumes of the

Natural History and exchanged their impressions of it. Haller's impressions were systematically unfavorable, Bonnet's were more subtle. As convinced as he was of the preexistence of germs, Bonnet was no less sensitive to the difficulty that hybrids represented. Haller remained imperturbable to the end, even to Buffon's announcement of the birth of hybrids between dogs and wolves.[39] Bonnet even defended the "Theory of the Earth" to Haller, who remained true to the biblical Flood,[40] and reminded Haller that after all it was he who had attacked Buffon most openly: "When it looks like he is not wrong, however, I am happy to say so."[41] On the winter torpor of swallows, Haller preferred to believe Klein rather than Buffon. Bonnet himself leaned visibly toward the idea of migration defended by Buffon.[42]

The point on which they agreed was in denouncing the anti-Christian tendencies of the philosophes. Bonnet complained that Jaucourt had plagiarized him in the *Encyclopédie*, without adding quotation marks and by replacing Bonnet's "adorable Wisdom" with "eternal laws." "A fanatic man who does not want to pronounce the name of the Creator," responded Haller. The same thing applied to Buffon and to the *Interprétation de la Nature* [Interpretation of Nature]. "These gentlemen do not know how to conceal their aim: they want to strip us of the strongest and the most popular demonstration we have of the existence of God." Bonnet agreed: he knew Diderot's *Interpretation:* "I have read few worse books. The Author is a Charlatan who dispenses bad drugs from a peddler's stand."[43] Bonnet was more hesitant than Haller on Buffon. Although the "History of the Pig" attacked final causes, Bonnet was satisfied to say: "I am not sure, but underneath I seem to glimpse a secret design to weaken one of the best proofs of God's existence." Haller's response was unqualified: "M. de Buffon reasons like an atheist and is not a physicist." God foresaw everything, and nipples on males "are no more useless than the breasts of girls whom God has foreordained to die as virgins."[44] Haller was very attentive to what Buffon wrote: "His discourse on Nature, and his ideas on imperfect animals, species of which have disappeared, display a very dangerous tendency."[45] Also, when Bonnet was troubled by the discourse "On the Degeneration of Animals," Haller had an immediate retort: "Assume that M. de Buffon is wrong. . . . These Philosophers believe nothing that we believe; to make up for it, they believe everything that their imagination creates, if it is advantageous to their cause."[46] Bonnet tried nevertheless to remain impartial: "Not all the modern *Epigenesists* are unbelievers: witness Needham and Paul. All the *Unbelievers* or *Infidels* are not epigenesists either: witness Voltaire, who blames Buffon and his partisans here, there, and everywhere, and who preaches about *Germs* from the rooftops."[47] Haller did not differentiate among the philosophers: he condemned "the arrogant splendor of d'Alembert, Buffon, and Diderot"[48] once and for all, and even all the French philosophers as a whole, starting with Descartes, "that poor observer, writer of fictions, author of an anthropogeny more impious than that of any Buffon, and even less reasonable."[49]

In 1764 Bonnet published his *Contemplation of Nature.* In it Buffon found himself criticized, although without being named, and with respect. "Systematic genius," "famous creator of animals" (apropos of Needham's observations), certainly. But "if Nature did not make him an Observer, she did enrich him with her noblest gifts and made him the most eloquent Man of his century. If he is not a Malpighi, or a Réaumur, he is a Plato, a Milton; and his Writings, full of fire and life, will tell Posterity that the Painter of Nature was not always the Draftsman."[50] Did Buffon read the book? We do not know. Be that as it may, in October 1765 Bonnet learned, through several intermediaries including the Duchess d'Anville and the Count de La Bourdonnaye, that "M. de Buffon will vote for M. Bonnet, if it is a question of naming him foreign associate to the Academy of Sciences." This was Bonnet's dream. He immediately wrote to La Bourdonnaye to explain to him that despite his criticisms, he had never stopped feeling "a lively admiration for the one whom I dared to combat, and whom I would have liked to praise ceaselessly." To this letter filled with praise, enthusiasm, and good will, La Bourdonnaye responded quickly in the same tone, guaranteeing Buffon's esteem for Bonnet: "I desire that you know of him, Sir; he is virtue itself, he is candor itself, he is sweetness itself; yet he is a man; therefore he could have been mistaken, no matter what zeal he has, like you, for the truth. I am entirely satisfied with the justice that you do to each other. He knows your feelings for him." In 1768 two letters from Horace-Bénédict de Saussure confirmed Buffon's good intentions.

In June 1769 Bonnet sent Buffon a copy of his *Palingénésie philosophique* [Philosophical Palingenesis], which had just appeared, accompanied by a letter in his eloquent and sentimental style, full of false modesty and, above all, with admiration for this "sublime, extensive, fertile genius, who knows so well how to see and paint the large picture of nature, bring together the most remote objects, grasp their diverse relationships, and deduce the most general results from them." Cleverly mixing flattery and sincerity, Bonnet praised the "Theory of the Earth," which he truly admired. He concluded: "We are therefore rivals and not enemies at all."

To Bonnet's great disappointment, Buffon did not answer except in very vague terms through a third person, and more than a year later. Evidently, he had not even opened *Palingenesis,* and if he read it, he probably had little enthusiasm for its confused metaphysics. He had, in fact, left the book and the letter in Paris. Bonnet was very hurt. With that art of complicated psychological construction of which intellectuals are so fond, he came to the conclusion that Buffon justified his criticisms by the fact that, in an article of the *Natural History,* he had attributed to Leeuwenhoek the discovery of the parthenogenesis of aphids, Bonnet's most famous discovery.[51] In reality, Buffon, had no reason to be interested in Bonnet's works, and the two men had nothing much in common, either in their vision of the world or in their scientific preoccupations. In the end, Bonnet was elected foreign associate in 1783. It is not known if Buffon voted for him.

In 1768 Voltaire published his *Singularités de la Nature* [Singularities of Nature]. "What a poor book!" Bonnet exclaimed. "There is practically not one single line that does not sound like a loudmouthed, ignorant schoolboy. And the admiring fools of this stupidity present this as a Masterpiece."[52] Bonnet was being a bit ungrateful, for Voltaire had published a very favorable review of the *Considerations on Organized Bodies*, where Bonnet found himself spoken of as an "ingenious and deep author" because he defended the preexistence of germs, while the theory of organic molecules and interior molds, presented by "an eloquent and very enlightened philosopher," was accused of "multiplying obscurities."[53] What began as a passing criticism would soon become a leitmotif. In 1765 Voltaire found himself opposed to Needham in a dispute over miracles, at the very moment when he learned, through Damilaville, that several philosophes including Diderot and d'Holbach had frankly opted for atheism, using none other than Needham's observations (approved by Buffon) to defend spontaneous generation and therefore the uselessness of a creative God to explain life. Voltaire thus launched himself on an unrestrained campaign against Needham, a campaign he would continue until his death.[54]

What Voltaire defended was a stable creationism, which could admit only that nature had changed only minimally since its creation. Buffon could not escape his criticism here. In 1767 Voltaire published the *Défense de mon oncle* [Defense of My Uncle], which attacked, first of all, the unity of the human species. Such a unity was impossible by definition, since it presupposed that the current races had appeared progressively and therefore that nature had changed. For the same reasons, he then attacked the "Theory of the Earth" and, in passing, organic molecules. "When I read forty years ago that shells from Syria had been found in the Alps, I said, I admit, in a rather mocking tone, that these shells had apparently been brought by pilgrims coming back from Jerusalem. M. de Buffon corrected me very sharply in his theory of the earth on page 281. I did not want to argue with him over shells; but I have retained my opinion." Voltaire had a good memory, and the reference was correct! He did add, however, "I know how to render justice to M. de Buffon's vast knowledge and genius, although I am strongly persuaded that the mountains are of the same date as our globe and all things. And I do not believe in organic molecules."[55] Buffon had no chance of ignoring this text, since the main points of the work were an attack on a Burgundian scholar Pierre-Henry Larcher, with whom Buffon had worked.[56]

Voltaire, who did not fear repetition, raised the same criticisms in *Singu-*

larities against the "illustrious and scholarly author of the *Natural History*,"
that "scholarly and eloquent academician." The same year, *Les colimaçons du
révérend père L'Escarbotier* [The Snails of the Reverend Father L'Escarbotier]*
took up these criticisms even more sharply. "When the Irish Jesuit named
Needham" (note that Needham was neither a Jesuit nor Irish) believed to
create "eels using ergotic wheat," "a geometer, a philosopher, a man who
has rendered great services to physics, and whose work, erudition, and
eloquence I have always esteemed, had the misfortune of being seduced by
this chimeric experiment" and hastened "to substitute his organic molecules for the evidence of germs." Yet "barely had the father of organic
molecules progressed halfway with his creation when the mother and
daughter [vinegar] eels disappeared.** M. Spallanzani, an excellent observer, has by observation proved the chimerical nature of these animals
born of corruption, just as reason has showed it to the mind. The organic
molecules disappear along with the eels into the nothingness from which
they came."57 All that was very prettily said, yet Buffon was not wrong to
think that he would have been wasting his time in answering "Voltaire's
foolish remarks." Here again, the opposition of philosophies came out into
broad daylight, but in this case it had existed from the very beginning. And
here, again, it was the Buffon of 1749 who was criticized, and whom Voltaire never tired of attacking. The questions raised by the *History of Quadrupeds* did not seem to have attracted Voltaire's attention. In a way, that is
understandable: these questions only had meaning within Buffon's intellectual framework. For Voltaire, the matter was closed.

The complex question remains of Buffon's relationship with Jean-
Jacques Rousseau.58 The two men had met in 1742, we remember, at the
home of Mme. Dupin de Francueil, but Rousseau, then unknown, had
probably not dared approach too closely the intendant of the Royal Botanical Garden. It seems to have been Diderot who brought them into closer
contact, and Rousseau then discovered behind the official personage an
independent spirit capable of understanding and even instructing him.
What is sure is that Rousseau read, admired, and widely used the first three
volumes of the *Natural History*. In 1753 in a letter he wrote but did not
send, Rousseau said to the journalist Fréron: "If all men were Montes-
quieus, Buffons, Duclos, etc., I would ardently desire that they cultivate all
the sciences, so that humankind would be a society of sages only."59 And
from the first page, his discourse *Sur l'origine et les fondements de l'inégalité
parmi les hommes* [On the Origin and the Foundations of Inequality among

*Ed. note: An *escarbot* is a snail (escargot), and Voltaire was clearly using the Reverend
Father's name satirically.
**Ed. note: Eels were believed to appear by spontaneous generation when vinegar spoiled.
"Mother" of vinegar is a ropy, gooey substance produced in the formation of vinegar from
wine.

Men] made references to the *Natural History of Man,* "one of those respectable authorities for philosophers, as [the authority] comes from solid and sublime reasoning that [philosophers] alone know how to find and feel."[60]

Rousseau in fact borrowed much from Buffon—not for his main ideas but for facts and also for a method—to the point that Formey, a member of the Academy of Berlin and collaborator on the *Encyclopédie,* wrote in 1756: "M. Rousseau is as much in his element as M. de Buffon is in his; he handles man the way that that philosopher handles Nature and the Universe; he frames hypotheses on Society as the Academician frames them on the Globes of the Universe and the origin of the Planets."[61] There was indeed an apparent similarity of method; the two authors wanted to reconstruct the past by starting with the present. A more precise analysis would show, however, that Rousseau reconstructed the state of primitive nature on very different foundations from those Buffon used in his reconstruction of the cosmos before the formation of the planets.

The two men saw each other during this period,[62] but we do not know how long this cordial relationship lasted. It is probable that Buffon was satisfied to observe from a distance as Rousseau's relationship with the encyclopedists, especially Diderot, became increasingly difficult. We do know, on the other hand, that the dialogue between the two men continued through their works, and here it seems to be Buffon who was especially impressed by Rousseau's ideas. I have often had the occasion of showing how very sensitive Buffon was to Rousseau's theses about man in nature, civilization, and society. He probably did not always agree: we have seen how he refuted Rousseau's ideas about man in the state of nature, how he judged the relationships between animals and man, in particular for apes, differently, and how he continued to state that man's mastery over animals was legitimate, and that civilization was a progress of humanity. It is no less true that Buffon often contradicted himself, that he seemed divided in his reactions, that he remained seduced by the purity of original nature, and that a great part of him he felt was in harmony with what was most profound in Rousseau's sensibility. Perhaps it was to Rousseau that Buffon should have dared show that independence of spirit and rejection of social conventions that he so carefully hid behind his official personage.

Rousseau continued to read the *Natural History,* whose volumes he constantly sought to obtain. When *Émile* appeared in 1762, readers easily recognized the advice that Buffon had given about the precautions to be taken with young children and even on education in general; Rousseau himself willingly recognized his debt. How the publication of *Émile* and the *Contrat Social* [Social Contract] upset Rousseau's life is known. The two books were condemned in Paris and then in Geneva. Obliged to flee France hurriedly and banned from staying in Geneva, he found himself in Môtiers when he learned that Daubenton wanted news of him "for M. and Mme. de Buffon."

Both of them hoped that he knew "how much they were very deeply interested in his fate, his situation, his well-being, and how much they wished to have good news of him." Rousseau was very moved by this solicitude: "May God bless both of them for being interested in this poor exile. Their kindnesses are one of the consolations of my life: let them know, I pray of you, that I honor them and love them with all my heart!" It was in this letter that Rousseau expressed the judgment of Buffon so often cited: "His Writings will instruct me and please me my entire life. I believe him to have equals among his contemporaries in his quality as a thinker and philosopher, but for his quality as a writer, I know of none [equal] to him. His is the most beautiful pen of his century."[63]

Buffon did not rest here. A few weeks later he wrote a letter to Rousseau that has been lost. Rousseau said that he was "extremely touched" by it but did not agree with the advice Buffon gave him, "not to incite M. de Voltaire to turn against him." "Does M. de Buffon want me to soothe this tiger thirsty for my blood?"[64] "I cannot swallow that M. de Buffon supposes that it is I who caused his hatred," he wrote a little later. Nonetheless he asked his correspondent, "Until I am in a state to write him, speak to him, I beg of you, of all the feelings for him with which you know I am filled."[65] A little later, persecuted by a pastor who rallied the population against him, Rousseau had to leave Môtiers. It was then, in the darkest moment of his distress, that he received the only letter from Buffon that we have:

"It is with a very great pleasure, Sir, that I have received the testimonies of your friendship; I would not have postponed thanking you for them if, near the same time, I had not learned that new misfortunes had befallen you and that you had left the city of Môtiers; your address was just given me, assuring me that you were safe in Neuchâtel. May God wish to calm your persecutors since He does not want to confound them! I have trembled a thousand times for your fate, I saw with pain that your priests are even more intolerant, more ferocious than ours; I thought that after the injustices that had been done to you in Paris, you would find justice and peace in Geneva as a well-deserved reward; your fellow citizens owed it to you; they owed you much more, for independently of the honor that you do to your homeland, you were sincerely and perhaps even too warmly attached to it; you were the victim of your love for truth, and even of your patriotic love; what a sad example! It can only cool virtue; I know that yours is sustained by great courage and that your soul is as firm as elevated; yet courage does not prevent suffering, and when it is for an unjust cause, it turns into indignation, and this feeling is still unpleasant: I love you, Sir, I admire you and I sympathize with you with all my heart."[66]

This letter is remarkable. First of all, because it is true. Persecuted by the devout and the governments of Paris and Geneva, rejected by the philosophers, Rousseau was truly alone, except for a small group of friends. That

Buffon, apparently so distracted and always very busy, had chosen this specific moment to express his admiration, compassion, and friendship was the proof of a very personal feeling, and of much more than a simple literary and philosophical admiration, which would in and of itself already be surprising. Buffon was decidedly not the conventional and cold mind that one might think.

In the spring of 1770 Rousseau slowly made his way back to Paris and passed by Montbard, where Buffon "gave [him] the kindest reception."[67] According to tradition, Rousseau went down on his knees at the entrance of the small building in the park where Buffon had installed his office. An inscription would for a long time recall the event:

> Passerby, bow down; it is before this retreat
> That the author of *Émile* fell at the great Buffon's feet.

A theatrical gesture, perhaps, as was fitting for the taste of the time, but the esteem was real and reciprocal between the two men.[68] It is an esteem that, to our eyes, reveals a too little known aspect of Buffon's personality. Later, it is true, Buffon changed his mind about Rousseau. He would say to Hérault de Séchelles, "I liked him fairly well; but when I saw the *Confessions,* I ceased admiring him. His soul revolted me."[69] Despite all his intellectual admiration, despite even a certain complicity in their challenge to conventional values, Buffon was not ready to follow Rousseau wholeheartedly. In the end, Rousseau remained very alone in his century.

In a way, so did Buffon. All the texts, criticisms, judgments, and commentaries of which I have spoken in this chapter cite the three volumes of 1749. The rare exceptions are primarily the "Theory of the Earth" and the theory of reproduction and of organic molecules. It was as if the long work of description, discussions, and methodological reflections to which Buffon had devoted himself in the twelve volumes of the *Natural History of Quadrupeds* had passed unnoticed. His style was spoken about, but not his ideas. In reality no one, neither among the naturalists nor among the philosophers, and even less so among the public, could truly follow the intellectual progression that led Buffon to open new paths for natural history. His thoughts were too personal, his proofs too fragile, his deductions too rash. People stopped at what was easiest to admire or criticize; they remained ignorant of what was most difficult to understand, and ignorant of what remains one of Buffon's most solid claims to posterity's admiration: the transformation of natural history.

Part Four

THE
ENTRANCE
INTO
HISTORY

CHAPTER XXI

An Untamable Energy

During the summer of 1766, Mme. de Buffon did not return to Montbard. She stayed in Paris to prepare their new house: Hotel Lebrun, rue des Fossés-Saint-Victor, "within range of the Royal Botanical Garden." "I gave up my lodging in order to enlarge the cabinets," Buffon explained: "Everything was crammed in, everything was suffering . . . for lack of space. We needed one hundred thousand livres to build. The king was not rich enough for that; his comptroller general made a decision that would cost only forty thousand livres . . . and he pays the rent of my house." Therefore "I have been honestly treated for inconvenience, but not *magnificently*, as one says in Dijon; and, to be honest, the motives of personal interest have no place here, . . . at the very most that will enrich the Cabinet, and that is enough for me; for I am satisfied with my wealth, however mediocre."[1] It must be believed that Buffon had the reputation, according to the shrewd people of Dijon, of being a good businessman. It was true that in the circumstances it was the interest of the Cabinet that had persuaded him.

Very quickly, therefore, workers guided by the architect La Touche were busy taking down partition walls, creating four new rooms and installing windowed cases decorated with sculpted woodwork. Starting in the summer of 1767 Daubenton moved his collections. He tried to establish a methodical classification without always succeeding: a beautiful zebra reigns in the last room, but "between the feet of this quadruped are different dried fish and reptiles." And "the ceiling is decorated with different large whole fish, such as Porpoises, Dogfish, Sharks." In short, there still was not enough room, and these juxtapositions, which surprise us, represent the taste of the time.[2]

While this work was being completed, Buffon suddenly had a serious worry: his wife's health. In the month of May, Mme. de Buffon had fallen from her horse and suffered "several fairly large bruises." She recovered, and the air of Paris did her good.³ What were the effects of her fall? In December 1767, she was still in Paris, very weak. Her husband joined her. Nonetheless, her health was not improving at all: "We are here in a state of affliction," wrote Buffon. "My wife is seriously ill, and with an illness that will continue for some time and unfortunately is always painful." To stay close to her, he put off his business trips. "I even admit to you that I am not without worry for the future," he wrote in March. She was brought back to Montbard, but the health bulletins became more and more alarming: her jaw was locked, perhaps by a tumor, and she was becoming weaker and weaker. "My poor invalid is almost always in the same state of despair and pain. . . . It has reached the point that I can no longer even leave her."⁴ Finally, sensing herself dying, Mme. de Buffon ordered her husband to leave for Paris, so as not to be present during her last moments. She died March 9, 1769, at the age of thirty-seven.

"It was at first a cruel wound, which today has degenerated into an illness that I regard as incurable and that I must accustom myself to endure as a necessary evil. My health has been altered by it and I have abandoned all my activities, at least for the time being."⁵ During these two years Buffon suffered greatly: "No one has been unhappier two years in a row: study alone has been my resource, and since my heart and my head were too ill to be able to apply myself to difficult things, I amused myself by caressing birds."⁶ Modesty probably prevented him from saying more. He had surely loved his wife, who had surrounded him with affection and admiration. He found himself alone and with the responsibility of a child to raise. He reacted as might be expected: he went back to work.

The Adventure of the Forges

In 1767 Buffon had thrown himself into a new enterprise: the construction of a forge in the village of Buffon. His curiosity was first aroused by the problem of heat, its propagation and cooling. We shall see in the next chapter the origin of this new passion, which would guide the rest of his work. "I had not thought of this last year," he wrote to De Brosses in January 1768. "But having busied myself during the summer and autumn with experiments on heat, and in particular on the action of fire on iron, I managed in the end to make an iron of as good as or even better quality than that of Sweden and Spain from our worst Burgundy mines. This discovery will certainly be useful to the state, and in order to receive some product from it myself, I decided to establish a forge, especially since I have enough wood."⁷

Where had Buffon done these experiments of 1767, of which we will speak again? "For one month . . . I have buried myself in my forge," he

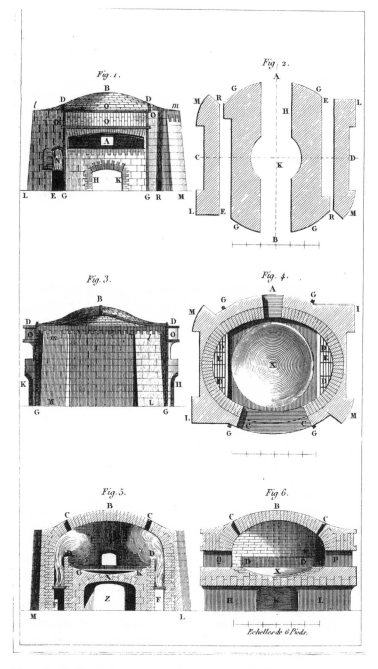

Plans for Buffon's forge. From *Oeuvres complètes*, IV, p. 118.

wrote in October 1767.[8] This could not have been the forges of Buffon, which did not yet exist. It was probably at Montbard, in a building called the Petit-Fontenet, where he had built a laboratory. Perhaps he carried them out with Rigoley, a master of a nearby forge. This entire part of Burgundy had been a region of great metallurgical activity since the Middle Ages. There were numerous pockets of ore close enough to the surface to be easily exploitable, and the forests furnished the wood necessary for the metallurgy of the time. By launching himself into this industrial enterprise, Buffon knew that he could profit from his neighbors' experience. He claimed, however, that he had invented a new method and hoped to get rich. As we shall see, success did not live up to his hopes.

After having obtained the necessary letters patent from the King's Council and, with the help of De Brosses, having registered them with the parlement and the chamber of accounts of Dijon,[9] he started to build his forges. The work progressed quickly but was not quite finished in the spring of 1770.[10] The buildings had been conceived on a grand scale. The largest part still exists and has been restored. Today it is the center of a museum of mining and metallurgical activity of the region, the forges are one of the finest monuments of industrial architecture of the eighteenth century.

The blast furnace is remarkable. It is approached by an "imposing facade," and a porch surrounded by two alcoves on either side, as if to shelter statues, which lead by a double-turned staircase with a bannister of wrought iron to the ground level, where the pig iron flowed out. Everything seems to have been built to allow noble visitors to watch the spectacle of this "Vulcan's cave" from above. Behind the blast furnace, two large buildings were devoted to the various activities for the production of iron. Other buildings received the ore brought in carts. Still others contained the lodgings of the permanent workers and the fine dwelling of the steward. Buffon reserved a pavilion for himself where he stayed during his visits.[11] The whole place was very impressive, and the enterprise employed up to four hundred workers. The hydraulic energy needed for the machines was furnished by the small Armançon River, whose course Buffon had altered and which turned two large wheels. It is difficult to imagine today, in the calm and solitude of the location, the feverish activity that took place there then, the flames and the smoke that frightened horses on the main road and the infernal noise of the tilt hammers striking the iron with the cadence of a machine gun. These forges, well worth a visit, were the most modern that could be built at the time. They unfortunately came during a period when metallurgy was being profoundly transformed, when wood would be replaced by coke. Buffon realized this, but too late.

Meanwhile, Buffon made and sold his iron energetically. If we can believe him, it was of course the best and the cheapest of the region. Still, out of habit, he sought to be subsidized by the government. At the request, he said, of the minister of the Navy and the duc de Choiseul-Praslin, he engaged in a series of experiments on the fabrication of naval cannons. He made them "with zeal, together with M. the viscount de Morogues, a very enlightened man," and sent his observations to the minister: "I do not know today the result and the success of all this," he wrote later, not without frustration. "The minister of the Navy having changed, I hear no more either about experiments or about cannons." He therefore published his results and his conclusions,[12] having taken the precaution of sending his bill of expenses to the minister.[13]

M. le Comte de Buffon

In February 1771 Buffon was in Paris when he fell suddenly and seriously ill, perhaps from an abscess of the intestines. His life was soon in danger and gossip spread rapidly. "M. de Buffon of the Académie Française, whose works assure immortality for him, is at the brink of death," wrote Bachaumont on February 16. "It will be a great loss for literature." His brother, prior of the abbey of Petit-Cîteaux, was urgently sent for, and he watched over him in the company of M. Laude, Buffonet's tutor. The physicians thought they had lost him. In the night of the sixteenth to the seventeenth, his condition suddenly improved, and he was considered saved. Again, the news was quickly known and, on the eighteenth, Bachaumont noted: "M. de Buffon is out of danger, and we are much relieved, since no one else could have continued his important and original work on natural history."[14] His convalescence was fairly rapid, but Buffon started to suffer from kidney stones and produced "bits of gravel" whose "passing caused him very sharp pains." He was not a very good invalid: "He constantly refused all sorts of remedies. He argued strongly with his doctors, and in the end believed nothing and did nothing." Above all, he did not like being kept on a diet. On April 2 he started to work again. He was finally able to return to Montbard on May 8.[15]

This violent but short illness, from which Buffon recovered all in all fairly rapidly, had unexpected consequences. As soon as he was believed to be dying, agitation for his replacement started at Versailles. Certain people wanted the position to go to the first physician to the king, to whom it had belonged before Du Fay. Others wanted it for themselves, and Buffon had given it so much brilliance that great lords put themselves into the running. Buffon, it is true, had two years earlier obtained from Louis XV the reversion of his position for his son as well as the transfer of two-thirds of the six thousand livres salary which he himself received. Such reversions were a common practice in the Ancien Régime, even for positions that assumed scientific knowledge, such as those of the professors of medicine at the University of Montpellier. Buffonet, however, was only seven years old, and no one knew if he would be able to fulfill the responsibilities required; in any case, for legal reasons he could not exercise it before reaching the "great majority"—twenty-five years of age—since legal accounting papers needed to be signed.

In the end, it was Charles-Claude de Flahaut, comte de La Billarderie d'Angivilliers, who obtained the promise of succession. A clever, intelligent courtier, this great lord was the director general of Buildings of the King: he was interested in chemistry and had put together a large collection of minerals. As soon as Buffon was informed, he protested. What ensued is unclear. The official act that assigned the succession to d'Angivilliers brought forward a demand from Buffon, who, considering that "his health, weakened by his work, no longer allows him to attend with the same care and the same activity" to the administration of the Garden, "implored" the king to give him d'Angivilliers as a successor. On the eve of his death, Buffon would state before a notary that he never had sent this supposed "petition." In fact, it was the fruit of laborious negotiations. By not dying, Buffon had embarrassed many people, foremost among them d'Angivilliers, who tried to get himself excused, as this letter attests:

"When M. the Dauphin thought of me for this position, I started by refusing to accept it for two reasons: the first, Sir, being that you have a son who should naturally gather the fruit of your labors. I was told that the age of M. your son was an insurmountable obstacle. . . . A second objection came from myself. I pointed out that I was not a scientist, that having only the superficial knowledge of a man of the world, I was not made for the position, even less so to replace M. de Buffon; I was told that it was not a position attached to the sciences; that it was solely an administrator's position. . . . On the verbal promise of the King, I asked that there be put on the page that, if M. your son devoted himself to the sciences, I would prefer him to have the reversion of the position that had been filled with such dignity by his father."

Buffon resigned himself to the inevitable. D'Angivilliers seemed to be an honest man, and there was a real danger of losing everything. The bookkeepers at Versailles had helped him to see the situation such as it was: "M. de Buffon should realize that it would be impossible for his son to obtain the approval for this position before twenty-five years of age. . . . In twenty years M. d'Angivilliers will be delighted to find rest [he was forty at the time]. Let M. de Buffon make sure, therefore, that his son be in a state to succeed him." The advice, after all, was wise.

Buffon learned very quickly also that the king would reward him handsomely for his compliance. As early as May 9, d'Angivilliers had sent him a rough draft of a communiqué for the *Gazette:* "M. de Buffon, having desired a successor for the position of Intendant of the Royal Botanical Garden, His Majesty named M. de La Billiarderie d'Angivilliers to it; and His Majesty, to give to this illustrious man a particular mark of his goodness, has established the holding of Buffon as a county and accorded the entrance to his Chamber to him." The letters patent "concerning the establishment of the land of Buffon as a county" were published in July 1772.[16] Georges-Louis Leclerc had become comte de Buffon.* By unpredictable ways, he had realized that dream of all bourgeois families of the Ancien Régime: to obtain nobility.

It is possible that this ennoblement made Buffon think about his son's future. The position of intendant of the Royal Botanical Garden was good, but the authentic mark of nobility was a military career, and many children in that sort of intermediary caste called "noblesse de robe" hastened to enter the army, to the great despair of the old traditional military nobles, who found that these young bourgeois lacked the true qualities of a soldier and spent more time in Paris than in their garrison towns. It was not until 1781 that the maréchal de Ségur signed the edict that required each officer to prove four quarterings** of nobility, but the idea was already in the air. Be that as it may, Buffon quickly busied himself finding noble ancestors *ab antiquo* in the Leclerc family, and finished by finding some Leclercs from Nivernais ennobled by Philippe de Valois in 1349, which was not too bad. And he just as quickly obtained his coat of arms: "Quartered, the I and IV of pure silver to the red band, charged with three silver stars; the II and III of azur with five small plaques of silver forming a chain."[17] Curiously, the arms were topped with the crown of a marquis and not a count.

Louis XV did more than ennoble Buffon: he decided to have a statue made and paid for it out of his personal privy purse. D'Angivilliers asked Pajou to execute it; first he did a model in terra cotta, then a bust in marble, which was exhibited in the Salon in 1773. Finally, the statue, almost three meters high, was finished in 1776 and placed in the Garden, where it remains. It is a statue "in the antique style," where the naturalist, half naked and half draped, appears to write the laws of nature on tablets of marble. Buffon, who knew nothing of it, discovered it upon his return from Montbard. It is possible to imagine his rapture. There were, however, a few difficulties with the inscription. At first it read, "Naturam amplectitur omnem" (He embraces all nature). But a bad jokester wrote below it: "He who embraces too much holds on badly." A new inscription was therefore engraved: "Majestati Naturae par Ingenium" (A genius equal to the majesty of nature). Pajou and then Houdon would later make several busts of Buffon.[18]

*Ed. note: A count clearly needs a county.
**Ed. note: This means that all four grandparents must be of noble birth.

While sculpture immortalized Buffon, poetry was not idle. In 1778 Ecou-chard Lebrun, called Lebrun-Pindare, again wrote an ode in honor of the great man, on the occasion of his illness. Here Envy demanded of the Fates the death of Buffon, that too fortunate genius:

> Black Divinities! a demigod braves us
> He has conquered Olympus and believes me his slave;
> His title of Immortal everywhere my eyes abrades:
> His image is my torture! and to enlarge him even more
>   A marble that I abhor
> Sanctions, my indignities and their odious traces.

The three Fates immediately "Directed toward Buffon their sinister flight." The hero will die. "Nature trembles at it." Two Fates are moved, but the third, Atropos, remains untouched.

> It was almost the end! Suddenly by Love aglow,
> From the base of the Elysium, a Shadow was thrown
> All in tears, and of Atropos kissed the toes.
> Yes, you see his Spouse, oh fatal Goddess!
>   Forgive me my tenderness,
> Forgive me my pain to suspend your blows!

Moved in turn, Atropos hesitates:

> The Fate, trembling, looks at her and sighs.
> Your tears, new Alceste, have saved your Spouse.

Immediately, Buffon lived anew, saved by the love of his wife. There was much invention, and even a few beautiful lines in it, but also long passages, and even Buffon advised Lebrun to shorten his poem, which had no fewer than twenty-nine stanzas.[19]

M. le Comte at Montbard

When it was known that M. de Buffon was healed, that he had also become M. le Comte, and that he would return home, the municipality of Montbard decided that "a cannon will be fired upon his arrival to show the town's esteem, the bourgeois militia under arms . . . will be present when he enters the town, and the Chamber will go as a body to do him honor."[20] They could hardly do less. When he returned there, however, Buffon found the inhabitants to be less docile. For a long time, the town remembered him as a harsh, demanding lord, greedy, and brutally claiming the least of his rights.

This "black legend" was born with the Revolution. In examining the original documents more closely, modern historians have come to a more balanced conclusion. In a time of "lordly reaction," Buffon had relatively few conflicts with the peasants who owed him seignorial rights. He did not have any with the inhabitants of the village of Buffon itself, but neither did he show himself to be very generous toward them. When their church was threatening to collapse, the villagers had to borrow 4,000 livres for repairs. Buffon agreed to loan them 2,641 livres, which was repaid fairly quickly, but he gave them nothing. Yet it was his "chaplain," Father Ignace, who was the curate of the parish. As Ignace said, however, "he loved money, because he knew its value and knew how to use it." Ignace was well placed to know this. In Buffon's account book, the following two entries occur: "It is owed me for the rental of the hall of Buffon whatever the R[everend] F[ather] Ignace Bougot can get for it, or *four livres* from Sir Tribolet, and more or less from the merchants who display their wares there." And: "It is owed me for the permission for the game of ninepins *three livres* per year, that the R. F. Ignace receives for me."[21]

A more interesting case is that of the village of Arrans, a hamlet of Montbard, whose inhabitants owed Buffon the tax of the *tierce,* or one sheaf for every thirteen. In 1776 one farmer refused to pay it. The following year, all refused. Buffon sued them. The villagers were condemned in 1780, then again upon appeal in 1785. Buffon's fermier général, Charles Humbert, a merchant and former master of the forges, suspected a plot, and the peasants finally admitted that it was their proprietors, all bourgeois from Montbard, who had ordered them not to pay the tierce.[22]

It was in Montbard that the resistance was found, and it was the bourgeois who resented the authority of a "lord" who was actually only one of them and who ruled either directly or through Daubenton. Indeed it seems that Buffon often intervened as a conciliator in local quarrels or as an intermediary between the city and the provincial authorities, in particular, the intendant of Burgundy. Because of his influence with the Court, Buffon was not treated like the others, and he several times placed his authority at the service of the town when it asked for help from the province for a hospice or for aid for the poor.[23]

Nevertheless, there were also lively conflicts. In 1772 the town had a house built for the curate. During construction, a part of the terraces that supported Buffon's park collapsed. Buffon asked the intendant to require the town to pay for the reconstruction. Mandonnet, the first alderman, got angry, and spoke some unfortunate words, saying that Buffon "was a terrible man, that his greediness was so great that if he could reach the eternal Father, he would take away his hat or his coat and that he was a tyrant and a usurper. If M. de Buffon had died at the time of his last illness, the town of Montbard would have gained a lot."[24] Buffon took his words seriously, attacked the alderman before the parlement, and even made the long trip to Dijon to "solicit" its judges, as was his habit.[25] Of course he won the lawsuit, and Mandonnet lost his position.

He always had other business to attend to; in 1767 he intervened in favor of the countrymen of Marmagne, a hamlet a half-league from Montbard, which the monks of Fontenay were in the process of drowning by lowering the spillway of their pond. "I take no other interest here than that of humanity," he wrote to Ruffey.[26] He could not say as much when he was involved in a lawsuit with the Ursulines of Montbard for a problem concerning the border of the woods.[27] In 1776 it was with the wood merchants that he had difficulties. They used the Armaçon River to float their wood in the direction of Paris, and Buffon, who had diverted the river to make his forges run, had to stop his machines to let the wood pass. He asked for compensation. The merchants refused. Buffon emphasized the national interest and "a series of useful experiments to the State." The merchants emphasized the necessity of supplying Paris, counting on being "supported by the town and the Provost of Merchants."[28] The argument was heated, and Buffon did not give up. His greed was probably exaggerated, but he was not a man to give up his "rights."

The major occupation remained the forges. For more than ten years, Buffon took direct charge of them. He even tried to use his influence at Court to obtain favorable arrangements from the government for himself and the other masters of forges. The technical aspect of the business still fascinated him, and he was in contact with other metallurgists like Grignon. Yet, after the initial period of experiments on a large scale and intellectual enthusiasm, he began to neglect the enterprise a little, unable to devote the necessary time to it: too much scientific work, too many preoccupations in the Royal Botanical Garden. And then, from the financial point of view, the operation was not a success. He said it publicly in 1783 in the *Natural History of Minerals:*

"I established, on my land of Buffon, a blast furnace with two forges. . . . All these constructions made on my own property and with my money have cost me more than three hundred thousand livres; I built them with care and economy; I then directed the entire handling of these factories for twelve years, I have never been able to obtain the interest on my investment of the twentieth denier [that is, 5 percent]; and after twelve years of experience, I leased all these factories for 6,500 livres; thus I do not have $2\frac{1}{2}$ percent from my funds. . . . I am only giving these facts to warn people who are thinking of building similar establishments against illusory speculations."[29]

He also took advantage of the occasion to denounce badly made regulations, which created "obstacles to the perfection of the art of forges in France."[30]

In fact, Buffon had leased his forges in 1777 to a certain Chesneau de Lauberdière and his wife, first for nine years, then at the end of the year "for nine more years, starting May 1, 1787, and ending May 1, 1796." In 1782, at the request of the interested parties, and "wanting to show them his satisfaction with their good management," Buffon prolonged the contract to 1803. The rent was 26,000 francs per year, but besides the forges the lease included the land, the fields, and "an annual cutting of 150 arpents [about 150 acres] of woods," destined in principle to feed the forge. A little later, Buffon agreed to two loans to Lauberdière and his wife of 30,000 and 31,000 livres and ceded to them the exploitation of his woods. Without Buffon's knowledge, Lauberdière immediately sold a cutting of 2,692 arpents to the wood merchants of Paris, left the forges, and finally fled "to the Islands," or the Antilles, taking the money with him. He was declared bankrupt and died "in the Islands" in 1787. Buffon tried to recover what he could. He was accused of going fiercely after Lauberdière's wife, who had stayed in France. There is every reason to believe that the two spouses were in it together and even that Buffon's notary, who served as his agent in business, was not above all suspicion in this affair.[31]

Be that as it may, Buffon lost more than 100,000 francs. He had not yet, however, finished with the forges. Starting in 1776, he became interested in metallurgy done with coal—or coke as we would say today.[32] The process had been invented in England, and it was needed in France to remedy a lack of wood, which was becoming drastic. Buffon helped with experiments done at his colleague Rigoley's establishment and he envisioned participating in the operation of a coal mine in Wassy, in the Haute-Marne.[33] In 1780 a company for the mining and preparation of coke was created under the auspices of Necker and Maurepas. It involved making coke for metallurgy. Buffon put 40,000 francs into it. Unfortunately, the business collapsed, through a lack of energetic direction and suitable coal. Another 40,000 francs lost.[34]

Among his contemporaries and even among historians, Buffon has long enjoyed the reputation of being a wise businessman. It is true that he succeeded in a few fruitful enterprises, and we shall see a few more examples. These dealings, however, had to do with land or buildings, with purchases and resales, and the state played a role in them. It was typically the operations of a bourgeois of the Ancien Régime who knew how to make useful contacts. When he entered into industrial or modern capitalist types of operations, Buffon always lost money. With the forges as elsewhere, his spirit of enterprise distinguished him from Burgundian landowners of his rank. He did not have any business acumen. Too absorbed in his scientific works and not taking the time to oversee his affairs, he too often put himself in the hands of incapable or doubtful intermediaries. This method-ical man, who did his accounts every Sunday morning and assiduously collected his annual rights of four livres, thus lost hundreds of thousands of livres through negligence and ineptitude in business.[35] Added to that, his eagerness for gain did not prevent him from being very generous, from helping many people, friends or strangers, in need and often anonymously. He was willing to give, but he wanted to do so freely.

Life was not sad at Montbard. Since the death of Mme. de Buffon, the house was run by Marie Blesseau. It was said that Buffon had noticed this young peasant when she was working in his park, and he had her come to the house "for less difficult tasks." The abbé Desvignes, curate of Montbard, was said to have condemned from the pulpit this guilty relationship, which he said would have greatly saddened Mme. de Buffon.[36] Be that as it may, Marie Blesseau reigned in the "chateau," directed the servants, watched over all the details of material life, and protected the work of her master, to whom she remained devoted until his death. She almost had an unex-pected rival, if it is true that Marie-Thérèse de Ruffey, the youngest daugh-ter of Buffon's old friend, had taken it into her head to marry this still brilliant widower. She was fifteen years old, and Buffon, after all, was only sixty-two! The young girl often came to Montbard with her mother. Noth-ing shows, however, that Buffon ever thought of remarrying. Marie-Thérèse therefore married another sexagenarian, before starting a tumultuous and famous affair with Mirabeau.[37]

There were many young women at Montbard, especially Mme. Nadault, Buffon's half-sister from his father's second marriage. She was thirty years younger than he, lively and charming, and very graciously did the honors of the house. For there were many visitors. Buffon did not change his schedule for them: he still arose very early in the morning, dressed, break-fasted frugally, and went to work in his office, where no one was authorized to disturb him. He joined his guests at two o'clock for "dinner." His cook was famous, for the master was a great gourmand. Thus "dinner" was a prolonged affair.

We have a late but very lively testimony about life at Montbard from Hérault de Séchelles, a young, elegant, refined, atheistic, and libertine magistrate who decided to visit Buffon in October 1785. The great man was then seventy-eight years old and was suffering cruelly from gravel. All the same he received his visitor, who found him "fresh as a child and calm as if healthy." "He dines, he likes to dine for a long time; it is at dinner that he places his mind and his genius aside; there he abandons himself to all gaieties, to all follies which go through his head. His greatest pleasure is to make naughty remarks, which is all the more agreeable because he retains his characteristic calm; his laugh and his old age form a sharp contrast to the seriousness and gravity that are natural to him, and these jokes are often so shocking that the women are forced to leave. In general, Buffon's conversation is very free.

"After his dinner, he hardly bothers with those who live in his chateau or the others who have come to see him. He goes to sleep for half an hour in his room, then he takes a walk, always alone, and at five o'clock, he returns to his office to go back to his studies until seven o'clock; then he returns to the salon, has his works read, explains them, and admires them." Hérault analyzed Buffon's vanity, for which he had been so often reproached, with much finesse. "He is also very interesting when he speaks of himself and he does so often with much self-praise. . . . It is not pride, it is not vanity: it is his conscience that one hears; he knows himself and does justice to himself. Let us therefore consent sometimes to have great men at this price. . . . His good faith contains something precious, original, old-fashioned, and seductive."

The clothing of the old man was also old-fashioned, and he reproached his son for wearing a fashionable tailcoat. He was always well groomed, having his hair curled every day and sometimes several times per day. Even when ill and in his dressing gown, he had great presence. Received at Montbard with a strong paternal affection, Hérault left very moved and filled with "a kind of veneration."[38]

Meanwhile, Buffonet grew up at Montbard, surrounded by his aunt and other young female relations. When he left for Paris, Buffon entrusted him to Mme. Guéneau de Montbeillard, who lived in Semur-en-Auxois and who had a son of the same age. He already had been given a tutor. It is probable that the child had never had much personal contact with an already aged, intimidating, preoccupied, and probably distant father. He was perhaps fairly difficult, to judge by the recommendations that Guéneau de Montbeillard sent to his wife when she had him in her care: "Insist especially that he take an honest tone with his inferiors, and that [these] inferiors be deaf, absolutely deaf, when he does not take an honest tone with them. It seems to me that the father has this most at heart. . . . The greatest service that one can render a child is to break him of bad habits."[39] It was perhaps because of these "bad habits" that there were so many tutors: four in ten years.

In 1773, at the age of nine, Buffonet entered the Collège de Plessis in Paris as a day student. He lived at the Royal Botanical Garden and always had a "governor" who oversaw his studies. In 1780 he turned sixteen, and Buffon was not very happy with him: "I would be obliged to you if you would exhort him not to abandon his studies," he wrote from Montbard to the abbé Bexon, his collaborator. "He does not sense the great wrong he is doing to himself by wasting his time."[40] The following year, Buffonet entered the French Guards. There was decidedly little likelihood that he would lean toward the sciences.

M. le Comte in Paris

Buffon's ennoblement had pleased his friends, but in 1771 a noble title was already no longer what it had been. Even Guéneau de Montbeillard, one of his most loyal supporters, said it in his own way:

> King Louis sought to honor your name
> By linking it to the beautiful title of count;
> Yet what title indeed is worthy of Buffon's name?

As for the philosophes, their reaction was mostly ironic. D'Alembert got in the habit of calling Buffon "the count of Tuffière": this was the name of a hero in a comedy by Destouches, *Les Glorieux* [The Glorious Ones]. Buffon had tried to heal the split: in 1774 he reconciled with Voltaire. This great undertaking was aided by Guéneau de Montbeillard and Mme. de Florian, the poet's niece. It was Buffon who took the first step by sending Voltaire the text of a "short note" that he intended to publish in the *Supplement* to the *Natural History*. Voltaire hastened to respond to these advances and wrote a fine letter to "Archimedes II," in which he even promised to come to Montbard. Buffon replied with a letter to "Voltaire I." "Ever since I can remember, you have had all [my esteem], but this is only a speck in the immense mass of the glory that surrounds you, whereas yours, Sir, is a diamond of the highest price for me."[41] The news of this literary event appeared in the gazettes, and was celebrated in lyrical verses.[42] In fact, it entailed what Voltaire himself called "a poorly darned patch": it is enough to look at the way Voltaire treated Buffon in his letters to d'Alembert or Condorcet to understand that for him it was pure comedy. Buffon, at least officially, respected the signed peace. In 1778 he published the promised note, where he apologized for not "having treated M. de Voltaire seriously enough" and for having mistreated the *Lettre italienne sur les coquilles* [Italian Letter on Shells]. "I admit that I would have done better to drop my opinion than to treat it as a joke, more so as it is not my tone, and is perhaps the only [joke] that is in my writings. M. de Voltaire is a man who, through the superiority of his talents, deserves the greatest respect." Buffon had not known that the *Letter* was Voltaire's.[43] And beginning in 1775, he took advantage of an academic discourse to praise his *Henriade* highly.

This discourse was specifically a call to peace among the men of letters: "Is the empire of opinion not vast enough so that everyone can live there in peace? Why make war? Oh, Sirs, we ask for tolerance, let us grant it therefore, and practice it to set an example."[44] At this date, even though the philosophes had become as powerful in the public's eye as in the Académie Française, peace hardly reigned, not even among themselves. In 1770 Saint-Lambert in his reception at the Académie had given a discourse on which Grimm commented ironically: "M. de Saint-Lambert is reproached for having praised everything and for having praised too much; but it was in the spirit of the institute. . . . The new academician did his service of flattery marvelously." Montesquieu and Voltaire deserved the flattery. But "the abbé de Condillac, M. Thomas, M. d'Alembert [also] had their portion of praise. I do not know by what fatality M. de Saint-Lambert forgot M. de Buffon, who is also one of the Forty."[45] It is true that Buffon had written, "They just received Saint-Lambert in the Académie Française. He is a poet without poetry, just as they previously had received the abbé de Condillac, who is a philosopher without philosophy. And it was Duclos who alone made these good choices."[46] It was also Duclos who had been responsible for d'Alembert's election.

Between d'Alembert and Buffon the hostility was irreconcilable. It was even more so since d'Alembert had gained entrance for Condorcet in the Academy of Sciences in 1769. D'Alembert and his protégé had a perfect understanding. They had the same passion for mathematics, which led Condorcet to the point of scorn for the "physics riffraff." They also had the same anticlerical passion and the same admiration for Voltaire, whom the two men visited in 1770. They shared the same political opinions as well: like Voltaire, d'Alembert and Condorcet were great admirers of Turgot, with whom they worked, and supported his theories on the free circulation of grain, while other philosophers—Diderot, Galiani, d'Holbach—defended Terray, the minister of Finance and the regulation of the wheat trade. After eighteen months, in 1771, Condorcet, then twenty-seven years old, was promoted to the rank of associate academician, and in 1772 d'Alembert manoeuvred to have him named permanent secretary of the Academy in the place of Grandjean de Fouchy, whom d'Alembert could not bear. Since d'Alembert himself had just been elected permanent secretary to the Académie Française, replacing Duclos, members wondered if he did not also want to rule the Academy of Sciences using Condorcet, and people began to worry about his "despotism."

Early in 1763 d'Alembert had encouraged the astronomer Jean-Sylvain Bailly to seek to obtain the same position. Bailly, who would pass into history as the unfortunate mayor of Paris at the beginning of the French Revolution, was a friend of Buffon's. D'Alembert's change of mind disappointed Bailly deeply. When in the beginning of 1773 Grandjean de Fouchy, pushed by d'Alembert, asked the minister, La Vrillière, for Condorcet to be named "his deputy and successor," the minister, instead of asking the Academy to designate someone freely, asked only "if M. de Condorcet was indeed qualified for that position." This abnormal procedure irritated many.

Meanwhile, in two months Condorcet had written eleven eulogies of academicians who had died between 1666 and 1699, eulogies that Voltaire and Turgot had, of course, admired, while Grimm wished for "a little more interesting style."[47] On March 6, 1773, Condorcet was elected by fifteen voices to six, but his nomination was no less resented as "an act of authority." The affair turned into an open scandal when Turgot came to power in 1774 and Condorcet was considered for a pension of six thousand livres taken from the funds allocated to the Academy for its experiments. In the beginning of 1775, the Academy designated four representatives to revise the eulogies. Condorcet feared being subject to a perpetual censor and, with d'Alembert and Turgot's help, tried to get the decision overturned. When Turgot named d'Alembert, Bossut, and Condorcet as inspectors of Navigation, and then Condorcet as inspector of the Mint in 1775, the cry went out that the "reign of Philosophy" was costing the public Treasury a fortune. Finally, on July 24, 1776, Grandjean de Fouchy resigned. Condorcet renounced his rights of succession and was unanimously elected permanent secretary on August 7.[48]

For Buffon, this was a defeat. He had been the "leader of harassment," as Condorcet said, and had supported Bailly as best he could. Condorcet did not have words harsh enough for him, at least in private. In his letters to Voltaire, he spoke of the "physical gibberish of the comte de Buffon . . . which must be let pass without saying a word, or without mockery." And it was left to Voltaire to answer by admiring "the skill necessary to pass for a superior mind when all he has shown the public is the size of a monkey's tail.*"[49] The tone is obvious, and also showed what the "reconciliation" of 1773 was worth.

Buffon, it is true, had all it took to be detested: aside from d'Alembert's old animosity and the resistance with which he opposed Condorcet's election, he was in Necker's party against Turgot. Necker was a partisan of the regulation of wheat commerce and protectionism in general, and Buffon, like his forge-owning colleagues, wanted their industry to be protected against foreign imports. "I have never understood anything of this poorhouse jargon of these charity seekers that we call economists," Buffon wrote to Necker in 1773, "nor of that invincible obstinacy of our ministers and assistant ministers for the absolute freedom of trade of this basic foodstuff." This letter was published by Grimm and not by chance. It did not go unnoticed.[50] In addition, Buffon had entirely personal reasons for defending Necker. Turgot detested Necker and did not hide it from him. He and his friends accused Necker of having stirred up the "flour war," those riots provoked by the rise in the price of bread. Condorcet became famous by taking up the accusation in two anonymous pamphlets against

---

*Ed. note: As in English, "tail" in French has a double and vulgar meaning, referring to the male generative organ.

the "Genevan usurer" that were so violent even his friends advised him to be more moderate.[51]

Things did not stop there, for d'Alembert led the campaign to have Condorcet enter the Académie Française. Once again, Bailly, supported by Buffon, presented himself as an alternative, and once again he was defeated. "The election of M. le marquis de Condorcet . . . is one of the greatest battles that M. d'Alembert has won against M. de Buffon," Grimm noted. He also observed that Condorcet's literary titles counted for little: one admissible work, "a thin collection of academic eulogies." All the rest had been published anonymously and showed a "decided atheism," except for the "vile libels" written against Necker: all writings that "certainly should have been seen by the Académie Française as reasons for exclusion. How many iniquities can the love of philosophy cover when it is carried to such a degree! It is like faith, which makes even more miracles than does charity." Nonetheless, "without a small betrayal by M. de Tressan," d'Alembert would not have won. Tressan had promised his vote to Buffon and to d'Alembert. "M. de Buffon, to whom M. de Tressan owes his position at the Académie, frankly believed himself able to rely on the word that he had been given. . . . M. d'Alembert . . . , a much better geometer than the French Pliny, . . . judged correctly that a verbal promise from the comte de Tressan was not a rigorous enough demonstration; consequently, he had the vote that he needed given in a conveniently sealed envelope." Condorcet was elected, sixteen votes to fifteen.[52] Buffon had decided not to attend the election: "I have no other way in which to avoid so many disagreeable things."[53] He barred his door to the comte de Tressan and decided never to set foot in the Académie again. Thus ended the academic activity of a man that Grimm, in 1771, had believed capable of being one of "those proud and free souls that disdains being in any party . . . and whose neutrality exposes them to the slander of both factions."[54]

The opinion of the philosophes, or at least of the faction that followed d'Alembert, was later written with a conceited complaisance by Marmontel, who was not the most inspired member but who owed a moment of celebrity to the persecutions of which Buffon had been the object:

"Buffon, with the Cabinet of the King and his *Natural History,* felt himself strong enough to elevate himself considerably. He saw that the encyclopedic school was in disfavor at the court and in the mind of the King; he feared being enveloped in the common wreck. . . . He cannot be blamed, but his retreat had yet another cause. Surrounded at his home by indulgent people and flatterers, and accustomed to an obsequious deference toward his systematic ideas, Buffon was sometimes unpleasantly surprised to find less reverence and docility among us. . . . Incontestably worthy, his pride and presumption were at least equal to his worth. Stimulated by adulation and placed by the multitude into the class of our great men, he was chagrined to see that the mathematicians, the chemists, the astronomers gave him only a very inferior rank among them; that the naturalists themselves were little disposed to put him at their head, and a few even reproached him for having written so sumptuously on a subject that required only a simple and natural style. . . . Buffon, ill at ease with his peers, therefore closeted himself at home with his ignorant and servile companions, going no longer to either Academy and working with the ministers to make his fortune and his reputation in foreign courts, from which, in exchange for his works, he received beautiful presents."[55]

Despite his obvious malevolence, Marmontel was right at least on one point: in the last years of his life, Buffon retired more and more from academic life and the disappointments that it brought him. He dedicated himself more and more to the Royal Botanical Garden and it was there that he organized his Parisian life.

Before abandoning the Academy of Sciences, he did have one last victory. In 1776 he had been introduced to a thirty-two-year-old former officer who had become a botanist, the chevalier de La Mark.* He was well known at the Garden, where he had been noticed by the teachers of botany. Convinced like Buffon of the impossibility of establishing a "natural" classification, and therefore very critical of Linnaeus's method, Lamarck had created a purely artificial method of classification, which allowed the rapid recognition of any plant. Interested, Buffon supported this unknown naturalist and had his *Flore française* [French Flora], a large work in three volumes printed at the Royal Press. More important, since Lamarck was without resources, he procured for him all the income from the sale.[56] Better yet, he obtained his entry to the Academy of Sciences. This, however, did not happen by itself. At the time of the election, Lamarck was ranked second on the list presented to the king. It is probable that Buffon intervened directly with the minister, for he reversed the order proposed by the Academy—something that rarely happened. Lamarck was therefore named assistant botanist, thus beginning a scientific career. He was nonetheless only half satisfied. He had wanted to put a long preliminary discourse at the beginning of his work, in which he would have presented all his ideas on physics, chemistry, and science in general. Buffon was opposed to this project. Through the intermediary of Daubenton, he asked René Just Haüy, a naturalist and crystallographer, to correct the preliminary discourse, which greatly upset Lamarck. The continuation of his relationship with Buffon would not be any happier, as we shall see, and it was perhaps because of these bad memories that Lamarck never acknowledged what he owed to Buffon's ideas.

*Ed. note: The spelling was later modified to Lamarck.

The Great Construction in the Royal Botanical Garden

In 1772 Buffon, tired of living in the Hôtel Lebrun, wanted to move closer to the Garden. He then bought a large house, today still called the the "house of Buffon," on the corner of the rue Buffon, which did not exist then, and the rue Geoffroy-Saint-Hilaire, then the rue du Jardin-du-Roi. He lived on the second floor only, the rest serving as storage for the unexhibited objects from the collection.

Seven years later, Buffon embarked on a campaign of acquisitions, which lasted until his death. He was perhaps taking advantage of favorable political circumstances. Necker was in power, and Buffon was linked to Mme. Necker by a tender friendship, which we will mention later. Thus it was Necker whom he addressed in 1779 to prevent the Company of Carriages of Paris from installing its stables in the Hôtel de Magny, very close to the Garden.[57] The Company withdrew its project, and the hotel stayed in the hands of its renter, the lord Verdier, a boarding house master with whom Buffon would have the worst difficulties. In 1780, under the direction of the architect Verniquet, the construction to enlarge the building that housed the Cabinet's collections was in progress, and starting in the beginning of the following year, Buffon acquired new lands and new buildings. First he secured the "house of the heirs of Father Lelièvre"; then, not much later, there was an exchange with the monks of Saint-Victor.[58]

The Garden was limited to the north by lands belonging to the abbey of Saint-Victor, which separated it from the Seine. The property of the Church was inalienable. Not being able to buy the land, Buffon decided to propose a trade to the monks. In 1779 he bought a piece of land close to the "house of Buffon" including a very large plot, the Patouillet field, which stretched between the Garden and the Bièvre stream. It was a marshy plot, but so was the monks'. All that remained was to convince the Victorins. Although the prior, Father Delaulne, agreed, the community resisted. Buffon called on the ministers to intervene. An arrangement was finally signed in 1781. Its official character was sanctioned by the letters patent of April 1782. The City of Paris participated in the dealings by abandoning a path that separated the Garden from the lands of Saint-Victor. From then on, the Royal Botanical Garden reached to the Seine.

All, however, was not over. There were tenants on the monks' land, the Bouillon ladies, who had built a few small buildings, which Buffon had had razed. The ladies demanded compensation. They were told that a decree from the Council in 1671 had forbidden all construction between the Garden and the Seine, even an extension. They were therefore without a case. But there was also a house belonging to the monks themselves, who lived in it and refused to leave. Buffon fixed a time limit for them. The monks did not budge. In the morning of the appointed day they were

awakened by a loud noise. It was workers sent by Buffon who were starting
to demolish the roof. It was raining torrentially that day. The monks disappeared. Their rage can be imagined.

While the lands of Saint-Victor were being converted by digging a pool
for aquatic plants, Buffon continued his acquisitions. In 1785 he bought a
small plot to the southwest, which today is the Alpine Garden. In 1786
Verniquet put the Labyrinth in place and built the gazebo that crowns it.
The following year Buffon bought the Hôtel de Magny for sixty thousand
livres; today it houses the administration of the Museum. The construction
of a lecture hall for anatomy based on Verniquet's plans was immediately
started on the land that was part of the property, and at the same time new
greenhouses were installed. The Hôtel de Magny was rented to M. Verdier,
who had boarders there and who was said by this fact to be a member of the
University. Verdier was compensated but not enough for his liking, and he
sued Buffon. His suit was rejected. His hour, however, came at the beginning of the French Revolution: in 1790 he associated himself with the
Bouillon ladies and presented a petition to the National Assembly. He
never really won, but the lawsuits he pursued for years prevented Buffon's
heirs from being reimbursed for the debts that the state had contracted
for him.[59]

The most surprising aspect for us in all these acquisitions is how they
were financed. Buffon paid for everything out of his own pocket: purchases, construction, changes to the Garden, and even the salaries of the
employees of the Cabinet of the King. Then he was reimbursed by the
Royal Treasury, which involved sending proof and evidence of the money
spent, as well as many procedures that needed to be followed. Buffon
himself did not go very often to Versailles: the trips tired him, so he sent his
collaborators. To André Thouin he gave this advice on how to behave
himself: "You will do me pleasure, my dear M. Thouin, to give the enclosed
letter to M. Dufresne [an important "assistant" in the financial office] in
person; and since it is to obtain money, of which we have a great need,
there would be no harm in bringing him a few flowers or a few shrubs,
which could only increase his good will."[60] One of his secretaries, Humbert-Bazile, left an exaggerated account of the magical power that a simple
letter could have on the comptroller general Joly de Fleury.[61] In reality,
things were more difficult, for the state of the Treasury was catastrophic
and construction was expensive, all the more so since it was necessary
constantly to reinforce the land, which was mined with catacombs.

Buffon did not make large acquisitions on his own authority. The purchase of the Patouillet field, for example, had been expressly authorized by
the minister. The same applied for construction plans, such as those for a
new amphitheater, which had to be accepted by the minister. It was, all the
same, necessary to advance the money. Buffon's account books allow us to

get an idea of the huge sums thus advanced, which the state reimbursed without interest and unhurriedly: nearly 470,000 livres for the year of 1786 alone, almost 250,000 for 1787.[62] Enormous sums, when one realized that Buffon, who was considered a rich man, earned about 80,000 livres per year.

Of course Buffon did not forget his own interests. He kept for himself certain lands that he had purchased. This land increased in value after the construction of a new road, which was named the rue de Buffon. In addition, by reselling to the state what he had bought, Buffon realized fairly good profits. The intendant's house, for which he had paid 12,000 livres in 1771, was resold to the state in 1778 for an annual "pension" of 5,600 livres, which corresponded to a sum of 80,000 francs of capital, or more than six times the purchase price. On the resale of the Patouillet field, he also made a good profit: 30,000 francs according to some, 200,000 according to others. These methods, which were common at the time, involved risks. The state paid very slowly, and could one even be sure that it would always pay? To become involved in such activities at a time when the public finances were known to be at the edge of bankruptcy required a lot of confidence or lack of realism. This was even more the case since Buffon, far from having the sums he used, took out more and more loans. His correspondence at this time was full of problems of reimbursements and notes to be honored, and he complained incessantly of being short of money. All this accounting remains obscure, yet it is unlikely that he borrowed money without interest. At Buffon's death, the state owed him, according to his son, more than 300,000 livres, of which only a small part was reimbursed. In all, the financial operations of the Garden were not much more successful than those of the forges.[63]

Buffon was not a philanthropist, but as Condorcet noted, "so many men separate their interests from the general interest that it would be unjust to be harsh to those who know how to join them." Condorcet, it is true, was well placed to speak about this. It is also true that at Buffon's death the Garden had almost doubled in area, the professors were comfortably lodged, the collections had more space at their disposal, and the amphitheater and the greenhouses offered new possibilities for teaching and research. In the fifty years of his "rule," Buffon had made the establishment one of the best in Europe, if not the best. The extraordinary success of the new Museum of Natural History would not have been possible if Buffon had not turned the Garden and the Cabinet into a magnificent research instrument.

At the end of his life, Buffon was certainly rich. First he had "pensions," an ambiguous term that includes both what we call a "salary" or "compensation" corresponding to a position and annuities made to him to reimburse the sums he advanced to the Treasury. In all he received close to

30,000 livres per year. There were also the revenues of the seigneury of Montbard, the lands, and the forges, which had been rented again, all of it representing close to 50,000 livres. Taking everything into account, at his death he had a revenue of 80,000 livres, to which it is necessary to add the returns from the *Natural History*. It was therefore a comfortable, if not immense, fortune. It put Buffon at the same level of the people of his milieu, the parlementarians and Burgundian landowners. It was not by any means comparable to the great fortunes of certain aristocrats of the time.[64]

During all these years, Buffon was so active that it is possible to forget his age. He was older than seventy when he embarked on these adventures, more than eighty when he finished with his projects. He suffered from gravel, or kidney stones, and the crises became more and more frequent and painful. The trips between Montbard and Paris were made in small stages. But nothing stopped him.

When Buffon lived at Montbard, he showered his Parisian collaborators with very precise instructions. At the Garden, André Thouin seconded or replaced him in everything. The son of the head gardener whom Buffon had hired in 1745, he was born in the Garden in 1747. At the time of his father's death in 1764, he was only seventeen years old, but that did not prevent Buffon from giving him the position, which he filled very well. He became an excellent botanist, entered the Academy of Sciences in 1786, and was one of the first professors of the Museum in 1793. During all these years of transformation, he watched over everything, carried out negotiations at Versailles, oversaw the construction, and prepared delicate transactions. He was helped by Lucas, the son of the "widow Lucas" of Montbard, who had so much presence that he was said to be Buffon's natural son. Buffon had him named bailiff at the Academy of Sciences, then made him his business agent in Paris. It was he who received the money from Montbard and gave it to André Thouin, who paid the bills. Every two weeks, Buffon did his accounts. He would do them until he died.

Buffonet's Misfortunes

Buffon remained preoccupied with his son's future. The army prepared men for anything, on the condition of their leaving it. In place of a formal and good scientific education, it was necessary to give the young officer an opportunity to know the world and make himself known. Buffon, therefore, decided to have Buffonet travel, but not alone. He assigned the Chevalier de Lamarck, his protégé, to him for a mentor. Lamarck was twenty years older than Buffonet and a former officer. The chevalier was of the old nobility and had a touchy personality: the relationship could have its difficulties.

Our travelers left Paris on May 12, 1781, in the direction of Holland. From there, they passed into Germany, arrived in Berlin in the month of August, then headed toward Prague and Vienna. In November they were in Munich. They were admirably received everywhere, by sovereigns as well as scholars. And everywhere they were honored with visits to collections of natural history, mines, and other scientific curiosities. All this fascinated Lamarck but bored Buffonet stiff, who became more and more difficult with his mentor. In Munich, after a particularly unpleasant prank, Lamarck had to complain to Buffon, who called the two travelers back home immediately. They arrived in Paris at the end of December.[65] Undiscouraged, Buffon sent his son to Russia the following year, with the mission of bringing his bust by Houdon and his admiring respects to Catherine II. This time, Buffon chose for his companion an officer of the French Guards. Upon his arrival at Saint Petersburg, the empress wrote to Buffon: "I will receive him as the child of a famous man, that is to say, without ceremony. He will sup this evening alone together with me."[66]

After spending six months in Saint Petersburg, he had to return, and it was said that the return trip alone cost 20,000 francs. In 1783, at the age of nineteen, the young comte de Buffon was named governor of Montbard. The following year, he made a brilliant marriage, which seemed to introduce the family definitively into the high aristocracy: he wed the daughter of the late marquis de Cépoy, Marguerite-Françoise, barely sixteen years old. The great names of France are found as witnesses to the contract: the maréchal-duc de Biron, commander of the French Guards, and the marquis de Sauzai, marshal of the camps and armies of the king. But there were also Grimm, Daubenton, the lieutenant of police Lenoir, and M. and Mme. Necker. The dowry was considerable: 450,000 livres in all. For his part, Buffon promised his son 20,000 livres in annual income. The fiancée was ravishing, and the couple had everything necessary to be happy.[67]

What Buffon probably did not know was that the young lady was already, like her mother, the object of the insistent attentions of the duc d'Orléans, the future Philippe-Egalité. Born in 1747, Louis-Philippe d'Orléans possessed one of the largest fortunes in France. Upon his father's death, he was able to pay off the debt of twenty million francs that he had accumulated during his youth without batting an eyelash. He spent freely on his horses, mistresses, and friends. Full of scorn for his cousin Louis XVI and harboring much animosity toward Marie Antoinette, he was the center of a brilliant circle of young "enlightened" aristocrats, and he certainly had political ambitions.

Between the two spouses, things became tense very quickly. In 1786 the young count left the French Guards to become a captain in the Chartres regiment, which was in fact dependent on the duc d'Orléans. It was an advance, but as if by chance, the new captain was sent to far-off garrisons such as Quesnoy, Philippeville, and Givet, and his wife stayed in Paris. It is true that she came for long stays in Montbard with her father-in-law, who treated her with affection. There, it was said, she often received visits from a young aristocrat, the duke of Fitz-James, whose carriage was driven by a strange postilion, who was none other than the duc d'Orléans. This detail is perhaps too fantastic to be true. Yet little by little the truth came out, Mme. de Buffon hid herself no longer, and her liaison with the duke became public knowledge.

On June 22, 1787, Buffon had the following letter delivered to his son:

"M. de Faujas, through friendship for me and for you, my dear son, has consented to carry my orders to you, which you must obey.

"1. Honor requires you, and I order you, to tender your resignation and to leave your regiment never to enter it again.

"2. You will leave immediately, saying that circumstances require you to do so, and you will give this same answer to everyone without other explanation.

"3. You will not go to Spa, and you will not come back to Paris before my return.

"4. You will travel where it pleases you. . . .

"5. These honest and necessary steps, far from harming your advancement, will serve it greatly. . . .

"6. Act entirely in accordance with the opinion of M. de Faujas for everything else. . . . If you need the three thousand francs that you should receive on August 5, I will give them to M. Boursier right now. You know that he must give back fifteen hundred francs at this same time to your former wife.

"These are, my very dear son, the absolute wishes of your good and tender father."[68]

It was as if David's Brutus were in the *Liaisons dangereuses.** This Roman-style letter was much admired in the nineteenth century, especially by Sainte-Beuve. Meanwhile, the son obeyed, but Buffon continued to watch over him. In September he wrote to Thouin: "Be moderate, I beg you, in agreeing to the requests that my son might make of you; I know too well your good will, which he could abuse."[69] While Mme. de Buffon became the official mistress of the duke, her husband, from then on separated from her, took up service again in the regiment of Angoumois. As his father had foretold, his "advancement" had not suffered.

People wanted to hold him responsible for his wife's behavior. He was accused of being less interested in her than in his horses and their grooms. The older Buffon was even accused of having wanted to seduce his daugh-

---

*Brutus, consul of Rome, had his two sons, who had conspired against the Republic, executed. This was the subject of a painting by Jacques-Louis David, exhibited in 1789 with great success.

ter-in-law. Buffonet was certainly a spoiled child, who threw money out the window and did not have much common sense. He had been crushed by his father's personality. He was the object of easy jokes: "The poorest chapter of the *Natural History*," said Rivarol. All that he was able to do was to erect a small column at the foot of the tower where his father worked in the park at Montbard. "Excelsae turri, humilis columna. Parenti suo, filius Buffon, 1785," said the Latin inscription: "To the high tower, the humble column. To his father, Buffon the son." "My son, that will honor you," said the father, who was "touched to the point of tears."[70] Probably, but how was Buffonet to become a man under these conditions? Buffon was not a bad father. He gave advice to his son more and more, worried about his health, and was preoccupied about his future up to his last moment. Yet he also feared the arrival of the boisterous boy at Montbard, which disturbed his tranquility. It was after the death of his father, and in hardship, that Buffonet would become a man.

He did not have much time in which to do it. When the Revolution arrived, he supported it and became a colonel of the 58th infantry regiment. In January 1793, as soon as the law allowed him, he divorced, and remarried a childhood friend, Betzy Daubenton, in October. Guyton de Morveau, who was for a time president of the Committee of Public Safety, was his witness. Arrested in obscure conditions at the beginning of 1794, accused of having participated in the "conspiracy of the Luxembourg prison," he was condemned to death and executed on the 22 Messidor, Year II. "Citizens, my name is Buffon": those were his last words, as if his death were a last homage to his father's glory.

One Last Love: Mme. Necker

At the Royal Botanical Garden, Buffon received many visitors, primarily royal ones. There was Prince Henry of Prussia, who offered him a fine porcelain cabinet. There was the archduke Maximilian, Marie-Antoinette's brother, to whom Buffon offered his *Natural History* and who answered gauchely, "I do not wish to deprive you of it." This tactlessness was repaired two years later by the emperor Joseph II: "M. le comte," he said to Buffon, "I have come seeking a copy of your works that the Archduke my brother left on your table." He also returned several times to chat familiarly with the naturalist. There was also the king of Denmark, who sent a magnificent collection of minerals. He was the "count of the North," son of Catherine II. The Garden was from then on a required stop on visits from foreign sovereigns.

Buffon also liked to receive guests in a more intimate manner. On Sundays he held his "salon." Here again, many women surrounded him. He loved them and they admired him. The Countess Grismondi came there during her infrequent trips to Paris; he called her "my sublime friend" because she had translated the ode by Lebrun-Pindare into Italian. The countess Fanny de Beauharnais was there more often, with her extravagant clothes. And then Mme. de Genlis, "governor" (she was very proud of this masculine title) of the children of the duc d'Orléans, a bluestocking and a pedagogue in a small way, but nevertheless friendly. She had declared war against the "modern philosophers," and had published a book, *La Réligion considerée comme l'unique base du bonheur et de la philosophie* [Religion Considered as the Sole Foundation of Happiness and Philosophy]. Buffon had congratulated her for it in a letter in which he called her "my noble daughter."[71] The letter was made public and the riposte was not long in coming. Under a cascade of false names, Rivarol published a new "dream of Athalia" that he attributed to Mme. de Genlis:

> It was during the repose of the night's work
> Buffon's image showed itself before me
> As at the Royal Botanical Garden pompously ready.
> His errors had not his pride defiled:
> He even still used that affected style
> Of which his work he took care to paint and decorate
> To avoid for years the inevitable outrage.
> Tremble, my daughter too worthy of me:
> It prevailed over you, Voltaire's party;
> I pity you for into his fearful hands falling,
> My daughter! By finishing these words appalling
> Natural history seemed to lower itself,
> And me, I extended my hands to help.
> But I found no more than a horrible mix
> Of reptiles, birds, and awful insects,
> Of dead quadrupeds dragged through the mire,
> Which Bexon and Guéneau fought over.[72]

The parody was good and not that mean, but Mme. de Genlis took it very badly. She admired Buffon greatly and remained loyal in her friendship to him.

Buffon's great friend, however, was Mme. Necker. D'Angivilliers, the badly accepted successor to Buffon, had in the end reconciled with him, and it was he who put Buffon in contact with Necker, then a rich Genevan banker in Paris. For Buffon, who always needed money, it was a useful acquaintance. Necker introduced him to his wife, and this was the beginning of a tender friendship. Suzanne Courchard, the orphaned daughter of a pastor, came to Paris to take care of children. She met Necker, whom she married, and aided his fortune by quickly making a place for herself in Parisian society. Pretty, intelligent, idealistic, she was also a Calvinist and a believer. The atmosphere of unbelief that permeated Paris shocked her. Her first discussions with Buffon were about religion and the naturalist remained reserved: "I protest to you, Madam, that I would esteem myself more if I could think as well on everything as do you. . . . Yet the first of all religions is to keep to one's own, and the greatest of all joys is to believe it to be the best. I have not taken any delight in those conversations where we were not quite in agreement."[73] Buffon nevertheless made an effort. A few months later, he sent her a "small written work" in which he strove to show that there was no contradiction between his "cosmogony" and the first verses of Genesis. He would later insert this small work in the *Epochs of Nature*. Often, he would make discreet allusions to their disagreement, apologizing for not being as "spiritual" as his correspondent.

Very quickly, however, Mme. Necker became "my very respectable friend," then "my adorable friend," even "my divine friend." Buffon was always careful to associate Necker with this friendship, but it was she he loved: "I love you and I will love you all my life." It was she he took as a confidant for all his worries, his tangled affairs, his problems with his health. It was to her alone that he dared to admit that he was aging, that his sight was becoming cloudy, and that his hand trembled when he wrote. It was for her that he discovered feelings that he had perhaps never known and that he felt he was incapable of describing: "The page, it seems to me, can only receive the product of the mind and not sensations of the soul. I feel them in wanting to paint for you those that are the dearest to me."[74] Nevertheless he tried: "Ah! God, it is not a feeling without fire, it is, on the contrary, a true heat of the soul, an emotion, a gentler but also livelier movement than that of all other passion; it is a delight without turmoil, a joy even more than a pleasure; it is a communication of a purer and yet more real existence than that of the feeling of love. . . . But for the intimate union of two souls, is it not necessary that they be of the same level, and can I flatter myself that mine rises as high as yours? I believe it does sometimes because I desire it, because you are my model, because I love and respect you beyond all that I have loved."[75]

Buffon did not have a pure mind, and he knew it. Mme. Necker was pretty and, as she had just been ill, he wrote her: "[My heart] has, in addition, feeling which is fond of your person. I cannot imagine that thinness, that loss of your alabaster roundness, without crying in despair; it is therefore not your soul alone that I love."[76] Yet he was virtuous from necessity: "I have also endeavored to make myself an *égide* [a shield] against the evil of absence. For that, I felt it necessary to separate the existence of one's soul from that of one's heart as much as possible; the heart's movements depend on the action of the senses and all tend toward love, those of the soul limit themselves to tender friendship. The first need to be nourished and maintained by the presence of the loved object, the second are exercised with more force and are even purer in absence. But in order to enjoy more fully this happiness, solitude is required, and it is necessary to say to oneself: I am only alone to love better, only to unite my soul with hers at all moments. Here is my most cherished illusion, do not make it fade away."[77]

Mme. Necker answered this adoration with a tender admiration and with constant attentions for her great man. She had not given up converting him, as he wrote again in 1785: "I feel the faculties of the spirit decrease with those of the body, and there lies the foundation for the difference of our opinions."[78] But that did not matter, after all. From afar, she watched over him. She sent him her most comfortable carriage so that the trip from Montbard to Paris would be less painful. She inquired about his health, she wrote long letters to Mlle. Blesseau, and the two women were joined in their ardent affection and their worries. They would also be at his death bed.

In the beginning of 1788, Buffon left Montbard. He had just had a serious medical crisis, following many others. His doctors advised against the trip. Since autumn, however, he had been restless. The work at the Garden was not progressing. The amphitheater was barely rising above ground. "It is nevertheless there . . . that it is necessary to put all our forces," he wrote to Thouin, "so that the school classes will not be interrupted, and so that the lessons of anatomy can be held in this new amphitheater."[79] There he could speed up construction. The indomitable old man left Montbard for the last time.

What was this man of eighty years thinking about, sitting in a jolting carriage under the worried eye of Mlle. Blesseau, who was furious with this senseless trip? Was he thinking of the tree nurseries of Montbard, from which perhaps came the trees lining the road? Of the forges his carriage passed, which had meant so many disappointments for him? Of his work, which was ending? The birds were done. Guéneau was dead. Also dead were the abbé Bexon and the young Daubenton. The minerals were done. The *Treatise on the Magnet* was in press, and it would be necessary to push the printing. One volume of the *Supplement* remained to be finalized. It was necessary to take care of it. Afterward? . . . Well! Faujas and Lacépède would continue his work and would do what he had not been able to do. But first, there were these buildings to raise, that architect to push, those workers . . . He alone could do it. And, in order to go faster, he made the horses trot on the terrible road that led from Fontainebleau to Paris.

CHAPTER XXII

The Great Visions of
a Passionate Genius

The Prolongations of the *Natural History*

By the time the *History of Quadrupeds* had been completed, the *History of Birds* was already begun. The initial project of a collection of "illuminated" plates grew into a work analogous to the *History of Quadrupeds*. The first volume appeared in 1770. Buffon, it will be recalled, had found in this work the sole remedy for the worries and pains that his wife's illness caused him. He did not write the last articles of volume I, however, although he signed them. In volume II, which came out in 1771, his participation was more minimal. Finally in volume III, in 1775, he revealed the name of his clandestine collaborator, whose style resembled his so closely that no one had noticed the difference: it was his old friend Guéneau de Montbeillard, and there was perhaps some irony in the way that Buffon remarked that "in his own way, he is one of those persons who, like Peacocks, are enthusiastically applauded by the Public and by the severest Judges." The official reason for Guéneau's collaboration was "a serious and long illness," which had "interrupted the course of my work for almost two years." And Buffon added this apology, typical of his temperament: "I could have produced, in the two years that I lost, two or three other volumes of the *History of Birds* without abandoning the project of the *History of Minerals,* with which I have occupied myself for several years."[1]

In fact, it was largely his new passion for "my dear minerals"[2] that turned Buffon away from the interminable details of the *History of Birds*. This does not mean that he left the control of the work to Guéneau. Not only did he constantly pressure him, for Guéneau was a bit lazy, but he also attentively read and corrected all his articles, returned them to Guéneau, and corrected them again. He even continued to write a good part of the volumes.

This collaboration lasted until volume VI. But as his widow later pointed out delicately, "M. de Montbeillard, in continuing his work on birds, always felt the difficulty of organizing himself in order to produce volumes on certain days. His independent genius could not march in step with anyone, not even with his intimate friend, with the great man with whom he was flattered to be associated. It was this that made him willingly accept the proposition of being solely responsible for the *Histoire des insectes* [History of Insects]."[3] Guéneau died in 1785, and the *History of Insects* did not see the light of day.

Buffon announced Guéneau's departure at the beginning of volume VII, which appeared in 1780. At the same time he pointed out that even though all the articles were now signed with his name, "a good deal of what they contain does not belong entirely to me." It was a new collaborator, the abbé Bexon, who furnished him with "all the nomenclatures and the majority of the descriptions" and even many "of the solid reflections and ingenious ideas that I have used, and for which I am indebted, came from him. It is a pleasure for me to express publicly my deep gratitude to him."[4] In fact Bexon had worked for Buffon since 1772. He was then twenty-four years old. He was a "small, humped, and deformed abbé, but with an open face, and eyes filled with expression," and he had encountered some difficulty in forcing open Buffon's door without letters of recommendation.[5] But he worked very well, knew how to capture something of Buffon's style, and soon became an indispensable collaborator. Like Guéneau, he was submitted to criticisms and corrections from the master,[6] but he never obtained the right of signature. Nonetheless, Buffon liked him, as the tone of his letters testifies, and was not ungrateful: he had him named in 1782 as preceptor of the Sainte-Chapelle in Paris, which was worth, in addition to certain privileges, a salary of eight thousand livres, which the "little abbé" greatly needed to support his mother and sister. Unfortunately, Bexon would not profit from it for very long. He survived the *Natural History of Birds* only by one year and died at the age of thirty-six, on February 15, 1784.

Three other collaborators are worth at least mentioning, although their relationships with Buffon were not always easy. The first, Sonnini de Manoncourt, a young officer fascinated with natural history, had lived in Guyana. Upon his return to France, he gave a large number of specimens to the Cabinet of the King, and even came to work at Montbard with Buffon in the winter of 1776–1777. He brought back a large amount of information on birds, and he is often cited in the articles that treat birds of South America. Buffon even financially sponsored a research trip to Egypt for him but soon cut off his means of support, for the young man's behavior was very disordered: he was "lost by debts and reputation." Nevertheless, Buffon ended up by forgiving him and helped him again financially.

He used his descriptions of Guyana at length in the *Epochs of Nature*.[7] After Buffon's death, Sonnini remained loyal to his memory, defending it against attacks by the new naturalists, and published a beautiful edition of the *Natural History*.[8]

Things were simpler with Edme-Louis Daubenton, called Daubenton the younger, a cousin to the "old" Daubenton, for whom Buffon had the place of "guardian and assistant demonstrator of the Cabinet of the King" created. It was he who watched over the work of the draftsmen and engravers for the *Natural History of Birds*. Buffon thanked him for it publicly at the beginning of the work, and at the same time he celebrated "the facility of M. Martinet's talent," the creator of the illuminated plates.[9]

Yet destiny seemed to work against Buffon's collaborators. Daubenton the younger died in 1786. Buffon replaced him with someone who would soon make a name for himself, the comte de Lacépède, who had begun his relationship with him in 1778; he entrusted him with the responsibility of writing the history of oviparous quadrupeds and snakes, giving him all the notes he had accumulated on the subject. Lacépède wanted to test his own wings, and decided to publish the work under his own name, as he was, after all, the author. The first volume appeared in the beginning of 1788 just before Buffon's death. Buffon, according to Mlle. Blesseau, was "very angry that M. de Lacépède had not spoken to him about it" and said that "it was a bad book."[10] After Buffon's death, Lacépède took on the task, apparently against the wishes of the deceased, to publish the last volume of the *Supplement*, which appeared in 1789. He would then publish a *Histoire de poissons* [History of Fishes] and a *Histoire des cétacés* [History of Cetaceans], which are considered the continuation of the *Natural History*. He continued to defend Buffon's memory with great concern.

In 1774 the first volume of the *Supplement* to the *Natural History* appeared. The very idea of publishing a "supplement" might seem strange: why not do a new "revised and expanded" edition? It is true that at the time "supplements" were common, and that the supplement to the *Histoire naturelle* was the counterpart to the *Encyclopédie*'s supplement: it was a symbol. The official reason, however, was so as "not to render my book too expensive for the public." In all the earlier reprintings, "there was not one changed word," and Buffon did not want to "make . . . all these editions superfluous."[11] Behind this pious motive, there was perhaps a more realistic consideration: we do not know how many copies of the old volumes were still in stock at Panckoucke. A new edition would have made them unsellable.

The subtitle of the first volume, "Des éléments" [On the Elements], clearly indicated the new orientation of Buffon's research. These seven "supplementary" volumes formed a collection of fairly odd texts. Volumes III, VI, and VII contain the "Additions" to the articles on the *History of*

Quadrupeds. Buffon here presented everything he had learned, especially through his correspondents, since the publication of the article. Aside from the long article on mules, of which I have already spoken, he did not renounce any of his essential ideas. He simply corrected his errors, added information, and thanked all those who had communicated with him. In particular, he reproduced the additions that "le Professeur Allemand" had added to the Dutch edition of the *Natural History.* Also on occasion, but rarely, Buffon defended himself from criticisms. Thus, "M. Vosmaër, an able naturalist and Director of the Cabinets of S.A.S. Mgr. the Prince of Orange," reproached him for two things in his description of sloths. Buffon rejected the first criticism: the fact had been confirmed by an eyewitness. "The second reproach is better founded. I very willingly admit that I made a mistake when I said that the unau and the ai did not have teeth, and I am not at all ungrateful to M. Vosmaër for having noticed this error, which came only from inattention. I like a person who notices an error as much as another who teaches me a truth, because an error corrected is indeed a truth."[12] Buffon also corrected or completed the plates. All that was a lot of work, requiring much attention. Even when he had collaborators, it was always Buffon who chose, decided, and finally said what he wanted to say. He corrected many of his former statements about the gigantic animals of Patagonia, the Hottentot's apron, and the giraffe. But again, nothing essential was changed.

Certain volumes of the *Supplement* serve as a bit of a grab bag, and one cannot help thinking that Buffon made money for himself by reselling to Panckoucke old texts at the price of twelve thousand livres per volume. The academic discourses, published in volume IV, did not have much to do with natural history. The famous "Discourse on Style" can be accepted, but the others? The same remark could be made about the academic theses, which Buffon also republished. The experiments on burning mirrors republished in volume I also fit in: they had an immediate relation to the research Buffon was doing on heat and light. But the old articles on the strength of wood, the eccentricities of ligneous layers, and the effect of frosts on plants, which take up a good part of volume II, have nothing to do with what he was doing at the time. But they made good copy.

The "Essay on Moral Arithmetic"

One text, however, is worth pausing over—the "Essay on Moral Arithmetic" published in volume IV in 1777. It is a composite text, where very old parts, anterior even to the *Natural History,* are found. It is here that the only known version of the article "On the Game of *franc-carreau*" of 1733 can be found, as well as the text of the letter written in 1730 to Cramer about the

Saint Petersburg paradox, as well as a long extract from the preface of the translation of Newton's *Method of Fluxions,* published in 1740, and themes already treated in the "First Discourse" of 1749. Thus the first impression is of a hodgepodge of old texts that Buffon did not want to lose, just as he had not wanted to let his academic discourses be lost.[13]

In reality, the text has a deep unity: it developed a theme that continued to preoccupy Buffon, or which at least was preoccupying him again, ever since he had thrown himself into new research and new experiments. This theme had a unifying effect that, as a first approximation, can be expressed in the following manner: how can mathematics be made to serve the study of reality? This question comes back to an essential problem, that of measure, as a recent study of this text has shown.[14] What can be measured and how? What can be measured in physics and also in ethics, in psychology, even in what we call today the social sciences? For it was not by chance that the "Essay" was immediately followed by a very long article titled "Les probabilités de la durée de la vie" [Probabilities of the Length of Life], an article based on statistical tables.

The last articles of the "Essay" (art. XXIX to XXXV) treat "Mesures géométriques" [Geometrical Measures]. Buffon judged it useful to set them apart in the table of contents of the volume, a sign of the importance he gave them. It was there that the fundamental problem lay: if we take a segment of a line as a unit of measure, we can measure the diameter of a circle exactly but not its circumference. Inversely, if our unit of measure is a segment of a circle, we can measure the circumference exactly but not the diameter. Buffon, remember, had already used this example in 1749. But "as soon as one has ceased to regard curves strictly as curves and has reduced them to be only what they are, in fact, in Nature, namely polygons whose sides are indefinitely small, all difficulties have disappeared" (art. XXXII). It is unlikely that a mathematician would consider infinitesimal calculus in such a way.

The same problem came up for the "moral" man. How to measure our fears, our hopes, our pleasures? In 1730, at the age of twenty-three, Buffon had considered our hopes to enrich ourselves. In 1777, at the age of seventy, he reflected on our fear of death. If we are reasonably sure of living to the following day, while we have one chance out of ten thousand of dying before then, a one in ten thousand probability does not affect us, and that is our unit of measure. If a phenomenon has ten thousand chances to one of occurring, we are "certain" it will occur. Let it be understood, certain means physical certitude, which is only a very high degree of probability.

This probability can be calculated only thanks to statistics. It is they, and they alone, that can allow the calculation of our "life expectancy," as we say today, at a given age. And this method can be applied to other physical phenomena, to all those that cannot be submitted to a simple analysis showing an immediate relationship of cause to effect. For death can be the effect of multiple causes, of so many causes that they cannot be analyzed. Many other phenomena are the same. It is therefore necessary "to see them such as they present themselves to us without paying attention to the causes, or rather without seeking causes for them. . . . As incomprehensible, as complicated as they seem to us, we will judge them as the most obvious and the simplest, uniquely by their results" (art. IV and V). Buffon gave only two examples: the probabilities of human life and a game with irregular dice. He could not then foresee the extent of the possibilities that would later be given to this method. For that matter, he did not yet clearly distinguish between the statistical study of a complex phenomenon and the phenomenological and "positivist" approach to a simple phenomenon, such as gravitation.

Condorcet, in his academic "eulogy," referred to this text, but hastened to repeat the criticisms he had already given.[15] It was undoubtedly difficult for him to recognize a predecessor in a man he disliked.

Toward a Unified Vision of Nature: Meditations and Experiments

If the "Essay on Moral Arithmetic" brings us back to the fundamental problems of knowledge, it is because these problems had taken on a new topicality for Buffon. Beginning in 1764, the "First View of Nature" showed a new direction in his thoughts, or rather a return to former preoccupations. This vast tableau placed earth in the immensity of the universe—this earth on which we live and which, in truth, is "barely recognizable among the other globes, and completely invisible from distant spheres." "A million times smaller than the Sun that lights it, and one thousand times smaller than other planets," its only advantage is its location, less frigid than Saturn or Jupiter, less burning hot than Venus or Mercury. On earth the heat of the sun is "gentle and fertile" and "hatches all the seeds of life."

This entire system maintained itself thanks to the balance of the two forces, attraction and impulsion. There are "thousands of luminous globes, placed at inconceivable distances," and "millions of opaque globes, circling around the first." Thus "it is from the very bosom of movement that the balance of the worlds and the repose of the Universe are born."

This universe is stable: these movements "will be eternal, unless the hand of the first Mover opposes and uses as much strength to destroy them as was necessary to create them." Along with Leibniz, and contrary to Descartes, Buffon believed that the universe, once created, survived by its own forces—unless he believed it to be eternal, like Aristotle. What is important, especially for us, inhabitants of the planet Earth, is that we are not at risk of losing solar heat. Buffon now had an explanation for this permanent blazing of the sun: it is the result of the motion of the planets and the comets that surround it: they "make up a wheel of a vast diameter whose axle carries all the weight, and which itself turning with speed must have heated, blazed, and spread the heat and light to the extremities of the circumference: as long as these motions last [and they will be eternal . . .] the Sun will shine. . . . This fertile source of light and life will never be exhausted, will never run dry."[16]

In this text, which records Buffon's thoughts in 1764, a few ideas must detain us: first of all, that of eternity or at least the stability of the universe in its entirety. Buffon does not change his thinking on this point. Next, this fact that the elevated temperature of the sun is a consequence of gravitation, of the "weight" of the heavenly bodies that turn around it. Whatever the worth of this analogy may be, and clearly it is not much, it guaranteed that the sun would shine forever. At the same time, it was a step toward the unification of the great forces that govern nature: heat is a consequence of attraction. It is also clear that in this text, heat did not interest Buffon much. The only two primitive forces of which he spoke were gravitation and impulsion. Finally, it was from the sun, and from it alone, that the "gentle heat" came that governed the earth and permitted life there. Buffon restated this in the following year: the temperature of the earth's globe "depends on its location, that is, on the distance it is placed from the Sun."[17] It was precisely on these last two points that Buffon's thoughts would undergo a radical change.

Buffon did not leave us the history of his ideas. "A few new ideas also came to me . . . of which I have sought to certify the worth and reality through experiments":[18] that was all he said to his readers. We are therefore reduced to guessing at the reconstruction, but we know that in the summer of 1767 Buffon threw himself into a series of experiments on heat, experiments he had not thought of one year earlier. What had happened between 1764 and 1767?

In 1765 an academician whom Buffon had known for a long time, Dortous de Mairan, read a memoir titled "Nouvelles recherches sur la cause générale du chaud en été et du froid en hiver" [New Research on the General Cause of Heat in Summer and Cold in Winter] to the Academy. We do not know if Buffon was present at this academic session, but the paper was published in 1767 in its *Memoirs*. Dortous de Mairan took up and developed ideas he had already expressed in the fourth edition of his *Dissertation sur la glace* [Dissertation on Ice], which appeared in 1749. Buffon had read the memoir of 1765, since he cited it.[19] Mairan showed that the difference in temperature between winter and summer in our climates did not correspond to the difference of sunshine. If the sun were the sole source of heat on earth, the winters should be much colder. He therefore concluded from this the existence "for the entire Earth" of a "fund of heat independent of the vicissitudes of the seasons." In other words, the earth had its own heat. Other facts, which Buffon listed in turn, supported this statement: the increase in temperature when descending into mines, what we call today "geothermal gradient," and the constant temperature of the depths of the oceans, where the solar light does not penetrate. It was therefore necessary to accept the existence of a "central fire" that Mairan did not know how to define, although, he said "many reasons persuade me that it comes from the internal structure of the Earth and the planets in general."[20]

The idea of a "central fire" was not new: it was found in Descartes and Leibniz. But Descartes thought that this fire had been buried under the piling up of "obscure matter," and Leibniz believed that it had been extinguished long ago. This was the point that Buffon criticized in 1749, as I have mentioned: "The great defect of this theory is that it does not apply to the present state of the Earth; it explains the past, and this past is so ancient and has left us so few vestiges that one can say anything that one wants about it."[21] Let us recall that Buffon himself did not establish any link between his hypothesis on the formation of the planets and his theory of the earth; the two texts were even contradictory.

But what was to be believed if it were established that the earth had its own proper heat? It can be assumed that Buffon did not entertain for an instant the idea of a "central fire." For him, the underground fires were superficial, caused by the flaring-up of "pyritic and combustible materials." The idea must have immediately come to him that this heat was what remained of the earth's heat at the moment that it had been torn from the sun. Since that event, the earth had cooled. How, and at what rate? If we could know that, we could date the birth of the earth. This meant that the entire history of the earth needed to be reviewed. Buffon thus undertook a long project that would end, in 1778, in the publication of one of his most important works, the *Époques de la Nature* [Epochs of Nature].[22]

The experiments of 1767, which Buffon would describe at length in the first volume of the *Supplement,* make sense only when seen from this point of view. Beyond these experiments and their results, to which we shall return, Buffon had to confront a new category of problems. He had to reflect on the very nature of the elements, light, heat, and fire.

In the first volume of the *Supplement* in 1774, he gives us the result of these reflections and an overall vision of the forces which act in nature. It has a general subtitle: "On the Elements." A detail to be noted is that the volume was presented at the same time as "serving as a follow-up to the Theory of the Earth and as an introduction to the history of minerals." Buffon also added in a note: "See in this work the article on the formation of the planets and afterward the articles on the epochs of Nature."[23] It was truly the meeting of two series of texts, two stages of Buffon's thoughts. He wrote two long chapters, the first one dedicated to light, heat, and fire, and the second to air, water, and earth. Aristotle's four elements, still widely accepted, were therefore found here, but light and heat were the more attentively studied.

Buffon started with a major statement: "The powers of Nature, as far as they are known to us, can be reduced to two primary forces, one that causes weight and one that produces heat. The force of impulse is subordinate to them."[24] What Buffon meant by "impulse" is not very clear. When a body rebounds after having struck another, it receives an "impulse." Since the rebound assumes "spring," or the elasticity of bodies in collision, and since this elasticity can be explained only by the attraction that maintains their cohesion, this type of impulse was a consequence of gravitation. But under this same name of impulse, Buffon also included "all the phenomena of living matter," and by "living matter" he included not only all that lives— organisms or organic molecules—but also light, fire, and heat, "in a word, all matter that seems inherently active to us." Thus understood, "it is principally through the means of heat that the impulse penetrates organized bodies." Which in fact, without saying so, was the same as bringing life back to the physical phenomenon of heat.

Light and heat radiate from a center. By generalizing, it was therefore possible to say that "living matter always tends from the center toward the circumference, while inorganic matter, on the contrary, tends from the circumference to the center; it is an expansive force that animates living matter, and it is an attractive force that inorganic matter obeys." The two forces "balance each other without ever destroying each other, and from the combination of the two equally active forces result all the phenomena of the Universe." Since these two forces are "general effects, one must not ask for the causes." They affect all bodies; to find a cause, it would be necessary to observe a body that is unaffected by them that would allow a comparison, which is impossible.

"This first reduction accomplished, it would perhaps be possible to do a second one, and to reduce the power of expansion itself to that of attraction, in such a way that all the forces of matter would depend on a single primary force: at least this idea seems to me worthy of the sublime simplicity of the plan according to which Nature operates." Centrifugal expansion was perhaps only a reaction to centripetal attraction. "From that moment on, the expansive force will no longer be a specific force opposed to the attractive force, but an effect that is derived from it, and that manifests itself every time bodies hit each other or rub against each other." The "primary molecules of matter" must be perfectly elastic, as are those that make up light. If they are found "in a situation of freedom," they are attracted by each other, enter into collision with great speed, and rebound with as much speed as they acquired at the moment of contact, which one must consider a true shock." Thus light and fire, examples of "living matter," "are not special forms of matter." They are simply ordinary matter, but in a state of "great division of parts," a state wherein the "primary molecules" regain freedom of their motion.

That light is matter was attested to by experience. A prism separates its "primary molecules," as Newton had shown. It is sensitive to gravity, "as it bends every time that it passes close to other bodies, and it is found within reach of their sphere of attraction." It can even "put fairly heavy bodies into motion." Hence, "all matter will become light as soon as its coherence is destroyed and it is divided into sufficiently small molecules" and these molecules are "freed." Conversely, "light can also be converted into all other matter by the adhesion of its own parts," for "all elements are convertible."

"After having shown that impulse depends on attraction, that the expansive force is the same as the attractive force become negative; that light, and even more so, heat and fire are only aspects of a common matter; that there only exists, in a word, one single force and one single matter always ready to attract or repel according to the circumstances: let us seek how, with this sole elasticity and this sole subject, Nature can vary her works infinitely."[25] Such was the program that Buffon laid out for himself.

One cannot help becoming lost in thought, even perplexed, when faced with these pages. It is of course possible to see here the "physical gibberish" denounced by Condorcet. This dream of the unification of nature, which has always haunted and continues to haunt physicists, can also be seen here. One can even find intuitions of genius here: was it not necessary to wait for Einstein and the great observations made just after World War I to show that light is indeed affected by gravity? Did not Buffon at least come close to an explanation of heat by the movement of molecules in matter?

It is easier to place this text in history. What is striking is its Newtonian roots. It was from Newton, from the *Principia* of 1687, even more from the "Queries" that conclude the *Opticks* of 1704, and especially from the subsequent editions, that Buffon drew his inspiration. Let me give two examples. "Query 1" is announced in the following manner: "Do not bodies act upon Light at a distance, and by their action bend its Rays; and is not this action, all other things being equal, strongest at the least distance?"[26] Do not be fooled by the interrogative form. Newton was convinced of this action simply because he considered light as a continuous emission of material corpuscles. If Buffon seems to be a precursor to Einstein, it is only because he followed Newton, and after the long episode of the wave theory of light, the twentieth century found itself obliged to admit that light is also made up of particles, photons. But what a gulf between the photon of the twentieth century—wave and particle at the same time, as de Broglie proved—and Newton's "corpuscles" or Buffon's "molecules"!

Newton also wrote in "Query 30": "The changing of Bodies into Light, and Light into Bodies, is very conformable to the Course of Nature, which seems delighted with Transmutations."[27] It was again from Newton that Buffon drew his considerations on light and heat, or on the glowing of opaque bodies that are heated. And it was yet again Newton who suggested the essential idea of connecting impulsion back to gravity: "And as in Algebra, where affirmative Quantities vanish and cease, there negative ones begin; so in Mechanicks, where Attraction ceases, there a repulsive Virtue ought to succeed."[28] More generally, what they have in common is the desire to give a purely physical explanation of chemical and even physiological phenomena. This attitude was common in the beginning of the eighteenth century, and it was also that of Stephen Hales, the other authority often cited by Buffon.

Through this loyalty to Newton's broad views and his boldest hypotheses, Buffon could seem completely anachronistic. Certainly, Newtonian mechanics was universally accepted, and his theory of light remained officially admitted, even if certain snipers like Goethe or Marat criticized it vigorously. A modest veil, however, had been thrown over the daring of the great man and his rash speculations. The chemistry of the middle of the century, with men like Rouelle or Venel, had followed Stahl rather than Newton and adopted the theory of phlogiston, which Buffon did not accept. These chemists insisted on the autonomy of their discipline, and rejected the tyranny of physicists. It was this independence that Venel declared in the article "Chimie" [Chemistry] in the *Encyclopédie:* "What is called chemical by so-called physicists . . . all this chemistry, I say, which is the most widespread, has the great defect of not having been discussed or verified in detail or by comparison of the facts; what Boyle, Newton, Keil, Boerhaave, etc. have written on these subjects manifestly bears the stamp of this inexperience." Buffon could be added to the list. And the article endeavored to show precisely how chemistry differed from physics, in its goal as in its method.

Reading this text from 1774, one cannot help but remember that Lavoisier was already at work and would revolutionize chemistry. It is nonetheless necessary to look at it more closely. Buffon was not completely isolated in his ivory tower. He had numerous contacts with chemists: Sage, Macquer—considered the greatest French chemist of his generation— Guyton de Morveau, parlementarian, chemist, and Burgundian. He followed their works carefully. The question of the elements, which he attacked with so much vigor, was, as it happens, a very topical problem. Were the four traditional elements of earth, water, air, and fire really completely distinct, or were there possible "transmutations," as Buffon felt, which basically would mean that there is only one type of matter? The problem was made much more complicated because the distinction between different elements and different states of the same matter was not always made. Was water vapor still water, or was it air? Was ice an "earth"? Buffon described with much precision the vaporization of water, its contraction when it is cooled, its sudden expansion when it is converted to ice, and even the phenomenon of "super cooling," which allows it to remain liquid below the freezing point in "a well closed and perfectly still tube," but he did not dare decide whether it was always the same "element" or if there was a "transmutation."[29]

The great Swedish chemist and mineralogist Wallerius (1709–1785), whose works Buffon must have known, had studied the question of the elements. In 1766 the German chemist Eller argued that there were only two elements, water and fire, and that air was only a combination of the two. Another well-known case of "transmutation" was that of water into "earth." When pure water was distilled, there was always a residue of "earth" at the bottom of the receptacle. In 1767 Charles Le Roy, author of the article "Evaporation" in the *Encyclopédie,* presented a memoir on this problem to the Academy. It was not, therefore, because he treated the elements that Buffon appears anachronistic. Was it then because of the way in which he treated them? At first glance, that could be.

The point, as Venel said, was that all these problems must be discussed in the light of experiments. This was precisely what Lavoisier did starting in 1768. Buffon did not seem to have known him personally, and we understand why: Lavoisier was Turgot's protégé, and therefore a friend of Condorcet. Lavoisier showed, by a precise experiment, that the earthy residue found after the distillation of water actually came from the glass of the receptacle itself. The result of this experiment was published in 1771 in the *Journal de physique* [Journal of Physics] of the abbé Rosier. Buffon did not seem to have known of it or paid it any attention. The irony of the situation is that Lavoisier's thinking, like Buffon's, grew out of the work of Stephen Hales, and that in 1772 Lavoisier wrote a text, which he kept to himself, where he admitted (almost like Buffon) that air mixed with fire could exist in air and (like Buffon) that air and fire could exist in a "free" state or a "fixed" state, that is, in combination with other bodies.[30] At the time no one, not even Lavoisier himself, could foresee where all this would lead. In fact, Lavoisier took up where Buffon left off. Lavoisier carried out experiments, whereas Buffon thought in the abstract. In the same way, if Buffon understood the difference between temperature and quantity of heat and came very close to the notion of "latent heat" proposed by the Scottish chemist Joseph Black (1728–1799), it was probably because both started from Newton. But it was Black and later Lavoisier and Laplace who would do the conclusive experiments.

Such is the conclusion one is tempted to draw from reading these pages, and one thinks of Buffon's remark to Macquer about the combustibility of diamond: "The best crucible is the mind." Buffon was not ignorant of the facts, but he was satisfied to gather and organize those that he knew, whatever their origin and their worth; and he often described them as he "saw" them—in his thoughts.

This conclusion is, however, largely false, first of all, because 400 of the 542 pages of the first volume of the *Supplement* are dedicated to accounts of experiments, of which the majority were carried out by Buffon himself or under his direction. But he set them apart, perhaps through a sort of obsolete elegance, or more likely because they were only one basis for his ideas, ideas that went well beyond them and attacked much more general problems, probably much too general for the taste of the younger generation of scientists. Because of the generality of the problems he discussed as much as through his style of exposition, Buffon was anachronistic. Instead of posing a limited problem and trying to resolve it through experiment, as was done more and more at this end of the century, he used his experiments as a springboard for a meditation on nature in general.

These experiments are nonetheless worth pausing for, especially those concerning "the progression of heat." In fact, he was studying times of cooling, but the very title assumes that, for Buffon, the speed of propagation of heat was the same in both directions, which seems logical to us, since the cooling of a body means the heating of the environment in which it is found. But let us not get ahead of ourselves. The first series of experiments concerned balls of iron of different diameters, from one-half inch to five inches. All came from the same forge and had been weighed to verify that their weight was proportional to their volume, that is, that all were of the same quality of iron. Each ball was heated until it was white, and the time of heating was measured. Then it was brought to a cellar where the temperature was constant, and the time of cooling was measured.

Here Buffon ran into a practical difficulty: a thermometer could not be used for these measurements. Buffon therefore "sought to ascertain two moments in the cooling, the first when the balls stopped burning, that is, the moment when one could touch them and hold them in one's hand for one second without burning oneself," the second when the cool ball had attained the temperature of the cellar, which the observer verified by touching the cool ball and a test ball which had not been heated. For the comparison to be more exact, "care had been taken that the cold balls be crude and similar to those that had been heated, whose surface was covered with small bumps produced by the action of the fire."[31]

The goal of these experiments, which it is said had been suggested to Buffon by an officer of the artillery, Potot de Montbeillard, was to verify a statement made by Newton, or rather two contradictory statements. In his *Principia*, Newton said that a larger globe would cool proportionally faster than a smaller globe. In the *Opticks*, he said the opposite. Buffon was persuaded that the statement from the *Opticks* was correct, and his experiments showed him to be right: the larger globes cooled slower proportionally, as the very precise tables of results that he gave us prove.

Buffon's method has been severely criticized, especially because of the absence of a thermometer. In fact—and we shall come back to this—his results were good. Was this because, as a spiteful local legend accused, Buffon employed women, who have a more sensitive skin, to take the measurements? More surprising for us, he was content to present his results in tables and did not try to give a mathematical formulation of the relationship between the diameter and time of cooling. On the other hand, he tried to explain the cooling itself and, faithful to his principles, explained it not by "contact with the ambient medium" but by "the expansive force that animates the parts of heat and fire, which chases them out of the body where they reside and pushes them directly from the center to the circumference."[32] Here is why the time of cooling measures the "progression" of heat: it was the "expansive force" he was measuring. Immediately and, by the way, following Newton, he asked himself the question of the time necessary to cool the earth or a comet. This question indicated both the true goal of his experiments and underlined the distance that separated his preoccupations from those of his contemporaries.

These preoccupations led him to a second series of experiments, which were perhaps even more interesting. Since the earth is not a globe of pure iron, it was necessary to compare the times of cooling of the different substances of which it is made. For that, Buffon had "a great number of spheres made, all of one inch diameter," of about thirty different materials, be they metals or minerals. Certain spheres that were too small were discarded. The others were heated in ovens, together or separately. To be sure of having the same temperature at the beginning of the cooling, Buffon put a sphere of tin, a metal that melts at a low temperature, next to them. As soon as the sphere of tin started to melt, the spheres were removed from the oven and placed on a table. This time, Buffon himself did the experiment: "I let them cool without moving them, trying fairly often to touch them, and at the moment they no longer burned my fingers, and I could hold them in my hand for a half second, I marked the number of minutes that had passed since they had been taken out of the fire; then I let them all cool to the current temperature," using the same method of control balls used for the balls of iron.[33] The experiments, which lasted close to six years beginning in 1767, had been repeated in all possible ways, and their results took up more than one hundred pages, before being put into a comparative table.

The conclusion was that it was not "proportional to their density . . . that bodies receive and lose more or less heat" but "by inverse reason of their solidity," the "solidity" being "the quality opposed to fluidity"[34] and, at least in metals, corresponding to their "fusibility."[35] This quality Buffon explained by the very structure of matter and by "the facility with which the particles of heat are able to separate [the] molecules of solid matter from one another."[36]

We shall leave aside here the experiments done in collaboration with Macquer and Guyton de Morveau on platinum, a recently discovered metal whose nature was still not known, or those that Buffon did on iron with Potot de Montbeillard and Tillet, with whom Buffon seemed to be in continual contact. Tillet used his position as inspector of the mint to supply Buffon with the sphere of pure gold needed for his experiments on cooling. All these experiments showed that Buffon did not work alone and that he was not behind the times. What is perhaps most remarkable is that a modern mineralogist took the results of the experiments on cooling seriously and found them to be highly accurate. For example, Buffon gave a density of 7.76 to iron, whereas the accepted density today is 7.8. More important, Buffon's classification of metals by the speed of cooling corresponds to the product of the density times the specific heat, a relation that Buffon did not know.[37] Thus his experiments, which have long been criticized without the results being seriously examined, have been restored.

These experiments, however, serve only as a point of departure for some of the ideas that Buffon offers in his two chapters on the elements. In these pages, there are a few ideas that he developed in his later works. Relying on the works of the German chemist Johann Heinrich Pott (1692–1777)— further developed by Jean d'Arcet (1725–1801), a professor at the Royal College whom he cited in a lengthy note—Buffon discussed the action of heat on minerals and observed that they were "vitrifiable": with the exception of calcareous rocks, they melted in fire. He returned later to this distinction, which raised the question of the origin of calcareous rocks. For Buffon, this was a case of "transmutation" of water into earth, but this transmutation could be accomplished only by an "animal filter." The mollusk shell "is certainly a terrestrial substance, a true stone." It was therefore these "animals with shells" that transformed the "particles of water" into stone. This was the origin of calcareous rocks. If one adds to this "all the combustible materials that come only from animal or plant substances," it will be seen "how much living Nature has worked for dead Nature, for here inorganic matter is only [that which is] dead." This did not prevent the essential mass of the earth from being made up of a "glassy-like substance," which strictly speaking formed the "terrestrial element."[38] If to these two categories of rocks were added "all those that had been altered by the fire of volcanoes or sublimed by the interior heat of the earth," a division of rocks into three fundamental categories was found: vitrifiable rocks, calcareous rocks, and rocks "altered" by fire, which anticipated our metamorphic rocks.[39]

In each of these major categories of rocks, there is an infinite variety of minerals, which were all products of nature's activities. Nature through the action of water and salts could produce all that chemistry produced using fire: "Only time is needed. . . . It only took time . . . and time is nothing for Nature and . . . is not lacking to her."[40] However, "if we want to form a correct idea of her processes in the formation of minerals, we must first go back to the origin of the formation of the earth." Buffon then gave a rapid sketch of what he would discuss in detail in the *Epochs of Nature*.[41] This sketch would complete his "Recherches sur le refroidissement de la Terre et des planètes" [Research on the Cooling of the Earth and Planets] in the second volume of the *Supplement,* to which we shall return.

The *Natural History of Minerals*

These fundamental reflections also inform the entire *Natural History of Minerals,* whose five volumes, along with a treatise on the magnet, appeared between 1783 and 1788. The outline of the work, at least overall, was inspired by the history of rocks, such as Buffon stated it: "As the order of our ideas must here be the same as that of the succession of time, and time can only be represented to us through . . . the succession of Nature's operations, we shall consider her first in the great masses that are the results of her first and greatest actions on the terrestrial Globe."[42]

Buffon therefore started with the "primitive glasses," then proceeded to calcareous rocks and to rocks of organic origin, then to volcanic materials before arriving at metals and discussing the endless series of minerals in which the fundamental materials were mixed. It is impossible here to go into the infinite detail of these articles, which number close to five hundred, where Buffon risked losing himself even more completely than in his articles on birds. He did not work alone. He was helped, in particular, by Faujas de Saint-Fond (1741–1819), who had abandoned a career as a lawyer to become a mineralogist and geologist. In 1778 Faujas published his *Recherches sur les volcans éteints du Vivarais et du Velay* [Research on Extinct Volcanoes of the Vivarais and the Velay], in which he established the volcanic origin of basalt (almost at the same time as Desmarest, of whom we shall speak later). It was perhaps to Faujas that Buffon owed his interest in rocks that had "undergone a second action by fire." He cited him at length in the article "Basalt." All the same, the whole represented an enormous and careful work, and it is admirable that Buffon's passion for his "dear minerals" could sustain him for the entire time of its writing.

Buffon explained very precisely the characteristics that he used to define and distinguish the minerals: the degree of "fusibility"; the presence or absence of calcination before fusion; whether they effervesced with acids;

the capacity to "spark or make fire against tempered steel," which indicated the body's hardness; interior texture; and finally, "the colors which show the presence of metallic particles."[43] Nothing in all this was new, but many of these distinctive characteristics are still used today. What is absent was actual chemical analysis. Its use was greatly discussed at that time by mineralogists. Werner (1749–1817), a professor at the School of Mining in Freiberg and probably the greatest mineralogist of the time, had published a *Traité des caractères extérieurs des minéraux* [Treatise on the Exterior Characteristics of Minerals], in 1773 which excluded chemical analysis. He changed his mind later, it is true.

Buffon was concerned with not mixing fields, with keeping the "operations of Nature" and those of "Art" separated: "The Naturalist, in treating minerals, must therefore limit himself to the objects Nature presents to him, and send back to the Artists all that Art produces."[44] Perhaps there was behind this caution the feeling that "art," that is, chemical analysis, created as many substances as it discovered. Lamarck had this feeling already, but he was able to express his ideas openly only in 1794; he must, however, have often felt in agreement with Buffon's way of thinking.

It was not that Buffon was ignorant of chemistry, or even of the most recent research. If he violently criticized a "Logomachy" that had "delayed the progress of the Sciences," and "this creation of *half-technical* and *half-metaphysical* new words, which from now on represent clearly neither the effect nor the cause," it was phlogiston he was taking on: why not call it simply "fixed fire"?[45] On the other hand, he often cited the *Dictionnaire de chimie* [Dictionary of Chemistry] that Macquer had published in 1766 and reedited in 1778. He reproduced a "Table of chemical nomenclature" that Guyton de Morveau published in 1782,[46] which included certain distinctive traits of the new nomenclature that would be proposed in 1787 by the *Méthode de nomenclature chimique* [Method of Chemical Nomenclature], the collective work of Guyton, Lavoisier, Berthollet, and Fourcroy. Buffon was a friend of Guyton's, and it was probably through him that he knew the works of Black and Scheele. He cited Berthollet and spoke with respect of "our knowledgeable Academician, M. de Lavoisier"[47]; as for Fourcroy, he named him professor at the Royal Botanical Garden upon the death of Macquer in 1784, perhaps following Guyton's advice. The other possible candidate had been Berthollet, but the choice was excellent, for Fourcroy was a remarkable professor. All this did not show an attachment for the theory of phlogiston, or ignorance of what was happening in chemistry, or hostility toward the new chemists.

Nevertheless, it is clear not only that the viewpoints were not the same, which would not have been serious, but that Buffon's fundamental ideas could not agree with the theories of the new chemistry. The viewpoints were not the same because chemists manipulated, analyzed, or combined bodies without being concerned with their historical "origins." Buffon was pleased to see that Lavoisier thought "that fixed air, or mephitic acid, is formed by the combination of air and fire," but specified that in his "general system on the formation of the earth" this fact meant that it was necessary to "regard [it] as the primitive acid from which all the others have come."[48] Lavoisier would have considered this, at best, as being without interest for him. Above all, as the new chemistry began to define the notion of "simple bodies"—that is, bodies that at the time could not be decomposed—it definitively distanced itself from the idea of classical elements, to which Buffon remained loyal, and therefore from any idea of "transmutation." On this point, too, Lamarck represented Buffon's thoughts well, when he later criticized the idea of "simple bodies": entities that are irreducible and that must consequently have existed since the moment of creation. Lamarck saw here the philosophical equivalent of preexistent germs, because, like Buffon, he posed the question of origins and wanted everything that exists to be the product of "operations of Nature." This preoccupation hardly interested the new chemists.

It was, however, a main preoccupation for Buffon, for whom mineralogy after all lay within the framework of a general history of the earth. For him, the infinite diversity of chemical and mineral bodies was the product of a history. It was the "more than perpetual movement" of nature, "helped by the eternity of time," that "produces, leads, brings all revolutions, all possible combinations. . . . Everything occurs, because through the work of time everything interacts."[49] In order to put order into this history, it was first necessary to establish an "arrangement of the minerals in a methodical table, drawn up according to the knowledge of their natural properties," and this was what Buffon gave us at the end of volume III.[50] In its main lines, however, that table followed the order of the "genesis of minerals," to borrow the title of the work's last article.[51] For it was only by recognizing "the genesis or filiation of minerals, that is, the march of Nature in her successive productions in the mineral kingdom," that it was possible "to arrange them henceforth in a less arbitrary and less confusing manner than has been done until now."[52] In this sense, the "Genesis of Minerals" is the exact counterpart to the chapter "On the Degeneration of Animals": both are concerned with basing a classification on genetic relationships.

Modern mineralogy has not abandoned this ambition. Without going into the details of all that separates modern knowledge from that of Buffon, and in particular of all that chemical analysis has allowed us to discover, let us note only that one of the notions not available to Buffon was that of "metamorphic rock." When he spoke of rocks that had undergone "a second action of fire," Buffon had in mind only volcanic rocks. He did not for an instant imagine the extent of transformations that rocks could have undergone in the depths of the earth through the actions of heat and pressure. For him, the earth was absolutely solid, and nothing more could have happened since the primary "vitrification." As in Aristotle, geological phenomena were only surface phenomena; it was water and it alone that was the agent of all the "transmutations." Buffon did not know either that calcareous rocks could be of chemical origin, without mollusk shells playing a part. Given this, it is clear that many of his interpretations and his "filiations" had to be erroneous, and that the least obsolete are those that concerned sedimentary rocks. He retains the merit of having stressed, more than anyone else, the study of these "filiations" as a necessary part of mineralogy.

On one point at least, which was much contested in his time, Buffon was correct: the vegetable origin of coal and "mineral oils." D'Holbach had defended this idea in the *Encyclopédie*, and Buffon adopted it fairly late, but he made himself the convinced defender of it, even though the question would only be settled in the middle of the nineteenth century.[53] It is also interesting to note that Buffon used geological rather than chemical arguments to support his views.

Containing an enormous mass of documentation borrowed from all possible sources, with numerous pages of detailed discussions written most often in the most prosaic style, the *Natural History of Minerals* was and remains a dry work. The main ideas that animate it had already been presented with much more brilliance in the *Epochs of Nature*. Nonetheless, this work of his old age remains typical of its author. Here—in the middle of these hundreds of pages containing the miniscule details that he had toiled to gather, present, and discuss—Buffon affirmed with enthusiasm the right to "grand views" and the prerogatives of genius.

What he wanted to defend in this context were his ideas on the origin of metallic mines. "Simple conjecture," one will say, and "one will probably revive the trivial objection so often repeated against hypotheses, by exclaiming that in good physics neither comparison nor system is needed." Nonetheless, "the goal of the Philosopher naturalist must . . . be to raise himself high enough to be able to deduce all the particular effects from a single general effect taken as a cause. But in order to see Nature in this large frame, it is necessary to have examined, studied, and compared her in all parts of her immense breadth; a certain genius, much study, a little freedom of thought are three attributes without which one will only disfigure Nature instead of representing her . . . and woe to those who do not believe it!" Woe especially to "those Writers who have no other merit than of protesting against systems, because they are not only incapable of making any, but perhaps even of understanding the true meaning of this word, which appalls or shames them; nevertheless all systems are only a rational combination, an ordering of things or of the ideas that represent them, and it is genius alone that can create this order."[54] Here Buffon was probably responding to criticisms that his *Epochs of Nature* had received. He could not more clearly state, for the benefit of these new scholars, what had been the ideal of his entire lifetime of work.

Buffon presented the *Treatise on the Magnet,* which appeared in 1788, as a "sequel to the *History of Minerals,*" which explains why it is about terrestrial magnetism in particular. In fact, the treatise itself takes up only 192 pages of the thick volume bearing this title. The rest is 366 pages of tables of declination of the magnetic needle, accompanied by eight large maps of the earth's oceans, where these observations were made. The work seems to have been conceived for its practical utility rather than its scientific content.

Its content, however, is not negligible. Although he had treated the major forces that act in nature in the first volume of the *Supplement,* Buffon had not spoken about electricity. The new treatise corrected this oversight, for Buffon immediately stated that magnetism was only an effect of terrestrial electricity. He added at once that electricity "has its origin" in the earth's heat. "Electrical fluid" is produced by "continual emanations of this interior heat." Since these emanations were stronger at the equator, "electrical currents"* were produced that head from the equator toward the poles. It was also the "condensation" of these currents that produced the northern lights, as Mr. Franklin and M. de Lacépède had explained. One can wonder if it was not Lacépède's article that gave Buffon the idea of writing this treatise.

Electricity was therefore a "fluid" like light and heat, but Buffon took advantage of the occasion to discuss the idea of "force," which can only possibly be applied rigorously to attraction and the "primary impulse." These forces were not "material substances," and we must give up transforming them, "to aid our imagination, into subtle materials, elastic fluids, truly existent substances." We must renounce ever knowing "the origin and the essence" of the "primitive force," for it "will be forever unknown to us, because this force in effect is not one substance, but a power that animates matter." Buffon, three pages later, explained that this primitive force "emanated from divine power" and that "attraction [is] essential to all material atoms."[55] Caution or contradiction? Regardless, the important idea was that of "force," which Buffon neither wanted to nor could exclude from his vision of nature.

To explain the production of subterranean electricity, Buffon spoke of all the "vitreous materials," "electrifiable by rubbing," mixed with conducting substances "isolated and charged with electric fluid," that the subterra-

*Ed. note: The first "electrical currents," as opposed to electrical discharges, were first produced in 1800, when Alessandro Volta constructed the first voltaic cell. It is also worth noting that André-Marie Ampère, who had read Buffon as a youth, used terrestrial electrical currents to explain terrestrial magnetism in 1820.

nean waters could link together. This must produce electrical discharges analogous to lightning, which sets fire to combustible materials. Buffon let himself be carried away by the analogy, to the point of attributing earthquakes and even some volcanic eruptions to electricity; this fact would explain the "very large quantity of ferruginous matters" found in lava and basalt.[56] He went so far in this direction that he was almost tempted to correct the theory of the earth that he had presented in the *Epochs of Nature,* and to hold that "volcanic mountains" like those of the Andes cordillera "had been projected or lifted by the force of lightning and subterranean fires," instead of being "primary mountains" as he had believed. There would thus have been volcanoes long before the deposit of pyritic materials, whose combustion explained for him the action of ordinary volcanoes.[57]

As for the actual magnet, Buffon was satisfied to gather accounts, but nowhere did he try to find an original explanation for magnetic force, except to affirm its electrical origin. He was satisfied to criticize those who, like Descartes or Euler, imagined the flux of "magnetic material."[58] He was also impatient to move on to technical questions about the fabrication of compasses, and especially to discuss the inclination and declination of compasses. Once again, it was terrestrial magnetism that interested him more than magnetism itself.

From the first volume of the *Supplement* in 1774 to the *Treatise on the Magnet* in 1788, Buffon's thoughts had been dominated by the history of the earth. He related this history in detail in the *Epochs of Nature,* which appeared in the spring of 1779. This great work brought together and summarized all the ideas that had preoccupied and even obsessed Buffon during the last twenty years of his life.

CHAPTER XXIII

Nature in History

"In civil History, we refer to titles, we search for ancient coins, and we make out antique inscriptions to determine the course of human revolutions and record the dates of changes of customs; it is the same in Natural History; it is necessary to mine the archives of the world, to pull old monuments from the entrails of the earth, to collect their ruins, and put together in one body of evidence all the signs of the physical changes that can let us go back to the different ages of Nature. This is the only way to establish a few points in the immensity of space, and to place a certain number of milestones on the eternal road of time."[1]

It was Buffon himself who read this text on August 5, 1773, before the Academy of Sciences, Arts and Belles-Lettres of Dijon, which on that day inaugurated its new home, the Hôtel de Pringles.[2] The great man's visits were rare, and Buffon had come to Dijon only to win his lawsuit against the alderman Mandonnet. But he was not looking down on his colleagues: he was offering them a preview of the first pages of the *Epochs of Nature,* which would appear five years later.

These first pages immediately indicated the direction of the work, which, as a historical work, was no longer a "natural history" but a history of nature. In a way, it recapitulated all that Buffon had published since 1749. In another way, it transformed and reorganized everything. It was certainly not the first occasion in which time intervened in Buffon's ideas. The discourse "On the Degeneration of Animals," which ended with the very title "Epochs of Nature," was already a description of the transformations of animal forms throughout the course of time. As soon as an action was attributed to nature and what existed was considered the results of her "operations," time, the great worker, was made to intervene.

Time, however, is not enough to make a history. A history supposes a beginning, perhaps an end, in any case an irreversible series of events. If we are ready today to admit that history does not have a "direction," this is not what was generally thought in the eighteenth century. For Christian thought, the history of humanity was a long march toward the accession to the kingdom of God. For many philosophers, like Turgot or Condorcet, it was the slow march of progress toward happiness. In both cases some force must be at work that made history "advance." For the Christians, it was God's grace; for the philosophers, it was reason. This conception of history as "oriented" toward an end appeared in the eighteenth century and became the conception of the nineteenth century. The history of nature, as Buffon would tell it to us, was also an oriented history. Unfortunately, it was the cooling of the earth that directed it, and it could have no other end than death.

The death in question was that of life on earth, and it must never be forgotten that the history of nature was the history of nature on earth. Other similar histories have occurred, are occurring, or will occur on other planets. The sum of these particular histories do not make up a history of the universe. These local episodes do not trouble the universal order any more than the majesty of the sea is troubled by the history of each wave that swells, breaks, and comes to die on the beach.

Eternity formed the backdrop of this history, but it was a history all the same, and that posed new problems of method. "It involves piercing the mist of time; to recognize by the inspection of existing things the former existence of annihilated things, and to go back to the historical truth of buried facts relying only on the strength of remaining facts."[3] "Historical truth" was not the same as physical truth. It was not observed but reconstructed. For this reconstruction, Buffon wanted to use "three great means: (1) the facts that can bring us closer to the origin of Nature; (2) the vestiges that we must consider witnesses to the earliest ages; (3) the traditions that give us some idea of subsequent ages."[4] We shall come back to the use of "traditions." For the rest, the methods were hardly vulnerable to attack. The "vestiges" were of course fossils. The "facts" were directly observed facts, which could only be explained by a former state of things.

Buffon enumerated five "facts," which actually boiled down to four: the shape of the earth, "raised at the equator and flattened at the poles, in the proportion that the laws of weight and centrifugal force require"; the interior heat of the earth; the "vitrifiable" nature of the rocks that make it up; the presence of sea shells "[existing] at altitudes up to one thousand five hundred and two thousand fathoms." Buffon knew the first "fact" in 1749. The second and the third were recent "discoveries," as we have seen. The fourth involved an important correction to what Buffon believed in 1749. He thought at the time that shells were found up to the peaks of the highest mountains. Even then he knew that La Condamine had not found any on the peaks of the Andes, but he had carried on regardless. The new theory would allow him to take this into account.

As for the "vestiges," they came from paleontology. They included the seashells that made up calcareous rocks, but especially the presence of fossils of lost species or species that no longer existed except in the warm seas in temperate climates. In the same way, more recent fossils of typical mammals of warm regions, like the rhinoceros or the elephant, were found in Europe and as far north as Siberia or Canada. All this was not in question in 1749. These are the facts that it was necessary "to link through analogies" in order to "form a chain that will descend from the top of the ladder of time to us."5

The conclusion seemed immediately evident: the earth had first been a globe of fused materials that slowly cooled. There was, therefore, a time when the temperate zones had a tropical temperature and could be inhabited by elephants and rhinoceroses. The facts thus ordered themselves, so to speak, and it was only to examine them in more detail that Buffon divided his account into six "epochs," to which he would add a seventh, the epoch of man.

"But before going further, let us be quick to head off a serious objection, which could even degenerate into an accusation. How, someone will say, can you bring together this great antiquity that you attribute to matter, with the sacred traditions that give the world only six or eight thousand years? As strong as your proofs may be, as founded as your reasonings may be, as obvious as your facts may be, are not those that are recounted in the sacred Book even more certain? Does not contradicting them mean abandoning God, who had the goodness to reveal them to us?"6 The question was bound to be asked. In 1749 Buffon was able to present his theory of the formation of the planets as a simple "hypothesis" and abandon it to theologians. Now he was telling a history and could not get out of it so easily. Moreover, as we have seen, the "Theory of the Earth" of 1749 did not permit asking the question of the age of the earth. The historical form of the new theory required asking the question, and Buffon attempted to answer it. Nevertheless, by bringing out only the difficulties that arose from chronology, Buffon allowed himself to leave others handily in the shadows. He would not have God intervene in either the formation of the earth and the planets or in the appearance of life. Better to attract attention to an objection he believed himself capable of dismissing.

We have every reason to think that for his defense he used here the "little written work" that he had sent to Mme. Necker in 1774, to reassure the worried conscience of his friend. Still he adopted a risky strategy; by making himself an exegete, he commented on the first verses of Genesis in his way, to show two things: that he could, according to that very text, assume a very long time between the creation of the world and its being put into order, and that the Bible "spoke only in a popular sense to the first men, who were still very ignorant." It was therefore inappropriate to seek scientific truths there. Buffon wanted to remain as orthodox as possible, and it was from a known exegete, the Benedictine Dom Calmet, that he borrowed the details of his argument, which did not reflect, we suspect, very deep biblical knowledge.[7] Let us not judge this commentary by modern criteria: today, any Catholic theologian would go much further than Buffon. Let us simply note that at least certain exegetes would not at the time have been too shocked by his argument, which was hardly revolutionary.

Still, theologians do not like laymen who come to teach them their profession under the pretext of being scholars. Galileo had suffered the consequences of this, and Buffon would suffer in turn—even more so, since unfortunate phrases escaped him. "I am afflicted each time someone abuses this great, this holy name of God; I am wounded each time man profanes it, and prostitutes the idea of the first Being, by substituting for it the phantom of his opinions. The more I have penetrated to the heart of Nature, the more I have admired and deeply respected her Author; yet a blind respect would be superstition: true religion, on the contrary, supposes an enlightened respect."[8] And he clearly let it be understood that it was necessary "to distance ourselves from the holy tradition . . . when *the letter kills,* that is to say, when it seems directly opposed to sane reason and the truth of the facts of Nature."[9] With a little bit more theological culture, Buffon could have relied on the authority of Saint Augustine, as Galileo had done. But that would probably not have succeeded for him any better than it had for Galileo.

Buffon therefore believed he had gotten rid of that "serious objection." To be more secure, he proclaimed his obedience in advance: "If this explanation, though simple and very clear, seems insufficient and even untimely to some minds too literally attached to the letter [of the Bible], I pray them to judge me by my intentions, and to consider my system in the *Epochs of Nature* as being purely hypothetical; it cannot harm revealed truths, which are so many immutable axioms independent of all hypothesis, and to which I have submitted and do submit my thoughts."[10] Perhaps he went a bit far there. In seeking to repeat the manoeuvre of 1753, he appeared to take the theologians of the Sorbonne as fools. In fact, they had learned to be careful.

The ground thus laid, Buffon could start his history. And since this was first of all the history of our earth, it was necessary to start at the moment when, with its sister planets, it had been formed from the incandescent mass of the sun.

The Birth of the Earth

Remember the hypothesis that Buffon offered in 1749: a comet had hit the sun obliquely and had torn off about $\frac{1}{650}$ of its matter. This "torrent" of liquid matter divided into spheres, each rotating rapidly. The centrifugal force created by this rotation had torn the material for their satellites from these spheres. The theory was, therefore, already there, and Buffon had nothing to add to it, except for his new explanation of the sun's' heat. The criticisms forthcoming in 1749 had apparently not moved him. The objections that could be made based on the laws of celestial mechanics, and which Buffon had discussed in 1749, here disappeared. What is more regrettable is that he did not seem up to date on the new ideas on comets. Lalande, as early as 1771, considered "that there is reason to believe that their substance has little density." Father Pingré, who was working on his *Cometography*, published in 1784, proved it even more precisely, but Buffon perhaps had no knowledge of his results. Be that as it may, the idea of a comet being dense enough to tear a part of its matter from the sun would quickly become unacceptable.

Although the theory remained the same, the viewpoint had changed: it was now a question of showing that the earth started by being a molten sphere, which must have immediately started to cool. This change in viewpoint risked bringing about a change in the theory's epistomological status. In 1749 Buffon could maintain that it was only a "hypothesis," which had nothing to do with the actual theory of the earth. Where did he stand now? If need be, it could be said that the explanation of the earth's *formation* remained a hypothesis, and that only its primitive molten state was a historical fact. This was perhaps what Buffon meant when he wrote: "It was . . . not that I affirmed or even positively claimed that our Earth and the Planets had necessarily and really been formed by the impact of a comet."[11] Or when he spoke of "this time that had preceded known time and was hidden from our view."[12] Given the narrow link that Buffon established between the planets' formation, the primitive molten state of the earth, and the continuation of its history, however, it was easy to be mistaken, and the "hypothesis" was generally considered a "theory." What changed everything was the existence of subterranean heat. This was the "vestige" that Leibniz lacked in order to make the link between the past and the present. Despite his assurances, Buffon did present this formation of the planets as the "First Epoch" of nature.

The First Moments of Cooling

Left to itself, the primitive earth was a globe of molten matter which turned rapidly on its axis. This rotational motion had two effects. First, the centrifugal force had removed the least dense matter, which formed the moon. Then it gave the earth the shape we know, that of a spheroid flattened at the poles. Progressively, the earth cooled and "solidified" to its center. This slow cooling produced the same effects on it that Buffon had observed in the forges on masses of metal when they cooled. "Holes, waves, wrinkles are formed on the surface of these masses; and below the surface, hollows, cavities, blisters are made." These "wrinkles" are very simply "primary mountains," of which "the highest . . . are in relation to the Earth's diameter only what one-eighth of a line is in relation to a diameter of a globe of two feet,"[13] or less than three-tenths of a millimeter for sixty-five centimeters. These "primitive mountains of the globe," the Andes cordillera, the high mountain range that runs the length of Africa from south to north, and the one that stretches from the Pyrenees toward the Alps, the Caucasus, Tibet, and China, were thus made up of "vitrifiable" rocks, and come from the very mass of the earth. They have existed since the beginning, and the peaks that we can still see today are therefore void of fossils. As for the "cavities" and "blisters," they are gigantic subterranean caverns, whose role we shall see later. And in this mass of cooled "glass," full of unevenness, cracks, and "blisters," the metals intermixed with the primitive mass are "sublimed" in vapors that were deposited in the crevices and made up the "first mines."

The globe of molten and then "consolidated" matter was from the beginning surrounded by an atmosphere made up of materials volatilized by the original heat. With the progression of cooling, "the chaos of the atmosphere started to separate: not only the waters, but all the volatile matters that the great heat had kept isolated and suspended, fell successively; they filled all the depths, covered all the plains, all the spaces that were found between the hills of the globe, and they even topped all those that were not excessively high."[14] Thus a "primitive ocean" was formed, from which at first only the peaks emerged, precisely where we do not find fossils.

"Let us try to envision for ourselves the prodigious effects that accompanied and followed this precipitous fall of volatile matter. . . . The separation of the element of air and the element of water, the shock of the winds and the floods that fell in vortices on a steaming earth; the purification of the atmosphere, which the Sun's rays could not penetrate before; this same atmosphere obscured by clouds of thick smoke . . . the continual boiling of alternatively fallen and condensed waters. . . . What motions, what storms must have preceded, accompanied, and followed the local establishment of each of these elements! . . . Agitated by the rapidity of their fall, by the action of the Moon on the atmosphere and on the waters already fallen, by the violence of the winds, etc.,"[15] these waters immediately started to attack the primitive mountains. "Glass slags" decomposed into clayey silt, immediately carried away by sea currents before depositing themselves on the bottom.

At the same time, the water eroded the vaults of underground caverns and rushed in, lowering the level of the primitive ocean. Life was already appearing in the warm waters, and shells started to accumulate at the bottom in banks of calcareous rocks which formed above banks of clay or sometimes between layers of clay. The emerged land became covered with plants, whose remains, carried by erosion, in their turn were deposited to form masses of coal. Among these materials, countless chemical combinations produced all the minerals we discover today. By attacking the "primitive mines" of metals and by transporting their debris, the waters created "secondary mines," which are found in sedimentary regions. Sea currents sculpted all these new lands, which as the water level progressively lowered, emerged and formed "secondary mountains." Little by little, the earth took on the surface that we know. Then the volcanoes appeared, born from accumulated combustible matter, which spontaneously ignited in contact with water. With the retreat of the primitive oceans, many were extinguished, like the volcanoes of Auvergne, from then on quite far from the sea. Each moment of this description was justified by a tight argument, based on geological, stratigraphical, geographical, and mineralogical considerations.

Such was, in its basic outline, the new theory of the earth that Buffon presented in 1778. As we can see, it drew upon many of the elements of the 1749 theory. Water and the phenomena of erosion, transport, and sedimentation continued to play important roles. Calcareous rocks were still of organic origin, and fossils were still the sign of the former presence of water. But these phenomena henceforth concerned only one part of the formation of mountains and rocks. Buffon introduced the distinction between "primitive mountains" and "secondary mountains," which had appeared among geologists of the middle of the century, a distinction that already assumed a chronological order. What disappeared, therefore, was the cyclic character of phenomena as described in 1749. It was no longer the present ocean that alternately covered the continents: instead, there was a "primitive ocean," which at first covered almost the entire earth and then slowly sank, allowing land to emerge, land that would never again be covered by water.

Linnaeus, we remember, had had the same idea, and Werner, at the very moment when Buffon published his *Epochs,* also defended the idea of a primitive ocean in which different mineral substances were found mixed. Nevertheless, this primary ocean posed a difficult problem, that of its progressive retreat, and the role of underground caverns was a somewhat ad hoc explanation. The true thread of this history was obviously the cooling, which imposed a linear chronology. It also raised the great problem of geological chronology; this was a problem that did not come up, as we have seen, in 1749.

The Age of the Earth

The biblical chronology was still the most often used to calculate the age of nature at the beginning of the eighteenth century. Without mentioning the difficulties that Chinese or Egyptian chronologies posed, the authors of theories of the earth found themselves in difficulty with such a short time line. Usually they evaded it only by recourse to cyclical theories. The boldest spoke vaguely of "millions of years," but their theories, by definition, did not allow them precise evaluations. Buffon had eluded the problem in 1749 by suggesting that the very slow geological phenomena he described could have occurred faster in the beginning. In the *Epochs of Nature*, he no longer had this escape, and to tell the truth, he could not hope to pass over the problem in silence, since it was the very idea of the progressive cooling of the earth that had inspired the work.

Recall the experiments on "the progress of heat in bodies," described at length in the "experimental part" of the first volume of the *Supplement*. As soon as he finished the first series of experiments on the cooling rates of iron balls of differing diameters, he extrapolated his results to calculate "with Newton how much time was necessary for a sphere as big as the Earth to cool" and obtained the following figures: 42,964 years and 221 days to touch the sphere without burning oneself; 96,670 years and 132 days for the sphere to arrive at "the point of its current temperature."[16] The apparent precision of the numbers came simply from the fact that they were the result of a calculation, and it would be immediately necessary to correct them to take into account the fact that the iron balls had cooled in the air, and that the earth had cooled in a vacuum, even if it was probable that the difference was "not considerable."[17]

But the earth is not an iron globe, and "in reality, the principal materials of which the earth is made, such as clays, sandstone, stones, etc., must cool in less time than iron."[18] From this came the second series of experiments that we have seen, on the comparative rates of cooling of all the substances that make up the earth. The following year, in volume II of the *Supplement,* Buffon presented other experiments, destined to measure the time necessary for the "consolidation" of different masses of iron and, by the same method of extrapolation, he obtained the figure of 1,342 years "for the consolidation . . . to the center" of a "sphere of iron as big as the Earth." This figure was too small if the heat of the earth's atmosphere at the beginning was taken into account. Buffon rounded it off to two or three thousand years: he thus obtained, for a sphere of iron, 100,000 years for the cooling to the current temperature, "without counting the duration of the first state of liquefaction, which further pushes back the limits of time, which seems to flee and grow as we seek to grasp it."[19]

Armed with these experimental results, Buffon finally attacked the calculation of the time of cooling for the present earth. He left out the metals, whose total mass was insignificant, and retained only "glass, sandstone, hard calcareous rocks, marbles, and ferruginous materials." He thus obtained the following figures: 2,905 years for the "consolidation"; 33,911 years before it was possible to touch it; 74,047 years to reach the current temperature. These were approximate numbers: each time Buffon added "around."

He was, nonetheless confident enough in his results to apply them immediately to the moon and the other planets of the solar system, for which he calculated the cooling times. He immediately encountered a difficulty: during its cooling, the earth was heated by the sun. In the beginning, it was even heated by the moon. The same situation existed for the moon, heated by the earth among other things, and for the other planets, in proportion to their distance from the sun, without forgetting that Jupiter was heated by its four satellites and Saturn by its five satellites and its ring at the time of their early incandescence. Buffon then embarked on one hundred fifty pages of interminable calculations, whose details he gave so that the reader could verify them—surely a rare temptation! The table recapitulates the results for the moon and the six planets. A "more exact Table" gives additional details. Let us simply retain the age of the earth: 74,832 years since its formation to its current temperature.[20]

Buffon focused his attention on one date in particular, that of "the birth of organized Nature on the globe of the Earth." He placed it "a few centuries" after the moment when the earth could be touched without burning oneself.[21] He knew that living organisms had been found in fountains of very hot water: therefore life could have started fairly quickly. What was true for the earth was also true for the other planets and their satellites. As soon as the temperature was or would be suitable, life appeared or would appear, as the laws of nature were universal.

Buffon brought his conclusions together in a table with an astonishing title: "Beginning, end, and duration of organized Nature on each Planet." We learn here that life disappeared from the moon a little over two thousand years ago, that it also had disappeared from Mars, but that it was currently "in full existence" on Mercury, Venus, Saturn, and a few satellites, which consequently "could be peopled like the terrestrial globe." On the other hand, Jupiter was still too hot and would have to wait 41,000 years for life to appear on its surface.[22] And, since the cooling continued, Buffon also estimated for each planet the time that life still had before it would be extinguished by the cold.

Let us come back to the earth. According to the *Supplement*, life had appeared here 35,983 years after its formation. In the *Epochs*, Buffon dated the establishment of the primary ocean "at thirty or thirty-five thousand years from the formation of the planets."[23] It was only at that moment, therefore, around forty thousand years ago, that sea mollusks could start to make their shells, which would form the calcareous rocks, and that plants, whose remains would form coal, started to grow on the emerged lands. In the chronology that Buffon had just established, all that must have taken place in twenty thousand years. All these phenomena were very slow. "This multiplication of plants and shells, as rapid as one can imagine it, could only have occurred over many centuries, since it produced such prodigious volumes of their remains. . . . How many centuries were necessary to produce all the calcareous matter of the globe's surface? And is one not forced to admit, not only centuries, but centuries of centuries, for these marine productions to be not only reduced to powder but also transported and deposited by water, in such a way as to be able to create chalks, marls, marbles, and calcareous stones? And how many more centuries need to be admitted for these same calcareous materials, newly deposited by water, to be purged of their superfluous moisture, then dried and hardened to the state in which they are today and have been for a long time?"[24]

It was at the very moment when he was writing the *Epochs* that Buffon became aware of the difficulty and immediately concluded that the chronology he had based on the speed of cooling was much too short. This can be seen thanks to a specific example, the rate of the sedimentation of clay. Observations made on the Normandy coasts allowed him to say that the deposit increased "by much less than five inches per year."[25] "In order to judge what happened, it is necessary to consider what is happening."[26] Even by assuming a more rapid rate of deposit, more than fourteen thousand years would be necessary for the formation of a clay hill one thousand yards in height. And above the clay, there is calcareous rock.[27] Twenty thousand years would not suffice.

Buffon had thus reevaluated his calculations, especially his estimation of "hidden causes" (here he borrowed Newton's expression "causae latentes") which delayed the "progress of heat" in larger spheres. He tried various hypotheses that gave him higher and higher results: from seven to eight hundred thousand years for the stabilization of the primitive ocean, and a million years for the current age of the earth. Then two and three million years, respectively, which multiplied the action of hidden causes by forty. This coefficient was perhaps too small: it was probably necessary to assign the earth a probable age of at least ten million years.

The manuscripts of the *Epochs of Nature* show traces of these hesitations.[28] The printed text returned to the figures Buffon had reached in 1775, and the earth once again was given an age of seventy-five thousand years. Why this timidity? It was not for fear of censure: for theologians, seventy-five thousand years was as scandalous as 10 million. It is necessary to seek the explanation in the fear of shocking the intellectual habits of his contemporaries. Buffon justified himself twice, both in the manuscript and in the printed copy. Is it not, one might ask, "adding a new cause of obscurity to the difficult things that you claim to explain, when such large numbers and inappropriate lengths of duration are used?"[29] "Is it easy, is it even possible to form an idea of all or of parts of such a long series of centuries?" And then he has to wonder, "Why does the human mind seem to lose itself in the length of the time rather than the expanse, or in the consideration of the weights or numbers? . . . Is it not that being accustomed to our short existence we consider one hundred years a long time, and have difficulties forming an idea of one thousand, cannot even imagine ten thousand years, or even conceive of one hundred thousand years?"[30]

Buffon did not want, therefore, to plunge his readers into the "dark abyss of time." "Although it be very true that the more we stretch time, the more we will near the truth and the reality of the use that Nature makes of it, it is nevertheless necessary to shorten it as much as possible to conform to the limits of our intelligence," he wrote in the manuscript.[31] In print, he was content to say, "Instead of pushing the limits of time back too far, I brought them forward as much as it was possible for me to do without obviously contradicting the facts recorded in the archives of Nature."[32]

It is not certain that when we today assign an age of 4.5 billion years to the earth, we have a clear idea of what that represents, and geologists or paleontologists sometimes have difficulty sharing their sense of time with laymen. In Buffon's time, biblical chronology had largely lost its influence, and geologists spoke freely of "long series of centuries." Buffon's originality lies in his refusing to remain vague and in attempting an estimate of the age of the earth based solely on scientific foundations. It is useless to underline the weaknesses of his method, and he was sufficiently conscious of them to challenge his first results, which relied only on the rate of cooling. Perhaps the most interesting thing is to see why he challenged them.

Geological arguments required him to discuss the results derived from a physical theory, and it is here of little importance that these results had little value. What deserves to be noted is that eighty years later the same debate arose again, involving two scientific personalities of the first importance: Darwin and Lord Kelvin, one of the greatest physicists of the time. In the interval, geologists had practically abandoned the hope of evaluating the age of the earth and were satisfied to hold that "geological times" were immense. With Lyell's *Principles of Geology,* whose first volume appeared in 1830, this immensity became practically infinite. Darwin, although a disciple of Lyell, needed a little more precision to defend his theory of evolution, which also required a very long time. In 1859, in the first edition of the *Origin of Species,* he stressed the slowness of the phenomena of erosion and sedimentation and estimated the time necessary for the erosion of a chalk hill in the Weald, a region in the south of England, at 300 million years.[33] Lord Kelvin answered this assertion by saying that the entire solar system could not have lasted that long. Without insisting, Darwin removed the passage from subsequent editions of his book.

The basis of Kelvin's calculations was precisely the rate of cooling, no longer of the earth alone, but of the sun itself. For him, solar heat came from the initial heat produced by the contraction of a primary nebula, according to Laplace's hypothesis. Armed with all the knowledge of thermodynamics, Fourier's mathematics, and numerous observations made since the beginning of the century, he concluded that the age of the earth could be evaluated at 98 million years. To be cautious, he accepted an estimate between 20 and 400 million years. In fact, it was necessary to wait for the beginning of our century and the discovery of radioactivity as a still-present source of heat to put an end to the debate.[34] What interests us here is to see that Buffon had understood the two sides of the problem. If publicly he preferred calculations based on physics, it can be believed that deep down he judged the requirements of geology better justified.

The Birth and History of Life

Let us come back to the primitive ocean at the moment when life appeared. Buffon was persuaded that life could appear in very hot water, but how? He did not explain it anywhere with precision. We know that he did not give up his theory of organic molecules. On the contrary, he soon gave them an additional role by making them the agents of the "representation" of minerals that come from organic substances.[35] Up to this time, organic molecules were for him a matter of fact. It was, however, possible to understand that they had appeared spontaneously. Here, Buffon indicated more specifically, although very rapidly, that they were the results of chemical combinations: "and since organic molecules are only produced by heat on their malleable matters, these were as well."[36] It was up to the reader to pick up this last phrase and draw the necessary conclusions.

With the existence of organic molecules the history of life started. In the historical account that made up the *Epochs*, we can hope to find a development of the ideas expressed in the discourse "On the Degeneration of Animals" as well as responses to questions that had not been answered until then. In particular, what was the origin of the "prototype" that served as a starting point? And what was the origin of the "interior mold" for those biological species within which degeneration had introduced so many varieties that it was necessary to consider them as forming "genera"? And how far would Buffon push his thoughts on "degeneration" and its effects at this time?

To our great surprise, we do not find much more said here on the main themes discussed in the last volumes of the *History of Quadrupeds*. It seems Buffon was completely a prisoner of the logic of a new system. He seemed to forget the results of his former ideas and moved on to new problems, which he resolved with varying success. What now seemed to dominate his thoughts was paleontology, or more specifically the question of the bones and teeth of large quadrupeds—mammoths, hippopotamuses, or rhinoceroses—that were discovered in Siberia and North America.

Organic molecules thus were produced by chemical combinations at the heart of the primitive ocean, which foreshadowed the "primary soup" of Haldane and even more so the primitive ocean of Oparin, both biologists of our century.* The elevated temperature and the abundance of "oily and ductile matter" permitted this production, which was no longer possible when the temperature fell, and that is why no more organic molecules are produced. But for them to form, the primitive ocean had to be cooled a bit. Where did this first happen? In the northern regions, and more specifically in the northeast of Siberia. It was there that life appeared; it was there that organic molecules spontaneously combined to give birth to complex organisms, both plant and animal.

"Nature was then in its first force, and it worked upon organic and living matter with a more active power in a warmer temperature."[37] These first living organisms had therefore not been simple and rudimentary beings, but, on the contrary, beings of very large size: giant ammonites in the seas, elephants and hippopotamuses much larger than their "current analogues." Buffon admitted without difficulty, in a Lucretian manner, that the combination of organic molecules had immediately produced very complex organisms. He did not seem to have been touched by the idea, which today seems obvious to us, that the first "productions of Nature" were necessarily very simple organisms, which became progressively more complex with time. This was Lamarck's idea later on, but Lamarck was an evolutionist and Buffon was not. Even so, this spontaneous formation of mollusks and giant mammals seems particularly archaic.

As the cooling continued, these first giant species had to leave their birthplace, since the lands to the north cooled earlier than the others, and descend toward warmer regions. Thus their remains are found at relatively

---

*The modernity of Buffon's ideas on the origin of life becomes curiously striking with the reading of the following text, published by the Soviet biologist Oparin in 1936: "As the temperature of the Earth had cooled off sufficiently to permit the formation of droplets of liquid water, torrents of boiling water must have poured down upon the earth's surface and flooded it, thus forming the primitive ebullient oceans. The oxygen and nitrogen derivatives of the hydrocarbons already present in the atmosphere were carried down by these torrents of rains and the oceans and sea, at the moment of the first formation, contained, therefore, the simplest organic compounds in solution" (Oparin 1953, pp. 108–9).

low latitudes. Yet even there, the heat became insufficient and these species disappeared. The disappearance of these creatures, however, left the organic molecules without any function, and they recombined in the north to give birth to a new category of species, adapted to a lower temperature. Fleeing, in their turn, from the cold, these beings, animal or vegetable, migrated southward. They were then found in tropical regions, until the time that the cooling condemned them in turn. A third generation, appearing in the same way, started its migration toward the south and is today found in the temperate regions. A fourth and last generation still lives in the lands of the north. They will leave one day also. "And who knows if with the passage of time, when the Earth will be cooler, new species will not appear, whose constitution will differ from that of the reindeer as much as the nature of the reindeer differs in that respect from that of the elephant?"[38]

This new vision of the history of life reinforced or perhaps brought to light a posteriori an old idea, that of the "land of origin" and the perfect adaptation of living beings to the climate in which they were born. Here the climate produced beings directly, that is, it directed the combinations of the organic molecules, and "climate," previously a quite complex idea, was now reduced to temperature alone. In 1775 Buffon expressed this idea strongly: "In all the places where the temperature is the same, one finds not only the same species of plants, the same species of insects, the same species of reptiles without their having been brought there, but also the same species of fish, the same species of quadrupeds, the same species of birds without their having gone there."[39] From this he drew the argument that the same temperature must have created the same species on other planets and their satellites as on earth. The relationship between temperature and living forms thus became a universal law of nature that was applicable to the entire solar system and even beyond.

All the animal or plant species were thus born in the lands of the north, where there was an abundance of free organic molecules, and had undertaken or would undertake a migration toward the south, pursuing the heat that eluded them. How then to explain the originality of South American fauna, a fact that Buffon was particularly proud of having discovered? The new system permitted a new explanation: species born in the north descended toward the south in the New World as in the Old, since at that time they were not separated. In America, however, animals and plants encountered the isthmus of Panama. At that point there was an impassible barrier because of the cold of the mountain peaks. South America was thus protected, so to speak, from these invasions and, "reduced to its own forces, gave birth only to weaker and much smaller animals than those which came from the North to populate our regions of the South."[40] This means that the "malleable matters" and organic molecules appeared later and at a lower temperature: nature there was "born late and had never existed there with the same force, the same active power as in the northern lands."[41] And if these animals, although smaller, resembled those of the Old World, it was because the laws that presided over the combination of organic molecules were the same everywhere.

This new explanation should have rid Buffon of the vehement criticisms directed at him by the young North American scientific community, vigorously represented in Paris by Benjamin Franklin and Thomas Jefferson, with whom Buffon remained in frequent and friendly contact. Remember that in 1761, comparing animals of the Old and the New Worlds, Buffon had stated that America, in its entirety, was colder and wetter than the old continent, and that the animals there were smaller than the species of the Old World, whether they be native or "degenerated." The new theory reserved only for South America the sad privilege of possessing these shrunken species: North America was now exempted from this flaw.

The American intellectual elite knew Buffon's works well, but it seems the *Epochs of Nature* was not read there with sufficient attention, since Thomas Jefferson, arriving in Paris in 1784 as the ambassador of the young United States, went to much trouble to convince Buffon of his former error. Together with Franklin, he tried to show that at equal latitudes America was neither colder nor wetter than Europe, and he had his friends send him hides of the largest American animals, in particular moose and bison, which he gave to the Royal Botanical Garden. In his *Notes on Virginia,* published in 1785, he gave detailed tables of comparisons of the weights and sizes of American and European species. He even hoped to find live mammoths in the wilderness of the West and made inquiries among the Indians, but with no success. In reality, and we shall return to this, it was the American man he was defending.[42]

This debate aside, it is easy to see what Buffon gained with his new method of explanation: the possibility of accounting for the gigantic fossilized bones found in the cold and temperate regions of the two continents, which seem to have captured his imagination. By introducing a correct idea, that these regions had previously known a warmer climate, he reached an elegant solution to a difficult problem. He did not, however, convince everyone, and several of his critics preferred to stay with Hannibal's elephants or that great vanished river that could have carried these bones from India. The strange thing is that Buffon arrived at a correct conclusion by starting with an error, one that Cuvier quickly pointed out: the mammoth, which Buffon wrongly identified with the elephant, and the "rhinoceros of Siberia were covered with thick fur and could withstand the cold."[43]

What disappeared in this new history of life is also apparent: it is the notion of "degeneration," which was no longer referred to here except in passing[44] and which no longer played a real role. Now, species did not change climates: they followed the one in which they were born in its slow displacement toward the equator. When the heat was no longer strong enough, they did not "degenerate," they disappeared. All the wealth of the analyses to which Buffon had paid so much attention in the *History of Quadrupeds,* all his ideas on the relationship of species and their evolution disappeared at the same time. The history of life that Buffon had sketched out earlier depended on geographical and climatic conditions, but it kept its autonomy in a physical environment that had, so to speak, no history. Here, the history of life was completely subordinate to that of the earth. This undoubtedly represented a unification of Buffon's thoughts[45] but it was also a terrible impoverishment.

More attentive to the determinism of nature's great laws, which governed the appearance of living forms as they governed everything else, and more preoccupied with the broad principles that had always dominated his conception of life, Buffon seems to have become less of a "naturalist" and to have forgotten all his observations on the diversity of animal forms. His "dear minerals" had made him lose the memory of the "quadrupeds" to which he had dedicated so many years of work.

The History of Man

What became of man in this new history of nature? It would be unthinkable not to have included him. But the logic of his new system required Buffon to ask at this point a difficult question, one that he had always carefully avoided, with regard to man as well as other species: when and how did man appear? This amounts to asking if man had the same origin as other animals or if he benefited from some special treatment. Buffon's answers were so unorthodox that they barely appear in the printed text. Fortunately, important parts of the manuscripts have been preserved; here is where we find Buffon's true thoughts, in the same place that we found his deep convictions on the age of the earth.

In the "First Discourse" of the *Epochs,* at the end of his commentary on Genesis, Buffon seemed ready to admit that man had been created only six or eight thousand years ago, "as the different genealogies of the human family do not indicate more."[46] Elsewhere, he was satisfied with saying, "We are persuaded; independent of the sacred Books, that man was created last." This reassuring statement, however, came at the end of a long discussion, in which Buffon presented all the arguments that might show that "the human species followed the same path and dates from the same time as the other species." To this opinion founded on "analogy," Buffon opposed the superiority of the human species and the fact that although humans knew how to defend themselves from the cold, they were impotent against great heat. He concluded, therefore, that God "had wanted to give all the time necessary for the Earth to consolidate, shape, cool, emerge, dry up, and finally reach that state of rest and tranquility where man could be the intelligent witness, the peaceful admirer, of the great spectacle of Nature and the marvels of Creation."[47]

The manuscript record, however, was in favor of "analogy": "Analogy, vestiges, and even traditions prove to us . . . that the human species followed the same path and dates from the same time as other species." Thus "without wanting to affirm it, we therefore presume that according to our hypothesis, the human species is as old as that of the Elephants; that being able to stand the same degree of heat and maybe an even stronger one, man will have been the first to penetrate southern regions."[48] The statement remained cautious, for there were "specific reasons" that "seemed to make it necessary to say" that man had been created at the end of this entire history. In fact, Buffon had already betrayed himself: in 1775, at the end of volume II of the *Supplement*, speaking of the degree of heat that living beings could stand, he had concluded with this surprising sentence: since certain species can stand strong heat, "and since Negroes are those in the human genus that are bothered least by the heat, should not one reasonably conclude that according to our hypothesis their race could be older than that of white men?"[49]

This bold conclusion is not found in the *Epochs of Nature*, but one sees that Buffon now accepted the existence of "races of giants so common earlier in Asia," and of which "that of the Patagonians is the only one left."[50] Buffon, it will be recalled, had refused to believe in these giant Patagonians. Now they became the equivalent in the human species of the gigantic elephants at the beginning of life.

All the same, Buffon did not forget that man is not an animal like the others. For him, "matter is led by the mind," and the manuscript went further than the printed copy, for it explained that man, having invented fire, clothes, and weapons, "soon found himself the master of the Earth's domain."[51] At least this is what the "Fifth Epoch" says, and we have no reason to doubt Buffon's sincerity.

We are thus surprised by the picture of these first men that he presented at the beginning of the "Seventh Epoch"—"witnesses of the Earth's convulsive movements, still recent and very frequent . . . trembling on an earth which trembled under their feet, naked in mind and body, exposed to the assaults of all the elements, victims of the fury of ferocious animals, to which they inevitably succumbed as prey," and keeping "a lasting and almost eternal memory of these misfortunes of the world." A memory that had inspired and continues to inspire so many myths and superstitions.[52] We are even more surprised to see that barely a few centuries later a "happy people" emerged "in the center of the continent of Asia," a people who, for perhaps three thousand years, cultivated the arts and the sciences, perfected astronomy, and rejoiced in peace and posterity. A people, alas! who succumbed to barbarous invasions, and whose knowledge darkened into oblivion, without leaving other traces than "practical methods" for the calculation of eclipses, which were not even understood by those who used them.[53]

In these two pictures, which so clearly contradict each other, it is necessary to see the influence of two recent works. The first was *Anecdotes de la Nature* [Anecdotes of Nature] by Nicolas-Antoine Boulanger. An inspector of bridges and roads, Boulanger died at the age of thirty-seven in 1759.[54] He had written two articles for the *Encyclopédie* and left several long manuscripts, which d'Holbach published, although not without rewriting them: *Recherches sur l'origine du despotism oriental* [Research on the Origin of Oriental Despotism] (1761) and *L'Antiquité dévoilée par ses usages* [Antiquity Unveiled through Its Customs] (1766). He also left a manuscript, which was long considered lost, the *Anecdotes of Nature;* it circulated and finally fell into Buffon's hands, who borrowed from it, notably in the *Epochs*—possibly the very idea of his commentary on Genesis and certainly his description of the mountain of Langres and the map that accompanied it.* Boulanger was, however, not interested in the history of the earth for itself. He was chiefly seeking to explain the traditions and myths of ancient man, including the biblical text, by means of the terrified memory that men had kept of the grandiose natural phenomena of which he had been the witness. It was from him that Buffon borrowed his image of the first men, "trembling on an earth which trembled under their feet," an image that hardly corresponded to the idea that Buffon had of man, and which very quickly disappeared from the text.

The second work was the *Histoire de l'astronomie ancienne* [History of Ancient Astronomy], which Jean-Sylvain Bailly had published in 1775 and which was followed two years later by his *Lettres sur l'origine des sciences et sur celle des peuples de l'Asie* [Letters on the Origin of the Sciences and of the Peoples of Asia]. We have seen that Buffon supported Bailly in his various academic candidatures. Bailly himself bragged about Buffon's favorable reception of his ideas. Indeed, starting in 1758 Buffon had spoken of the "centuries of light" in the first ages of humanity, which were followed by a "revolution of darkness."[55] Bailly was also inspired by Court de Gébelin and an entire contemporary movement that sought the memory of a primitive revelation in the ancient traditions.[56] This preoccupation was as foreign to

---

*Ed. note: Nicolas-Antoine Boulanger (1722–1759) was a French engineer who carefully examined the strata in northeastern France and concluded that they had been formed by the remains of marine shelled animals. His map of these strata is the map referred to.

Buffon as that of Boulanger, but Bailly brought him confirmation of an idea he had held for some time.

The last "Epoch" was more oriented to the future than to the past, and it was during this epoch that the history of humanity resumed, after the "revolution of darkness," with the slow conquest of nature by man. Here again, the theme of "degeneration" disappeared, as well as the "varieties of the human species." The pitiful portrait of the American man that Buffon had given in 1761 in his discourse "Animals Common to both Continents" also disappeared. This discourse, however, had left its mark. Cornelius de Paw, in his *Recherches sur les Américains* [Research on the Americans], published in 1768, and the abbé Raynal in his *Histoire philosophique et politique des deux Indes* [Philosophical and Political History of the Two Indies], published in 1770, took up the theme of a "degeneration" of the Indians, victims of a hostile physical environment whose action could even affect the European populations that had emigrated to the New World.[57]

Here again, one can understand the violent reaction of the American colonies against a European prejudice of which they felt themselves personally the victims. When they crossed the Atlantic, they discovered, in London as in Paris, that their presence was enough to turn a conversation to this theme. As they saw it, Buffon was responsible for this. Here again, Jefferson made himself the strong advocate not only of the colonists settled in the New World but especially of the Indians of North America; he lauded their courage, moral and family virtues, loyalty in friendship, intelligence, and even eloquence, which he favorably compared to that of Cicero and Demosthenes. The differences that could be noticed between them and the *Homo sapiens Europaeus* came from those of life style but did not imply any innate inferiority.[58] Jefferson attacked Buffon because he admired him and considered him an adversary worthy of being discussed and someone he needed to convince. Here again, he did not seem to notice that Buffon had taken himself out of the discussion, as he had done with the North American fauna. It is true that he did so only through his silence and that he had not publicly renounced ideas that were perhaps still his.

To repeat, the "Seventh Epoch of Nature" was more oriented to the future than to the past. Man's superiority over animals and his power over nature are familiar themes to which we need not return. Now a new question arose: was man capable, if not of stopping, at least of slowing down the inexorable cooling that condemned all life to a distant but certain death? Buffon thought so. He was even persuaded that he knew how: "Cleansing, clearing, and populating a country gives it heat for several thousand years." The proof is that "Paris and Quebec are at about the same latitude: Paris would therefore be as cold as Quebec if France and all the regions surrounding it were as lacking in men, as covered with woods, and as bathed by waters as are the neighboring lands of Canada."[59] The further proof was that in Cayenne it was less cold in the cleared lands than in the forest.[60] Man, who could tame wild animals, create wheat, breed new flowers and fruits, who had known how to perfect nature, could still fight to survive. But would he understand this?

For it was himself that he had to change first. "And what could he not do for himself, I mean for his own species, if the will were always directed by the intelligence? Who knows to what point man might perfect either his mental or his physical nature?"[61] "Six hundred centuries were necessary for Nature to construct her great works, to cool the earth, to fashion its surface and arrive at a tranquil state; how many will be necessary for men to arrive at the same point and stop annoying, agitating, and destroying one another? When will they realize that the peaceful enjoyment of their native lands is enough for their happiness?"[62] Spain, even England, "this so sensible, so deeply thinking people," wanted to have colonies. To what end? The war of independence in the United States was in full swing. Buffon was against France's intervention, for he thought correctly that she would meet her ruin there.

One must go even further: "Is there a single nation that can pride itself at having obtained the best possible government, one that would be to make all men, if not equally happy, at least less unequally unhappy, while watching over their preservation, the sparing of their sweat and their blood through peace, through the abundance of food, through the ease of life and reproduction: this should be the moral goal of every society that is seeking to improve itself." Buffon did not propose a political program, and the prospect of a revolution hardly appealed to him: "I see a large movement coming, and no one to lead it," he would later say. He was satisfied to ask for less injustice, and especially peace and prosperity. The moment had perhaps come. It was with this hope that a work on the history of nature ended: man has admired for too long "great talents in the art of doing evil" and has let himself be captivated for a long time by "those who have amused him." But finally, and this is the book's last sentence, "after an overly long use of these two means of false honor and sterile pleasure . . . he has recognized that his true glory is science and peace his true happiness."[63] The message would barely be heard.

"One of the Most Sublime Novels, One of the Most Beautiful Poems . . ."

The publication of the *Epochs* was long awaited. Everyone knew that Buffon had worked on it for a long time, and manuscript copies were already circulating. Even though it bore the date 1778, the work, which made up the main part of volume V of the *Supplement*, was not for sale until

April 10, 1779.[64] Meister, who succeeded Grimm in the *Literary Correspondence,* immediately gave an account of it, but with caution: "If the system established in this work does not seem to be equally solid to all its readers, it will at least be admitted that it is one of the most sublime novels, one of the most beautiful poems that philosophy has ever dared to imagine."[65] Curious praise for a scientific work. The weak point of the system was the theory of the formation of the planets. Everyone would agree on this.

The same caution appeared in the *Journal encyclopédique* [Encyclopedic Journal] and the *Journal des Savants,* where the astronomer Lalande had the difficult task of reconciling the respect due the author and skepticism with regard to the system. It is useless to deal here with all the reviews that the work provoked[66]: suffice it to say that all the scientific commentaries showed the same caution, not only with respect to "cosmogony" but for the system in general. They explicitly or implicitly reproached Buffon for having wanted to describe nature's "general laws" without paying enough attention to the "specific laws." It was these "specific laws" that the science of the time sought patiently to elucidate. Buffon's science already belonged to an obsolescent era in the history of thought. The breadth of the views and the generality of the design of the work often hid the numerous precise remarks that it contained from the reader of the time. Only later would they be discovered and their value appreciated: the theory of "lost species," for example, was one of the most criticized of the time.

The liveliest attacks came from the defenders of religion. The abbé Grosier, the editor of the *Journal de littérature, des sciences et des arts* [Journal of Literature, the Sciences, and the Arts], who defended the Flood, took a malicious pleasure in publishing a letter that denounced Buffon's borrowings from Boulanger's manuscript. The abbé Royau denounced the work to the Sorbonne and started to attack it in *L'Année littéraire* [The Literary Year]. The Sorbonne became involved with the book in November 1779.[67] The theologians remained cautious: Buffon was famous, he had often declared his respect for the Creator in his works, and he had not participated in the *Encyclopédie.* Still, it was necessary to insist that "his general principles on the manner of understanding Scripture" were not acceptable, and that his "different epochs . . . had no relationship to the different days of creation, neither for the order of time nor for the circumstances of the facts."[68] Buffon, who had at first gone to Montbard to let the storm pass and had ordered his friends not to answer Royau, contacted the Sorbonne to ask that its criticisms be communicated to him. After that he asked Bexon to prepare an answer.

Bexon did so with the utmost ingenuity, but without result. Finally, Buffon signed a new retraction prepared by the Sorbonne, which very closely resembled the one he had signed in 1751. Again, he promised to publish it at the beginning of his next volume.[69] This time he did not keep the promise. The Faculty, triumphant, published all the documents on the entire business in 1780 in the form of a Latin brochure. There was no formal censure. It was said that the minister Amelot and the king himself had intervened with the Sorbonne. Buffon's errors, nevertheless, were publicly revealed and his retraction published.

Astonishingly, no one paid attention to it, and the brochure was forgotten for a long time by all but a few specialists, before a modern historian looked into the problem. In fact, these theological debates did not interest anyone anymore in 1780, and the defenders of orthodoxy who attacked the *Epochs*—the abbé Barruel and the abbé de Feller, whose works, printed in Luxemburg, were forbidden in France—based the main points of their attacks on the scientific weaknesses of the work: they knew that more people would listen to them on these grounds. As for Buffon himself, he apparently said to Hérault de Séchelles in 1785, "When the Sorbonne picked petty quarrels with me, I had no difficulty in giving it all the satisfaction that it could desire: it was only a mockery, but men were foolish enough to be contented with it."[70] He probably did say this. Did not Buffon write to Guéneau de Montbeillard on November 15, 1779, "I do not think that this affair will have any other regrettable consequence than that of causing a stir and perhaps requiring me to give the same foolish and absurd explanation that they made me sign thirty years ago"?[71] As for the "ignorant scholar whom prejudice blinds," and "those Writers whose only claim to notice is opposing systems," Buffon settled his score with them, we recall, in the *History of Minerals* in 1783.

Buffon consoled himself for their criticisms with the praises and gifts received from Catherine II. Enchanted to see that Buffon's theory made her empire the cradle of life and humanity, the empress, always aware of events in Paris through Grimm, sent Buffon a collection of sumptuous furs and all the medals of her reign, in gold, which he displayed at the Royal Botanical Garden and which the curious evaluated at forty thousand livres. She also sent him her portrait on a gold snuffbox encrusted with diamonds and an even rarer piece, a gold chain found in excavations in Siberia, the probable remains of the contented and enlightened people who, according to Buffon, had lived at an earlier time in these regions. The philosophes were jealous, as Marmontel told us earlier, and Buffon thanked the empress with a lovely letter, in which he wished her full success in her undertakings against the Ottoman empire, "that stagnating part of Europe," and expressed the desire to see "beautiful nature and the arts descend a second time from the North to the South under the standard of its powerful genius."[72] And he sent his son to give her his bust by Houdon.

The *Epochs of Nature* contradicted the scientific spirit of the time too openly to be taken seriously. Gradually, however, Buffon's main ideas had a measurable effect on people's minds. His cosmogony was universally crit-

icized, but in 1795 Laplace presented a new one, which was more respectful of celestial mechanics but no less a cosmogony for all that. In 1812
Cuvier published his *Discours sur les révolutions du globe* [Discourse on the
Revolutions of the Globe], which obviously was intended as a revised and
corrected version of the *Epochs of Nature*. We can probably no longer say
today what Flourens wrote in 1860: "All the efforts that our current geologists make and all that future geologists will ever make . . . will never be
more than a constant reworking and endless perfecting of the *Epochs of
Nature*."[73] But it is true that Buffon had asked questions and opened paths
that would be those of the nineteenth century.

The influence of the model proposed by Buffon in the *Epochs of Nature*
was not restricted to the natural sciences alone. The end of the eighteenth
century was a moment when the history of humanity and the philosophy of
history interested intellectuals more and more, particularly in Germany
but also in England and Scotland. For the majority of these thinkers, the
history of humanity necessarily followed the history of nature and posed
the same problems of method. History was no longer concerned with
studying wars or political conflicts but rather with rediscovering the course
of all of humanity by revealing the forces at work and the slow processes of
transformation. It has recently been shown[74] how Herder in Germany or
Ferguson in Scotland transposed the spirit of Buffon's method—as elaborated and perfected by naturalists like Blumenbach and Sömmering—in
order to adapt it to the needs of a new history, whose development would
be one of the most important aspects of the nineteenth century in Europe.
Thus in his last works, Buffon had at the same time been a witness to and
an actor in the deep intellectual transformation that marked the passage
from the eighteenth to the nineteenth century.

The End of a World

In the beginning of April 1788, Buffon had a new attack, more serious than the preceding ones. The kidney stones were making him suffer terribly, and it was feared he would not live much longer. He himself had no illusions. He wanted to see the Garden that had been his kingdom one last time. Under the timid spring sun he was able to take a few steps there, leaning on two servants and covered with the furs from Catherine II. Soon he was incapable of getting up. He refused all remedy and all food: "I feel myself dying," he said. For the first time, he surrendered. His son had come back to him. Mlle. Blesseau did not leave him, and Mme. Necker spent her days at his bedside.

"When I become dangerously ill and I feel my end approaching, I will not hesitate to send for the sacraments. One owes it to the public cult," he apparently said to Hérault de Séchelles.[1] On April 11, Father Ignace arrived at Montbard. The burlesque Capuchin ridiculed by Hérault de Séchelles had regained all the dignity of his ministry. He heard the invalid's confession. On April 15, as Buffon suffered and weakened increasingly, he gave him extreme unction and had communion brought to him.

Mme. Necker, who was present, described his last moments. "He spoke to Father Ignace and said to him in a very anxious manner, 'Someone give me the good God quickly! Quickly! Quickly!' But the *porte-Dieu** had not arrived; the invalid intensified his request, even adding a certain impatience; finally Father Ignace gave him communion and M. de Buffon repeated during the ceremony, 'Give it then! But give it then!' This terrible spasm of death was then partly calmed." A few hours later, "his pulse gradually diminished, his mouth remained open, his extremities cooled, he squeezed

---

*Ed. note: The priest who brings the viaticum (the Eucharist).

the hand of Mlle. Blesseau several times; his breathing became almost imperceptible and at forty minutes past midnight, he gave his last sign."[2] It was April 16, 1788.

Of what was he thinking, this dying man who received the last sacraments of his Church? We will never know: the last moments of a human life concern God only. What had he believed during his lifetime? Was he a Christian, a Deist, an atheist? This has been much debated.[3] There are a hundred ways to believe or not believe. Is it legitimate to define a religious thought or sensibility as if it were not primarily the expression of an intellectual temperament? In the same way, is it possible to study a "scholar" or a "writer" as if he were not the same man who thinks, who seeks, who writes? Such examinations are useful, and we have already made them.[4] Here, it is the deep forces of a personality and a temperament that we would like to reach before taking leave of an extraordinary man. This is a riskier attempt, since it entails finding a unity that perhaps did not really exist.

The first difficulty is remembering that the man whose life was ending had not always been the majestic elderly man of 1788, nor even the man of fifty years, sure of himself and of his glory, whose portrait Drouais has left us. Houdon's beautiful busts show us a Buffon of more than sixty years of age, and we risk forgetting the young man, his appetites and his ambitions, his strength of character as well, when he launched himself on a career whose success no one could have imagined. Although we are looking for something that could show the unity of his personality and his work, we cannot forget that both evolved over time. Between his friendship with the duke of Kingston and his fervent love for Mme. Necker, many things changed.

What strikes us first with Buffon is his attachment to reality, to what could be touched and felt, to what gave the immediate feeling to the body and the mind of being here and the possibility of enjoying it: the body of a woman, the earth on which we walk and which we feel beneath our feet, the food we eat, the houses in which we live. He bought lands, but close to home. He liked to build for himself or for the Royal Botanical Garden, because buildings are tangible things. And reality for him was primarily that on which he could act. Managing a forest, "making water run," overseeing lumberjacks or unskilled laborers—these were ways of transforming reality, harnessing it, mastering it. Knowledge and possession had the same goal: action.

But action supposes knowledge. It is reality that it is necessary to know. Even though the sense of sight is the "sense of knowledge," it is touch that assures us of the reality of things. If we cannot really imagine millions of years, Buffon tells us, that is because years are not like gold pieces that can be touched. He had not known how to refute Berkeley and his "immaterialism," but he never took him seriously. Matter is there. Dead or living, mineral or organic molecule, matter is what counts, it is in matter and through matter that everything is accomplished. The logic of mathematical abstractions or classifications exists only in the human mind. It does not account for reality. In this sense, Buffon was deeply materialist, through instinct rather than through doctrine. He was the heir of those generations of country folk who had turned the earth and reaped wheat to make bread. Like them, he battled against things, patiently, unwearingly.

In this daily victory, in this hard-won mastery of the elements and forces of nature, he placed all of man's grandeur and dignity. Yet despite all the power he gave to human reason, when it concerned understanding it he was more interested in its physiological roots than in its metaphysical nature. It seems doubtful to me that he believed in the "spirituality" of the soul in the sense of Christian theology or in the existence of the soul as a distinct "substance" in Descartes's meaning. Beginning with the letters of his youth that he wrote to Dutour and ending with his letters to Mme. Necker, taking into account all the analyses of the *Natural History,* everything leads us to think that he held reason, that "divine ray," to be the highest expression of physiological organization. In any case, reason was not possible without either matter or organization. At least on this point, it seems difficult to believe that Buffon was as Christian as was necessary at that time.

It is not enough to touch reality to be sure that it is there. One must try to understand it. Reality for Buffon was at the same time opaque and multiple. It was opaque, first, because we know only matter's exterior such as our senses reveal it to us, and it is in the interior, at the heart of matter, that important things happen. "In the interior" means in the density of the mass and according to the logic of nature, who "does not take a single step that is not in all directions," whereas the logic of our mind is linear. This is why Newtonian gravity was a sort of revelation for Buffon. The force of attraction penetrates to the core of matter; it acts on the mass, and not on the surfaces like the laws of traditional mechanics. Because of it, Buffon felt himself correct in suggesting "penetrating forces" that permit the assimilation of the organic molecules, and he was again following Newton when he studied heat, that other "penetrating force," and its "progress in bodies." We will never grasp these active forces in the intimacy of their action or even conceive of their nature. But we can at least imagine them and, in certain cases, measure their effects.

Reality was also multiple, and that is the lesson of the *Natural History.* Here we are dealing with forms, and therefore it was first necessary to describe them. Buffon wrote hundreds of pages, as patient, as technical, as dry, and as meticulous as the description of the experiments on heat and the infinite calculations that follow it. To see in Buffon a "pompous" writer, the "great phrase maker" of whom d'Alembert spoke with disdain, it is necessary to ignore at least three-quarters of the *Natural History*. Aside from a few "borrowed ornaments" that decorate the descriptions of birds, he does not seek after effect. The vocabulary is one of absolute precision: no one would claim that "redistillation," "testa," or "scoria" are noble words. In reading the text, this becomes obvious. Still, it is necessary to read it. And Buffon never considered himself exempt from writing these interminable pages, simply out of respect for this reality that he wanted to understand.

For a physicist or a philosopher of science today, this conception of reality as a raw and immediate given is naive, even "prescientific," as Gaston Bachelard has said. It probably is, and Buffon's physics did not follow the slow evolution of science that started to make itself felt only at that time. His realism, which warned him against the use of mathematics in his physical theories, prevented him from seeing how physics could create its own subject and finally reduce nature to a game of equations. But this realism no doubt helped him in his work as a naturalist.

As a naturalist he studied a reality that might have seemed irreducible: the multiplicity of living forms. Strictly speaking, the physiologist can isolate a problem, prepare an experiment, and create a laboratory "reality." The naturalist does not have this possibility: all beings are present in their own complexity and in the complexity of the relationships that link them to their physical environment and to other living beings. Description was essential, but it was only the first step of research. To collect animal or mineral species was not an end in itself.

More precisely, Buffon refused to consider these forms irreducible givens because no structure for him was an element of the essential order of the world. In order to deal with the structure of the solar system or the structure of animals it was necessary to try to understand them, that is, to know how they were formed. Here Buffon's ambition was close to Descartes', except that he refused for a long time to look for absolute beginnings. It was only in the *Epochs of Nature* that he would explain the origin of organic molecules and the "prototype" of each species. Meanwhile it was in the genesis of forms, the "degeneration" of animals, or the "genesis" of minerals that he sought to introduce order into the infinite diversity of things.

For it was necessary to introduce order. That Buffon had a passion for order in everything—in his schedule, his accounts, his papers, and his life no less than in his study of nature—was such an obvious aspect of his temperament that his contemporaries noted it. He wanted an order, but not just any order; he wanted a true and legitimate order. Buffon wished there to be an order in society, and in passing he did define a few rules that should preside over such an order. Respect for the established religion is one of them, and he observed it all his life. Yet he knew that this social order was not satisfying. Although he submitted to it on the outside and played the game to the point of accommodating himself perfectly to the France of the Ancien Régime and using all its devices, he did not give up his freedom of mind. He knew the weaknesses of the system and sometimes denounced them. As Mlle. Blesseau said, "no one has ever been able to take credit for having controlled him."[5] In the hierarchical society in which he lived, he knew how to carve out a place for himself, without excessive qualms or dishonoring servility. He used institutions as he found them and did not seek to change them because it was none of his business and because he did not have a great deal of confidence in human wisdom. It was within the environment of society and through it that he found the freedom that Rousseau vainly sought in solitude.

The order of nature is surely more perfect than that of society so long as we do not wish to introduce our feelings and our values into it. There are unhappy animals, there are predators and prey, and death is the necessary price of life. The individual is only a link in the chain of the species. At least animals are free and independent. Only man spreads terror and servitude in nature. When Buffon is "painting" the deer's beauty, the lion's nobility, the horse's servitude, the ass's misfortunes, even the poor peasant's or black slave's misery, the tone rises, the sentence finds its cadence, and a firm and measured eloquence reveals the controlled emotion of the writer. This is an intellectual emotion, so to speak, one that is wary of easy sentiment. Thus there is no ranting, no emphasis, but almost a new breadth of view that makes the reader feel that the tone is changing and makes him want to read the text aloud. The punctuation, which marks a breath for the reader more than a logical articulation of reasoning, suggests that Buffon—incessantly correcting his sentences with a taste for perfection that was already celebrated during his lifetime—had it pass by his "muzzle," as Flaubert said (but as Buffon would never have).

One must, however, go beyond this first level of emotion, even if it is an intellectual one, and the search for order must be pursued. Once again, it was a matter of finding a true order, based on the profound reality of things, and not on their superficial resemblance as happens in the classifiers' order. The reality of living beings was their organization, their physiology, their behavior, everything that allowed them to live and reproduce, to make their species survive, whose very existence was part of nature's order. Interacting with this persistence of the living and cosmic forces, there is the "degeneration" that modifies the form without interrupting the genealogies and allows the establishment of a real classification, that which reunites in the "genus" all those beings that have a common ancestor.

The search for order would soon spread beyond living nature to the entire cosmos, locating its ultimate and unique origin in the force of gravity and in the secondary forces of heat, electricity, and impulse, which are derived effects. This evolution of Buffon's ideas, already apparent in 1764, became more intense in the following years. Perhaps more sure of his genius, he left the familiar path of natural history and embarked on the conquest of the cosmos. Very understandably, he extended the power of active nature to the entire universe such as he observed it on the earth and started hunting for a universal order. This fundamental order, an order of forces and laws, maintained the universe in equilibrium, where everything leads us to believe that it is eternal or at least could be. This equilibrium, which he had first discovered between living species, he now found in the cosmos: an equilibrium constantly broken and reestablished, a balance of moving forces as opposed to a balance of death and immobility, yet a balance that nothing could threaten.

Assured of the existence of this unchanging order on the scale of the universe, Buffon could then abandon the earth to history, that is, to disequilibrium, to the irreversible change of which cooling was at the same time the cause and the symbol. In a way, the cooling continued the work of "degeneration." Where did this negative vision of time come from, this mythical vision of the radiant perfection of beginnings and the inevitable decadence? It is difficult to say, since nothing in what we know of Buffon's philosophy required this orientation of history. It was probably from the depths of his being, in what lies beyond our reach, that the root of this pessimism is to be found.

Life thus was born one day on earth, and one day it will be extinguished. It had, so to speak, lost its autonomy, it was now entirely subject to great cosmic forces. It was no longer, as Buffon had believed in 1749, a separate being: its appearance and disappearance were no more than necessary consequences of universal laws. And that is why it was also abandoned to history. Certain species had disappeared a long time ago, for Buffon had, like Aristotle, believed for quite some time that the existence of species was an element in the order of nature. They remained so, but only to the extent that the forces that gave birth to them and the laws that had presided over the formation and the combination of organic molecules were the same everywhere; and thus it could be believed that the same species, including the human species, had existed or will exist on other planets, even in other solar systems. Actual earthly species are no more than the local and fleeting manifestation of universal laws. If Buffon abandoned them to history, to extinction and death through cold, it was because in the infinite universe other planets in turn would receive and shelter life. The order of the universe is beyond that of history.[6]

Where did this order come from? Although it is clear that it is man who discovered it, he is not the author of it. He discovered it, moreover, at the price of a fundamental ignorance: man does not know and will never know what the force of attraction is. Matter here guards its secret. Does the order that man has succeeded in discovering come from God, or is it simply the order of nature such as it is, an already established order with no metaphysical guarantee? "I have always named the Creator; but we need only remove this word and, of course, put in its place the power of Nature," Buffon apparently said to Hérault de Séchelles.[7]

This is not so easy. One cannot replace God by nature in the famous "Prayer" that concludes the "First View of Nature": "ALMIGHTY GOD! whose sole presence sustains Nature and maintains the harmony of the laws of the Universe; YOU who from the unmoving throne of the Empyrean see all the heavenly spheres roll under Your feet without collision and without confusion; WHO from the bosom of repose, reproduces in every instant their immense movements, and alone in a deep peace regulates this infinite number of heavens and worlds; grant, grant calm to the Earth."[8] Curiously, it was for man that Buffon prayed to God. It was not necessary for nature: God knew what he had to do. Yet no one forced Buffon to write this prayer, which was more Deistic than Christian. He launched it like a bottle in the ocean toward the One who, if He reigned over nature, should not forget man.

For the scholar, it was essential that there be an order in nature, but its metaphysical origin was a secondary problem. Nevertheless, if Buffon, as we believe, changed his mind several times about the existence of God— God the creator and governor of the universe and its laws—this was because he had changed his opinion on the very existence of the order of nature. In 1749, overwhelmed by the diversity of beings and the anarchical wealth of forms, he thought that "all that can be is," and that nothing really permitted discovering an order or a God. In 1778, convinced he had discovered the order of nature, he wrote in the *Epochs of Nature:* "The more I have penetrated the bosom of Nature, the more I have admired and deeply respected its Author."[9] For "Nature is the system of laws established by the Creator for the existence of things and for the succession of beings."[10] Later, when Buffon was immersed in the history of minerals, God's image as guarantor of the order of the universe seems to have dimmed. Only the ferment of matter, creating all forms by chance, remained: "Everything works because with time everything collides with everything."[11] But it is true that here we are far from the cosmos and lost again in the infinite diversity of forms. If Buffon believed in God, it certainly was neither the Christian's God nor the God of Linnaeus or Réaumur, the meticulous artisan of living forms. It was the Legislator who created matter, gave it His laws, and left nature to do the rest, that is, carry out the wishes of her Author.

Whether the order of the universe came from God or from nature, it always inspired admiration, respect, and moving eloquence. It is useful to evoke here the famous "Discourse on Style," for such was the traditional name for Buffon's discourse before the Académie Française. "Style is only the order and the movement that one puts in one's thoughts." Order comes first; nature herself gave the example of it: "Why are the works of Nature so perfect? It is because each work is a whole and because Nature works according to an eternal plan from which she never deviates." To impose an order on his discourse and to discover the order of nature was to show the same power of thought, it was to affirm that man could be the interpreter of nature. To write a natural history was still in that century to write a literary work.[12] This did not mean "embellishing" nature, but understanding her and painting her as she is. "A beautiful style is indeed only so by the infinite number of truths it presents." Scientific discovery alone was not enough to immortalize a scholar: discoveries fall into the public domain and belong to everyone. On the contrary, "style is the man himself." It was found in the scientific work, which no one could borrow.

If the tone, which "is just the suitability of style to the nature of the subject," could become sublime, it was when the "Philosopher . . . speaks of the laws of Nature, of beings in general, of space, of matter, of motion and of time, of the soul, of the human mind, of feelings, of passions."[13] We can recognize from this list at least a few of the themes of the *Natural History*, those of the most famous "discourses" that Buffon had not yet written, the two "Views of Nature" or the "First Discourse" of the *Epochs of Nature*, all those where the majesty of nature awakened emotion in its painter.

Order is an abstract notion, and it is in the contemplation, or rather in the vision, of nature and the cosmos that this order can be grasped. When Buffon's thoughts turned to the universe, a vision took over his thought. The paradox, which is not really one, is that Buffon was never so visionary as when he described what he had not seen or would never see. It was with the eyes of the mind, with a purely intellectual vision, that he "saw" the "penetrating forces" and the "interior mold," hidden in the very bosom of matter. Later he wrote those descriptions of the desert of Arabia or the boggy forests of Guiana that I have cited: countries that he had never seen and that he knew only through texts. What gave them their power, what moved Buffon himself when he wrote these pages, was the savagery of these inhuman countrysides, which made immediately palpable the horror and the grandeur of which nature was capable. Later still, in what Buffon specifically called his "View of Nature," were two large visions that opposed each other, that of living nature and that of the infinite universe.

This visionary power was never as strong as in the *Epochs of Nature*, when Buffon described scenes that no human eye had ever seen, the torrent of molten matter escaping from the sun, which would form the planets, the barely cooled earth smoking from the torrential waters that fell from its atmosphere and quickly boiled away in thick vapors, volcanoes erupting, and gigantic earthquakes caused by the collapse of underground caverns.

This vision extended toward the future, toward that ever cooler earth where man and even all living nature would fight for their survival. For as the ocean continued to recede, as plants transformed inorganic matter into combustible matter, a reserve of heat accumulated. Still, the inexorable cooling would doubtless have the last word. The image of approaching death, for Buffon as an old man, was associated with the ultimate death of life on earth:

"It is with regret that I leave these interesting objects, these precious monuments of old Nature, that my own old age does not leave me the time to examine sufficiently to draw the consequences that I glimpse. . . . Others will come after me. . . . They will weigh the losses and the gains of this globe whose own heat is ceaselessly being exhausted, but which in compensation receives all the fire that resides in the remains of organized bodies. . . . They will see at the same time that . . . the diminishing of the waters, joined to the multiplication of organized bodies can retard only for a few thousand years the invasion of the entire globe by ice and the death of Nature through cold."[14] This was the ultimate vision of inescapable death, a tragic vision that sealed the ancient yet fleeting history of life and of man.

The day after his death, his body was opened. Following the deceased's orders, his heart was given to Faujas de Saint-Fond. His skull was sawed open and it was found that the brain was "of a slightly larger size than that of ordinary brains." He had fifty-seven stones in his bladder. On April 18, the funeral took place in the parish of Saint-Médard. Fourteen liveried horses, nineteen servants, sixty ecclesiastics, and thirty-six choirboys led the procession. "His funeral rites," read the *Mercure,* "were of a splendor rarely accorded to power, opulence, dignity. A numerous gathering of distinguished people, academicians, and men of letters accompanied the funeral procession and were united in this solemn homage rendered to the memory of a man of genius. Such was the influence of this famous name that twenty thousand spectators waited for this sad procession, in the streets, in the windows, and almost on the rooftops, with that curiosity that the people reserve for princes."[15] The death certificate was signed by the family members and representatives of the Academies, including Condorcet. That evening, the body left for Montbard, where it was interred on April 20. During the Revolution, the coffin would be opened and the lead would be taken from it to make bullets, which earned the municipality of Montbard a dry reprimand from the Committee of Public Instruction.

After the burial came the academic praise. Vicq d'Azyr, elected to Buffon's place at the Académie Française, did not mix criticisms with his admiration for the scholar and the writer. The most awaited discourse was that of Condorcet at the Academy of Sciences. Everyone knew of the hostility that had existed between the two men. The preparation was difficult: "Here I am busy with another charlatan, the great Buffon," wrote Condorcet to Mme. Suard. "The more I study him, the more I find him empty and bombastic. Fortunately that one had a lot of spirit, lucky insights, and a great talent for writing; thus I can avoid overly displeasing his admirers, without dishonoring myself in the eyes of educated people."[16] Nevertheless, either out of fear of opinion or because of intellectual honesty, Condorcet emerged with honor. Skillfully mixing criticism and praise, underlining the weakness of "systems" but also the value of erudition and description, finding honorable excuses for the scholar's faults, which still allowed him to mention them, Condorcet at least avoided admiring only his style, refrained from considering Buffon only a talented popularizer, and lastly placed him on the same level as Aristotle and Pliny. In the *Epochs of Nature* he could, however, see only a false hypothesis defended with stubbornness; he had understood nothing of how revolutionary the idea of a history of nature could be.

For many naturalists of the new generation, Buffon had simply "cleared the horizon," as was said of Victor Hugo a century later. "This time, the comte de Buffon is dead and buried," the young Cuvier noted simply.[17] The reaction was not long in coming, and the cult of Linnaeus was celebrated, whose work, it was said, had remained unknown in France because of Buffon. The "black legend" began. It would have a long life.

The fairest word was probably that of Meister in the *Literary Correspondence:* "M. le comte de Buffon died Wednesday, April 16, at two o'clock in the morning. If he has survived all his systems, his genius will more surely survive all those that have risen and are still rising from their superb ruins. He just closed the door on the most beautiful century that could ever honor France."[18] On August 8, 1788, Louis XVI convoked the Estates General.

Notes

The references to Buffon's works are to the original edition. Different collections have been designated by the following abbreviations.

HN Histoire naturelle, générale et particulière, Paris, Imprimerie royale, 1749–1767, 15 vols., in quarto.

HNO Histoire naturelle des oiseaux, Paris, Imprimerie royale, 1770–1783, 9 vols., in quarto.

Supplément Histoire naturelle, générale et particulière . . . Supplément, Paris, Imprimerie royale, 1774–1789, 7 vols., in quarto.

EN Époques de la Nature, appearing in volume 5 of the *Supplément*, 1778.

HNM Histoire naturelle des minéraux [and *Traité de l'aimant*], Paris, Imprimerie royale, 1783–1788, 5 vols., in quarto.

MAS Histoire et mémoires de l'Académie des Sciences, published annually starting in 1699.

CHAPTER I

1. For this part of Buffon's biography, I have essentially used Bourdier 1952a, Bertin 1952, and Hanks 1966, where one finds the traditional sources for what is known about Buffon's life. For information about the Leclerc family, I have also used Nadault de Buffon's notes from his two editions of correspondence with Buffon (*Correspondance* [hereafter *Corr.*], 1860 and 1971), even though they cannot always be trusted.

2. *Histoire naturelle de l'homme*, "De la vieillesse et de la mort," HN, II, pp. 602–3.

3. On intellectual life in Dijon and the role that the parlementarians played, especially Bouhier, as well as the Collège de Godrans, see Bouchard 1930. For the scientific teachings in Jesuit colleges, see Dainville 1964.

4. To Richard de Ruffey, 1729, in *Corr.*, 1971, I, pp. 3–4.

5. According to the Hickman papers, he and Kingston lived in Dijon, after a brief stop at Montbard, at least from March 7 to August 6, 1729, then in Angers from August 24 to January 15, 1730. See Fellows and Milliken 1972, p. 41.

6. Quoted by Monod-Cassidy 1941, pp. 40–1, from whom I borrow information on Kingston.

7. All of these quotations are from letters to Ruffey, respectively April 2, 1731 (*Corr.*, 1971, I, pp. 9–10), January 20, 1732 (I, p. 12), January 22, 1731 (I, p. 8), and November 5, 1730 (I, p. 6).

8. The quotations are taken from *Essai d'arithmétique morale* of 1777 (*Suppl.*, IV, p. 75 ff.) where in a footnote Buffon reproduced his October 3, 1730, letter to Cramer, that we know only because of this publication. It is also found in Buffon 1954, p. 465, in Weil 1961, pp. 115–8, and in Binet and Roger 1977, p. 49.

9. Note 1 in Weil 1961, p. 116.

10. *Pensées*, Lafuma, 136, Brunschvicg, 139, in Pascal 1963. We know that Pascal began his study of probability precisely with games of chance. This was the standard perspective in Buffon's era.

11. October 25, 1732, *Corr.*, 1971, I, pp. 17–8.

CHAPTER II

1. For information on the old Academy of Sciences, see Maindron 1888 and in particular Hahn 1971 (which includes a large bibliography). The chronological list of the academicians from 1666 to 1793 is found in Saigey 1873.

2. A detailed analysis of the dissertation is in Hanks 1966.

3. I am quoting the "Essai d'arithmétique morale," § 23 (*Suppl.*, IV [1777], p. 95+). This text is also found in Buffon 1954, p. 471+, and in Binet and Roger 1977, pp. 60–6.

4. Cf. Pascal 1963, p. 62. The arithmetic triangle indeed allows the calculation of the binomial coefficients for a given power from the preceding power. For information on the role of Pascal's triangle at the end of the seventeenth century, see Boyer 1959, pp. 203–4.

5. Newton knew the triangle of Pascal through the intermediary of Wallis, and had used it several times. Cf. Westfall 1980, pp. 37, 115, 118–9. My hypothesis about Pascal's role in the meeting of the minds of Buffon and Newton seems reasonable to Westfall (personal communication).

6. *Pensées*, Lafuma, 82, Brunschvicg, 44, in Pascal 1963, p. 504.

7. According to Hanks (1966, p. 60), this part of the text was added in 1777; this is possible but not certain. Hanks also opposes Buffon to Pascal but mentions only their optimism or pessimism. These are in reality two visions of man and the world that are radically opposed.

8. See Miles and Serra 1977, in particular pp. 3–28.

9. Quoted by Hanks 1966, p. 42 n. 46.

10. Letter to Bouhier, October 9, 1736, in Monod-Cassidy 1941, pp. 250–1.

11. The text appears in *Corr.*, 1860, I, pp. 213–4.

12. Quoted in ibid.

13. Ibid., I, p. 218. Nadault de Buffon, the editor of his correspondence, was the descendant of Buffon's half-sister.

14. Quoted in ibid., I, p. 217.

15. A precise analysis of Buffon's work as a translator appears in Hanks 1966, pp. 77–87.

16. Most of this preface is reproduced in Buffon 1954, pp. 5–6.

17. Hales 1961, p. 94.

18. About the role of theory in Hales's work, see Hanks 1966, pp. 73–7, which, however, appears to me a bit severe.

19. Quoted in Roger 1971, p. 205 n. 242.

20. Quoted by Hanks 1966, pp. 71–2.

CHAPTER III

1. *Corr.*, 1971, I, p. 34.

2. "Sur la Nature des Oiseaux," HNO, I (1770), p. 19.

3. *Corr.*, 1971, I, p. 32.

4. Letter written to Bouhier, September 19, 1739, in Monod-Cassidy 1941, p. 341.

5. Cf. Helvétius 1981, I, p. 40.

6. *Corr.*, 1971, I, pp. 33–4.

7. Quoted by Gascar 1983, p. 77.

8. Buffon to Le Blanc, February 22, 1738, in *Corr.*, 1971, I, p. 35.

9. For information about this entire affair, see Monod-Cassidy 1941, pp. 42–3, 296, and notes pp. 439, 511.

10. *Confessions*, book VII, in Rousseau 1959, p. 292.

11. Cf. Hanks 1966, p. 259.

12. *Corr.*, 1971, I, p. 38. For Voltaire's views on Newton, see Pomeau 1956, pp. 187–90, 203–6, and Roger 1971, pp. 735–6.

13. Letter of October 27, 1740, in Helvétius 1981, I, p. 50.

14. Letter of October 3, 1739, ibid., I, p. 40.

15. Ibid., I, p. 41.

16. Cf. Fellows 1956.

17. "He often asked me of news from you," wrote Buffon to Le Blanc June 23, 1750, in *Corr.* 1971, I, p. 72.

18. Letter of May 23, 1755, ibid., I, p. 101.

19. Ibid., I, p. 16.

20. Ibid., I, p. 40.

21. Quoted by Bourdier 1952, pp. 48–9.

22. February 16, 1736, in Weil 1961, pp. 121–2.

23. On the history of Newton's text, see Hanks 1966, pp. 250–2, and Westfall 1980, p. 226 ff.

24. Cf. Margaret Jacob 1976. On the argument between Newton and Leibniz, see Hall 1980.

25. Cf. Alexandre Koyré, *Du monde clos à l'univers infini*, 1962.

26. Letter to Mersenne of January 28, 1641, quoted in Costabel 1985, p. 38. I thank P. Costabel, who pointed out to me the texts of Descartes and Leibniz that I quote here and enlightened me on this debate.

27. *Excerpta* (notes taken by Leibniz on Descartes' papers), in Descartes 1967, XI, p. 656.

28. Letter to Varignon, April 14, 1702, in Leibniz, *Mathematische Schriften*, IV, p. 98.

29. Cf. the letter to Cramer of January 21, 1731, in Weil 1961, pp. 108–9. Buffon indicated here that he had read Fontenelle "with great attention eighteen months ago," that is, in the middle of 1729.

30. Cf. Hanks 1966, pp. 113–8, and especially Brunet 1931. About the debate in general in the eighteenth century, see Boyer 1959, chapter VI.

31. All the quotations are from Fontenelle 1727, "Préface," except the last, which is from Section II, p. 30.

32. Buffon 1954, pp. 448–9.

33. About this argument and Buffon's position, see Hanks 1966, pp. 122–5, and Boyer 1959.

34. About this dissertation, see Hanks 1966, pp. 122–5.

35. *Corr.*, 1971, I, p. 177.

36. Ibid, I, pp. 177–8. There are several versions of this incident, of which the date is uncertain.

37. Cf. Hanks 1966, pp. 156–68.

38. Cf. ibid, p. 183.

39. Cf. ibid, pp. 193–213.

40. *Corr.*, 1971, I, p. 36.

41. Cf. Hanks 1966, pp. 127 ff, 253–7.

42. This activity of the English aristocrats is currently the object of important research by Mme. de Blomac, who was kind enough to inform me of it.

43. Cf. Monod-Cassidy 1941, p. 522 n. 4 to letter LXXXIII.

44. Cf. Hanks 1966, pp. 131–2.

45. *Buffon*, 1952, pp. 186–7.

46. Ibid., p. 189. The letter is from January 6, 1739. The most daring passages of the text have been carefully altered in ink (perhaps by Dutour himself) and could be deciphered only through infrared photography.

CHAPTER IV

1. Cf. Hanks 1966, pp. 134–5.

2. Letter from Le Blanc to Bouhier, June 15, 1739, in Monod-Cassidy 1941, p. 335.

3. *Corr.*, 1971, I, pp. 41–2.

4. *Eloge de M. du Fay*, in Fontenelle 1742, VI, pp. 668–9. The eulogy of Du Fay was the last that Fontenelle, then eighty-one years old, gave. He was then replaced in his functions of permanent secretary by Grandjean de Fouchy.

5. *Corr.*, 1860, I, p. 231, also *Corr.*, 1971, I, p. 42. Note of Nadault de Buffon. This note is the only source that mentions Hellot's intervention. According to the chemist Geoffroy, Du Fay would have designated several possible successors. According to Le Blanc, Maurepas was the one who spontaneously chose Buffon (see below). In any case, Buffon did not have the time to intervene personally in his nomination.

6. Fontenelle 1742, VI, p. 669.

7. Cf. Genet-Varcin and Roger 1954, p. 516 nn. 12, 13.

8. Letter from Le Blanc to Bouhier, in Monod-Cassidy 1941, p. 337. The letter is dated July 9, which is a problem, since the decision was made only on the 25th,

and Le Blanc wrote, "It is by this same post that I am writing to send him, without his expecting it, the Provisions for the position of Intendant of the Royal Botanical Garden." If the "Provisions," that is, the official nomination, were ready, it is possible that it was August 9, not July 9. Buffon probably did not receive the letter before the 12th.

9. Ibid. The Court was at Compiègne, where Louis XV liked to stay.

10. On the history of the Royal Botanical Garden, see François 1952 and above all Laissus 1986.

11. See Pintard 1943, pp. 195–200.

12. Quoted in François 1952, p. 105.

13. See Howard 1983, where the inventory of Guy de la Brosse's library and laboratory after his death is found.

14. See the beautiful eulogy of Fagon by Fontenelle in Fontenelle 1742, VI, p. 34 ff.

15. *Eloge de M. du Fay*, ibid., VI, pp. 664–5.

16. Cf. Goerke 1966, p. 126.

17. Cf. Larson 1971, p. 59.

18. A complete list of the papers and reports presented by Buffon to the Academy is found in Hanks 1966, pp. 275–81.

19. Republished in Binet and Roger 1977, pp. 137–49.

20. Published with commentary in Hanks 1961.

21. "La Dioptrique, Discours huitième," in Descartes 1987, p. 119.

22. To obtain the vast mirrors needed by modern telescopes, small thin mirrors of which the curvature can be regulated are juxtaposed.

23. Letter to Guyton de Morveau, June 26, 1772, in *Corr.*, 1971, I, p. 217. Guyton de Morveau, a chemist and Burgundian parlementarian, wanted to take up these experiments again, and Buffon proposed giving him the settings that he still had.

24. Letter of July 1, 1747, in Mme. du Châtelet 1958, II, p. 157. Mme. du Châtelet heard the reading of the first paper at the Academy, but could not attend the experiment and regretted it.

25. Cf. his article "*Scala Graduum Caloris*" (first published in 1701), in Newton 1958, pp. 259–64.

26. Letter of April 6, 1750, in Hanks 1966, pp. 263–6.

27. See *Corr.*, 1860, I, pp. 417–9, where the poem from De Ruffey is found.

28. All the texts cited are found in the *Mémoires de l'Academie des Sciences* for the year 1745. The *Mémoires* appeared several years late. Because of the importance of the debated questions, the Academy decided to publish all the papers cited here in the volume of the year 1745, which would in fact appear in 1749. For the quarrel seen from Clairaut's point of view, see Brunet 1952, chapter VI.

29. Letter from Clairaut to Euler, September 11, 1747, in Euler 1980, p. 177.

30. *MAS*, 1745.

31. Letter from d'Alembert to Euler, July 20, 1749, in Euler 1980, p. 302.

32. *MAS*, 1745.

33. "In the argument that I had more than two years ago with Clairaut on the subject of the movement of the apogee of the moon, I defended Newton and his commentators; Clairaut has since retracted his arguments, and having suppressed the part of his paper which directly attacked the commentators, I was also obliged

to suppress what I had written to uphold the theory, and there is printed in the volume of 1745 only that which looks at the law of attraction in general. I'm writing this to you so that R. P. Jacquier sees that I wish greatly his friendship." Buffon to Le Blanc, June 23, 1750, in *Corr.*, 1971, I, p. 71. Jacquier was one of the commentators in question.

34. "Réponse aux réflexions de Mr. de Buffon sur la loi d'Attraction et sur le mouvement des apsides," *MAS*, 1745.

35. These considerations have continued to occupy historians of mathematics, who see here one of the first definitions of the idea of a function.

36. *MAS*, 1745.

37. See the letter to Cramer of July 26, 1749, in Speziali 1955, pp. 226–7.

38. Letter of June 9, 1749, in Euler 1980, pp. 186–7.

39. See François 1952, pp. 107–8.

40. Letters to MM. de Madières, August 29 and 30, 1739, in Michaut 1931, pp. 17–8.

41. Letters from January 4, 1742, and February 10, 1747, in *Corr.*, 1971, I., pp. 47, 50.

CHAPTER V

1. On the mechanical philosophy of the seventeenth century, see Dijksterhuis 1950.

2. *Nouvelles de la République des Lettres*, edited by Pierre Bayle, August 1684. On this fashion, the role of microscopy, the development of entomology, and the entire evolution of the ideas described here, see Roger 1971, Part 2, chap. I.

3. *Eloge de M. Géoffroy*, quoted in ibid., pp. 165–6.

4. Cf. ibid., pp. 242–4.

5. Réaumur 1732, IV, xxix.

6. Pluche 1764, I, ix–x.

7. The expression is Voltaire's. On the fad of natural history and "cabinets of curiosities" in the eighteenth century, see Mornet 1911.

8. A map of M. Bouvet's expedition, drawn by M. Buache in 1739. "Theory of the Earth," *HN*, I, 72, note *f*. Buffon and Grandjean de Fouchy had presented a report on this map before the Academy on September 5, 1739. See Hanks 1961, p. 278.

9. Cf. Roger 1971, pp. 542–3.

10. Letter of July 18, 1741, which was published in the *Philosophical Transactions*.

11. The text of the sealed envelope was published in *Corr.*, 1971, I, pp. 54–6. It began: "February 9, 1746, I started a *Traité de la Génération* [Treatise on Generation], which is now completely finished."

12. The announcement of the *Journal des Savants* is reproduced in *Corr.*, 1860, I, pp. 243–4.

13. All these details are taken from the correspondence, in particular letters to Cramer and the abbé Le Blanc, in Weil 1961, pp. 129–33, and *Corr.*, 1971, I, pp. 56–9.

14. Certain academicians were prompt to underline the fact. See below, Chapter XIII.

CHAPTER VI

1. "Premier Discours," *HN*, I (1749), pp. 5–6.
2. Ibid., p. 4.
3. Ibid., p. 51.
4. Ibid., pp. 9–10.
5. Ibid., p. 11.
6. Ibid., p. 18.
7. Ibid., pp. 38–9.
8. Ibid., p. 13.
9. Ibid., p. 24.
10. Ibid., p. 52.
11. Ibid., pp. 53–4.
12. Cf. Michel Foucault 1966. The author's efforts to make Buffon enter into his picture of natural history in the eighteenth century do not seem very convincing to me.
13. "Premier Discours," pp. 31–2.
14. On the idea of the "chain of being" and its history, see the classic work of Lovejoy 1936. We will come back to Buffon's evolution of ideas on this topic: see below, Chapter VIII.
15. "Premier Discours," p. 12.
16. Ibid., pp. 34–5.
17. Ibid., pp. 32–3.
18. Ibid., pp. 36–7.
19. Pluche 1732, II, ix–x.
20. For information about these modern debates, see Tassy 1987.
21. "Premier Discours," p. 30.
22. Ibid., p. 51.
23. Ibid., p. 55.
24. Ibid., p. 57.
25. Ibid., pp. 58–62.
26. Ibid., p. 62.

CHAPTER VII

1. Unless otherwise indicated, the majority of this history is taken from Roger 1962 (introduction), Gohau 1988, and Ellenberger 1988.
2. On the birth of the theory of the earth in the seventeenth century, see Roger 1973.
3. On the epistomological and religious signification of Descartes' theory, see ibid.
4. First published in Latin, then rewritten and published in English in 1684. The English text was republished in Burnet 1965.
5. I am taking up here a distinction clearly set forth by Gohau 1988.
6. This forgotten author has been recently rediscovered; cf. Ellenberger 1976.
7. The work has recently been republished in *Corpus général des philosophes français* by the Librairie Fayard.

8. Linnaeus 1749, II, pp. 430–59. Republished and translated into French in Linnaeus 1972.

9. Bourguet 1729, p. 50.

10. Pluche 1732, III, p. 528.

11. "Preuves," art. II, *HN*, I, p. 178.

12. Ibid., pp. 178–9.

13. "Preuves," art. V, ibid., p. 203.

14. Ibid., p. 202.

15. "Théorie de la terre," ibid., p. 99.

16. Ibid., p. 96.

17. Ibid., pp. 69–70.

18. Kenneth Taylor, communication to the French Committee of History of Geology, February 24, 1988.

19. "Théorie de la terre" *HN*, I, pp. 94–5.

20. Ibid., p. 77.

21. "Preuves," art. XIII, ibid., p. 457.

22. "Théorie de la terre," ibid., pp. 95–7.

23. "Conclusion," ibid., p. 612.

24. "Théorie de la terre," ibid., p. 124.

CHAPTER VIII

1. Genesis 1:1–2. I am using the translation used by Buffon himself at the end of the "Premier Discours" of the *Epoques de la Nature*. See below, Chapter XXIII.

2. On the history of the notion of creation in Christian thinking, see, among others, Vigoroux 1889, Sertillanges 1945, Guelly 1963, and McMullin 1985.

3. Cf. Busco 1924.

4. On this question, see Roger 1973.

5. On this quotation, see the note on p. 66.

6. Newton 1966, II, p. 544.

7. See letter to Bentley, December 10, 1692, in Newton 1961, pp. 233–6.

8. *Lettres Philosophiques*, 15th letter. On Voltaire's attitude in regard to science, see Roger 1971, pp. 732–48.

9. "Preuves de la Théorie de la Terre," article I, "De la formation des planètes," *HN*, I, p. 131.

10. Ibid., p. 153.

11. Laplace 1836, II, pp. 547–8.

12. Buffon wrote 1683 by mistake. In 1749, the same year that the first three volumes of the *Natural History* appeared, a complete treatise by Leibniz on the theory of the earth was published under the same title of *Protogea*, a treatise that Buffon did not know about when he wrote his own theory.

13. "Preuves de la Théorie de la Terre," article V, *HN*, I, p. 196.

CHAPTER IX

1. On the history of the theories of the "generation of animals" of the sixteenth and seventeenth centuries, see Roger 1971, which is entirely devoted to this question, and the other references given in this and the following chapters.

2. Letter IX, in Fontenelle 1742, I, pp. 322–3.

3. "Remarques sur un nouveau monstre," in *Mémoires de l'Académie des Sciences,* 1740, p. 612.

4. Bourguet 1729, pp. 165–6.

5. *Corr.,* 1971, I, p. 40, letter of February 8, 1734 (which Nadault de Buffon erroneously dated 1739).

6. Chapter V, *HN,* II, pp. 163–4.

7. Chapter I, ibid., pp. 10–11.

8. I will come back to the originality and the importance of this definition of the species; see below, chapter XIX.

9. Chapter I, *HN,* II, p. 17.

10. See Roe 1988.

11. Chapter IV, *HN,* II, p. 58.

12. Ibid., p. 59.

13. Ibid., p. 68.

14. Ibid., pp. 68–9.

CHAPTER X

1. "Histoire des Animaux," chapter II, *HN,* II, pp. 32–3.

2. Ibid., p. 28.

3. Ibid., p. 27.

4. Ibid., p. 28.

5. Ibid., pp. 20–1.

6. Ibid., pp. 22–3.

7. Cf. the essay "*Sur les mesures*" presented to the Academy on December 23, 1738, and taken up again on May 15, 1743. See *MAS,* 1741 (published in 1744), pp. 219–21. Cf. Hanks 1966, pp. 277–8. The text shows up again in the "Essai de l'arithmétique morale" of 1777.

8. "Histoire des Animaux," chapter II, *HN,* II, p. 22.

9. Ibid., p. 37.

10. This error was committed in an otherwise interesting article (Farber 1972). The author translated "relativement à cette qualité" by "as analogous to this power" (p. 263). This misunderstanding led him to discover in Buffon's thinking contradictions and a development that were not there. Farley (1974) seems to have had the same misunderstanding (p. 23).

11. "Histoire des Animaux," chapter II, *HN,* II, p. 36.

12. Ibid.

13. Letter to Bentley of February 25, 1693, in Newton 1961, p. 254.

14. Newton 1966, II, p. 546.

15. Ibid.

16. "Histoire des Animaux," chapter III, *HN,* II, p. 50.

17. Ibid., p. 52.

18. Bourguet 1729, pp. 164–5.

19. "Histoire des Animaux," chapter III, *HN,* II, p. 43.

20. Ibid., p. 49.

21. Ibid., chapter IV, p. 54.

22. Ibid., p. 59.

23. Ibid.

24. On the ideas thus discussed on the subject of "spermatic animalcules," see Castellani 1965, chapter X, and 1973, p. 48, as well as Farley 1982, pp. 16–25.

25. "Histoire des Animaux," chapter V, *HN*, II, p. 113.

26. Ibid. Malpighi said exactly "a white or ashy spherical body, like a *mole.*" A "mole" here is an unorganized mass. On Malpighi and his work, see Adelmann 1966.

27. "Histoire des Animaux," chapter VI, *HN*, II, pp. 168–70.

28. On Needham, see Roger 1971, pp. 494–520, Roe 1981, 1982, and 1988, and Mazzolini and Roe 1986.

29. Cf. Sloan 1988a, which I follow here. On English microscopes of the eighteenth century, see Bradbury 1967 and Sloan's study.

30. "Histoire des Animaux," chapter VI, *HN*, II, p. 198.

31. Ibid., § IV, p. 180.

32. Ibid., § VI, pp. 182–3.

33. Ibid., § VIII, p. 185.

34. Castellani 1973, p. 52.

35. Cf. Sloan 1988a.

36. "Histoire des Animaux," chapter VI, § XXVI, *HN*, II, p. 203. It is very likely, as Frank Bourdier (1952b, p. 168) suggests, that it was this experiment that Sève had represented on the band decorating the first page of the "Histoire des Animaux." It is in this engraving that the double microscope of which I have spoken is found.

37. See Roe 1988.

38. "Histoire des Animaux," chapter VIII, *HN*, II, pp. 276–8.

39. Bazin 1741, p. 21.

40. "Histoire des Animaux," chapter VIII, *HN*, II, pp. 260–3.

41. Ibid., p. 265.

42. Ibid., p. 257.

43. Ibid., p. 263.

44. Ibid., chapter IX, p. 320.

45. Ibid., chapter VI, § XLIV, p. 222.

46. Ibid., chapter X, pp. 351–4.

47. Ibid., chapter XI, pp. 368–71.

48. Ibid., pp. 372–4. One can wonder how Buffon knew Leibniz's thoughts on the *Analysis situs,* as the texts that mention them were only published in the nineteenth century by Gerhardt.

CHAPTER XI

1. "Premier Discours," *HN*, I, p. 12. The *Histoire naturelle de l'Homme* has been republished by M. Duchet (Buffon 1971) with the supplements from 1777 and a very interesting introduction. The chapter "De la vieillesse et de la mort" has been republished by J.-L. Binet and J. Roger 1977. On the anthropology of the philosophes of the eighteenth century, see Duchet 1971, 2d part, chapter I, and the very original essay by Claude Blanckaert 1988.

2. *Histoire des Animaux,* book I (Aristotle 1783, I, p. 11).

3. On man's place in nature according to Linnaeus, see Broberg 1983. From this very shrewd analysis, it emerges that a Christian such as Linnaeus was as disposed to reduce man to modesty as was a materialistic philosophe, if not more.

4. "Premier Discours," *HN,* I, p. 39.

5. Ibid., p. 48.

6. Ibid., p. 12.

7. On the quarrel over the soul of animals in the seventeenth and eighteenth centuries, see the classic book by Rosenfield 1941.

8. *Dictionnaire de Trévaux,* 1752, article "Anthropologie." Cf. Roger 1979.

9. "Premier Discours," *HN,* I, p. 30.

10. *Histoire naturelle de l'Homme,* chapter I, *HN,* II, p. 430.

11. Ibid., p. 432.

12. Ibid., p. 430.

13. Ibid., pp. 434–5.

14. Ibid., pp. 432–3.

15. Ibid., p. 434.

16. On this philosophy, see Berkeley 1948.

17. *Lettre sur les aveugles,* in Diderot 1978, (OC, IV), p. 44.

18. *Histoire naturelle de l'Homme,* II, chapter I, p. 434.

19. *Principes de philosophie,* 1st part, § 13, in Descartes 1974, II, 2, p. 30.

20. *Histoire naturelle de l'Homme,* "Du sens de l'ouïe," *HN,* III, p. 349. The text was taken from the *Histoire de l'Académie des Sciences,* 1703, p. 18. It is therefore scientifically indisputable. But it is by Fontenelle . . .

21. *Histoire naturelle de l'Homme,* "Des sens en général," *HN,* III, pp. 360–1.

22. Ibid., pp. 362–3.

23. Ibid., p. 363.

24. Ibid., "Du sens de la vue," *HN,* III, pp. 305–34.

25. "Discours quatrième: Des sens en général," in Descartes 1974, VI, p. 109. Notice that Buffon took the title of Descartes' chapter.

26. "Discours sixième: De la vision," ibid., pp. 130–40.

27. Diderot, for example, referred immediately to it in his "Lettre sur les aveugles." Voltaire had already cited Descartes in 1738 and had taken up his comparison of inverted vision in the case of the sensations of a blind man who holds two crossed sticks (*Éléments de la philosophie de Newton,* Chapter VI).

28. *De la recherche de la vérité,* book I, chapter VI–IX, in Malebranche 1962, pp. 79–120.

29. Ibid., chapter IX, pp. 119–20.

30. Voltaire used it, even though he cited it only to criticize it. *Éléments,* Chapter VIII.

31. On this controversy, see Duchesneau 1973, chapter VI.

32. *An Essay concerning Human Understanding,* book II, chapter IX, § 8 in Locke 1740, pp. 54–5.

33. Ibid., chapter XXI, § 5, p. 103.

34. *Éléments,* chapter VII. Voltaire mistakenly gave the date of 1729, which was then accepted by all the French philosophers who did not verify it. Buffon was the

only one who went back to the original memoirs of Cheselden, which appeared in the *Philosophical Transactions* and in *The Tatler*, and to which he gave the references ("Du sens de la vue," *HN*, III, p. 314).

35. *Essai*, section VI, in Condillac 1792, I, pp. 191–213.

36. *Lettre sur les aveugles*, in Diderot 1978, p. 6. Also see Torlais 1936, pp. 251–2.

37. Diderot 1978, p. 17. For the relationship between Diderot and Buffon at this time, see Roger 1963.

38. In the chapter "Du sens de l'ouïe," he wrote, "Until that day [in the month of June 1749]." *HN*, III, p. 351.

39. "Du sens de la vue," *HN*, III, p. 318.

40. *HN*, III, pp. 352–3.

41. *HN*, II, p. 432.

42. Cf. Diderot 1978, p. 81, note 34. Commentary of Yvon Belaval.

43. *Dioptrique*, "Discours Cinquième," in Descartes 1974, VI, p. 129.

44. Articles XI and XII of the "Propositions . . . qui ont paru répréhensibles à MM. les Députés de la Faculté de Théologie de Paris." Published by Buffon in *HN*, IV, pp. ix, x. See below, chapter XIII.

45. "De la nature de l'homme," *HN*, II, p. 436.

46. Ibid., pp. 436–7.

47. Ibid., pp. 437–8.

48. See Darwin, in particular, *The Descent of Man*, 1881, chapters III and IV.

49. On the evolution of attitudes in England, see Keith Thomas 1983.

50. On this use of a human vocabulary for animal behavior, see Roger 1988a.

51. "De la nature de l'homme," *HN*, II, p. 438.

52. Ibid., p. 440.

53. Ibid., pp. 440–2.

54. Ibid., p. 444.

55. *HN*, III, pp. 364–70.

CHAPTER XII

1. Cf. Ph. Aries 1960.

2. "De l'enfance," *HN*, II, p. 445.

3. Ibid., pp. 447–9.

4. Ibid., pp. 451–2.

5. Ibid., pp. 454–5.

6. Ibid., pp. 459–61. Buffon came back to the problem of wet-nurses on p. 474.

7. Ibid., pp. 472–3. Cf. "Sur l'acroissement successif des enfants," in the *Supplément*, volume IV (1777), pp. 376–83, where Buffon reported the growth, measured every six months, of the son of his friend Guéneau de Montbeillard. These observations, which are the first done methodically on the subject, have been confirmed by modern anthropologists. Cf. Scammon 1927 (communication from Professor R. Dailly).

8. "De l'enfance," *HN*, II, pp. 471–2.

9. Ibid., pp. 474–5.

10. Ibid., p. 477.
11. "De la puberté," *HN*, II, pp. 478–9.
12. Ibid., pp. 485–8.
13. Ibid., p. 487.
14. The probable source was Louis de Lacaze, who had just published his *Specimen novi medicinae conspectus* in 1749. On these questions, see R. Rey 1987.
15. "De la puberté," *HN*, II, p. 481.
16. Ibid., pp. 492–500.
17. Ibid., pp. 501–2.
18. See Roger 1971, p. 706.
19. "De la puberté," *HN*, II, pp. 502–3.
20. Ibid., pp. 503–5.
21. Ibid., pp. 507–8.
22. "De l'âge virile," *HN*, II, p. 518.
23. Ibid., p. 554.
24. Ibid., pp. 540–5.
25. Ibid., pp. 519–37.
26. Ibid., pp. 587–8.
27. Ibid., p. 555.
28. Ibid., p. 546.
29. On this chapter, see J.-L. Binet and J. Roger 1977 and especially C. Milanesi 1989. On the physiology of the eighteenth century in general, see F. Duchesneau 1982. On the place that Buffon occupied there, see Duchesneau 1988.
30. "De la vieillesse et de la mort," *HN*, II, p. 558.
31. Ibid., p. 568.
32. Ibid., p. 569.
33. Ibid., p. 570.
34. Ibid., p. 567.
35. Ibid., p. 570.
36. Ibid., p. 571.
37. Ibid., p. 572.
38. Ibid., pp. 572–3.
39. Ibid., p. 579.
40. Ibid., p. 580.
41. Ibid., p. 581.
42. Ibid., p. 583.
43. Ibid., p. 584.
44. *Essais*, book III, chapter XII, "De la Physionomie."
45. "De la vieillesse et de la mort," *HN*, II, p. 584.
46. See Milanesi 1989 for the study and the bibliography of this question.
47. "De la vieillesse et de la mort," *HN*, II, pp. 587–8.
48. Ibid., p. 602.
49. Ibid., pp. 602–3.
50. "Variétés," *HN*, III, pp. 401–3.
51. Ibid., p. 502. The Chacrelas are cited p. 399, the albino Papuans p. 406, the Indians of the Isthmus p. 500, and the problem is considered again in its entirety

pp. 500–503. The main signs of albinism were correctly described, but Buffon could obviously go no further.

52. Ibid., p. 473. The Hottentots were famous and had often been described with horror.

53. On the Lapps, ibid., p. 373; the Kalmuks, p. 380; the inhabitants of Senegal, pp. 457–8; the Circassian women, p. 435.

54. On the Lapps, pp. 374–6; the Tartars, pp. 380–1; the Bengalis, p. 412.

55. Cf. Roger 1971, pp. 87–8, 215–6.

56. Cf. ibid., pp. 733–4, 739.

57. Cf. *Supplément,* volume IV (1777), p. 462.

58. "Variétés," *HN,* III, pp. 481–2, 453.

59. Ibid., p. 483.

60. Ibid., pp. 446–7.

61. On the Hottentots, ibid., p. 471; on the squashing of the nose and lips, p. 459.

62. On the Maldivians, p. 416; the Persians, p. 421; the Moors, p. 448.

63. Ibid., pp. 481–2.

64. On the Ostiaks, pp. 378–9; the Peuls, p. 456.

65. Ibid., pp. 528–9.

66. Ibid., pp. 526–7.

67. Ibid., p. 483.

68. "Exposé des causes," ibid., p. 446; "Discussion sur la noirceur des Nègres en particulier," pp. 522–7; final citation, pp. 523–4.

69. Books XIV–XVII. As Bodin before him, Montesquieu distinguished among three major climatic zones: cold, temperate, and hot. It is to be noted that Montesquieu attempted a physiological explanation for the action of heat on customs, which he saw as a consequence of its action on fibers. Cf. Montesquieu 1955, II, pp. 169–86, and *Esprit des Lois,* book XIV, chapter II. Buffon carefully refrained from offering this explanation.

70. "Variétés," *HN,* III, pp. 509–10. Later Buffon would more easily admit the existence of the "giant Patagonians," for reasons that we shall see.

71. Ibid., p. 515.

72. Ibid., pp. 529–30.

73. "Variétés," *HN,* III, p. 528.

74. Ibid., p. 462.

75. Ibid., pp. 468–70.

76. *Esprit des Lois,* book XV, chapter V, in Montesquieu 1955, pp. 220–1.

CHAPTER XIII

1. Cf. *Corr.,* 1971, pp. 60–72.

2. "Nouvelles littéraires," in *Correspondance littéraire,* I, p. 336.

3. *Mémoires de Trévoux,* October and November 1749; March and May 1750.

4. *Lettre sur quelques écrits de ce temps,* January 15 and 22, 1750.

5. *Nouvelles ecclésiastiques,* February 6 and 13, 1750, pp. 21–4, 25–7. The author is Jacques Fontaine, known under the name of abbé de La Roche.

6. Letter of June 23, 1750, in *Corr.*, 1971, I, p. 70.

7. Quoted in Bourdier 1952a, p. 28.

8. Letter to the president De Brosses, July 14, 1760, in *Corr.*, 1971, I, p. 114.

9. Cf. Stengers 1974 and Mengal 1989. In 1779, risking being condemned again by the Sorbonne, Buffon envisioned giving "an explanation as silly and absurd as the first *that I was forced to sign* thirty years ago." Letter to Guéneau de Montbeillard, November 15, 1779, in *Corr.*, 1971, I, p. 440. Emphasis added.

10. *Corr.*, 1971, I, pp. 78–9.

11. *HN*, IV, pp. v–xvi.

12. *Mémoires de Trévoux*, December 1753.

13. Quoted by Piveteau 1952, p. 127, and Mengal 1989.

14. Duhamel 1754, p. 137.

15. Cf. *Mémoires* by Bachaumont 1780, III, p. 150 (March 3, 1767). Marmontel paid the price for this repentance from the Sorbonne.

16. Quoted by Piveteau 1952, p. 128.

17. Cf. the correspondence Bettinelli–Voltaire quoted in Fellows 1970, p. 60.

18. The best study on Réaumur remains Torlais 1936, despite a few errors in details.

19. Cf. Roger 1971, pp. 379–83.

20. On July 20, 1740, Father Bignon, librarian of the king and president of the Academy of Sciences, invited Réaumur to dine at his house with Buffon. We do not have Réaumur's answer, but Bignon wrote him on August 9: "I did not know and I would not have even guessed what you have told me about M. de Buffon. The esteem that you yourself, Sir, have inspired in me for his character should not have allowed me to assume that he was capable of forgetting how much time is necessary for all kinds of things." Quoted in Torlais 1936, pp. 213–4. Buffon probably showed too much ambition. Bignon was powerful and Réaumur's denunciation could harm Buffon's academic career.

21. Letter to Seguier, May 25, 1749, in Réaumur 1886.

22. Letter to Ludot, May 3, 1751, ibid. See also the letters to Trembley, November 8, 1749, and to Bonnet, December 10, 1751 (in Réaumur 1943, pp. 329–30, 362–3). Trembley himself was more subtle: he was interested in the theory of the earth but condemned the theory of generation, founded on "a chancy hypothesis." He concluded similarly to Réaumur: "If his work is widely swallowed, I fear that he will do wrong to natural History by bringing to it the taste for hypotheses." Letter to Bentink, January 9/20, 1750. Cited in Réaumur 1943, p. 330.

23. Cf. Roger 1971, p. 692 n. 60.

24. *Journal* cited in Réaumur 1943, p. 362 n.

25. Letter to Séguier, cited in Réaumur 1943. See also the letter to Bonnet, December 10, 1751, and to Trembley, December 31, 1751.

26. For a more complete analysis of *Lettres*, see Roger 1971, pp. 692–702.

27. First letter, pp. 6–9.

28. Ibid.

29. *Mémoire pour servir à commencer l'histoire des araignées aquatiques* (Paris, 1749). The paper, and the title itself, are typical of Réaumur's style.

30. Volume II, fifth letter, pp. 62–3.

31. Ibid., pp. 59–60.

32. See above, chapter V, p. 80.

33. Letter to F.-M. Zanetti of May 11, 1750.

34. On the problems that this criticism of Needham posed, see Roger 1971, pp. 697–8.

35. April 2, 1752, p. 53.

36. "Nouvelles littéraires," June 28, 1751, in *Correspondance littéraire*, III, pp. 160–3.

37. "Nouvelles littéraires de la France," August 15, 1751, taken up in *Les Cinq Années littéraires*, III, pp. 160–3.

38. *Recueil de différents traités de physique*, III, p. xvi.

39. Lelarge de Lignac would continue his career as a polemicist, publishing in 1756 a *Suite des lettres à un Amériquain*, where he attacked Buffon and Condillac and lay into Helvétius and the Jesuits later. His work as a philosophe has retained the attention of historians: see Le Goff 1863, and Lewis 1951.

40. Letter to the abbé Le Blanc, March 21, 1750, in *Corr.*, 1971, I, p. 68.

41. Letter to M. Doussin, July 27, 1751, ibid., I, p. 83.

42. For a more complete analysis of this text, see Roger 1971, pp. 705–8.

43. On the later evolution of Haller's ideas, see Roger 1971, pp. 705–12, Roe 1981, and Haller 1983, which show the increasing hostility toward Buffon and French thought in general.

44. Cf. the letter to Séguier of December 9, 1751, in Réaumur 1886, p. 89.

45. Letter to Le Roy, August 23, 1759, in Helvétius 1981, II, p. 264.

46. For a more detailed analysis of the *Observations*, see Roger 1971, pp. 687–91.

47. On this entire aspect of Voltaire's philosophy and his ideas on the sciences, see Roger 1971, pp. 732–48.

48. *Preuves de la Théorie de la Terre*, art. VIII, *HN*, I.

49. Letter to Le Blanc, June 23 and October 22, 1750, in *Corr.*, 1971, I, p. 72.

50. *Correspondance littéraire*, IV, p. 80. Cited in Helvétius 1981, II, p. 83, as well as d'Alembert's opinion. Voltaire himself wrote to Helvétius: "You have unfortunately given a pretext to all enemies of philosophy" (ibid., p. 286). Buffon likewise evoked the persecution of Helvétius and the *Encyclopédie* when he gave advice of caution to the president De Brosses (May 5, 1760, in *Corr.*, 1971, I, p. 112). Nadault de Buffon claimed that Mme. Helvétius "was as linked to Mme. de Buffon as Helvétius was with the naturalist" (ibid., p. 112 n. 3), but we do not know on which document he based this statement.

51. Cf. Hankins 1970, p. 80.

52. Letter of September 21, 1749, cited ibid., pp. 76–7. D'Alembert reacted all the more violently because he felt that Buffon represented a movement of reaction against the tyranny of mathematics, a movement in which Diderot participated. Cf. Belavel 1952.

53. Cf. letters to Formey December 6, 1750, and to Le Blanc, April 24, 1751 (*Corr.*, 1971, I, p. 74, 79–80).

54. *Encyclopédie*, I, p. xiv.

55. Ibid., p. vii.

56. Ibid., p. xxxi.

57. Cf. Roger 1971, pp. 465–8.

58. June 20, 1751, in *Corr.*, 1971, I, p. 82.

59. On the intellectual relations between the two men, see Roger 1971, pp. 596–601, and Roger 1963.
60. Cf. above, chapter XI.
61. Cf. Vartanian 1989.
62. To the president De Brosses, February 16, 1750 (*Corr.*, 1971, I, p. 65).
63. To Le Blanc, March 21, 1750 (ibid., p. 68).
64. December 12, 1752 (ibid., p. 88).
65. *Lettres sur quelques écrits de ce temps*, October 13, 1753, in Fréron 1749, XI, p. 289.
66. *Correspondance littéraire*, October 1, 1753, in Grimm 1877, II, pp. 285–6.
67. "*Homo Duplex*," in "Discours sur la nature des animaux," *HN*, IV, p. 76.

CHAPTER XIV

1. November 23, 1753, in *Corr.*, 1971, I, p. 97.
2. Program of the trip announced to Guéneau de Montbeillard on September 18, 1752. Ibid., I, pp. 86–7. On the return date to Paris, see the letters to La Condamine (December 4, 1752) and to Ruffey (December 12), ibid., I, pp. 88–9.
3. To Ruffey, November 21, 1759, ibid., I, p. 111.
4. Cf. ibid., I, p. 132 and n. 4.
5. Cf. *Corr.*, 1860, I, pp. 339, 342–4, and the letter to Mme. Guéneau, May 2, 1766, in *Corr.*, 1971, I, p. 150.
6. Cf. Fellows 1970, pp. 35–8, which destroyed this legend.
7. Letters of May 5 and July 14, 1760, in *Corr.*, 1971, I, pp. 111–4.
8. To Guyton de Morveau, March 1762, ibid., I p. 124.
9. To Ruffey, March 13, 1762, ibid., I, p. 127. On the affairs of the clergy, the letter to Le Blanc, October 22, 1750, ibid., I, p. 72.
10. To Guyton de Morveau, March 1762, ibid., I, p. 124. The statement came from Rabelais.
11. To Ruffey, February 14, 1750, ibid., I, pp. 63–4.
12. Cf. *Corr.*, 1971, II, p. 168, no. 1.
13. Cf. *HN*, VI (1756), "Avant-Propos," p. v; "Le loup," VII (1758), pp. 51–2; "Le renard," ibid., pp. 81–2; "Le hérisson," VIII (1760), p. 30.
14. Cf. the letter to Ruffey, July 22, and August 15, 1752, in *Corr.*, 1971, I, pp. 83–5.
15. On the abbé Nollet, see Torlais 1954. On the history of electricity in the eighteenth century, see Heilbron 1966 and 1982. Relying on Torlais, Heilbron has a bad opinion of Buffon in general.
16. On the origin of this article, see Heilbron 1982, p. 188.
17. Cf. Heilbron 1982, pp. 183–95, and Falls 1938. The description of the existing theories is found in Heilbron.
18. July 22, 1752, in *Corr.*, 1971, I, p. 84.
19. D'Argenson 1859, VIII, pp. 59–60 (June 16, 1753).
20. Cf. the letter to Ruffey, July 4, 1753, in *Corr.*, 1971, I, p. 90. On the events of this election, see, besides d'Argenson, Grimm, *Corréspondance littéraire*, July 1, 1753, in Grimm 1877, II, pp. 261–2, *Corr.*, 1860, I, pp. 267–9 and *Corr.*, 1971, I, p. 90, note.

21. To Le Blanc, October 22, 1750 and April 24, 1751, in *Corr.*, 1971, I, pp. 73 and 79.

22. To Ruffey, July 4, 1753, ibid., I, p. 91.

23. *Corréspondance littéraire*, September 1, 1753, in Grimm 1877, II, pp. 275–9.

24. Letter read to the Academy of Dijon on November 5, 1753, cited in *Corr.*, 1860, I, p. 287.

25. Letter to Sophie Volland, December 1760.

26. July 14, 1760, in *Corr.*, 1971, I, p. 114.

27. To Le Blanc, March 23, 1761, ibid., I, p. 122 (on his failure at the Academy); May 23, 1755, ibid., I, p. 101 (on Voltaire).

28. On d'Alembert's election, cf. *Corr.*, 1860, I, pp. 296–7. On d'Alembert's activity to have philosophers enter the Académie Française, see Grimsley 1963, p. 95. On his role of "party chief" of the Academy of Sciences, see Hankins 1970, p. 138. On the criticisms that provoked his "despotic reign" in the two Academies, see Grimsley 1963, pp. 105–6.

29. Letter to Loppin de Gemeaux in De Brosses 1929, p. 274. On this entire affair, see Voltaire's correspondence in Voltaire, *Oeuvres Complètes*, ed. Th. Besterman, volumes XIX–XXIV.

30. De Brosses to Voltaire, December 20, 1759, ibid., XXI, p. 54.

31. De Brosses to Loppin de Gemeaux, June 1, 1760, in De Brosses 1929, p. 281.

32. De Brosses to Voltaire, November 1759, in Voltaire, *Oeuvres Complètes*, XIX, pp. 444–5.

33. Voltaire to De Brosses, ibid., XXIV, pp. 43–6.

34. Cf. *Corr.*, 1971, I, p. 171, note by Nadault de Buffon.

35. De Brosses to Loppin, March 26, 1761, in De Brosses 1929, pp. 283–4.

36. March 7, 1768, in *Corr.*, 1971, I, pp. 170–1.

37. Cited in *Corr.*, 1860, I, p. 387.

38. De Brosses to Loppin, January 14, 1769, in De Brosses 1929, pp. 312–3.

39. Cf. *Corr.*, 1971, I, p. 66.

40. Cf. ibid., I, p. 68, note by Nadault de Buffon.

41. To Le Blanc, September 22, 1765, ibid., I, p. 138.

42. Condorcet 1792, p. 46.

43. *Corr.*, 1971, I, p. 60.

44. Raitière 1981, p. 95.

45. Ibid., p. 91.

46. Ibid., p. 106.

47. Ibid., p. 102.

48. Ibid., p. 101.

49. Ibid., p. 104.

50. Ibid., pp. 91–2, 96–7.

51. Ibid., p. 110.

52. Ibid., pp. 95–6.

53. Ibid., pp. 106–7.

54. Ibid., p. 94.

55. Ibid., p. 97.

56. Letter to Charles Bonnet, March 14, 1754, cited in Torlais 1936, p. 243.

57. For this entire affair, I followed Torlais 1936, pp. 380–8.

58. On this other dark affair, see Huard 1951.

59. Cf. Farber 1972, p. 269.

60. Cited by Torlais 1936, p. 386. The entire letter is found in Haller 1983, p. 203.

61. Raitière 1981, pp. 101–2.

62. Ibid., p. 107.

63. Ibid., p. 99.

64. See letter to Gueneau, March 20, 1762, in *Corr.*, 1971, I, p. 128.

65. Cf. Bourdier 1952b, pp. 168–9.

66. Cf. Heim 1979, which gives the measurements of skeletons found in the Buffon family's burial place in Montbard.

67. Cited in *Corr.*, 1971, I, pp. 120–1.

68. Ibid.

69. The complete text of this ode is found in *Corr.*, 1860, I, pp. 307–10.

70. *Corr.*, 1971, I, p. 116.

71. All the names cited here appear in the correspondence from 1747 to 1755, almost all in letters to Le Blanc. In the letters to Ruffey or De Brosses, almost all the topics relate to Burgundy and Burgundians.

72. Morellet 1988, pp. 128–9. The "recitations" of which Morellet spoke were outlines of the article "On Nature, First View," appearing in 1764 in volume XII of the *Histoire naturelle*. They must therefore be dated around 1763.

73. Morellet 1988, p. 143.

74. Letter from Voltaire to La Harpe, September 4, 1771, cited in Badinter 1988, p. 47.

75. "Avant-Propos," *HN*, VI, p. iii.

76. Buffon to Daubenton, February 29, 1764, in Michaud 1931, pp. 21–4.

77. Daubenton to Buffon, 1767, ibid., pp. 30–1.

78. Letter to Ruffey, February 5, 1766, in *Corr.*, 1971, I, pp. 145–6.

79. Cf. the letter from Rousseau to Panckoucke, October 16, 1771, in Rousseau 1924, XX, pp. 95–6.

80. Cf. *Corr.*, 1971, I, pp. 128–9.

81. To Ruffey, February 5, 1766, ibid., I, pp. 145–6.

82. Montmirail to Daubenton, January 10, 1764, in Michaud 1931, pp. 20–1.

83. February 5, 1766, *Corr.*, 1971, I, p. 145.

CHAPTER XV

1. "De la Nature. Première Vue," *HN*, XII (1764), pp. iii–iv.

2. Ibid., p. v.

3. "De la Nature, Seconde Vue," *HN*, XIII (1765), p. ii.

4. "Les animaux carnassiers," *HN*, VII (1758), pp. 3–6.

5. Ibid., pp. 6–7.

6. "Le rat," ibid., pp. 278–9.

7. "Le lièvre," *HN*, VI (1756), pp. 247–8.

8. "Les animaux domestiques," *HN*, IV (1753), p. 173.

9. "Le lièvre," *HN*, VI, pp. 248–50.

10. On Laplace, see in particular *Exposition du système du monde* (1796) and the *Mécanique céleste* (1799–1805). The famous book by Adam Smith, *The Wealth of Nations*, appeared in 1776. On the notions of balance and "compensation" in the thought of that period, see Svagelski 1981.

11. "Les animaux sauvages," *HN*, VI (1756), p. 60.

12. Ibid., pp. 55–6.

13. Ibid., p. 61.

14. "Sur la nature des oiseaux," *HNO*, I (1770), pp. 49–51.

15. "Le cerf," *HN*, VI (1756), p. 63; "Le chevreuil," ibid., pp. 198 ff.; and "Le lion," IX, pp. 7–8.

16. "Le cheval," *HN*, IV (1753), p. 175.

17. "Les animaux sauvages," *HN*, VI, pp. 61–62.

18. "Les animaux carnassiers," *HN*, VII (1758), pp. 3–4.

19. "Le bœuf," *HN*, IV (1753), p. 440.

20. "Les animaux domestiques," ibid., p. 169.

21. "L'âne," ibid., p. 391.

22. "Le cheval," ibid., p. 175.

23. "Le buffle," *HN*, XI (1764), pp. 295–6.

24. "Le bœuf," *HN*, IV, pp. 440, 448.

25. "Le buffle," *HN*, XI, p. 320.

26. "Le héron commun," *HNO*, VII (1780), pp. 342–3.

27. "L'unau et l'aï," *HN*, XIV (1765), pp. 40–1.

28. "Les animaux domestiques," *HN*, IV, p. 170.

29. Ibid., p. 171.

30. Ibid., pp. 171–2.

31. Ibid., p. 170.

32. "Le chien," *HN*, V (1755), p. 195.

33. "Les animaux domestiques," *HN*, IV, p. 171.

34. "Le chien," *HN*, V, p. 195.

35. "La chèvre," ibid., p. 60.

36. "De la Nature. Première Vue," *HN*, XII (1764), pp. xi–xiv.

37. Ibid., p. xi.

38. Genesis 1:26 and 2:15.

39. "Les animaux domestiques," *HN*, IV (1753), p. 173.

40. "Le chien," *HN*, V (1755), p. 196.

41. "De la Nature. Première Vue," *HN*, XII (1764), pp. xiv–xv.

42. "Le chameau et le dromadaire," *HN*, XI (1764), pp. 220–1.

43. "Le kamichi," *HNO*, VII (1780), pp. 336–7.

44. Letter to Le Blanc, February 22, 1738, in *Corr.*, 1971, I, p. 34.

45. "De la Nature. Première Vue," *HN*, XII (1764), p. xiii.

46. "Le kamichi," *HNO*, VII, pp. 335–6.

47. "Premier Discours," *HN*, I, p. 12.

48. "De la nature de l'Homme," *HN*, II, p. 443.

49. "Sur la nature des animaux," *HN*, IV (1753), p. 3.

50. Cf. Condillac 1987 and chapter XX below.

51. Réaumur 1734, V, p. 388. The entire problem is studied in detail in pp. 379–93.

52. "Sur la nature des animaux," *HN*, IV (1753), pp. 90–5.
53. Letter to Charles Bonnet, March 14, 1754, cited in Torlais 1936, p. 243.
54. "Sur la nature des animaux," *HN*, IV, p. 104.
55. Ibid., pp. 93, 99–100.
56. Ibid., pp. 100–1.
57. Ibid., pp. 22–3.
58. Ibid., p. 6.
59. Ibid., pp. 38–9.
60. Ibid., pp. 6–9.
61. Ibid., pp. 34–5.
62. Ibid., pp. 39–40.
63. Ibid., pp. 40–1.
64. "Sur la nature des animaux," *HN*, IV (1753), pp. 35–6.
65. "Les animaux carnassiers," *HN*, VII (1758), p. 11.
66. "Sur la nature des animaux," *HN*, IV, pp. 41–4.
67. "Les animaux carnassiers," *HN*, VII, pp. 11–12.
68. "Sur la nature des animaux," *HN*, IV, pp. 51–61.
69. Ibid., pp. 61–7.
70. Ibid., pp. 68–9.
71. "Les animaux carnassiers," *HN*, VII (1758), p. 7.
72. "Sur la nature des animaux," *HN*, IV, pp. 78–89.
73. Ibid., pp. 101–8.
74. Beside in the texts cited in preceding notes, these objections are found in the fifth objections of *Méditations,* presented by Gassendi.
75. *Les Passions de l'âme,* art. 23 ff. (Descartes 1963, III, pp. 970 ff. [Adam-Tannery, XI, pp. 346 ff.]).
76. "Les animaux carnassiers," *HN*, VII (1758), pp. 10–19.
77. On the history of the concept of reflex, see Canguilhem 1955.

CHAPTER XVI

1. "Sur la nature des animaux," *HN*, IV (1753), pp. 33–34.
2. On this entire discussion, see Mauzi 1960.
3. On the origins of this theme, see Azouvi 1985, who insists more on the psychophysiological aspects than on the moral and religious aspects.
4. "Sur la nature des animaux," pp. 81–2.
5. Ibid., p. 77.
6. Ibid., pp. 73–4.
7. Ibid., p. 72.
8. Ibid., pp. 72–3.
9. Ibid., p. 75.
10. See above, chapter XIII, p. 200–1.
11. "Sur la nature des animaux," pp. 44–5.
12. Ibid., p. 45.
13. Ibid.
14. Ibid., p. 69.

15. Ibid., p. 78.
16. Ibid., pp. 80–1.
17. "Sur la nature des animaux," *HN,* IV (1753), p. 47.
18. Ibid., pp. 84–85.
19. Hérault de Séchelles 1890, p. 12. On Hérault de Séchelles' visit to Montbard, see below, chapter XXI.
20. "Sur la nature des animaux," passim, and in particular, pp. 41, 44, 47, 85.
21. *An Essay concerning Human Understanding,* book II, chapter XI, § 11–13, in Locke 1740, pp. 66–3.
22. Ibid., p. 27.
23. Ibid., pp. 41–2.
24. "Nomenclature des singes," *HN,* XIV (1766), pp. 32–3.
25. "Sur la nature des animaux," pp. 52–3.
26. Ibid., p. 84.
27. Ibid., pp. 67–8.
28. Ibid., p. 70.
29. *De l'origine de l'inégalité,* end of the first part, in Rousseau 1964, pp. 159–60. On the natural catastrophes that forced man to unite, see the second part, ibid., pp. 165 and 168.
30. "Les animaux carnassiers," *HN,* VII (1758), pp. 28–9.
31. Ibid., p. 29.
32. "Sur la nature des animaux," *HN,* IV, p. 110.
33. "Nomenclature des singes," *HN,* XIV (1766), p. 32.
34. Ibid., pp. 35–6.
35. Ibid., pp. 31–2.
36. "Les animaux carnassiers," *HN,* VII (1758), pp. 30–1.
37. "Nomenclature des singes," *HN,* XIV, p. 36.
38. Ibid., p. 38.
39. *Lettres philosophiques.* Letter V, in Le Roy 1802, pp. 86–7. See below, chapter XX.
40. *Philos. Zool.,* first part, chapter VIII, in Lamarck 1809, pp. 349–52. Lamarck used several of Buffon's major ideas, without citing him.
41. "Nomenclature des singes," *HN,* XIV, pp. 36–7.
42. "Sur la nature des animaux," *HN,* IV, p. 96.
43. Ibid., p. 97.
44. Ibid., p. 96.
45. "Les animaux domestiques," *HN,* IV (1753), pp. 172–3.
46. "Les animaux carnassiers," *HN,* VII (1758), p. 28.
47. Ibid., pp. 33–4. This idea would be taken up by the astronomer Bailly, then by Buffon in the *Epoques de la Nature.* See below, chapter XXIII.
48. "Le lièvre," *HN,* VI (1756), p. 248.
49. "Animaux communs aux deux continents," *HN,* IX (1761), pp. 104–5.
50. Ibid., p. 110.
51. "Les animaux carnassiers," *HN,* VII (1758), p. 27.
52. On "the entrepreneur," see the remarkable study by Claude Blanckaert 1988.
53. "Le castor" *HN,* VIII (1760), p. 282.
54. Ibid., p. 284.

55. Ibid., p. 285.
56. Ibid., p. 294.
57. Ibid., p. 283.
58. "Les animaux sauvages," *HN*, VI (1756), p. 62.

CHAPTER XVII

1. "Premier Discours," *HN*, I, p. 29.
2. "Plan d'Ouvrage," *HNO*, I, 1770, xiii–xiv. For the solution to this problem, see below, p. 287.
3. "Le petit gris," X, p. 118. Except for specific indications, this reference and those that follow refer to the *Histoire naturelle, générale et particulière* (1749–1767), 15 vols. in quarto, specifically to the *Histoire des quadrupeds*. References to the *Histoire naturelle des oiseaux* are indicated by the initials *HNO*.
4. "Le porc-épic," XII, pp. 404–6.
5. "Le buffle," XI, p. 316.
6. "Les animaux carnassiers," VII, pp. 34–5.
7. "Le buffle," XI, pp. 248–9.
8. "La chauve-souris," VIII, p. 118.
9. "Le surmulot," VIII, p. 206.
10. "Le sarigue ou opossum," X, pp. 290–1.
11. "Le tamanoir, le tamandua et le fourmiller," X, pp. 159–60.
12. "L'élan et le renne," XII, pp. 109–10.
13. "Les gazelles," XII, pp. 219–20.
14. "Le furet," VII, pp. 213–4.
15. "L'ondatra et le desman," X, p. 2.
16. "L'hippopotame," XII, pp. 45–6.
17. Animaux communs aux deux continents," IX, pp. 112–3.
18. "Le buffle," XI, pp. 284–6.
19. Ibid., pp. 291–2.
20. "L'élan et le renne," XII, pp. 80–5.
21. "Animaux communs aux deux continents," IX, p. 122.
22. "Le surmulot," VIII, p. 206.
23. "Les tigres," IX, pp. 54–5.
24. "Animaux de l'ancien continent," IX, p. 78.
25. "Le hamster," XIII, p. 125.
26. "Le raton," VIII, p. 337.
27. "L'hippopotame," XII, pp. 24–33.
28. "Le hamster," XIII, pp. 117–25.
29. "De la description des animaux," IV, pp. 120–6.
30. Ibid., pp. 127–30. Daubenton would take up the same ideas in 1795 in his lessons at the École Normale in the year III.
31. "Le palmiste, le barbaresque, et le suisse," X, pp. 129–30.
32. "Le cerf," VI, p. 63.
33. "Le chevreuil," VII, pp. 198–9.
34. "Le lion," IX, p. 8.

35. "Le tigre," IX, p. 130.
36. "Le grand aigle," *HNO*, I, pp. 79–81.
37. "La chauve-souris," VIII, pp. 113–5.
38. Ont he rabbit's "cacecotrophie," see Gallouin 1982.
39. "Le cheval," IV, pp. 189–97.
40. Ibid., pp. 175–7.
41. "L'éléphant," XI, pp. 4–17.
42. *Lettres philosophiques*, in Le Roy 1768, "Seventh Letter."
43. "La taupe," VIII, p. 82.
44. "Sur la nature des oiseaux," *HNO*, I, p. 4.
45. "Le chat," VI, pp. 3–4.
46. "Le chien," V, p. 186.
47. "Plan de l'Ouvrage," *HNO*, I, p. XV.
48. "Le hérisson," VIII, p. 29.
49. "Le tigre," IX, p. 133.
50. "Le renard," VII, pp. 79 and 82.
51. "Le furet," VII, p. 211.
52. "Le loup," VII, p. 52.
53. Ibid., p. 76.
54. "Le daim," VI, pp. 171–2.
55. "Addition à l'article du 'Cheval,'" in *Supplément*, III (1776), pp. 47–8.
56. "Le renard," VII, p. 81.
57. "La marte," VII, p. 188.

CHAPTER XVIII

1. "Premier Discours," *HN*, I, pp. 3–11.
2. "Nomenclature des singes," *HN*, XIV (1766), pp. 17–8.
3. Cited and commented on by Daudin 1926b, p. 73.
4. "Les tatous," X (1763), p. 202. As in the notes of the preceding chapter, this reference and those that follow refer to the *Histoire naturelle, générale et particulière*; the *Histoire naturelle des oiseaux* is abbreviated by the initials *HNO*.
5. "Premier Discours," I, p. 13.
6. "La musaraigne," VIII (1760), p. 57.
7. "Le cochon," V (1755), pp. 100–1.
8. "Les phoques," XIII (1765), p. 330.
9. "Premier Discours," I, p. 12.
10. I am here following the remarkable article by Barsanti 1988.
11. On the "great chain of being" in the eighteenth century, see Lovejoy 1936.
12. "Sur la nature des oiseaux," *HNO*, I (1770), pp. 45–6.
13. "Le cayopollin," *HN*, X (1763), p. 352.
14. "La bubale," XII (1764), p. 295.
15. "Le cariama," *HNO*, VII (1780), pp. 325–6. On all these ways of establishing these "relationships," see Barsanti 1988.
16. "Le cochon," V (1755), pp. 102–3.

17. "Les toucans," *HNO,* VII (1780), pp. 109–10.
18. "Le cochon," V (1755), pp. 103–4.
19. "Nomenclature des singes," *HN,* XIV (1766), pp. 21–3.
20. The phrase comes from Barsanti 1988.
21. Cf. Guyenot 1941, p. 139.
22. "L'âne," IV (1753) pp. 379–81.
23. Ibid.
24. "Nomenclature des singes," *HN,* XIV, pp. 28–9.
25. "Le cheval," IV, pp. 215–6.
26. See Mayr 1982, pp. 265 ff.
27. "Variétés dans l'espèce humaine," *HN,* III, p. 528.
28. "Le cheval," IV (1753), p. 249.
29. "Les animaux sauvages," VI (1756), p. 60.
30. *Oratio de Telluris habitabilis incremento,* in Linnaeus 1749, vol. II (1751), p. 441; French translation in Linnaeus 1972, pp. 30–1.
31. "Le chameau et le dromadaire," *HN,* XI (1764), pp. 217–8.
32. "Le lion," IX (1761), p. 2.
33. "Animaux communs aux deux continents," ibid., p. 102.
34. Ibid., p. 107.
35. Ibid., p. 106.
36. Ibid., p. 110.
37. "Animaux du nouveau monde," ibid., p. 96.
38. "Le cheval," IV, pp. 216–7.
39. Ibid., pp. 217–9.
40. "La brebis," V (1775), p. 4.
41. "La chèvre," ibid., pp. 59–60.
42. Ibid., pp. 60–1.
43. *Émile, ou de l'éducation* (1762), beginning of book I.
44. "Le mouflon et les autres brebis," *HN,* XI (1764), pp. 363–4. Modern naturalists still put the sheep and the mouflon together in the subfamily *Ovinae.*
45. "De la dégénération des animaux," XIV, pp. 317–9.
46. "Le chameau et le dromadaire," XI, pp. 228–32.
47. "Le bufle," ibid., p. 327.
48. Ibid., p. 328.
49. Ibid., p. 308.
50. Ibid., p. 307.
51. Ibid., pp. 294–5.
52. Ibid., pp. 295–6.
53. Ibid., p. 296.
54. "Animaux communs aux deux continents," *HN,* IX (1761), p. 97.
55. Ibid., pp. 101–2.
56. "De la dégénération des animaux," *HN,* XIV (1766), p. 316.
57. Ibid., pp. 326–7.
58. "Animaux communs aux deux continents," IX, p. 103.
59. "De la dégénération des animaux," XIV, pp. 329–30.
60. "Animaux communs aux deux continents," IX, p. 103.

61. See below, chapter XXIII, p. 420.
62. "Histoire naturelle de l'homme. Variétés dans l'espèce humaine," *HN*, III, p. 530. See above, chapter XII.
63. "De la dégénération des animaux," XIV, pp. 327–8, emphasis added.
64. "Variétés dans l'espèce humaine," III, p. 530.
65. Ibid.
66. "Les animaux sauvages," *HN*, VI, pp. 59–60.
67. "Animaux communs aux deux continents," *HN*, IX (1761), p. 127.
68. Ibid., p. 126.
69. "Animaux communs aux deux continents," IX, p. 127.

CHAPTER XIX

1. Foucault 1966, p. 86.
2. Stearn, in Blunt 1986, p. 332. On the problems of classification in general and of animals in particular in the eighteenth century, see Daudin 1927b, Guyénot 1941 and Sloan 1972. On Linnean classification, see Stearn.
3. Linnaeus 1743, p. iii.
4. "Genre," *Encyclopédie.*
5. Daubenton, "Classe," ibid.
6. Cf. Malesherbes 1971, I, p. 9.
7. On these questions, see Daudin 1926b.
8. Cf. Leroy 1957, pp. 188–90.
9. Linnaeus 1743, § VI, p. iv.
10. Malesherbes 1971, I, p. 41.
11. Ibid., p. 47.
12. *Essay,* book III, chap. III, § 10, in Locke 1740, p. 195.
13. Malesherbes 1971, I, pp. 59–60; 55 and 74.
14. Cf. Flourens 1844, pp. 2–15, and even Daudin 1927, pp. 128–31. See, however, Guyénot 1941, p. 66. On this attitude, see Sloan 1976, p. 356.
15. Cf. Stearn, in Blunt 1986, p. 332.
16. See especially Farber 1972, Sloan 1976, and Barsanti 1984.
17. Cf. Sloan 1976, p. 358, and Stafleu 1971.
18. Without claiming to give an exhaustive list see "L'âne" (IV, 1753, pp. 381–4); "Les animaux carnassiers" (VII, 1758, pp. 20–2); "Le rat" (VII, pp. 279–80); "Les animaux communs aux deux continents" (IX, 1761, pp. 121–5); "La roussette" (X, 1763, p. 55 n.); "Le mouflon" (XI, 1764, pp. 365–75); "La girafe" (XIII, 1765, pp. 7–9); "La mangouste" (XIII, pp. 154–5).
19. "Exposition des distributions méthodiques des animaux quadrupèdes," IV (1753), pp. 142–68.
20. Cf. Sloan 1976, pp. 360–1.
21. See above, chapter VI.
22. "Animaux communs aux deux continents," IX (1761), pp. 123–4.
23. *Essay,* book III, chap. II, in Locke 1740, pp. 190–2.
24. Cf. Sloan 1972.
25. Ibid.

26. *Essay,* book III, Ch. VI, § 37, in Locke 1740, p. 221.
27. Ibid., § 38.
28. Ibid., chap. III, § 17, p. 198.
29. Cf. Sloan 1972, pp. 44–51.
30. Cited in Mayr 1982, pp. 256–7.
31. "Histoire des Animaux," Chap. I, *HN,* II, pp. 10–11.
32. Ibid., chap. II, p. 18.
33. "L'âne," *HN,* IV (1753), pp. 385–6.
34. For a precise analysis of this question in Aristotle and Harvey, see the remarkable article by Sloan 1987.
35. "La chèvre," V (1755), pp. 61–4.
36. "Le pécari," X (1763), p. 22; "Le chien," V (1755), pp. 210–3.
37. "Animaux de l'ancien continent," IX (1761), pp. 66–7.
38. "Le bufle," XI (1764), p. 329.
39. "Le mouflon," XI (1764), pp. 364–5.
40. "Le bouquetin," XII (1764), pp. 155–6.
41. "Le daim," VI (1756), p. 167.
42. "Le chevreuil," VI, p. 201.
43. "Le renard," VII (1758), pp. 81–2.
44. "Le zèbre," XII (1764), p. 3.
45. "Le rat," VII (1758), p. 279.
46. "Le hamster," XIII (1765), p. 117.
47. "Premier Discours," I, p. 21.
48. Ibid., p. 22.
49. Ibid., p. 38.
50. Ibid., p. 40.
51. Cf. Larson 1971, pp. 84–5.
52. "Animaux communs aux deux continents," IX (1761), p. 122.
53. "L'âne," IV (1753), p. 378.
54. Ibid., pp. 381–2.
55. Cf. Sloan 1976, p. 373 n. 70.
56. "L'âne," IV, p. 382.
57. "Système de la nature," § XLV, in Maupertuis 1756, II, 148*–149* (because of a printing error, the number of the pages 145 through 160 are repeated. In the second series, the printer added an asterisk to the number of the page).
58. "L'âne," IV, p. 383.
59. Ibid., pp. 383, 386–91.
60. Ibid., pp. 390–1. On this text and all the questions discussed here, see Russel 1982, pp. 24–7.
61. Ibid., p. 386.
62. Ibid., p. 384.
63. "Le rat," VII (1758), pp. 279–80.
64. "La roussette," X (1763), pp. 55–6, n.
65. "Animaux communs aux deux continents," IX (1761), p. 121.
66. "La mangouste," XIII (1765), p. 154.
67. Ibid., pp. 154–5, n.
68. "La girafe," ibid., pp. 7–8.

69. "Le mouflon," XI (1764), p. 369.
70. Ibid., p. 370.
71. "Animaux communs aux deux continents," IX (1761), p. 99, emphasis added.
72. Ibid., p. 127, emphasis added.
73. "Les mulets," *Supplément,* III (1776), pp. 1–38. Announced in "La chèvre," V (1755), p. 63.
74. "Dégénération des animaux," *HN,* XIV (1766), pp. 336–47.
75. Ibid., p. 337.
76. Ibid., p. 350.
77. "Les mulets," *Supplément,* III, pp. 8–14.
78. "Le serin des Canaries," *HNO,* IV (1778), p. 11.
79. Ibid., p. 25.
80. "Dégénération des animaux," *HN,* XIV (1766), p. 353.
81. "Animaux communs aux deux continents," IX (1761), p. 119.
82. "Les mulets," *Supplément,* III (1776), p. 29.
83. "De la Nature. Seconde Vue," XIII (1765), p. ix.
84. "Dégénération des animaux," XIV, p. 335.
85. "Avertissement," XII (1764), p. j.
86. "De la Nature. Première Vue," XII, p. iii.
87. Ibid., p. iv.
88. Cf. Sloan 1987, pp. 103–6.
89. "Animaux communs aux deux continents," IX (1761), p. 121.
90. "Dégénération des animaux," XIV, pp. 358–63.
91. Ibid., p. 363.
92. Ibid., pp. 364–73.
93. Ibid., p. 352.
94. "Les mulets," *Supplément,* III (1776), pp. 32–4.
95. Letter to Daubenton, January 10, 1764, in Michaud 1931, p. 20.
96. "Plan de l'ouvrage," *HNO,* I (1770), pp. xx–xxi.
97. "Les coucous étrangers," *HNO,* VI (1779), p. 356.
98. "Plan de l'ouvrage," *HNO,* I, pp. xxiii–xxiv.
99. "Les mulets," *Supplément,* III (1776), pp. 33–4.
100. "Dégénération des animaux," XIV, (1766), p. 374.

CHAPTER XX

1. *Correspondance littéraire,* September 1, 1753, in Grimm 1877, II, p. 279.
2. Ibid., October 1, 1753, pp. 285–91.
3. Ibid., November 1, 1755, III, pp. 112–3.
4. Ibid., November 1, 1756, III, pp. 301–6.
5. Ibid., August 15, 1759, IV, pp. 131–4.
6. September 1, 1759, IV, pp. 136–9.
7. Ibid., December 15, 1759, IV, pp. 163–71.
8. Ibid., March 15, 1762, V, pp. 55–9.

9. Ibid., July 1, 1764, VI, pp. 22–9.

10. Ibid., December 1, 1759, IV, p. 158.

11. On Grimm, Diderot and the *Correspondance littéraire*, see J. Schlobach, in Diderot 1975, XIII, pp. xvi–xix.

12. Cf. Roger 1971, pp. 599–600.

13. Cf. Diderot 1975, IX, p. 28. The introduction and the notes of this edition show very clearly Buffon's influence in this text.

14. Ibid., Pensée XII, pp. 36–7.

15. Ibid., Pensée LI, p. 84.

16. Ibid., Pensée LVIII, § 2, p. 94.

17. Cf. Diderot 1975, XII, p. 128. On this philosophy of Diderot's, see the introduction to the *Rêve de d'Alembert* by Jean Varloot (ibid., pp. 25–73), and Roger 1971, pp. 654–69.

18. *HN*, IV (1753), pp. 384–6. The *Encyclopédie* erroneously gives pp. 784 ff.

19. See Rey 1987, in particular the second part.

20. See, for example, the article "Le lion" and "Le tigre."

21. *Correspondance Littéraire*, December 1, 1754, in Grimm 1877, II, pp. 438–44.

22. Ibid., pp. 203–4.

23. Ibid., III, p. 112.

24. This analysis is found in the introduction by F. Dagognet in a recent republication of the text: Condillac 1987, pp. 9–131.

25. *Correspondance littéraire*, November 1, 1755, in Grimm 1877, III, p. 112.

26. Ibid., January 15, 1769, VIII, p. 252.

27. Cf. Letter from Condillac to Rousseau, September 7, 1756, in Rousseau 1924, II, p. 332.

28. Cf. Rousseau 1967, IV, pp. 423–6.

29. Fifth letter, in Le Roy 1768, p. 92.

30. Ibid., seventh letter, p. 162.

31. Ibid., "Lettre sur une critique," pp. 129–30.

32. Ibid., "Lettre [. . .] sur l'homme," p. 186.

33. On Bonnet's critique of Buffon, his relationships with Haller, Needham, and Spallanzani, and all the history of this debate, see Roger 1971, pp. 708–24, which must be supplemented by Mazzolini and Roe 1986, Haller 1983, Bonnet 1971, Roe 1981a and 1981b.

34. Chap. XI, § 177, in Bonnet 1762, I, p. 176.

35. Ibid., § 175, p. 172.

36. Ibid., chap. VIII, § 135, p. 118 n.

37. Bonnet to Haller, October 22, 1762, in Haller 1983, p. 301. The correspondence with Malesherbes is found in Bonnet 1948, pp. 213–6.

38. Bonnet to Haller, December 4, 1762, in Haller 1983, p. 309.

39. Bonnet to Haller, January 10, 1777; Haller to Bonnet, January 23, 1777. Ibid., pp. 1251–4.

40. Letters of May 12 and June 18, 1769, pp. 818–26.

41. Bonnet to Haller, May 25, 1769, p. 822.

42. Haller to Bonnet, July 7, 1771; Bonnet to Haller, July 20, 1771, pp. 947–50.

43. Bonnet to Haller, June 10, 1757, p. 103; Haller to Bonnet, July 6, 1757, p. 105; Bonnet to Haller, July 22, 1757, p. 107.

44. Haller to Bonnet, August, 1761, pp. 243–4; Haller to Bonnet, October 5, 1761, p. 246.

45. Bonnet to Haller, November 16, 1765, p. 446.

46. Bonnet to Haller, November 6, 1767; Haller to Bonnet, November 8, 1767, pp. 685–7.

47. Bonnet to Haller, January 20, 1770, p. 856.

48. Haller to Bonnet, December 8, 1760, p. 228.

49. Haller to Bonnet, November 7, 1773, p. 1106.

50. Section VIII, chap. XVII, in Bonnet 1782, II, p. 204. I here respected the emphasis that Bonnet made in capital letters.

51. This entire correspondance and Bonnet's reactions are found in his *Mémoires autobiographiques* (Bonnet 1948, pp. 303–9).

52. Bonnet to Haller, January 20, 1770, in Haller 1983, p. 856.

53. *Gazette littéraire de l'Europe* April 4, 1764.

54. Cf. Roger 1971, pp. 740–4.

55. *Défense*, chap. XIX, in Voltaire 1978, pp. 228–30.

56. Cf. the introduction by J.-M. Moureaux, ibid., p. 43.

57. *Les Colimaçons*, "Dissertation d'un physicien de Saint-Flour."

58. On this problem, see Fellows 1960, taken up again in Fellows 1970, pp. 33–53, and Starobinski 1964.

59. Rousseau 1924, II, p. 58.

60. *Œuvres Complètes* (Rousseau 1964, III, p. 195).

61. Quoted in Starobinski 1964, p. 137.

62. Cf. Rousseau to Jalabert, December 16, 1754, in Rousseau 1924, II, p. 119.

63. Letter to Du Peyrou, November 4, 1764, ibid., XII, pp. 24–5.

64. Letter to Du Peyrou, January 31, 1765, p. 272.

65. Letter to Du Peyrou, February 7, 1765, p. 325.

66. October 13, 1765, ibid., XIV, pp. 195–6. *Corr.*, 1971, I, pp. 140–1.

67. Letter to M. de Tourette, July 4, 1770, in Rousseau 1924, XIX, p. 344.

68. *Corr.*, 1860, I, p. 402.

69. *Voyage à Montbard*, Hérault de Séchelles 1890, p. 19.

CHAPTER XXI

1. Letter to De Brosses, September 1, 1766, in *Corr.*, 1971, I, p. 153.

2. Cf. Y. François 1952, pp. 111–2.

3. Letter to Daubenton, May 21, 1766, in Michaut 1931, p. 29.

4. See the letters to Guéneau de Montbeillard, De Brosses, and other friends between December 15, 1767, and January 10, 1769, in *Corr.*, 1971, I, pp. 164–79.

5. Letter to Ruffey, April 5, 1769, ibid., p. 180.

6. Letter to De Brosses, September 29, 1769, ibid., p. 185.

7. Letter to De Brosses, January 1768, ibid., p. 169.

8. Letter to Guéneau de Montbeillard, October 8, 1768, ibid., p. 161.

9. Letter to De Brosses, March 7, 1768, ibid., p. 170.

10. Letter to De Brosses, May 28, 1770, ibid., p. 190.

11. On the history and the functioning of the forges, see Benoit and Rignault 1989. On the archeological research that followed, see Benoit and Peyre 1984.

12. "Observations et expériences faites dans le but d'améliorer les canons de la Marine," in *Supplément*, II (1775), pp. 81–110. The quotations are found on pp. 84–5.

13. To the duc de Choiseul-Praslin, Minister of the Navy, December 18, 1769, in *Corr.*, 1971, I, pp. 186–7. Choiseul-Praslin left this ministry in December 1770.

14. February 16 and 18, 1771, in Bachaumont 1780, V, pp. 219–20.

15. All the correspondence on this illness is found in *Corr.*, 1860, I, pp. 388–94.

16. This correspondence and the official documents are found in ibid., I, pp. 403–12.

17. Ibid., pp. 463–5.

18. Cf. Bourdier 1952b.

19. To Lebrun, March 3, 1778, in *Corr.*, 1971, I, pp. 385–6. The complete text of the ode is found in *Corr.*, 1860, II, pp. 288–92.

20. Deliberation quoted in ibid., I, p. 394.

21. Quoted in *Corr.*, 1860, I, p. 347.

22. Cf. J. Bart and P. Bodineau 1988.

23. Cf. ibid.

24. Quoted in *Corr.*, 1860, I, p. 429.

25. Cf. *Corr.*, 1971, I, p. 239.

26. Ibid., I, p. 154.

27. Ibid., I, p. 164.

28. Ibid., I, pp. 314–5 and *Corr.*, 1860, I, p. 257.

29. "Du fer," *HNM*, II (1783), pp. 409–10 n.

30. Ibid., p. 411.

31. See *Corr.*, 1971, I, p. 426, and II, pp. 265, 364–5.

32. Ibid., I, p. 315.

33. Ibid., I, pp. 397, 414.

34. See F. Bourdier and Y. François 1952, pp. 212–8.

35. This is the conclusion of Fortunet, Jobert, and Woronoff 1988.

36. Bourdier 1952a, p. 33.

37. Cf. *Corr.*, 1860, I, pp. 375–6.

38. Cf. Hérault de Séchelles 1890.

39. Letters from Guéneau de Montbeillard to his wife, in *Corr.*, 1860, I, pp. 421–4.

40. July 9, 1780, in *Corr.*, 1971, II, p. 23.

41. November 12, 1774, ibid., pp. 271–4.

42. Cf. Bachaumont, December 23, 1774, and January 17, 1775, in Bachaumont 1780, VII, pp. 254, 266–7.

43. *Supplément*, V (1778), pp. 285–6.

44. Answer to M. le Maréchal, duc de Duras, May 15, 1775, ibid., IV (1777), pp. 43–4.

45. *Corréspondence littéraire*, July 15, 1770, in Grimm 1877, IX, p. 85.

46. Letter to De Brosses, May 12, 1770, in *Corr.*, 1971, I, p. 189.

47. *Corréspondence littéraire*, February 1773, in Grimm 1877, X, p. 197.

48. On this entire history, see Baker 1967, Baker 1975, pp. 35–47, and Badinter 1988.

49. Quoted in Badinter 1988, pp. 119–20.

50. *Corr.*, 1971, I, p. 244, and Grimm 1877, X, p. 313 (November 1773).

51. Cf. Badinter 1988, pp. 126–32 and Baker 1975, pp. 61–2.

52. Grimm 1877, XIII, pp. 83–4 (February 1782).

53. To Mme. Necker, January 20, 1782, in *Corr.*, 1971, II, p. 96.

54. Grimm 1877, IX, p. 308 (May 15, 1771).

55. Quoted in *Corr.*, 1860, I, pp. 254–5.

56. Cf. letter to Amelot, the man responsible for the Royal Press, October 19, 1777, in *Corr.*, 1971, I, pp. 356–9.

57. Letter to Mme. Necker, August 30, 1779, ibid., I, pp. 436–7.

58. Cf. letters to Thouin of September 15, 1780, to July 20, 1781, ibid., II.

59. On all these operations, see François 1952, pp. 113–24, and *Corr.*, 1860, II, pp. 356–7, 463–4, 522–7. About the Verdier affair, see ibid., pp. 591–4.

60. July 20, 1781, in *Corr.*, 1971, II, p. 66.

61. Quoted in *Corr.*, 1860, II, pp. 374–5.

62. Cf. *Corr.*, 1971, II, pp. 356–9 nn.

63. Cf. Fortunet, Jobert, and Woronoff 1988.

64. Cf. ibid.

65. Cf. Delange 1984, pp. 96–104.

66. Quoted in *Corr.*, 1860, II, p. 393.

67. Cf. ibid., II, pp. 490–3.

68. *Corr.*, 1971, II, pp. 344–8. On this entire business and the correspondence of the different participants, see *Corr.*, 1860, II, pp. 563–75.

69. September 27, 1784, ibid., II, p. 361.

70. Hérault de Séchelles 1890, pp. 7–8.

71. March 21, 1787, in *Corr.*, 1971, II, p. 340–2.

72. Quoted in ibid. The complete text is found in Grimm 1877, XV, pp. 170–2 (December 1787).

73. March 22, 1774, in *Corr.*, 1971, I, pp. 263–4.

74. July 23, 1779, ibid., I, pp. 428–9.

75. July 18, 1781, ibid., II, pp. 62–5.

76. October 1, 1781, ibid., II, p. 82.

77. December 26, 1783, ibid., II, p. 221.

78. July 1, 1783, ibid., II, p. 286.

79. September 27, 1787, ibid., II, pp. 360–1.

CHAPTER XXII

1. *HNO*, III (1775), "Avertissement," pp. i–ii.

2. Letter to Guéneau de Montbeillard, January 1781, in *Corr.*, 1971, II, p. 37.

3. "Biographie de M. Guéneau de Montbeillard," by Mme. Montbeillard, in *Corr.*, 1860, I, pp. 335–41. Citation, p. 41.

4. *HNO*, VII (1780), "Avertissement," p. ii.

5. Cf. Humbert-Bazile, in *Corr.*, 1860, II, pp. 275–6.

6. An example is found in Bourdier 1952a, pp. 36–8.

7. On Sonnini, see Anderson 1970, 1973, and 1974.

8. Cf. Corsi 1983, pp. 37–40.

9. *HNO*, I (1770), "Plan de l'ouvrage," p. viii.

10. Letter from Mlle. Blesseau to Faujas de Saint-Fond, June 12, 1788, in *Corr.*, 1860, II, p. 643.

11. "Introduction à l'histoire des minéraux," in *Supplément*, I (1774), p. 144.

12. *Supplément*, III (1776), pp. 290–1.

13. About the *Essai*, see Brunet 1931; Fréchet 1954, pp. 438–46 (with the text); Roger in Binet and Roger 1977, p. 25 ff (with the text) and Coumet 1988.

14. Coumet 1988.

15. Cf. Baker 1975, pp. 241–2.

16. *HN*, XII (1764), pp. v–ix.

17. "De la Nature. Seconde vue," *HN*, XII, p. vi.

18. "Introduction à l'histoire des minéraux. Partie expérimentale," in *Supplément*, I (1774), p. 143.

19. "Supplément à la théorie de la Terre. Partie hypothétique, Premier Mémoire," in *Supplément*, II (1775), p. 372.

20. *Dissertation sur la glace* (1749), p. 56, quoted in Buffon 1962, p. xxvii.

21. "Preuves de la théorie de la Terre," art. V, *HN*, I, p. 196. See above p. 167.

22. On the genesis of the text, see "Introduction" in Buffon 1962, pp. xv–xxxi.

23. "Des éléments," in *Supplément*, I (1774), p. 36 n.

24. Ibid., pp. 1–3.

25. Ibid., pp. 4–19.

26. *Opticks*, III, 1st part, in Newton 1952, p. 339.

27. Ibid., p. 374.

28. Ibid., p. 395.

29. "Des éléments," in *Supplément*, I (1774), pp. 102–4.

30. Cf. Guerlac 1975, pp. 68–75.

31. "Introduction à l'histoire des minéraux. Partie expérimentale," in *Supplément*, I, pp. 145–8.

32. Ibid., p. 159.

33. Ibid., p. 176.

34. Ibid., pp. 169–71.

35. Ibid., p. 289.

36. Ibid., p. 291.

37. Cf. Leclaire 1988.

38. "Des éléments," in *Supplément*, I, pp. 106–8.

39. Ibid., p. 121.

40. Ibid., p. 116.

41. Ibid., pp. 122–4.

42. *HNM*, I (1783), p. 1.

43. Ibid., pp. 41–2.

44. Ibid., p. 43.

45. "Du soufre," *HNM*, II (1783), pp. 112–4.

46. "Des sels," ibid., pp. 160–2.

47. Ibid., pp. 167–8.

48. Ibid.
49. "Du soufre," ibid., p. 107.
50. *HNM*, III (1785), pp. 609–36.
51. *HNM*, IV (1786), pp. 433–48.
52. Ibid., p. 448.
53. "Du charbon de terre," *HNM*, I, p. 427 ff. On the history of the problem, see Buffon 1962, introduction, p. lvi n. 2.
54. "Du fer," *HNM*, II (1783), pp. 344–6.
55. *Traité de l'aimant*, pp. 2–7.
56. Ibid., pp. 11–4, 23–4.
57. Ibid., pp. 29–31.
58. Ibid., art. II, pp. 90–91.

CHAPTER XXIII

1. *Époques de la Nature*, "Premier Discours," in *Supplément*, V (1778), p. 1. My references to the text of the *Époques* (indicated by *EN*) refer to this edition, whose pagination is reproduced in my edition (Buffon 1962). The references to the manuscript (which is reproduced in this edition), to the Introduction, or to the notes of this edition, refer to the actual pagination of the edition.
2. Cf. Buffon 1962, introduction, pp. xxxi–xxxii.
3. *EN*, "Premier Discours," p. 5.
4. Ibid.
5. Ibid., pp. 5–28.
6. Ibid., pp. 28–9.
7. Cf. Buffon 1962, introduction, pp. ci–civ.
8. *EN*, "Premier Discours," p. 29.
9. Ibid., p. 35.
10. Ibid., p. 39.
11. *EN*, "Première Époque," p. 53.
12. Ibid., p. 55.
13. *EN*, "Deuxième Époque," p. 71.
14. *EN*, "Troisième Époque," p. 93.
15. Ibid., pp. 95–6.
16. "Introduction à l'histoire des minéraux," Partie expérimentale, Prémier mémoire, in *Supplément*, I (1774), pp. 157–8.
17. Ibid., p. 159.
18. Ibid., p. 166.
19. "Partie expérimentale. 8e mémoire," ibid., II (1775), pp. 33–6.
20. "Supplément à la théorie de la Terre, Partie hypothétique. Recherches sur le refroidissement de la Terre et des planètes," ibid., p. 404.
21. Ibid., p. 369.
22. Ibid., pp. 513–4.
23. *EN*, "Troisième Époque," p. 93.
24. Ibid., pp. 115–6.
25. *EN*, "Première Époque," p. 69.

26. *EN,* "Troisième Époque," p. 115.
27. *EN,* "Première Époque," pp. 69–70.
28. Cf. Buffon 1962, introduction, pp. lxiii–lxvii.
29. *EN,* "Première Époque," manuscript, in Buffon 1962, p. 40.
30. *EN,* "Première Époque," pp. 67–8 (printed).
31. Ibid., manuscript, in Buffon 1962, p. 40.
32. *EN,* "Première Époque," p. 70 (printed).
33. *On the Origin of Species,* chap. IX, in Darwin 1964, pp. 282–7.
34. Cf. Burchfield 1975, pp. 32–6 and chap. VI, and Badash 1989.
35. "De la figuration des minéraux," *HNM,* I (1783), pp. 4–13.
36. *EN,* "Cinquième Époque," pp. 185–6.
37. *EN,* "Troisième Époque," p. 99.
38. *EN,* "Cinquième Époque," p. 176.
39. "Partie Hypothétique. Premier mémoire," in *Supplément,* II (1775), p. 510.
40. *EN,* "Cinquième Époque," p. 177.
41. Ibid., p. 179.
42. On this controversy, see Roger 1989.
43. Cited in Buffon 1962, p. 273 n. 45.
44. *EN,* "Premier Discours," p. 27.
45. Cf. Hodge 1988.
46. *EN,* "Premier Discours," p. 35.
47. *EN,* "Cinquième Époque," pp. 187–9.
48. Ibid., manuscript, in Buffon 1962, pp. 159–61.
49. "Partie Hypothétique. 2e mémoire," in *Supplément,* II (1775), p. 564.
50. *EN,* "Sixième Époque," p. 213 and "30e note justificative."
51. *EN,* "Cinquième Époque," p. 187, and corresponding manuscript in Buffon 1962, p. 160.
52. *EN,* "Septième Époque," p. 225.
53. Ibid., pp. 228–34.
54. On N.-A. Boulanger, see Roger 1953 and Hampton 1955. On the borrowings and the differences between the two thoughts about early man, see Buffon 1962, introduction, pp. lxxvi–lxxviii.
55. "Les animaux carnassiers," *HN,* VII (1758), p. 34.
56. Cf. Buffon 1962, introduction, pp. lxvii–lxviii.
57. On De Paw and Raynal, see Duchet 1971, in particular pp. 202–6.
58. Cf. Roger 1989.
59. *EN,* "Septième Époque," p. 240.
60. Ibid., pp. 241–2.
61. Ibid., p. 253.
62. Ibid., p. 258.
63. Ibid., pp. 253–4.
64. On the details of the edition, see Buffon 1962, introduction, pp. xxxvii–xl.
65. April 1779, in Grimm 1877, XII, p. 237.
66. The details can be found in Buffon 1962, introduction, pp. cxxix–cxliv.
67. For a very detailed analysis of this history, see Stengers 1974, pp. 109–24.
68. Quoted ibid., pp. 114–5.
69. Ibid., pp. 120–1.

70. Hérault de Séchelles 1890, p. 26.
71. Letter of November 15, 1779, in *Corr.*, 1971, I, pp. 439–40.
72. Letter of December 14, 1781, ibid., II, pp. 88–92.
73. Flourens 1860, p. 75.
74. Cf. Reil 1988.

EPILOGUE

1. Hérault de Séchelles 1890, p. 26.
2. This account from Mme. Necker is found in *Corr.*, 1860, II, pp. 612–4.
3. Cf. Piveteau 1952, and Buffon 1962, introduction, pp. xciv–cxiv.
4. Cf. Buffon 1962, introduction, pp. lxxxiii–cxxvii.
5. Letter to Faujas de Saint-Fond, June 12, 1788, in *Corr.*, 1860, II, p. 643.
6. Cf. Roger 1988.
7. Hérault de Séchelles 1890, p. 26.
8. "De la Nature. Première Vue," *HN*, XII (1764), pp. xv–xvi.
9. *EN*, "Premier Discours," p. 29.
10. "De la Nature, Première Vue," *HN*, XII, p. iii.
11. "Du soufre." *HNM*, II (1783), p. 107. On this evolution, see Buffon 1962, introduction, pp. cvii–cxii.
12. Cf. the interesting study by Reynaud 1988, which unfortunately mentions Buffon little. Buffon the writer and especially the "Discours sur le style" have been the object of a large number of commentaries. We will only mention here Bruneau 1954; Buffon 1962, introduction, pp. cxiv–cxxvii, and especially the remarkable pages of Otis Fellows in Fellows and Milliken 1972, pp. 148–70.
13. Discourse given at the Académie Française, in *Supplément*, IV (1777), pp. 3–12.
14. "Des pétrifications et des fossiles," *HNM*, IV (1786), pp. 172–3.
15. Quoted in *Corr.*, 1860, II, p. 615.
16. August, 1788, in Condorcet 1988, pp. 240–1.
17. Letter to Pfaff, June, 1788 (?), in Cuvier 1858, p. 49.
18. April, 1788, in Grimm 1877, XV, p. 249.

Bibliography

For the bibliography of editions by Buffon and works dedicated to him through 1954, see Genet-Varcin and Roger, "Bibliography of Buffon," in Buffon 1954, pp. 513–75.

WORKS BY BUFFON

My references are to the original edition:

Histoire naturelle, générale et particulière, Paris, Imprimerie royale, 1749–1767, 15 vols. in-4°. (Designated by the abbreviation *HN.*)

Histoire naturelle des oiseaux, Paris, Imprimerie royale, 1770–1783, 9 vols. in-4°. (Designated by the abbreviation *HNO.*)

Histoire naturelle générale et particulière . . . , Supplément, Paris, Imprimerie royale, 1774–1789, 7 vols. in-4°. (Designated by *Supplément.* The *Époques de la Nature,* which are part of volume V, are designated by the abbreviation *EN.*)

Histoire naturelle des minéraux, Paris, Imprimerie royale, 1783–1788, 5 vols. in-4°. (Designated by the abbreviation *HNM.*)

PARTIAL EDITIONS

Buffon 1954, *Œuvres philosophiques de Buffon,* edited by Jean Piveteau, Paris, Presses Universitaires de France, Corpus général des philosophes français, 1954. This excellent edition is out of print.

Buffon 1962, *Les Époques de la Nature,* critical edition, with introduction and notes by J. Roger, Paris, Éditions du Muséum, 1962, reedited 1988.

Buffon 1971, *De l'homme* introduction by M. Duchet, Paris, Maspero, 1971.

See also Jean-Louis Binet and Jacques Roger, *Un autre Buffon,* Paris, Hermann, 1977.

CORRESPONDENCE

Correspondance inédite, collected and annotated by Henri Nadault de Buffon, Paris, Hachette, 1860, 2 vols. (Designated in the notes by the abbreviation *Corr.* 1860).
Correspondance générale, collected and annotated by H. Nadault de Buffon, Geneva, Slatkine Reprints, 1971, 2 vols. (Reprints of volumes XIII and XIV of the edition of the *Œuvres complètes* published by Lanessan, Paris, A. Le Vasseur, 1884–1885, 14 vols. Designated by the abbreviation *Corr.* 1971.)
See also below Franck Bourdier and Yves François, *Buffon,* Paris, 1952; Gustave Michaut, "Buffon administrateur et homme d'affaires," in *Annales de l'université de Paris-IV, 1931;* Françoise Weil, "La correspondance Buffon-Cramer," in *Revue d'histoire des sciences,* 1961.

GENERAL WORKS

Adelmann, Howard B. *Marcello Malpighi and the Evolution of Embryology,* 5 vols. Ithaca, Cornell University Press, 1966.
Anderson, Elizabeth. "Some possible sources of the passages on Guiana in Buffon's *Époques de la Nature,*" I: *Trivium,* V (1970), pp. 72–84; II: *Trivium,* VI (1971), pp. 81–91.
———. "More about possible sources of the passages on Guiana in Buffon's *Époques de la Nature,*" I: *Trivium,* VIII (1973), pp. 83–94; II: *Trivium,* 9 (May 1974), pp. 70–80.
———. "La collaboration de Sonnini de Manoncourt à l'*Histoire naturelle* de Buffon," in *Studies on Voltaire and the Eighteenth Century,* 1974, CXX, pp. 329–59.
Argenson, René-Louis de Voyer, marquis d'. *Journal et Mémoires du marquis d'Argenson,* 9 vols., published by E. J. B. Rathery, Paris, Vve J. Renouard, 1859–1867.
Ariès, Philippe. *L'enfant et la vie familiale sous l'Ancien Régime,* Paris, Plon, 1960.
Azouvi, François. "Homo duplex" in *Gesnerus,* 1985, vol. 42, pp. 229–44.
Bachaumont. *Mémoires secrets pour servir à l'histoire de la république des lettres . . . ,* 36 vols. London, John Adanson, 1780–1789.
Badash, Lawrence. "The Age-of-the-Earth Debate," in *Scientific American,* August 1989, pp. 78–83.
Badinter, Elisabeth, and Robert Badinker. *Condorcet. Un intellectuel en politique,* Paris, Fayard, 1988.
Baker, Keith M. "Les débuts de Condorcet au secrétariat de l'Académie royale des Sciences, 1773–1776," in *Revue d'histoire des sciences,* 1967, XX, 3, pp. 229–80.
———. *Condorcet. From Natural Philosophy to Social Mathematics,* Chicago, University of Chicago Press, 1975.
Barsanti, Giulio. "Linné et Buffon: deux visions différentes de la nature et de l'histoire naturelle," in *Revue de synthèse,* 1984, 3° série, nos. 113–114, pp. 83–111.
———. "Buffon et l'image de la Nature: de l'échelle des êtres à la carte géographique et à l'arbre généalogique," in *Buffon 88,* ed. Jean Gayon, Paris, Librairie Philosophique J. Vrin, 1992, pp. 255–96.
Bart, Jean. "Georges-Louis Le Clerc, Seigneur de Buffon et autres lieux . . . ," in *Buffon 88,* pp. 29–37.

Bazin, Gilles-Auguste. *Observations sur les plantes,* Strasburg, J. R. Doulssecker, 1741.

Belaval, Yvon. "La crise de la géométrisation de l'univers dans la philosophie des Lumières," in *Revue internationale de philosophie,* 1952, no. 21, pp. 337–55.

Benoît, Serge, and Philippe Peyre. "L'apport de la fouille archéologique à la connaissance d'un site industriel. L'exemple des forges de Buffon," in *L'Archéologie industrielle en France,* 1984, no. 9.

Benoît, Serge, and Bernard Rignault. "La grande forge de Buffon," in *Buffon, 1788–1988,* Paris, Imprimerie nationale, 1988.

——. *La Grande Forge de Buffon. Guide de visite,* Dijon, 1989.

Berkeley, George. *Works,* 9 vols., ed. A. A. Luce and T. E. Jessup, London and New York, T. Nelson, 1948–1957.

Bertin, Léon. "Buffon homme d'affaires," in *Buffon* 1952, pp. 87–104.

Binet, Jacques-Louis, and Jacques Roger. *Un autre Buffon,* Paris, Hermann, 1977.

Blanckaert, Claude. "La valeur de l'homme: l'idée de nature humaine chez Buffon," in *Buffon 88,* pp. 583–600.

Blunt, Wilfrid. *Linné, le prince des botanistes,* translation from the English by F. Robert, Paris, Belin, 1986.

Bonnet, Charles. *Considérations sur les corps organisés,* Amsterdam, M.-M. Rey, 1762, 2 vols.; 2d ed., Paris, Fayard, 1985.

——. *Contemplation de la Nature,* new ed., Hambourg, Virchaux, 1782, 2 vols. (1st ed.: 1764; the additions of 1782 are clearly marked).

——. *Mémoires autobiographiques,* Paris, Vrin, 1948.

——. *Lettres à M. l'abbé Spallanzani,* ed. C. Castellani, Milan, Episteme editrice, 1971. (See also Haller 1983 below.)

Bouchard, Marcel. *De l'humanisme à l'Encyclopédie. L'esprit public en Bourgogne sous l'Ancien Régime,* Paris, Hachette, 1930.

Bourdier, Franck. "Principaux aspects de la vie et de l'œuvre de Buffon," 1952a, in *Buffon,* 1952, pp. 15–86.

——. "Buffon d'après ses portraits," 1952b, ibid., pp. 167–80.

Bourdier, Franck, and Yves François. "Lettres inédites de Buffon," 1952, ibid., pp. 181–224.

Bourguet, Louis. *Lettres philosophiques sur la formation des sels,* Amsterdam, François l'Honoré, 1729.

Bowler, Peter J. "Preformation and preexistence in the seventeenth century: a brief analysis," *Journal of the History of Biology,* vol. 4, no. 2, 1971, pp. 221–44.

——. "Bonnet and Buffon: theories of generation and the problem of species," ibid., vol. 6, no. 2, 1973, pp. 259–81.

Boyer, Carl B. *The History of the Calculus and its Conceptual Development,* New York, Dover, 1949, 2d ed., 1959.

Bradbury, S. "The quality of the image produced by the compound microscope: 1700–1840," in Bradbury, S., and G. l'E. Turner, eds., *Historical Aspects of Microscopy,* Cambridge, W. Heffer and Sons, 1967, pp. 151–73.

Broberg, Gunnar. "*Homo sapiens.* Linnaeus' classification of man," in *Linnaeus: The Man and His Work,* ed. Tore Frängsmuyr, Berkeley and Los Angeles, University of California Press, 1983, pp. 156–94.

Brosses, président De. *Lettres du Pdt De Brosses à Ch.-C. Loppin de Gémeaux,* ed. Yvonne Bézard, Paris, Firmin-Didot, 1929.

Bruneau, Charles. "Buffon et le problème de la forme," in *Buffon*, 1954, pp. 491–99.

Brunet, Pierre. "La notion d'infini mathématique chez Buffon," in *Archeion*, vol. 13, 1931, pp. 24–39.

———. *La vie et l'œuvre de Clairaut*, Paris, Presses Universitaires de France, 1952.

Buffon, Paris, Muséum national d'histoire naturelle, 1952.

Buffon 88. Actes du Colloque International, Paris, Montbard, Dijon, June 14–22, 1988, Paris, Vrin, 1992.

Burchfield, Joe D. *Lord Kelvin and the Age of the Earth*, New York, Science History Publications, 1975.

Burnet, Thomas. *The Sacred Theory of the Earth*, introduction by Basil Willey, Carbondale, Ill., Southern Illinois University Press, 1965 (Latin original: 1681; English original: 1684).

Busco, Pierre. *Les Cosmogonies modernes et la théorie de la connaissance*, Paris, Alcan, 1924.

Canguilhem, Georges. *La Formation du concept de réflexe aux XVII et XVIII siècles*, Paris, Presses Universitaires de France, 1955.

Carter, Richard B. *Descartes' Medical Philosophy: The Organic Solution to the Mind-Body Problem*, Baltimore, Johns Hopkins University Press, 1983.

Casini, Paolo. "Buffon et Newton," in *Buffon 88*, pp. 299–308.

Castellani, Carlo. *Dal mito alla scienza. La storia della generazione*, Milano, Longaneri, 1965.

———. "Spermatozoan biology from Leeuwenhoek to Spallanzani," *Journal of the History of Biology*, vol. 6, no. 1, 1973, pp. 37–68.

Changeux, Pierre. *L'homme neuronal*, Paris, Fayard, 1983.

Châtelet, marquise du. *Les Lettres de la marquise du Châtelet*, published by Th. Besterman, 2 vols., Geneva, Institut et Musée Voltaire, 1958.

Condillac, Étienne Bonnot de. *Œuvres philosophiques*, 4 vols. Parme, 1792.

———. *Traité des animaux*, introduction by François Dagognet, Paris, Vrin, 1987.

Condorcet, Marie-Jean-Antoine Caritat, marquis de. *Éloge historique de M. le comte de Buffon*, Aux Deux Ponts, Sanson, 1792.

———. *Correspondance inédite de Condorcet et de Mme Suard*, edited, presented, and annotated by Elisabeth Badinter, Paris, Fayard, 1988.

Corsi, Pietro. "Models and analogies for the reform of natural history: Features of the French debate, 1790–1800," in *Lazzaro Spallanzani e la biologia dell' Settecento*, ed. G. Montalenti and P. Rossi, Florence, Olschki, 1982, pp. 381–96.

———. *Oltre il mito. Lamarck e le scienze naturali del suo tempo*, Bologne, Il Mulino, 1983. English trans.: *The Age of Lamarck*, Berkeley and Los Angeles, University of California Press, 1988.

———. "Buffon sous la Révolution et l'Empire," in *Buffon 88*, pp. 639–48.

Costabel, Pierre. "Descartes et la mathématique de l'infini," *Historia scientiarum*, no. 29, 1985, pp. 37–49.

Cuvier, Georges. *Lettres à C. M. Pfaff, 1788–1792*, Paris, Masson, 1858.

Dainville, François de. "L'enseignement scientifique dans les collèges des Jésuites," in *Enseignement et diffusion des sciences en France au XVIII siècle*, ed. René Taton, Paris, Hermann, 1964, pp. 27–65.

Darwin, Charles. *The Descent of Man, and Selection in Relation to Sex,* 2 vols., London, John Murray, 1871.

———. *On the Origin of Species,* facsimile of the first edition, Cambridge, Mass., Harvard University Press, 1964.

Daudin, Henri. *De Linné à Lamarck. Méthodes de la classification et idée de série en botanique et en zoologie (1740–1790),* Paris, Alcan, 1926.

Daumas, Maurice. "La Description des Arts et Métiers de l'Académie des Sciences et le sort de ses planches gravées en taille douce," *Revue d'histoire des sciences,* vol. 7, 1954, pp. 163–71.

Dechambre, Ed. "L'article des Chiens dans l'*Histoire naturelle,*" in *Buffon,* 1952, pp. 157–66.

Delange, Yves. *Lamarck,* H. Nyssen ("Actes Sud" collection), 1984.

Descartes, René. *Correspondance avec Arnauld et Morus,* with an introduction and notes by Geneviève Rodis-Lewis, Paris, Vrin, 1953.

———. *Lettres à Thomas Morus,* ed. G. Rodis-Lewis, Paris, Vrin, 1953.

———. *Œuvres philosophiques,* 3 vols., ed. F. Alquié, Paris, Garnier, 1963–1973.

———. *Œuvres,* 11 vols., published by Ch. Adam and P. Tannery, Nouvelle présentation, Paris, Vrin-CNRS, 1974.

Deslandes, André-François Boureau. *Recueil de différens traitez de physique et d'histoire naturelle propres à perfectionner ces deux sciences,* 3 vols., Paris, J. F. Quillau fils, 2d ed., 1748–1753.

Diderot, Denis. *Œuvres complètes,* Paris, Hermann, 1975, in-4°, in press.

Dijksterhuis, E. J. *The Mechanization of the World Picture Pythagoras to Newton,* Princeton, Princeton University Press, 1986 (first edition in Dutch, 1950).

Donovan, Arthur. "Buffon and chemistry," in *Buffon 88,* pp. 387–95.

Duchesneau, François. *L'Empirisme de Locke,* La Haye, Martinus Nijhoff, 1973.

———. "Buffon et la physiologie," in *Buffon 88,* pp. 451–62.

Duchet, Michèle. *Anthropologie et histoire au Siècle des lumières,* Paris, Maspero, 1971.

Duhamel, abbé Joseph-Robert-Alexandre. *Lettre d'un philosophe à un docteur de Sorbonne sur les explications de M. de Buffon* (March 5, 1754), Strasburg, G. Schmouk, n.d.

Ellenberger, François. "A l'aube de la géologie moderne: Henri Gautier," part II: "La théorie de la Terre d'Henri Gautier," *Histoire et Nature,* 1976, nos. 9–10, 1976–1977.

———. "Les sciences de la Terre avant Buffon," in *Buffon 88,* pp. 327–42.

Euler, Leonhardt. *Correspondance,* vol. V, Bâle, Birckhäuser, 1980.

Falls, William F. "Buffon, Franklin, et deux Académies américaines," in *Romantic Review,* February 1938, pp. 37–47.

Farber, Paul Lawrence. "Buffon and the concept of species," *Journal of the History of Biology,* vol. 5, no. 2 (Fall 1972), pp. 259–84.

———. "Buffon and Daubenton: divergent traditions within the *Histoire naturelle,*" *Isis,* 1975, vol. 66, no. 231 (March 1975), pp. 63–74.

———. "Research traditions in 18th century natural history," in *Lazzaro Spallanzani e la biolgia del settecento,* ed. G. Montalenti and P. Romi, Florence, Olschki, 1982, pp. 397–403.

Farley, John. *The Spontaneous Generation Controversy from Descartes to Oparin,* Baltimore, Johns Hopkins University Press, 1974.

——. *Gametes and Spores. Ideas about Sexual Reproduction, 1750–1914,* Baltimore, Johns Hopkins University Press, 1982.

Fellows, Otis. "Voltaire and Buffon: clash and conciliation," *Symposium,* vol. 9, no. 2 (Fall 1955), pp. 222–35. (Reprinted in Fellows 1970, pp. 22–32.)

——. "Buffon and Rousseau: aspects of a relationship," *PMLA,* vol. 75, no. 3 (June 1960), pp. 184–96. (Reprinted in Fellows 1970, pp. 33–53.)

——. "Buffon's Place in the Enlightenment," *Studies on Voltaire and the 18th century,* vols. 24–27, 1963, pp. 603–629, Geneva, Institut et Musée Voltaire. (Reprinted in Fellows 1970, pp. 54–71.)

——. *Problems and Personalities,* Geneva, Droz, 1970.

Fellows, Otis E., and Stephen F. Milliken. *Buffon,* New York, Twayne Publishers, 1972.

Flourens, Pierre. *Des manuscrits de Buffon,* Paris, Garnier, 1860.

Fontenelle, Bernard Le Bovier de. *Éléments de la géométrie de l'infini,* Paris, Imprimerie royale, 1727.

——. *Œuvres de Monsieur de Fontenelle,* 5 vols., Paris, Brunet, 1742.

Force, James E. *William Whiston: Honest Newtonian,* Cambridge University Press, 1985.

Fortunet, Françoise, Philippe Jobert, and Denis Woronoff. "Buffon en affaires," in *Buffon 88,* pp. 13–28.

Foucault, Michel. *Les Mots et les choses,* Paris, Gallimard, 1966.

François, Yves. "Buffon au Jardin du Roi (1739–1788)," in *Buffon,* 1952, pp. 105–24.

Fréchet, Maurice. "Buffon comme philosophe des mathématiques," in *Buffon,* 1954, pp. 435–46.

Fréron, Élie. *Lettres sur quelques écrits de ce temps,* 13 vols. Paris, 1749–1754. (later became: *L'Année littéraire,* 292 vols., Paris, 1754–1790).

Gallouin, François. "Ruminant ou coprophage?" *Cahiers rationalistes,* no. 376 (March 1982), pp. 161–80.

Gascar, Pierre. *Buffon,* Paris, Gallimard, 1983.

Gasking, E. *Investigations into Generation, 1651–1828,* Baltimore, Johns Hopkins University Press, 1967.

Genet-Varcin, E. and Roger, Jacques. "Bibliographie de Buffon," in *Buffon,* 1954, pp. 513–75.

Gilibert, J. E. *Abrégé du système de la Nature de Linné. Histoire des Mammaires ou des Quadrupèdes et Cétacés,* Lyon, Fr. Matheron, 1802, An X.

Goerke, Heinz. *Carl von Linné. Artz, Naturforscher, Systematiker,* Stuttgart, Wissenschaftliche Verlaggesellschaft m. b. H., 1966.

Gohau, Gabriel. *Histoire de la géologie,* Paris, La Découverte, 1987.

——. "La 'Théorie de la Terre' de 1749," in *Buffon 88,* pp. 343–52.

Grimm, Friedrich Melchior, baron de. *Correspondance littéraire, philosophique et critique par Grimm, Diderot, Raynal, Meister,* ed. Maurice Tourneux, 16 vols., Paris, Garnier, 1877–1882.

Grimsley, Ronald, *Jean d'Alembert 1717–1783,* Oxford, Clarendon Press, 1963.

Grinevald, Paul-Marie. "Les éditions de l'*Histoire naturelle,*" in *Buffon 88,* pp. 631–37.

Groult, Martine, Pierre Louis, and Jacques Roger, eds., *Transfert de vocabulaire dans les sciences*, Paris, Editions du C. N. R. S., 1988.

Guelluy, Robert. *La Création*, Tournai, Desclée, 1963.

Guerlac, Henry. *Antoine-Laurent Lavoisier, Chemist and Revolutionary*, New York, Scribner's Sons, 1975.

Guyenot, Émile. *Les Sciences de la vie aux XVII et XVIII siècles. L'idée d'évolution*, Paris, Albin Michel, 1941.

Hahn, Roger. *The Anatomy of a Scientific Institution: The Paris Academy of Sciences, 1666–1803*, Berkeley and Los Angeles, University of California Press, 1971.

Hales, Stephen. *Vegetable Staticks. Foreword (. . .) by M. A. Hoskin*, London, Macdonald, and New York, American Elsevier, 1969. First published 1727.

Hall, A. Rupert. *Philosophers at War: The Quarrel between Newton and Leibniz*, Cambridge University Press, 1980.

Haller, Albrecht von. *Réflexions sur le système de la génération de M. de Buffon*, Geneva, Barillot et fils, 1751.

———. *The Natural Philosophy of A. von Haller*, ed. Shirley A. Roe, New York, Arno Press, 1981.

———. *The Correspondence between Albrecht von Haller and Charles Bonnet*, ed. Otto Sonntag, Bern, Hans Huber, 1983.

Hampton, John. *Nicolas-Antoine Boulanger et la science de son temps*, 1955, Geneva, Droz, and Lille, Giard, 1955.

Hankins, Thomas L. *Jean d'Alembert: Science and the Enlightenment*, Oxford, Clarendon Press, 1970.

———. *Science and the Enlightenment*, Cambridge, Cambridge University Press, 1985.

Hanks, Lesley. "Buffon et les fusées volantes," *Revue d'histoire des sciences*, vol. 14, no. 2 (April–June 1961), pp. 137–54.

———. *Buffon avant l'"Histoire naturelle,"* Paris, Presses Universitaires de France, 1966.

Heilbron, J. L. "A propos de l'invention de la bouteille de Leyde," *Revue d'histoire des sciences*, vol. 19 (1966) pp. 133–42.

———. *Elements of Early Modern Physics*, Berkeley and Los Angeles, University of California Press, 1982.

Heim, Jean-Louis. *Les Squelettes de la sépulture familiale de Buffon à Montbard (Côte-d'Or). Étude anthropologique et génétique*, Paris, Éditions du Muséum national d'histoire naturelle, 1979.

Helvétius, Claude-Adrien. *Correspondance générale*, 2 vols., ed. Alan Dainard, Jean Olsoni, and Peter Allan, Toronto, University of Toronto Press, Oxford, Voltaire Foundation, 1981–1984.

Hérault de Séchelles, Marie Jean. *Voyage à Montbard*, with preface and notes by F.-A. Aulard, Paris, Librairie des Bilbliophiles, 1890.

Hodge, Jonathan. "Two cosmogonies ('Theory of the Earth' and 'Theory of Generation') and the unity of Buffon's thought," in *Buffon 88*, pp. 241–54.

Howard, Rio. *La Bibliothèque et le laboratoire de Guy de La Brosse au Jardin des Plantes à Paris*, Geneva, Droz, 1983.

Huard, Georges. "Les planches de l'*Encyclopédie* et celles de la *Description des Arts et Métiers* de l'Académie des Sciences," *Revue d'histoire des sciences*, special edition, "L'*Encyclopédie* et le progrès des sciences et des techniques," 1951, pp. 238–49.

Jacob, Margaret C. *The Newtonians and the English Revolution, 1689–1720*, Ithaca, Cornell University Press, 1976.

Koyré, Alexandre. *Du monde clos à l'univers infini*, Paris, Gallimard, 1973 (*From the Closed World to the Infinite Universe*, 1st English ed.: 1957).

Laissus, Yves. "Le Jardin du Roi," in Y. Laissus and J. Torlais, *Le Jardin du Roi et le Collège Royal dans l'enseignement des sciences au XVIII siècle*, Paris, Hermann, 1986, pp. 287–341.

Langaney, André. *Les Hommes, passé, présent, conditionnel*, Paris, Armand Colin, 1988.

Laplace, Pierre-Simon de. *Exposition du système du monde*, 2 vols., Paris, Bachelier, 1836 (6th ed.).

Larson, James L. *Reason and Experience: The Representation of Natural Order in the Work of Carl von Linné*, Berkeley and Los Angeles, University of California Press, 1971.

Leclaire, Lucien. "L'*Histoire naturelle des minéraux*: Buffon géologue universaliste," in *Buffon 88*, pp. 353–69.

Le Goff, F. *La Philosophie de l'abbé de Lignac*, Paris, Hachette, 1863.

Leibniz, Gottfried Wilhelm. *Mathematische Schriften*, ed. Gerhardt, 7 vols., Berlin, Halle, 1849–1863.

Lelarge de Lignac, abbé Joseph-Adrien. *Lettres à un Amériquain sur l'Histoire naturelle, générale et particulière, de M. de Buffon*, 5 vols. in 12°, Hamburg, 1751.

———. *Suites des Lettres à un Amériquain sur les IV et V volumes de l'Histoire naturelle de M. de Buffon et sur le Traité des animaux de M. l'abbé de Condillac*, 4 vols. in 12°, Hamburg, 1756.

Lenay, Charles. *Enquête sur le hasard dans les grandes théories biologiques de la deuxième moitié du XIX siècle*, unpublished thesis, Université de Paris-I Panthéon-Sorbonne, 1989.

Le Roy, Charles. *Lettres sur les animaux*, Nuremberg, 1768.

———. *Lettres philosophique sur la perfectibilité des animaux, avec quelques lettres sur l'homme*, Paris, 1802.

Lewis, Geneviève. "L'innéité cartésienne et sa critique par Lelarge de Lignac," *Revue des sciences humaines*, January–March 1951, pp. 30–41.

Linné, Carl von. *Amoenitates Academicae*, 9 vols., Holmiae, Laurentius Salvius, 1749–1785.

———. *L'Équilibre de la Nature*, trans. Bernard Jasmin with introduction and notes by Camille Limoges, Paris, Vrin, 1972.

Locke, John. *An Essay concerning Human Understanding*, in vol. I of *The Works of John Locke, Esq.*, 4th ed., 3 vols., London, 1740.

Lovejoy, Arthur O. *The Great Chain of Being*, Cambridge, Mass., Harvard University Press, 1936 (2d ed. New York, Harper and Row, 1960).

McMullin, Ernan, ed. *Evolution and Creation*, University of Notre Dame Press, 1985.

Maindron, Ernest. *L'Académie des Sciences*, Paris, Alcan, 1888.

Malebranche, Nicolas. *Œuvres complètes*, vol. I: *De la recherche de la vérité*, books I–III, Paris, Vrin, 1962.

Malesherbes, C. G. de Lamoignon de. *Observations sur l'Histoire naturelle générale et particulière de Buffon et Daubenton*, 2 vols., Paris, Charles Pougens, 1798 (reprinted Slatkine Reprints—Geneva, 1971).

Maupertuis, Pierre-Louis Moreau de. *Œuvres*, 3 vols., Lyon, J. M. Bruyset, 1756.

Mauzi, Robert. *L'idée de bonheur au XVIII siècle*, Paris, Armand Colin, 1960.

Mayr, Ernst. *The Growth of Biological Thought. Diversity, Evolution, and Inheritance*, Cambridge, Mass., Harvard University Press, 1982.

Mazzolini, Renato G., and Shirley A. Roe. *Science against the Unbelievers: The Correspondence of Bonnet and Needham 1760–1780*, Oxford, Voltaire Foundation, 1986.

Mémoires pour servir à l'histoire des sciences et des beaux-arts, dits Mémoires de Trévoux, 265 vols. in-12°, Trévoux, then Paris, 1701–1767.

Mengal, Paul. "La psychologie de Buffon à travers le traité 'De l'homme,' " in *Buffon 88*, pp. 601–12.

Merian, J. B. *Sur le Problème de Molyneux*, Paris, Flammarion, 1984.

Michaut, Gustave. "Buffon administrateur et homme d'affaires," *Annales de l'université de Paris-IV*, 1931, pp. 15–36.

Milanesi, Claudio. *Morte apparente e morte intermedia. Medicina e mentalità nel dibattito sull'incertezze dei segni della morte (1740–1789)*, Rome, Instituto della Enciclopedia Italiana, 1989.

Miles, R. E., and J. Serra. *Geometrical Probability and Biological Structures: Buffon's 200th Anniversary*, Berlin, Heidelberg, and New York, Springer Verlag, 1977.

Monod-Cassidy, Hélène. *Un voyageur-philosophe au XVIII siècle: l'abbé Jean-Bernard Le Blanc*, Cambridge, Mass., Harvard University Press, 1941.

Montaigne, Michel de. *Essais*, book III, 2d vol., in *Œuvres complètes*, 6 vols., ed. J. Plattard, Paris, F. Roches, 1931–1932.

Montesquieu, Charles-Louis de Secondat, baron de. *De l'esprit des lois*, 4 vols., ed. J. Brethe de La Gressaye, Paris, Les Belles Lettres, 1955.

Morellet, abbé. *Mémoires de l'abbé Morellet*, introduction and notes by J.-P. Guicciardi, Paris, Mercure de France, coll. "Le Temps retrouvé," 1988.

Mornet, Daniel. *Les Sciences de la nature en France au XVIII siècle*, Paris, Armand Colin, 1911.

Needham, John Turberville. *Nouvelles Observations microscopiques*, Paris, L. E. Ganeau, 1750.

Newton, Sir Isaac. *Optics*, New York, Dover, 1952.

———. *The Correspondence of Isaac Newton*, ed. H. W. Turnbull, vol. III: *1688–1694*, Cambridge, Cambridge University Press, 1961.

———. *Principia*, Motte's translation revised by Cajori, 2 vols., Berkeley and Los Angeles, University of California Press, 1966.

Nouvelles ecclésaistiques ou Mémoires pour servir à l'histoire ecclésiastique, n.p., then Utrecht, 1713–1803, in-4° (undiscovered until 1791).

Nouvelles littéraires, etc. de la France, published by Pierre Clément, Paris, 1748–1753, in-4° (reedited in *Les Cinq Années littéraires*, 4 vols. in-16°, La Haye, Goss Press, 1954).

Oparin, A. I. *Origin of Life*, trans. Sergius Morgulis, New York, Dover, 1953, 2d ed. (original Russian edition published in 1936).

Pascal, Blaise. *Œuvres complètes*, Paris, Seuil, 1963.

Pintard, René. *Le Libertinage érudit dans la première moitié du XVII siècle*, Paris, Boivin, 1943.

Piveteau, Jean. "La pensée religieuse de Buffon," in *Buffon*, 1952, pp. 125–32.

———. "Introduction à l'œuvre philosophique de Buffon," in *Buffon*, 1954, pp. VII–XXXVII.

Pluche, abbé Noël. *Le Spectacle de la Nature*, 9 vols. in-12°, Paris, Les frères Estienne, 1732–1750. (I have used an edition published from 1764 to 1770.)

Pomeau, René. *La Religion de Voltaire*, Paris, Nizet, 1956.

Raitière, Anna. "Lettres à Buffon dans les 'Registres de l'Ancien Régime' (1739–1788)," *Histoire et Nature*, no. 17/18, 1981, pp. 85–148.

Réaumur, René-Antoine Ferchault de. *Mémoires pour servir à l'histoire des insectes*, 6 vols. in-4°, Paris, Imprimerie Royale, 1734–1742.

———. *Art de faire éclore des œufs et d'élever en toute saison des oiseaux domestiques*, 2 vols. in-12°, Paris, Imprimerie royale, 1749.

———. *Lettres inédites*, ed. G. Musset, La Rochelle, 1886.

———. *Correspondance inédite entre Réaumur et Abraham Trembley*, ed. M. Trembley, Geneva, Georg et Cie, 1943.

Reill, Peter H. "Buffon and historical thought in Germany and Great Britain," in *Buffon 88*, pp. 667–79.

Rey, Roselyne. *Naissance et développement du vitalisme en France, de la deuxième moitié du XVIII siècle à la fin du Ier Empire*, Unpublished thesis, Université de Paris-I, 1987.

———. "Buffon et le vitalisme," in *Buffon 88*, pp. 399–413.

Reynaud, Denis. *Problèmes et enjeux littéraires en histoire naturelle au XVIII siècle*, Unpublished doctoral thesis, Université Lumière, Lyon-II, September 1988.

Roe, Shirley A. *Matter, Life, and Generation: 18th Century Embryology and the Haller-Wolff Debate*, Cambridge, Cambridge University Press, 1981.

———. "Buffon and Needham: diverging views on life and matter," in *Buffon 88*, pp. 439–50.

Roger, Jacques. "Un manuscrit perdu et retrouvé: les 'Anecdotes de la Nature,' de Nicolas-Antoine Boulanger," *Revue des sciences humaines*, July–September 1953, n.s., fasc. 71.

———. "Diderot et Buffon en 1749," in *Diderot Studies IV*, ed. Otis Fellows, Geneva, Droz, 1963, pp. 221–36.

———. *Les Sciences de la vie dans la pensée française du XVIII siècle*, Paris, Armand Colin, 2d ed., 1971.

———. "Buffon et la théorie de l'anthropologie," in *Enlightenment Studies in Honour of Lester G. Crocker*, ed. A. J. Bingham and V. Topazio, Oxford, Voltaire Foundation, 1979, pp. 253–62.

———. "Note sur le vocabulaire de la socio-biologie," in *Transfert de vocabulaire dans les sciences*, ed. Martine Groult, Pierre Louis, and Jacques Roger, Paris, CNRS, 1988a, pp. 179–86.

———. "La Nature et l'Histoire dans la pensée de Buffon," in *Buffon 88*, pp. 193–205.

———. "Buffon, Jefferson et l'homme américain," in *Histoire de l'anthropologie: hommes, idées, moments*, ed. A. Ducros and J. Hublin, Special no. of *Bulletins et Mémoires de la Société d'anthropologie de Paris*, n.s., t. 1, no. 3–14, (1989), pp. 57–66.

Rosenfield, Leonora Cohen. *From Beast-Machine to Man-Machine: The Theme of Animal Soul in French Letters from Descartes to La Mettrie*, New York and Oxford, Oxford University Press, 1941.

Rousseau, Jean-Jacques. *Correspondance générale*, 19 vols., ed. Théophile Dufour, Paris, Armand Colin, 1924–1933.

——. *Œuvres complètes*, 1959, vol. I: *Les Confessions. Autres texts autobiographiques*, ed. B. Gagnebin and M. Raymond, Paris, Gallimard, Bibliothèque de la Pléiade.

——. *Œuvres complètes*, 1964, vol. III: *Du contrat social. Écrits politiques*, Paris, Gallimard, Bibliothèque de la Pléiade.

——. *Correspondance complète*, ed. R. A. Leigh, Geneva, Les Délices, 1967.

Russel, E. S. *Form and Function: A Contribution to the History of Animal Morphology*. Chicago, University of Chicago Press, 1982 (1st ed.: London, J. Murray, 1916).

Saigey, Émile. *Les Sciences au XVIII siècle. La physique de Voltaire*, Paris, Germer-Baillière, 1873.

Scammon, R. E. "The first seriation study in human growth," *American Journal of Physical Anthropology*, vol. 10 (1927), pp. 329–36.

Sertillanges, O. P., A. D. *L'idée de création et ses retentissements en philosophie*, Paris, Aubier, 1945.

Sloan, Philip R. "John Locke, John Ray, and the problem of the natural system," *Journal of the History of Biology*, vol. 5, no. 1 (Spring 1972), pp. 1–53.

——. "The idea of racial degeneracy in Buffon's 'Histoire Naturelle,' " *Studies in 18th Century Culture*, vol. 3, Cleveland, Case Western Reserve University Press, 1973.

——. "The Buffon-Linnaeus controversy," *Isis*, vol. 67, no. 238 (September 1976), pp. 356–75.

——. "The impact of Buffon's taxonomic philosophy in German biology: the establishment of the biological species concept," *Proceedings of the 15th International Congress of the History of Science*, 1978, pp. 531–8.

——. "Buffon, German biology, and the historical interpretation of biological species," *British Journal for the History of Science*, vol. 12, no. 41 (1979).

——. "From logical universals to historical individuals: Buffon's idea of biological species," in *Histoire du concept d'espèce dans les sciences de la vie*, Paris, Fondation Singer-Polignac, 1987, pp. 101–40.

——. "Organic molecules revisited," in *Buffon 88*.

——. "L'hypothéticalisme de Buffon. Sa place dans la philosophie des sciences du XVIII siècle," in *Buffon 88*, pp. 207–22.

Smith, Edwin Burrows. "Jean-Sylvain Bailly: Astronomer, Mystic, Revolutionary, 1736–1793," in *Transactions of the American Philosophical Society*, n. s., vol. 44, part. 4 (September 1954), pp. 427–538.

Speziali, Pierre. "Une correspondance inédite entre Clairaut et Cramer," in *Revue d'histoire des sciences*, vol. 8, no. 3 (July–September 1955), pp. 193–237.

Stafleu, F. A. *Linnaeus and the Linnaeans: The Spreading of Their Ideas in Systematic Botany, 1735–1789*, Utrecht, 1971.

Starobinski, Jean. "Rousseau et Buffon," in *J.-J. Rousseau et son œuvre. Problèmes et recherches*, Paris, Klincksieck, 1964, pp. 135–47.

Stengers, Jean. "Buffon et la Sorbonne," in *Études sur le XVIII siècle*, Bruxelles, Éditions de l'Université de Bruxelles, 1974.

Svagelski, Jean. *L'idée de compensation en France, 1750–1850*, Lyon, L'Hermès, 1981.

Tassy, Pierre. *L'Ordre et la diversité du vivant*, Paris, Fayard-Fondation Diderot, ed. 1987.

Taylor, Kenneth. "Buffon's 'Époques de la Nature' and geology during Buffon's later years," in *Buffon 88*, pp. 371–85.

Thayer, H. S. *Newton's Philosophy of Nature: Selections from His Writings*, New York, Hafner, 1953.

Thomas, Keith. *Dans le jardin de la Nature*, trans. C. Malamoud, Paris, Gallimard, 1985. (Translated from *Man and the Natural World: Changing Attitudes in England, 1500–1800*, Harmondsworth, Eng., Penguin Books, 1983.)

Torlais, Jean. *Réaumur. Un esprit encyclopédique en dehors de l'Encyclopédie*, Paris, Desclée De Brouwer, 1936.

——. *Un physicien au Siècle des lumières: l'abbé Nollet*, Paris, SIPUCO, 1954.

Trousson, Raymond. *Jean-Jacques Rousseau*, 2 vols., Paris, Tallandier, 1989.

Vartanian, Aram. "Buffon et Diderot," in *Buffon 88*, pp. 119–33.

Vigouroux, F. *La Cosmogonie mosaïque d'après les Pères de l'Église*, Paris, Berche et Tralin, 1889.

Vincent, Jean-Didier. *Biologie des passions*, Paris, Éditions Odile Jacob, 1986.

Voltaire, François-Marie Arouet. "Voltaire's Correspondence," in *Œuvres complètes*, ed. Théodore Besterman, Geneva, Institut Voltaire, 1968, then Voltaire Foundation, Banbury, then Oxford (in press).

——. *La Défense de mon oncle*, ed. José-Michel Moureaux, Geneva, Slatkine, and Paris, Champion, 1978.

Weil, Françoise. "La correspondance Buffon-Cramer," *Revue d'histoire des sciences*, vol. 14., no. 2 (April–June 1961), pp. 97–136.

Westfall, Richard S. *Never at Rest: A Biography of Isaac Newton*, Cambridge, Cambridge University Press, 1980.

Index

Académie Française, 7, 187, 212–17, 336, 367–68; and philosophes, 197
Academy of Inscriptions, 186–87
Academy of Sciences, 12, 30, 192, 196; administration of, 218–19; Buffon as treasurer of, 58; expeditions of, 111; *Mémoires*, 56, 142, 387, 451n28; and philosophes, 197, 367–69
Acosta, José de, 179
Adanson, Michel, 219, 291
Alembert, Jean le Rond d,' 224, 339, 366–67, 427, 462n52; and Académie Française, 212, 214–15; and *Encyclopédie*, 197–99; and universal attraction, 55–58
America. *See* New World
Angivilliers, Charles-Claude de Flahaut d,' 358–59
Animals: behavior of, 274–75, 282–87, 328; biology of, 116; domestication and degeneration of, 299–304; and dreams, 246; exotic, 272–74, 277; and feelings, 245, 247–48, 251, 282–83, 304, 341; and final causes, 292–93; geographic distribution of, 275–76, 298, 324; and history, 266; horse, 233, 281, 286, 300–301, 320–21; humanizing of, 159–60; and intelligence, 242–44, 247–48, 342; interior sense of, 244, 248; and language, 342; and love, 327–28; as machines, 243–44; mammals, 289; marsupials, 292; and memory, 244–46, 342; New World, 298–99, 304–6, 331–33, 415–16, 420; nobility of, 301; and passions, 282; and sensation, 243–45, 248;

sheep, 301–2, 316; and society, 265–66, 285–86; soul of, 162, 240–49, 282, 341. *See also* Degeneration; *Natural History of Quadrupeds;* Species
Année littéraire, L,' 422
Anthropology, 152
Anthropomorphism, 283, 286, 304
Arcet, Jean d,' 395
Archimedes, 52–53
Argenson, Pierre de Voyer, Marquis d,' 25, 80, 187, 191, 212
Aristotle, 26, 70, 86–87, 116, 127, 148, 151, 314, 322, 386; view of earth, 94
Arnauld, Antoine, 122
Astronomy. *See* Celestial mechanics; Solar system
Augustine, Saint, 66, 107, 119, 405
Azyr, Vicq d,' 433

Bachaumont, 357
Bailly, Jean-Sylvain, 367, 419
Barruel, abbé, 423
Bazin, Gilles-Auguste, 146
Belidor, Bernard Forest de, 41
Belon, Pierre, 293
Benoît de Maillet, 98–99
Bentley, Richard, 109
Berkeley, George, 37–38, 154, 157
Bernard, Samuel, 31
Bernoulli, Nicolas, 12
Berthollet, 397
Bexon, abbé, 382, 422
Black, Joseph, 392
Blaisot, Georges, 4–5

Blesseau, Marie, 364, 380, 383, 425–26, 429

Bodin, Jean, 179

Bonnet, Charles, 87, 121–22, 125, 189, 220–21, 256, 338; *Considérations sur les corps organisés,* 342–46

Bonnier de La Mosson, 61

Botany, 50, 309; and classification, 71, 310. *See also* Plant physiology

Bouguer, Pierre, 192

Bouhier, President, 7–8, 16–17, 22, 43, 78, 125

Boulanger, Nicolas-Antoine: *Anecdotes de la Nature,* 419–20, 422

Boulduc, Gilles-François, 15, 59, 60

Boulduc, Simon, 15

Bourguet, Louis, 24, 78, 100, 102, 129

Boyle, Robert, 107

Brisson, 220–21, 309, 320, 324

Brongniart, Adolphe, 141

Brosse, Guy de La, 48–49

Brosses, Charles de, 7, 43, 45, 189, 214–17, 354, 356; *Dissertations sur les dieux fétiches,* 208–9

Brown, Robert, 141, 144

Bruhier, 172

Bruno, Giordano, 35

Buffon, Georges-Louis Leclerc, comte de: aggressiveness of, 9, 92, 324–25; ambition of, 13–14, 253; on biblical Flood, 100–101; as businessman, 363–64; bust of, 359, 423; on childhood, 5–6; on Clairaut, 451–52n33; on classification, 85–86; as count, 357–60, 366; death of, 425–26, 433–34; and description, 427–28; as diplomat, 20, 23–24, 45–46, 61, 80, 218; education of, 6–10, 77; emotional life of, 34; estate of, 5, 20, 22–24, 360–63, 373–74; and evolution, 322–23; as exegete, 405; and experimentation, 26–27, 52–53, 393; and free thought, 41–43; friendships of, 9–10, 32, 340; on happiness, 264–65; independence of, 22; on infinity, 35–37; intellect of, 35–37, 83–84, 149–50; as intendant of Royal Botanical Garden, 58–61; lack of meticulousness, 18–19; literary talents of, 287, 336, 429; on love, 252–53, 379–80; as man of action, 28–29; marriage of, 205–8; materialism of, 159, 162, 192, 341, 427; and mathematics, 8, 34–38, 41, 58; menagerie of, 210–11; on Negroes, 181–82; and observation, 142, 190; and order, 428–31; in Paris, 29–30; as paternal

lord, 24, 360–65; on patience, 4; as pioneer of genetics, 130; poetry about, 222–23, 360; on politics, 209–10, 420–21; portrait of, 222; procedure of, 127–28; psychology of, 11–12, 200–201; and rationalism, 427; realism of, 36–38, 425–28; relationship with mother, 5; and religion, 6–7, 18, 43, 110, 379, 425–26, 431; and skyrockets, 52, 78, 112; social life of, 28, 223–24, 364–65, 377–79; style of, 339; transfer within Academy, 44–45; as translator, 25–26, 34, 154, 385; unorthodox ideas of, 110, 126; as unrealistic, 226–27; visionary power of, 432–33; on Voltaire, 32; *Works: De la Nature,* 264; *Discourse on Style (Discours sur le style),* 213, 224, 384, 431–32; "Dissertation sur les causes du strabisme," 52; "Dissertation sur les couleurs accidentelles," 52; *Essay on Moral Arithmetic (Essai d'arithmétique morale),* 12, 17, 38, 52, 384; *Nomenclature des singes,* 258; "On the Game of franc-carreau" ("Mémoire sur le jeu de franc-carreau"), 17–19, 34, 384–85; "Prayer," 431; "Proofs," 93, 95–96, 102–3; "Réflexions sur la loi d'attraction," 56; "Sur les mesures," 38; "Sur les probabilités de la vie," 51–52; *Treatise on the Magnet,* 380, 400–401. *See also individual works*

Buffon, Georges-Louis-Marie (Buffonet), 207–8, 358–59, 365, 374–77

Buffon, Mme. de, 353–54

Buffonet. *See* Buffon, Georges-Louis-Marie

Burgundy parlement, 5, 7, 14, 215

Burnet, Thomas, 95, 97

Catherine II, 375, 423

Celestial mechanics, 106–9, 406, 424; and law of attraction, 53–58; Newton's model, 53–55

Celsius, Anders, 98–99

Cépoy, Marguerite-Françoise de, 375–77

Chain of beings, 87–88, 291–92, 429

Châtelet, Marquise de, 16, 32–33, 53, 196

Chemistry, 24, 391–92; and medicine, 48–49; and minerals, 397–98

Chesneau de Lauberdière, 363

Chevallier, François, 8

Childhood, 164–66, 173, 257–60

Chirac, Pierre, 50

Christianity. *See* Religion

Church of England, 42

DATE DUE

OC 10 '05			

Demco, Inc. 38-293